THE ANALYTICAL AND
NUMERICAL SOLUTION OF
ELECTRIC AND MAGNETIC FIELDS

THE ANALYTICAL AND NUMERICAL SOLUTION OF ELECTRIC AND MAGNETIC FIELDS

K. J. Binns
*University of Liverpool,
UK*

P. J. Lawrenson
*Switched Reluctance Drives Ltd
Leeds, UK*

C. W. Trowbridge
*Vector Fields Ltd,
Kidlington, UK*

JOHN WILEY & SONS
Chichester · New York · Brisbane · Toronto · Singapore

Copyright © 1992 by John Wiley & Sons Ltd,
Baffins Lane, Chichester,
West Sussex PO19 1UD, England

Other Wiley Editorial Offices

John Wiley & Sons, Inc., 605 Third Avenue,
New York, NY 10158-0012, USA

Jacaranda Wiley Ltd, G.P.O. Box 859, Brisbane,
Queensland 4001, Australia

John Wiley & Sons (Canada) Ltd, 22 Worcester Road,
Rexdale, Ontario M9W 1L1, Canada

John Wiley & Sons (SEA) Pte Ltd, 37 Jalan Pemimpin #05-04,
Block B, Union Industrial Building, Singapore 2057

The front cover illustration shows a computer model for the 12 MeV Superconducting
Cyclotron designed and built by Oxford Instruments Ltd, UK, for medical
applications using software provided by Vector Fields Ltd, Oxford.

1073312-4

British Library Cataloguing in Publication Data

A catalogue record for this book is available from the British Library
ISBN 0 471 92460 1

Typeset in 10/12pt Times by Thomson Press (India) Ltd, New Delhi
Printed in Great Britain by BPCC Wheatons Ltd, Exeter

To our many colleagues and
collaborators past and present

CONTENTS

Preface   **xiii**

1   Basic Field Theory   **1**

1.1   Introduction   1
1.2   Quasi-static electromagnetic field equations   3
        1.2.1   Magnetic vector potential   5
        1.2.2   Magnetic scalar potential   6
        1.2.3   Electric vector potential   7
        1.2.4   **H** field formulations   8
1.3   Static fields   8
        1.3.1   Magnetostatics   8
        1.3.2   Electrostatics   8
        1.3.3   Current flow   9
1.4   Integral equation formulations   9
        1.4.1   Volume magnetization integral equation method   9
        1.4.2   The boundary element method   9
1.5   Interface and boundary conditions   10
        1.5.1   Interface condition   10
        1.5.2   Equivalent pole and charge distributions   12
1.6   Calculation of global quantities   13
        1.6.1   Capacitance   13
        1.6.2   Inductance   14
1.7   Forces   15
        1.7.1   Line sources   15
        1.7.2   Distributed sources   15
        1.7.3   Total force acting on a boundary   16
        1.7.4   Force distribution over a boundary   17
1.8   Field equations in partial differential form   17
1.9   Summary   18
References   19

2   Images   **21**

2.1   Introduction   21
2.2   Plane boundaries   22
        2.2.1   Single plane boundary·   22
        2.2.2   Parallel plane boundaries   25
        2.2.3   Intersecting plane boundaries   27
        2.2.4   Inductance of parallel bus-bars near an iron surface   29
2.3   Circular boundaries   30

	2.3.1	Charge or current near a circular boundary	31
	2.3.2	Doublets: circular cylinder in a uniform field	33
2.4	General considerations		39
References			42

3 The Solution of Laplace's Equation by Separation of the Variables

43

3.1	Introduction		43
3.2	Circular boundaries		44
	3.2.1	The solution of Laplace's equation in circular-cylinder coordinates	44
	3.2.2	Iron cylinder influenced by a current	48
	3.2.3	The screening effect of a permeable cylinder	50
	3.2.4	The force between rotor and stator conductors in a cylindrical electrical machine	53
	3.2.5	Specified distribution of potential or potential gradient on the perimeter of a circular boundary	55
3.3	Rectangular boundaries		57
	3.3.1	Solution of Laplace's equation in Cartesian coordinates	57
	3.3.2	The semi-infinite strip and the rectangle	58
	3.3.3	Pole profile in the inductor alternator for a sinusoidal flux distribution	62
3.4	Conclusions		65
References			66

4 Fields with Distributed Currents: Poisson and Diffusion Equations

69

4.1	Introduction		69
4.2	Current-carrying conductors in free space		70
	4.2.1	Basic method: vector potential of a line current	70
	4.2.2	The field of a rectangular bus-bar	70
	4.2.3	The force between parallel rectangular bus-bars	73
	4.2.4	Filamentary conductors of arbitrary three-dimensional shape: Biot–Savart law	76
	4.2.5	Boundary effects: use of the image method	80
4.3	Solution of Poisson's equation in single and double series		80
	4.3.1	Introduction	80
	4.3.2	Single series solutions: Rogowski method	81
	4.3.3	Double series solutions: Roth's method	86
	4.3.4	Scope of the methods	94
4.4	Time-dependent fields: eddy currents		95
	4.4.1	Introduction	95
	4.4.2	Treatment of time variation	96
	4.4.3	Harmonic time variation with spatially fixed fields	97
	4.4.4	Harmonic time variation with travelling fields	102
	4.4.5	General time dependence: transients	107
4.5	Three-dimensional fields		109
	4.5.1	Introduction	109
	4.5.2	Diffusion equation in rectangular coordinates	110
	4.5.3	Diffusion equation in cylindrical polar coordinates	110
References			112

5 Conformal Transformation: Basic Ideas

117

| 5.1 | Conformal transformation and conjugate functions | | 117 |
| | 5.1.1 | Conformal transformation | 117 |

	5.1.2	The solution of Laplace's equation	120
	5.1.3	The logarithmic function	122
5.2	Approaching the solution		124
	5.2.1	Choice of origin	124
	5.2.2	Multiple transformations	125
	5.2.3	Field maps	125
	5.2.4	Scale relationships between planes	126
	5.2.5	Conservation of flux and potential	126
	5.2.6	Field strength	126
5.3	The bilinear transformation		127
	5.3.1	Mapping properties	128
	5.3.2	The cross-ratio	132
	5.3.3	The magnetic field of currents inside an infinitely permeable tube	133
	5.3.4	Capacitance of and voltage gradient between two cylindrical conductors	135
5.4	The simple Joukowski transformation		138
	5.4.1	The transformation	138
	5.4.2	Flow round a circular hole	140
	5.4.3	Permeable cylinder influenced by a line current	141
5.5	Curves expressible parametrically: general series transformations		143
	5.5.1	The method	144
	5.5.2	The field outside a charged, conducting boundary of elliptical shape	145
	5.5.3	General series transformations	146
	5.5.4	Field solutions	147
References			148

6   Polygonal Boundaries   149

6.1	Introduction		149
6.2	Transformation of the upper half plane into the interior of a polygon		149
	6.2.1	The transformation	149
	6.2.2	Polygons with two vertices	152
	6.2.3	Parallel plate capacitor: Rogowski electrode	155
	6.2.4	The choice of corresponding points	158
	6.2.5	Scale relationship between planes	161
	6.2.6	The field of a current in a slot	163
	6.2.7	Negative vertex angles	168
	6.2.8	The forces between the armature and magnet of a contactor	170
	6.2.9	A simple electrostatic lens	173
6.3	Transformation of the upper half plane into the region exterior to a polygon		175
	6.3.1	The transformation	175
	6.3.2	The field of a charged, conducting plate	178
6.4	Transformations from a circular to a polygonal boundary		180
	6.4.1	The transformation equations	181
	6.4.2	The field of a line current and a permeable plate of finite cross-section	183
6.5	Classification of integrals		185
References			186

7   General Considerations   189

7.1	Introduction		189
7.2	Field sources		189
	7.2.1	Infinite boundaries	190
	7.2.2	Finite boundaries	194
	7.2.3	Distributed sources	195

7.3   Curved boundaries   196
      7.3.1   Rounded corners   196
      7.3.2   Curvilinear polygons   198
7.4   Angles not multiples of $\pi/2$ 199
7.5   The use of elliptic functions   200
      7.5.1   Elliptic integrals and functions   202
      7.5.2   Two finite charged plates   204
      7.5.3   Elliptic integrals of the third kind   207
      7.5.4   The occurrence of elliptic functions   208
7.6   Numerical methods   208
      7.6.1   Classes of integral that arise in numerical evaluation   210
7.7   Non-equipotential boundaries   215
      7.7.1   Boundary value problems of the first kind   215
      7.7.2   Boundary value problems of the second and mixed kinds   217
      References   217

8   Computational Modelling—Basic Methods   **221**

8.1   Introduction and engineering objectives   221
8.2   Mathematical models   224
      8.2.1   Differential equations   224
      8.2.2   Integral equations   226
      8.2.3   Finite difference methods   227
8.3   The finite element method   231
      8.3.1   Elements   231
      8.3.2   Variational method   232
8.4   Method of weighted residuals   233
      8.4.1   Collocation methods   234
      8.4.2   Galerkin method (global)   236
      8.4.3   Comparison of weighting functions and generalization   239
8.5   Concluding remarks   241
      References   242

9   Two-dimensional Static Linear Problems   **245**

9.1   Discretizing Poisson's equation by finite elements   245
      9.1.1   Galerkin method and local basis functions   246
      9.1.2   Element continuity and element coefficients   248
9.2   Assembling the matrix   251
9.3   Boundary conditions   253
9.4   Linear algebra and solving the system equations   255
9.5   Summary   258
9.6   The solution of Poisson equation by boundary elements   260
9.7   Weighted residual and boundary elements   261
      9.7.1   Boundary element coefficients and system matrix   264
      9.7.2   Multiple regions and open boundary problems   269
9.8   Alternative boundary integral methods   274
9.9   Axisymmetric form of Poisson's equation   277
9.10 Scalar potential solution   278
9.11 Vector potential solution   280
9.12 Analytic transformations to improve the accuracy near the axis   281
9.13 Concluding remarks   281
      References   286

10   Non-linear Effects in Two-dimensional Fields   289

10.1   Introduction   289
10.2   Saturation effects   289
    10.2.1   Introduction   289
    10.2.2   Simple iterative methods   291
    10.2.3   The Newton–Raphson methods   295
10.3   Permanent magnets   302
10.4   Anisotropy   304
    10.4.1   Permanent magnets with anisotropy   304
    10.4.2   Laminated structures   306
10.5   Summary   309
References   310

11   Two-dimensional Time-dependent Fields   313

11.1   Introduction   313
11.2   Equations for low frequency fields   313
11.3   Numerical solution of steady state AC problems   314
    11.3.1   Time harmonic solutions   314
    11.3.2   Skin effect problems and the surface impedance boundary condition   319
11.4   Time-dependent formulations   320
    11.4.1   Spatial and time discretization   320
    11.4.2   Time marching schemes   320
    11.4.3   Non-linearity   323
11.5   Second order elements   325
11.6   Modelling the external circuit   329
11.7   Motional effects   332
    11.7.1   Modifications to field and finite element equations   332
    11.7.2   Upwinding   334
    11.7.3   Petrov–Galerkin method—one-dimensional example   336
    11.7.4   Extension to two dimensions   338
    11.7.5   Examples   339
11.8   Summary   339
References   340

12   Three-dimensional Problems   343

12.1   Introduction   343
12.2   Scalar potential formulations   344
    12.2.1   The reduced scalar potential approach   344
    12.2.2   The two-potential approach   345
    12.2.3   Finite element solutions   346
    12.2.4   The two-scalar potential formulation   347
    12.2.5   The reduced scalar potential formulation   348
    12.2.6   3D elements and shape functions   349
    12.2.7   Examples from magnetostatics   353
    12.2.8   Discussion on field cancellation   357
12.3   Integral methods for three-dimensional problems   359
    12.3.1   Magnetization and polarization formulation   359
12.4   Integral methods and parallel processing   364
12.5   Magnetic vector potential formulations   365
    12.5.1   Coulomb gauge   366
    12.5.2   Lorentz gauge   367

	12.5.3	Examples	373
	12.5.4	Ungauged solutions	377
12.6	Comments on vector potential formulations	379
12.7	**H** formulation using edge elements	379
12.8	Summary	380
	References	382

13 Electromagnetic Software Environment   **385**

13.1	Introduction	385
13.2	Elements of a CAD system	386
13.3	Pre-processing and mesh generation	387
	13.3.1 Semi-automatic	387
	13.3.2 Manual	388
	13.3.3 Mapping methods	390
	13.3.4 Automatic mesh generation	391
13.4	Solution processor	396
	13.4.1 Field equations	396
	13.4.2 Methods for solving the algebraic system of equations	398
	13.4.3 Gaussian elimination revisited and remarks on conditioning	399
	13.4.4 Iterative methods	402
	13.4.5 A comparison of methods	405
13.5	Error estimation	406
	13.5.1 Error estimators for finite element calculations	407
	13.5.2 Local error estimation using interpolation theory	410
13.6	Open boundary problems	411
13.7	Mesh adaptation solutions	414
13.8	Post-processing and optimization	414
13.9	Summary	418
	References	419

Appendix 1	**A11**	**Potential and flux functions**	**423**
	A1.1	Potential and flux functions	423
	A1.2	The magnetic field of line currents	424
	A1.3	Conjugate functions	427
Appendix 2	**The Field Inside a Highly Permeable Rectangular Conductor**		**433**
Appendix 3	**Table of Transformations**		**437**
	A3.1	Transformations to the upper half-plane	438
	A3.2	Other transformations	451
Appendix 4	**Useful Vector Identities and Integral Theorems**		**453**
	A4.1	Vector identities	453
	A4.2	Integral theorems	453
Appendix 5	**Quadrature Rules for Numerical Integration**		**455**
Appendix 6	**Uniqueness of Scalar Potentials**		**457**

| **Bibliography** | **459** |

| **Index** | **465** |

PREFACE

This book has been designed within a single volume to make accessible to students, to researchers and to design and development workers the full range of classical and modern methods for the solution of electric, magnetic, some thermal and other similar fields. It deals with one, two and three space dimensions, with linear, non-linear and anisotropic media, and with static and 'low'-frequency time variation (and with the special case of microwave-frequency characteristic impedances). In mathematical terms this equates to solution methods for the Maxwell equations when displacement current is negligible—and the derivative partial-differential equations most commonly known as the Laplace, Poisson and diffusion equations. In application terms it relates to understanding, analysis and design in a long and growing list of areas: electrical machines, recording and audio, electromagnetic forging, particle physics, non-destructive testing, transport, medical equipment and testing, electromagnetic compatibility, superconductivity, printing and copying, power transmission and distribution, electron optics, fusion and others.

The treatment inevitably reflects the fact that the subject is a mathematical one but special effort has been made, consistent with soundness and sufficiency, to simplify the presentation and to emphasize the practical utility of the ideas and techniques. Similarly, special effort has been made to bring out the physical significance of the mathematics and partly to this end, and partly to try to make the text as useful as possible, a good number of examples has been presented in some detail.

The reader is presumed to have a relatively modest knowledge of electromagnetic (or other) fields and of partial-differential equations such as would certainly be contained in a first or second year undergraduate course in engineering, physics or mathematics. The book can therefore be taken up at any time as appropriate to a particular scheme of undergraduate study or thereafter, and used as an adjunct to other studies; or, of course, it can be taken up in an independent, self-contained way. The final level reached is such as to present a secure picture of the state-of-the-art of the subject, together with comments on current research areas and likely future computational environments.

Throughout the book numerous examples are included to bring out the physical significance of the mathematics and the practical considerations involved in implementing the solutions. Additionally, a full bibliography and references are provided so that the reader, as appropriate to need, can establish greater familiarity with fundamental concepts, be directed to a very large number of already existing solutions and applications, and be introduced to those ideas and techniques which remain the subject of current research and development.

In preparing the book the authors were able to build on the experience of an earlier text by two of the present authors (KJB and PJL) on very much the same subject and with many of the objectives outlined above. Solution techniques, particularly in the numerical area, have advanced enormously over recent years and the appropriate balance and role of the book have shifted very significantly. This is most obviously reflected in the changed proportions devoted to numerical methods—less than one sixth originally to more than one half now and coupled with the increase in the book size by 60%. To devotees of numerical methods it may seem that the change in proportions could be even more marked. However, the classical methods, albeit in modern guises, remain the most useful in a surprising number of situations. Moreover they provide both essential fundamental background and, along with application knowledge dependent upon experience, general insight and understanding which are invaluable in the sound application and assessment of numerical models and solution techniques.

In structure the book falls into two parts, the first part dealing with analytical methods, Chapters 2–7 inclusive, and the second part with numerical methods, Chapters 8–13 inclusive. Chapter 1 provides a general introduction to the scope and applicability of the book, outlines the basic field concepts and equations using vector notation and in a style judged to be suited to the needs of the numerical methods, and reviews some of the considerations involved in the derivation of physical quantities, such as force and inductance, from the field solutions.

The analytical section begins with a treatment of the method of images in Chapter 2, providing a very easy starting point which is helpful in establishing a good feel for basic boundary conditions and field symmetries, and also is useful in solving a variety of practical problems. Basic solutions using Fourier series (single and multiple) are established in Chapter 3 in the context of the Laplace equation with attention restricted to two-dimensional fields with rectangular and circular boundary shapes, i.e. the simplest and most relevant for practical application. Chapter 4 is substantial and could equally well have been presented as at least two separate chapters. The rationale, however, is that it provides a progressive and coherent extension from the solution of the Laplace equation. First the zero right-hand side is replaced by a constant to form the two-dimensional Poisson equation and solve the fields due to spatially distributed (but time invariant) current or charge; then the constant RHS is replaced by the time derivative to form the diffusion equation and solve induced (eddy) current-type fields; and finally these treatments are extended to deal with three-space dimensions, culminating in powerful, but little-known, general solutions for cylindrical and rectangular geometries. Treatments of distributed field sources by integration using the Biot–Savart law, analytical for two-dimensional sources, and quasi-analytical, but powerful and general, for three-dimensional, filamentary sources are presented also.

The next three chapters are dedicated to the subject of conformal transformation. Whilst this is not so important as it once was, and it does not lead into the discrete numerical procedures in the way that the earlier methods do, it none the less remains of considerable practical value in a number of areas. Strikingly too it is the only analytical method capable of dealing with highly complex boundary shapes unrestricted by the need to work within the constraints of particular coordinate systems. For these reasons and, in the absence of an alternative up-to-date treatment, it was felt proper to deal reasonably comprehensively with the subject emphasizing some general results and the power of the method when implemented using numerical quadrature.

The remaining six chapters deal with the full range of numerical techniques in a thoroughgoing and modern way. This treatment is progressive throughout. Chapter 8 lays the foundations covering the basic numerical models in the space coordinates, finite-element methods and weighted residuals. In Chapter 9 solutions for two-dimensional, linear, time-invariant fields, including integral methods and use of scalar and vector potentials are discussed, and sufficient detail is given to enable the reader to develop his own software for finite-element solution of the Poisson equation. Chapter 10 extends the modelling and solution methods to include non-linearities, e.g. magnetic saturation, anisotropy and permanent-magnet materials, and laminated structures. Time-dependence is introduced in the next chapter including the use of implicit methods in a complex variable for time-harmonic fields and implicit and explicit methods for fully time-dependent fields including transient ones. Methods for treating moving conductors are also included. Next in Chapter 12 the extension to three-space dimensions is dealt with including the various single and mixed potential methods, and also integral methods. Additionally, the crucial importance of 'bench-mark' problems to validate three-dimensional solutions is emphasized. In all chapters care is taken to bring out the computational and software considerations, and the whole of the final chapter is devoted to the software environment and its crucial effects on capability and usability.

The latter includes pre-processing and mesh generation, solution processing, a brief survey of some of the more important techniques of linear algebra in solving large sparse sets of equations, error estimation, automatic mesh adaptation and post-processing and optimization.

In conclusion, we wish to express our thanks and indebtedness to many people particularly those, who as colleagues or students, have worked with us in our various roles and our various employments and have contributed so much over many years to the understanding and experience which we have drawn upon here.

1

BASIC FIELD THEORY

1.1 INTRODUCTION

The electrical and magnetic fields which are central to many aspects of electrical and electronic engineering and to physics are all governed by the set of equations, defining the curl and divergence of the field quantities and known as Maxwell's equations, together with equations defining the continuity of the field quantities. Depending on particular physical conditions—essentially whether the fields are independent (or not) of time and, if they are dependent on time, whether the time variation is at a 'high' or a 'low' frequency—the nature of the field can be classified into the three types. In the first, when the fields are wholly invariant with time, or when the rate of change with time is sufficiently low as to be treated as invariant for practical purposes they are described respectively as static or 'quasi'-static. In the second, when the rates of change are low, induced or 'eddy' currents exist; and in the third, rates of change are sufficiently high that displacement current and wave behaviour characterize the fields.

In this book, with the exception of a special case mentioned below, attention is given to the static (and quasi-static) and to the eddy-current types of field. In the first case of static fields in a uniform medium, the Maxwell equations reduce to a single partial-differential equation known as Poisson's equation. This equation fully describes static fields, including the effects of distributed field sources. In the absence of distributed sources, it reduces to the simpler form known as Laplace's equation. In the second case, with time-varying fields, Maxwell's equations reduce to the diffusion equation. For steady harmonic conditions this yields an implicit (phasor) solution for the field variation, but more generally needs to be solved as an explicit function of time.

There are also two special cases for which simplified solutions of the above type are useful: in the first, eddy-currents are sufficiently strong that no flux can penetrate the conducting medium so that a Laplacian field model can be used for the region outside the medium; and for some radiation phenomena, such as the determination of the characteristic impedance of transmission lines, Laplacian solutions can also be used.

The methods of solution of the partial differential equation may be analytical or numerical and reflecting this the book is divided into two parts, the first dealing with the analytical and the second with numerical methods. In the first class, a solution takes the form of an algebraic function into which the values of the parameters defining the particular problem can be substituted. Generally speaking, analytical solutions are most likely to be successful in the case of fields which are, or can be (by the choice of co-ordinate system), treated as two-dimensional, the properties of the media occupying the field are linear, and the time variation is relatively 'simple'. Otherwise, numerical solutions, which take the form of a paricular set of numerical values of the function describing the field for one particular set of values of the physical parameters become appropriate (if not essential). Numerical methods can, of course, also be used in cases amenable to analytic solution and the decision as to which method should be used is dependent on available resources, the pre-existence (or not) of solution routines and experience—all aspects which will be explored later in the book. Analytical methods have, of course, provided the classical approach for at least 80 years, but, more recently, within the last 30 years and particularly the last 10 or 15, numerical methods have been developed alongside digital computers in very powerful forms. Despite this, it remains true that in a good number of cases analytical methods are still the most effective and, of course, they provide general insights which cannot easily be extracted from numerical procedures.

The analytical methods discussed in the first part of this book include the image method which provides a particularly economical approach to the fields of discrete sources and has conceptual significance, direct methods yielding solutions in the form of convergent series, conformal transformation methods, which are probably the most efficient means of solution for those two-dimensional fields to which the method applies. The numerical methods include those of finite differences, finite elements, integral and time-dependent solutions, both by time-stepping and use of complex variable formulations. As indicated, numerical methods provide not only a much more powerful approach to three-dimensional situations, but also to those involving distributed media particularly media with non-linear properties.

Analogous fields

In many aspects of engineering and physics there are physical phenomena which are directly analogous to electric and magnetic field phenomena. Amongst these are the

Table 1.1 Analogous quantities for different scalar potential fields.

Quantity	Electrostatic	Electric current	Magnetostatic	Heat flow	Fluid flow	Gravitational
Potential	Potential V	Potential V	Potential Ω	Temperature	Velocity potential	Newtonian potential
Potential gradient	Electric field strength E	Electric field strength E	Magnetic field strength H	Temperature gradient	Velocity	Gravitation force
Constant of medium	Permittivity ε	Conductivity σ	Permeability μ	Thermal conductivity	Density	Reciprocal of gravitation constant
Flux density	Electric flux density D	Current density J	Magnetic flux density B	Heat flow density	Flow rate	
Source strength	Charge density ρ_e	Current density J	Pole density ρ_m	Heat source density	Density of efflux	Mass density
Field conductance	Capacitance C	Conductance G	Permeance Λ	Thermal conductance		

flow of heat in conducting media and the flow of an inviscid liquid. For example, the temperature distribution between two boundaries having a constant temperature difference between them, or the distribution of the stream function of an ideal fluid passing between these boundaries, is identical in form with the voltage distribution between the same boundaries having a constant electric potential difference. Thus a solution to one problem of a particular physical type is directly applicable to other problems of different types, and methods developed in this book for electric and magnetic fields apply in a similar way to the other fields mentioned above. Table 1.1 shows the equivalence of quantities for the different physical types of field. In addition to the ones tabulated, consideration is given in the book to magnetic fields within regions of distributed current, and it is of interest to note that this type of field is analogous, for example, to that of fluid flow with vorticity.

1.2 QUASI-STATIC ELECTROMAGNETIC FIELD EQUATIONS

In this chapter the basic equations describing electromagnetic fields are presented, without detailed explanations, to remind the reader of the concepts and nomenclature that are used extensively in the later chapters. For the purpose of introducing the field equations it is convenient to begin with the elementary model problem shown in Fig. 1.1 in which a volume of conducting material Ω_c, with magnetic permeability μ, electric permittivity ε and electrical conductivity σ, bounded by a surface Γ_c, is contained within a volume of free space Ω bounded by a surface Γ which may be extended to infinity if required. The global region may also contain one or more prescribed conductor sources presumed not to intersect Ω_c. This configuration arises in a very large number of applications of practical importance, care is needed if Ω contains multiply-connected regions, see section 12.6.

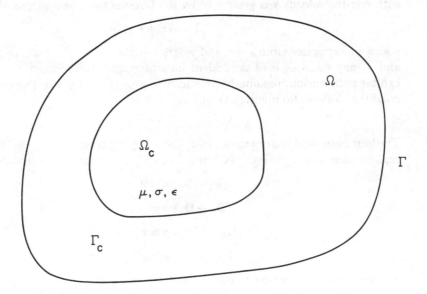

Fig. 1.1 Simple model configuration of arbitrary shape.

If the wavelengths of any time-varying fields are large compared with the physical dimensions of the problem, the displacement current term in Maxwell's equations is negligible compared with the free current density **J** and there is no radiation (see Stratton, p. 277). This regime means that the pre-Maxwell field equations, the so called *quasi-static* case, where Ampère's law is a good approximation, are applicable. In this situation the field equations can be approximated by:

$$\nabla \cdot \mathbf{D} = \rho_c \quad \text{(Gauss's law)}. \tag{1.1}$$

$$\nabla \cdot \mathbf{B} = 0. \tag{1.2}$$

$$\nabla \times \mathbf{E} = \frac{\partial \mathbf{B}}{\partial t} \quad \text{(Faraday's law)}. \tag{1.3}$$

$$\nabla \times \mathbf{H} = \mathbf{J} \quad \text{(Ampère's law)}. \tag{1.4}$$

D is the electric flux density and **E** is the electric field strength. **B** is the magnetic flux density and **H** the magnetic field strength. ρ_c is the free charge density and **J** the free current density at any point in the region.

The field vectors **D** and **E** and also **B** and **H** are related by the properties of the materials at any point in the field region. These are often referred to as the *constitutive properties* of the material and are given by:

$$\mathbf{D} = \varepsilon_r \varepsilon_0 \mathbf{E} \tag{1.5}$$

$$\mathbf{B} = \mu_r \mu_0 \mathbf{H}, \tag{1.6}$$

where ε_r and μ_r are the relative permittivity and permeability respectively and ε_0 and μ_0 are the primary electric and magnetic constants respectively. However, the symbol μ will also be used to denote the product $\mu_r \mu_0$ (i.e. $\mu = \mu_r \mu_0$) as is common practice. Equally ε will often be used as an alternative to $\varepsilon_r \varepsilon_0$. The current density in a conductor moving with relative velocity **v** is generated by the Lorentz force and is given by:

$$\mathbf{J} = \sigma(\mathbf{E} + \mathbf{v} \times \mathbf{B}), \tag{1.7}$$

which incorporates Ohm's law and where σ is the material conductivity. In practice μ and σ may often be field dependent quantities and, furthermore some materials will exhibit both anisotropic and hysteretic effects (see Chapter 10). The current continuity condition follows from eqn (1.4), and is:

$$\nabla \cdot \mathbf{J} = 0. \tag{1.8}$$

The four basic field vectors must satisfy the following conditions at the interfaces between regions identified by subscripts 1 and 2 of different material properties;

$$(\mathbf{B}_2 - \mathbf{B}_1) \cdot \mathbf{n} = 0, \tag{1.9}$$

$$(\mathbf{D}_2 - \mathbf{D}_1) \cdot \mathbf{n} = \omega, \tag{1.10}$$

$$(\mathbf{H}_2 - \mathbf{H}_1) \times \mathbf{n} = \mathbf{K}, \tag{1.11}$$

$$(\mathbf{E}_2 - \mathbf{E}_1) \times \mathbf{n} = 0, \tag{1.12}$$

where **K** and ω are the surface current and charge densities respectively. These conditions are referred to as *boundary or interface conditions*. Equation (1.9) expresses the continuity of magnetic flux across a boundary and eqn (1.10) is the equivalent form in terms of electric flux. Equation (1.11) results from Ampère's law and eqn (1.12) expresses the continuity of the tangential component of electric field strength. This is discussed in detail in section 1.5.

1.2.1 Magnetic vector potential

Since the field vector $\mathbf{B}$ satisfies a zero divergence condition, see eqn (1.2), it can be expressed in terms of a vector potential $\mathbf{A}$ as follows:

$$\mathbf{B} = \nabla \times \mathbf{A} \tag{1.13}$$

and then, from eqns (1.3) and (1.13), it follows that,

$$\nabla \times \left(\mathbf{E} + \frac{\partial \mathbf{A}}{\partial t} \right) = 0. \tag{1.14}$$

Hence, by integration, one obtains

$$\mathbf{E} = -\left(\frac{\partial \mathbf{A}}{\partial t} + \nabla V \right), \tag{1.15}$$

where V is a scalar potential. Neither $\mathbf{A}$ nor V are completely defined since the gradient of an arbitrary scalar function can be added to $\mathbf{A}$ and the time derivative of the same function can be subtracted from V without affecting the physical quantities $\mathbf{E}$ and $\mathbf{B}$. These changes to $\mathbf{A}$ and V are the so-called gauge transformations (see Stratton, p. 23), and the uniqueness of the solution is usually ensured by specifying the divergence (or gauge) of $\mathbf{A}$ together with the necessary boundary conditions. Thus in region Ω_c the field equations in terms of $\mathbf{A}$ and V are as follows (see section 12.5):

$$\nabla \times \frac{1}{\mu} \nabla \times \mathbf{A} + \sigma \left(\frac{\partial \mathbf{A}}{\partial t} + \nabla V \right) = \mathbf{J} \tag{1.16}$$

$$\nabla \cdot \sigma \left(\frac{\partial \mathbf{A}}{\partial t} + \nabla V \right) = 0 \tag{1.17}$$

and, in the global region where $\sigma = 0$ and $\nabla \times \mathbf{H} = \mathbf{J}_s$, eqn (1.2) reduces to

$$\nabla \cdot \mu \nabla \phi = 0, \tag{1.18}$$

where ϕ is the reduced magnetic scalar potential with $\mathbf{H} = \mathbf{H}_s - \nabla \phi$ for a source field $\mathbf{H}_s$. If μ is a constant then eqn (1.18) is known as Laplace's equation. See also section 1.8.

At points just inside a conductor, eqns (1.10) and (1.12) imply.

$$\mathbf{J}_n = -\sigma \left(\frac{\partial \mathbf{A}}{\partial t} + \nabla V \right) \cdot \mathbf{n} = 0 \tag{1.19}$$

Hence,

$$\frac{\partial \mathbf{A}_n}{\partial t} + \frac{\partial V}{\partial \mathbf{n}} = 0 \tag{1.20}$$

at conductor surfaces, and at interfaces, across which the conductivity changes from σ_1

to σ_2, eqn (1.10) further implies that,

$$\sigma_1\left(\frac{\partial \mathbf{A}}{\partial t} + \nabla V_1\right)\cdot\mathbf{n} = \sigma_2\left(\frac{\partial \mathbf{A}}{\partial t} + \nabla V_2\right)\cdot\mathbf{n}. \tag{1.21}$$

The question arises whether eqns (1.16) and (1.17) are sufficient for a solution to be found. It is clear that eqn (1.17) is a consequence of taking the divergence of eqn (1.16) and is not, therefore, independent. Some investigators have obtained unique solutions to eqns (1.16) and (1.17), as they stand, by using finite elements (see section 12.5), but they showed that the uniqueness depended upon the particular procedures used. Although this approach has been adopted by a number of workers it leaves a flexibility of the system unused. As has already been stated, it is necessary to specify the divergence (gauge) of $\mathbf{A}$ and appropriate boundary conditions to ensure uniqueness. The two commonest conditions are the Coulomb ($\nabla\cdot\mathbf{A} = 0$), and the Lorentz ($\nabla\cdot\mathbf{A} = \mu\sigma V$) gauges. Consideration of these is given in detail in section 12.5.

The use of the magnetic vector potential for 2D problems is widespread and for the very common limiting case of infinitely long models, where the current flows parallel to the z-axis, there will only be one component of $\mathbf{A}$ involved, namely A_z, and this quantity will depend upon x and y only, that is the solution is effectively two-dimensional. Since, by definition, A_z is now independent of the z-coordinate the Coulomb gauge is automatically imposed and V is independent of x and y and the solution is effectively two-dimensional.

1.2.2 Magnetic scalar potential

As has already been noted the free space region, Ω, shown in Fig. 1.1 can be modelled using a scalar potential, see eqn (1.18), and furthermore, static electric and static magnetic fields can likewise be modelled in all regions. For example, the magnetic field $\mathbf{H}$ can be partitioned into two fields namely, the field generated by the prescribed sources $\mathbf{H}_s$ and the field arising from induced magnetism in ferromagnetic materials $\mathbf{H}_m$. Thus,

$$\mathbf{H} = \mathbf{H}_m + \mathbf{H}_s \tag{1.22}$$

and, since from eqn (1.4) $\nabla \times \mathbf{H}_m = 0$, it follows that,

$$\mathbf{H} = -\nabla\phi + \mathbf{H}_s, \tag{1.23}$$

where ϕ is called the *reduced* scalar potential [1], and by definition for conductor source regions with current density $\mathbf{J}_s$ the source field is given by,

$$\mathbf{H}_s = \frac{1}{4\pi}\int_\Omega \mathbf{J}_s \times \nabla\left(\frac{1}{R}\right)d\Omega, \tag{1.24}$$

where $R = |r' - r|$ is the distance from the source point r' to the field point r, eqn (1.24) is known as the Biot–Savart law. See Stratton, p. 232.

In many cases, this can be integrated to give an analytic expression for $\mathbf{H}_s$; for complicated current paths, the expression can be integrated by a combination of analytic and numerical quadrature. The permanent magnet sources can be represented by a modified form of the constitutive relation, eqn (1.6), of the form,

$$\mathbf{B} = \mu(\mathbf{H})(\mathbf{H} - \mathbf{H}_c), \tag{1.25}$$

where μ is a non-linear function of $\mathbf{H}$ and is in general a tensor, and $\mathbf{H}_c$ is the coercive field for the material. In 'soft' magnetic materials, the coercive field intensity is normally assumed to be zero.

The governing equation for the reduced scalar potential, ϕ, is obtained by taking the divergence of eqn (1.25), i.e.

$$\nabla \cdot \mu \nabla \phi = \nabla \cdot \mathbf{H}_c + \nabla \cdot \mu \mathbf{H}_c + \nabla \cdot \mu \mathbf{H}_s, \tag{1.26}$$

where eqn (1.26) is now a generalization of eqn (1.18). Whilst direct solutions of eqn (1.26) are possible in magnetic materials, the two parts of the field $\mathbf{H}_m$ and $\mathbf{H}_s$ tend to be of similar magnitude but opposite in direction, so that cancellation occurs in computing the field intensity $\mathbf{H}$, giving a loss in accuracy (see section 12.2.1). This loss is particularly severe when μ is large. Fortunately, for regions where there are no conductor sources the total field $\mathbf{H}$ can be represented by a scalar potential since, in this case, $\nabla \times \mathbf{H} = 0$ and it follows that

$$\mathbf{H} = -\nabla \psi, \tag{1.27}$$

where ψ is known as the total scalar potential. The governing equation for regions without currents is given by,

$$\nabla \cdot \mu \nabla \psi = \nabla \cdot \mu \mathbf{H}_c. \tag{1.28}$$

Both eqns (1.26) and (1.28) are forms of the generalised Poisson equation and may be linear or non-linear depending upon eqn (1.25). See also section 1.8. It is clear that the total scalar potential should be used to avoid cancellation errors, but unfortunately it cannot represent the whole problem since in regions where there are currents this potential is multivalued (see Stratton, p. 254). This difficulty can become acute when three-dimensional solutions are required.

1.2.3 Electric vector potential

Since the current density $\mathbf{J}$ satisfies the divergence condition, eqn (1.8), it can be expressed in terms of a vector potential $\mathbf{T}$ as follows:

$$\mathbf{J} = \nabla \times \mathbf{T}, \tag{1.29}$$

with

$$\mathbf{H} = (\mathbf{T} - \nabla \phi), \tag{1.30}$$

where ϕ is a magnetic scalar potential. The corresponding field equations in terms of the electric vector potential $\mathbf{T}$ are obtained from eqns (1.3), (1.29) and (1.30) and are given by

$$\nabla \times \frac{1}{\sigma} \nabla \times \mathbf{T} + \mu \frac{\partial}{\partial t} (\mathbf{T} - \nabla \phi) = 0 \tag{1.31}$$

and

$$\nabla \cdot \mu (\mathbf{T} - \nabla \phi) = 0 \tag{1.32}$$

for the case where σ is piecewise constant, and in the global region where $\sigma = 0$ and $\nabla \times \mathbf{H} = 0$, eqn (1.32) limits to eqn (1.18) as before. There have been a number of successful implementations of this approach for the numerical solution of three-dimensional eddy current problems, [2, 3 and 4].

1.2.4 H field formulations

The corresponding field equation in terms of the field intensity **H** is obtained directly from eqns (1.1), (1.2) and (1.3) and is given by:

$$\nabla \times \nabla \times \mathbf{H} + \sigma \frac{\partial}{\partial t}(\mu \mathbf{H}) = 0 \qquad (1.33)$$

however, in the global region where $\sigma = 0$ and $\nabla \times \mathbf{H} = 0$, eqn (1.52) limits to eqn (1.18) as before. A direct application of classical finite elements, using nodal basis functions, in the conducting regions leads to difficulties. This is because the physical interface conditions, eqns (1.11) and (1.12), do not emerge naturally when a variational or weighted residual method is applied to eqn (1.33). The implications and solution to this problem in a numerical context will be addressed in section 12.7.

Equation (1.33) is an example of the diffusion equation and limits to the Laplace equations for time-invariant conditions.

1.3 STATIC FIELDS

The preceding formulations allow the following three cases describing static fields to be distinguished. These fields are all described by a potential function which satisfies Poisson's equation. In order for the potential to be guaranteed unique, boundary conditions must also be specified (see Appendix 6).

1.3.1 Magnetostatics

In two dimensions the vector potential, equation, reduces to the single component equation as follows:

$$\nabla \cdot \frac{1}{\mu} \nabla A_z = -\mathbf{J}_s. \qquad (1.34)$$

The full vector form required for the 3D statics case is not efficient for most problems whereas the alternative scalar forms of section 1.2.2 are very appropriate, i.e.

$$\nabla \cdot \mu \nabla \psi = \nabla \cdot \mu \mathbf{H}_c. \qquad (1.35)$$

1.3.2 Electrostatics

For electrostatic problems the field eqns (1.1, 1.2) reduce directly to the Poisson equation to:

$$\nabla \cdot \varepsilon \nabla V = -\rho, \qquad (1.36)$$

where V is the electrostatic potential and ρ is the charge density.

1.3.3 Current flow

For static steady state current flow the continuity eqn (1.8) reduces to:

$$\nabla \cdot \sigma \nabla V = 0, \tag{1.37}$$

where σ is the electrical conductivity.

1.4 INTEGRAL EQUATION FORMULATIONS

The integral equation forms of the governing field equations are often a viable alternative to the differential forms. There are many different types that can be derived, see for example [5, 6, 7], however in this chapter only two formulations appropriate to static problems will be considered.

1.4.1 Volume magnetization integral equation method

For example, in magnetostatics the magnetization vector $\mathbf{M}$ given by:

$$\mathbf{M} = (\mu_r - 1)\mathbf{H} \tag{1.38}$$

can be used instead of $\mathbf{H}$ in eqn (1.22) to derive an integral equation over all ferromagnetic domains of the problem [5].

Since,

$$\mathbf{H}_m = -\frac{1}{4\pi} \nabla \int_\Omega \mathbf{M} \cdot \nabla \left(\frac{1}{R}\right) d\Omega \tag{1.39}$$

see Stratton, p. 229, and using eqns (1.22) and (1.38) the following integral equation results:

$$\mathbf{M}(r) = (\mu_r - 1)\left[\mathbf{H}_s(r') - \frac{1}{4\pi} \nabla \int_\Omega \mathbf{M}(r') \cdot \nabla \left(\frac{1}{R}\right) d\Omega\right]. \tag{1.40}$$

1.4.2 The boundary element method

The general boundary element method requires either the function or its normal derivative to be given at all boundary points.

The classic boundary element method (BEM) first applied to mechanical problems [8] and later to electromagnetics [9] is more general in that bounding surfaces with assigned boundary conditions can be included, but now additional complexity arises because normal derivative values of potential have to be defined. The method is based on Green's theorem to obtain solutions inside defined volumes in terms of surface values of potential and the normal derivative of potential (see section 9.7).

If in region Ω, with surface Γ there exists two differentiable scalar functions u and v, then Green's second theorem can be written as follows:

$$\int_\Gamma \left(u \frac{\partial v}{\partial n} - v \frac{\partial u}{\partial n}\right) d\Gamma = \int_\Omega (u\nabla^2 v - v\nabla^2 u)\, d\Omega. \tag{1.41}$$

If u is a singular solution of Laplace equation (e.g. $u = 1/R$ in 3D and $u = -\log R$ in 2D) and if ϕ, a magnetic scalar potential say, is associated with v then for linear problems eqn (1.69) becomes,

$$4\pi\phi = -\int\left(\frac{1}{R}\frac{\partial\phi}{\partial n} - \phi\frac{\partial(1/R)}{\partial n}\right)d\Gamma, \tag{1.42}$$

i.e. given ϕ or $\partial\phi/\partial n$ on Γ, ϕ is uniquely defined in Ω. Problems with many regions with or without sources can be solved by applying eqn (1.42) to each region simultaneously using total scalar potential ψ where there are no sources and the reduced potential ϕ elsewhere see section 9.7.2. The additional equations required at the interfaces, where neither the potential nor its derivative are known, are supplied by the interface.

1.5 INTERFACE AND BOUNDARY CONDITIONS

1.5.1 Interface condition

When a field has regions with different electric or magnetic properties, separate functions are often used to describe the fields in the different regions. It is simple to establish the relationship between these functions at the boundaries, and this is done here for an interface between two regions in a magnetic field. Directly analogous equations exist for electric fields.

Consider an elemental length bd of the interface between two regions of relative permeability μ_1 and μ_2. Flux passes from one region to the other and a pair of parallel flux lines passing through b and d are considered to bend at the boundary (Fig. 1.2). The lines ab and cd are normal to the flux lines ϕ' and ϕ'' respectively. Now unless the current density on the boundary itself is infinite (a case discussed later), the work done in taking unit pole round the typical infinitesimally small path $abcda$ is zero. But ab and dc are equipotential lines, and so the work done from b to c plus the work done from d to a is zero. Hence if H_1 and H_2 are the field strengths on either side of the interface.

$$-H_1(bc) + H_2(da) = 0, \tag{1.43}$$

the negative sign denoting movement in a direction opposite to that of H_1. If θ_1 and

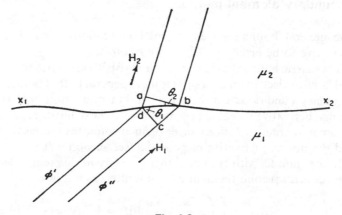

Fig. 1.2

θ_2 are the angles made by dc and ab with db, then

$$da = db \sin \theta_2 \quad \text{and} \quad bc = db \sin \theta_1,$$

and combining these with (1.43) gives

$$H_1 \sin \theta_1 = H_2 \sin \theta_2. \tag{1.44}$$

Hence, in the absence of an interface current, the tangential components of H on each side of an interface are equal. Again, since flux is continuous, if B_1 and B_2 are the densities on each side of the boundary,

$$B_1(dc) = B_2(ab),$$

and so

$$B_1 \cos \theta_1 = B_2 \cos \theta_2. \tag{1.45}$$

Thus the normal components of flux density are equal on each side of the boundary.

It can be seen by combining eqns (1.44) and (1.45) that

$$\frac{H_1}{B_1} \tan \theta_1 = \frac{H_2}{B_2} \tan \theta_2,$$

and so

$$\frac{\tan \theta_1}{\tan \theta_2} = \frac{\mu_1}{\mu_2}. \tag{1.46}$$

If the permeability is infinite on one side of the boundary, then θ is zero on the other side, which means that flux enters an infinitely permeable (equipotential) surface at right angles.

In the case when current of infinite density (as in a current sheet) is distributed along the boundary surface, the normal components of flux density on the two sides of the boundary are still equal, but the tangential components of field strength are discontinuous by an amount equal to the magnitude of the surface density S; i.e.,

$$-H_1 \sin \theta_1 + H_2 \sin \theta_2 = S. \tag{1.47}$$

For any given problem the relationship connecting the field vectors on the two sides of an interface, eqn (1.45) and eqn (1.44) or (1.47) are expressed in terms of the appropriate derivatives of the functions ψ and ϕ (or defining the fields). In terms of the potential functions ψ_1, and ψ_2, eqns (1.44) and (1.45) become

$$\frac{\partial \psi_1}{\partial s} = \frac{\partial \psi_2}{\partial s}, \tag{1.48}$$

where s is distance measured tangential to the interface, and

$$\mu_1 \frac{\partial \psi_1}{\partial n} = \mu_2 \frac{\partial \psi_2}{\partial n},$$

where n is distance measured normal to the interface. In terms of ϕ they become

$$\frac{1}{\mu_1} \frac{\partial \phi_1}{\partial n} = \frac{1}{\mu_2} \frac{\partial \phi_2}{\partial n}$$

and

$$\frac{\partial \phi_1}{\partial s} = \frac{\partial \phi_2}{\partial s}. \qquad (1.49)$$

These equations are used in terms of the particular coordinate system appropriate to a particular problem.

When current is distributed on one or both sides of the interface, the above boundary conditions are conveniently expressed in terms of vector potential. Then, since the normal components of flux density are equal on the two sides of a boundary, the tangential gradients of the vector potential function are of equal magnitude and A replace ϕ in eqn (1.49). Also, as mentioned in the example of a cylindrical conductor, the vector potential function is made continuous across a boundary.

1.5.2 Equivalent pole and charge distributions

For a number of purposes—in particular the calculation of forces on boundaries (see the next section) and the derivation of image solutions (see Chapter 2)—it is best to consider the effect of a boundary is being due simply the charges, poles, or currents which lie along the boundary line. The following discussion of such surface distributions is restricted to magnetic fields and pole distributions, but, of course, electric fields and charge distributions are analogous. Surface currents can also be used as an alternative to poles, but the two representations are equivalent and surface poles are easier to handle.

Consider a region 1 of permeability $\mu_1\mu_0$ separated from a region 2 of permeability $\mu_2\mu_0$ by a boundary of arbitrary shape (Fig. 1.3). The distribution of poles, which when lying along the boundary line, gives the same effect on the field in region 1 as does the presence of the interface, is to be found. It is convenient to consider the boundary as consisting of an infinitely thin region of permeability μ_0 in which the pole distribution lies. At any point on the boundary, let H_n be the normal component of the applied field strength, i.e. the field strength due to all the field sources which may be in either or both of the regions in the absence of polarized media. The effect of the polarized media can be accounted for by a normal component of field H'_n, at the boundary, and considered to act in the same direction as H_n in region 1.

Thus the resultant normal field at a point on the boundary is $H_n + H'_n$ in region 1 and $H_n - H'_n$ in region 2, and since the normal component of flux is continuous across

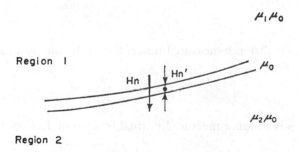

Fig. 1.3

the boundary it is necessary that

$$\mu_1(H_n + H'_n) = \mu_2(H_n - H'_n)$$

or

$$H'_n = \left(\frac{\mu_2 - \mu_1}{\mu_2 + \mu_1}\right) H_n. \tag{1.50}$$

The surface pole density ϱ_s which gives rise, in a region of permeability μ_0, to the component H'_n, and so to the effect of the boundary, is found simply. Since flux passes equally in each of the two directions normal to the boundary,

$$\varrho_s = 2\mu_0 H'_n, \tag{1.51}$$

and so, eliminating H'_n between eqns (1.50) and (1.51), ϱ_s is related to the applied field by

$$\varrho_s = 2\mu_0 \left(\frac{\mu_2 - \mu_1}{\mu_2 + \mu_1}\right) H_n. \tag{1.52}$$

For the calculation of force discussed in the next section, it is convenient to express the pole density in terms of the *resultant* field strength in region 1, H_{n1}. This equals $H_n + H'_n$, and from eqns (1.50) and (1.52) it is seen that

$$\varrho_s = \frac{\mu_0}{\mu_2} (\mu_2 - \mu_1) H_{n1}. \tag{1.53}$$

1.6 CALCULATION OF GLOBAL QUANTITIES

1.6.1 Capacitance

The capacitance between two conducting surfaces is given by the ratio of total flux common to the surfaces to the potential difference between them. Consequently it may be evaluated conveniently from the flux and potential functions. If ψ_1 and ψ_2 are the potential functions of the two conductors, and if ϕ' and ϕ'' are the values of the flux functions for the lines bounding the mutual flux, then the capacitance C is given by

$$C = \frac{\phi' - \phi''}{\psi_1 - \psi_2}. \tag{1.54}$$

This relationship is valid even when more than two conductors are present. When there are only two conductors (one of which may be at infinity) the flux between them is equal to the charge q on either, and

$$C = \frac{q}{\psi_1 - \psi_2}. \tag{1.55}$$

In the case of two charged concentric cylinders the potential function has been shown to vary as

$$\frac{q}{2\pi\varepsilon_0\varepsilon} \log r,$$

and so, if the boundaries have radii of r_1 and r_2, the difference in potential function between the boundaries is

$$\frac{q}{2\pi\varepsilon_0\varepsilon}\log r_1 - \frac{q}{2\pi\varepsilon_0\varepsilon}\log r_2.$$

Therefore the capacitance between the cylinders is

$$q\bigg/\left(\frac{q}{2\pi\varepsilon_0\varepsilon}\log\frac{r_1}{r_2}\right),$$

which equals

$$\frac{2\pi\varepsilon_0\varepsilon}{\log r_1/r_2}. \tag{1.56}$$

1.6.2 Inductance

The concept of inductance is most useful in analysing the effects in a circuit of the magnetic fields of currents changing in magnitude with time. Though a discussion of its significance involves a consideration of induced e.m.f. and stored energy and so is beyond the scope of this book,[†] a definition is given in mathematical form and its use is demonstrated.

Consider a conductor of arbitrary cross-section, as shown in Fig. 1.4 in a magnetic field described by the vector potential function A. The field may be due to current in the conductor itself, giving rise to what is termed *self*-inductance, or to currents in other conductors, giving rise to *mutual* inductance. Let δS be any elemental area of the conductor and let the vector potential function of the field at the position of the element be A. Then the inductance L, associated with all the flux linking the conductor but bounded by the flux line, $A = A_0$, is defined by

$$L = \frac{1}{IS_0}\int (A_0 - A)\,\mathrm{d}S, \tag{1.57}$$

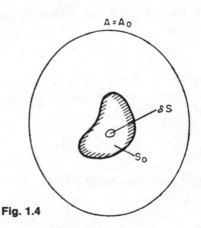

Fig. 1.4

[†]For a full discussion the reader should consult the book by Carter.

where I is the current giving rise to the field and where S_0 is the total area of the conductor, the integration being performed over this area.

For a rectangular conductor the integration is conveniently performed using cartesian coordinates with axes chosen to be parallel to the sides of the conductor. Then

$$L = \frac{1}{IS_0} \int_{x_1}^{x_2} \int_{y_1}^{y_2} (A_0 - A)\,dx\,dy, \qquad (1.58)$$

where x_1 and x_2 are the limits of the conductor on the x-axis and y_1 and y_2 are those on the y-axis.

In evaluating inductance analytically it is simply necessary to determine the function A for the field, to substitute in eqn (1.57) and to perform the integration. If the field solution is known in the form of a map of flux lines, the conductor is divided up into finite elements, each bounded by flux lines; and then taking a mean value of A for each element the sum of the terms $(A_0 - A)\delta S$, one for each element, is obtained.

Equation (1.47) has been used by Billig [11] and is used, in section 4.3.3, to determine analytically the inductance of a rectangular conductor in a slot.

1.7 FORCES

The force acting on a line source in a field is found simply, but the calculation of forces acting on boundaries is often difficult. There are several methods available for calculating *total* force on a boundary, and two of the most useful ones are discussed. These are reasonably simple to apply and give solutions for most fields of practical interest. For general reading on the calculation of force the authors recommend the paper by Carpenter [12] and the books by Moullin, Stratton, Carter, and Maxwell. The calculation of force *distribution* is briefly mentioned in Maxwell and reference [12], but the application to practical problems is not discussed, except by Carter [13] and Hammond [14]. In section 1.7.4 is given a simple method applicable to equipotential boundaries.

1.7.1 Line sources

In any physical problem sources have a finite cross-sectional area, but in most calculations it is sufficiently accurate to treat them as line sources. The force per unit length experienced by a line source in a field (of the same kind as that of the source) is given by the product of the field strength (or in the case of a line current flux density) at the position of the source due to all the other influences, and the strength of the source. For a line charge q per unit length in an electric field of strength E, the force is Eq; for a line pole in a magnetic field it is Hm; and for a line current in a magnetic field it is Bi.

1.7.2 Distributed sources

The force experienced by a field source, which is distributed over an area, can be calculated by considering the source as being made up of line elements and by summing the forces on these elements. The field strength at the position of an element is calculated as that due to all the field influences except the source considered. If the field of the other elements of the source is taken into account, the force calculated includes a contribution from he internal forces in the source and these sum to zero when the total external force is calculated.

1.7.3   Total force acting on a boundary

Simple expressions for the total force acting on a boundary are derived in terms of the magnetic field; analogous relations exist for the electric field.

Equivalent pole distribution

The first method is based on a consideration of the force exerted on the surface pole distribution which accounts for the influence of the boundary on the external field. The method does not give the force *distribution* for a (practical) piece of iron but, since the external field is truly represented, it does give the correct value for the total force. The required surface pole distribution is given by eqn (1.53) in terms of H_{n1}.

Now the normal component of force per unit length F_n acting on the surface poles is given by

$$F_n = H_n \varrho_s,\tag{1.59}$$

where H_n is the applied normal field strength. The tangential component of force per unit length F_T is given by

$$F_T = H_{T1} \varrho_s,\tag{1.60}$$

where H_{T1} is the tangential field strength. From eqn (1.50) remembering that $H_{n1} = H_n + H'_n$ it is simply shown that

$$H_n = H_{n1} \left(\frac{\mu_2 + \mu_1}{2\mu_2} \right).\tag{1.61}$$

Hence, substituting for ϱ_s from eqn (1.53) in eqn (1.59) gives

$$F_n = \tfrac{1}{2}\mu_0 (1 - \mu_1^2/\mu_2^2) H_{n1}^2,\tag{1.62}$$

and in eqn (1.60) gives

$$F_T = \mu_0 (1 - \mu_1/\mu_2) H_{n1} H_{T1}.\tag{1.63}$$

The total force on the iron is given by integration of these functions over the whole of the boundary surface.

Virtual work

The mechanical force acting on a magnetic material can be determined from the effect, on the energy balance, of a small displacement dx in the direction of the force. The equations given below are derived, for example, in Chapter 2 of the book by Seely.

If the potential difference between two iron surfaces is assumed constant and the flux passing between them changes by an amount $d\phi$ for a displacement dx, the force in the

x-direction is given by

$$f_x = \frac{1}{2}\psi\frac{d\phi}{dx}. \tag{1.64}$$

If the flux is assumed constant and the potential changes by an amount $d\psi$, the force is then given by

$$f_x = \frac{1}{2}\phi\frac{d\psi}{dx}. \tag{1.65}$$

The assumption of constant potential difference or flux is made as convenient for the analysis. Equation (1.64) is used later (see section 6.2.8) in the analysis of the force on the armature of a contactor.

1.7.4 Force distribution over a boundary

The evaluation of force distribution is in general extremely difficult but, when the boundaries are so highly permeable that it can be assumed that H is zero inside the iron boundary, a simple method may be used. The magnetic forces act only on the surface poles, and since $\mu_2 = \infty$, eqns (1.62) and (1.63) become

$$F_n = \tfrac{1}{2}\mu_0 H_n^2$$

and

$$F_T = 0. \tag{1.66}$$

Hence the force, which is normal to the boundary at any point, is obtained by integrating the square of the field strength along the boundary.

1.8 FIELD EQUATIONS IN PARTIAL DIFFERENTIAL FORM

The field equations for magnetostatic and electrostatic fields can be expressed as partial differential equations as can the diffusion equation which describes time-varying fields at 'low' frequency, i.e. negligible displacement current. Equation (1.36) which describes the electrostatic field in terms of the (electrostatic) potential V can be written in the form:

$$V^2 V = -\rho/\varepsilon \tag{1.67}$$

which is called Poisson's equation. For all points at which the charge density ρ is zero, the equation clearly reduced to

$$V^2 V = 0 \tag{1.68}$$

which is called Laplace's equation.

Equations analogous to (1.67) apply to the magnetostatic field, in linear media, so that the equation in terms of vector potential, equivalent to eqn (1.67), is then:

$$\nabla^2 \mathbf{A} = -\mu\mathbf{Js} \tag{1.69}$$

which for the two-dimensional case reduces to

$$\nabla^2 A_z = -\mu J_{sz} \tag{1.70}$$

for current flow in the z-direction. Care needs to be taken in modelling magnetic fields due to current sources, because the potential is multivalued, see sections 3.2.1, 8.2 and Appendix 2 for example.

The diffusion equation describes the time-varying magnetic fields, such as those of eddy currents, and can be expressed for two-dimensional problems as:

$$\nabla^2 \mathbf{A} = \mu \frac{\partial \mathbf{A}}{\partial t} \tag{1.71}$$

in terms of the vector potential. It can also be written in terms of field strength **H**, see eqn (1.33), or flux density **B** instead of **A**.

Each of these field equations can be expressed in a wide range of coordinate systems as convenient for solution. This form of expression is essential for the achievement of a practical analytical solution and the systems most commonly encountered are Cartesian or circular cylinder coordinates. (See Chapters 3 and 4.)

The three-dimensional Cartesian form of eqn (1.67) for example is:

$$\frac{\partial^2 V}{\partial x^2} + \frac{\partial^2 V}{\partial y^2} + \frac{\partial^2 V}{\partial z^2} = -\frac{\rho}{\varepsilon} \tag{1.72}$$

which reduces to

$$\frac{\partial^2 V}{\partial x^2} + \frac{\partial^2 V}{\partial y^2} = -\rho/\varepsilon \tag{1.73}$$

in two dimensions.

In two-dimensional circular-cylinder coordinates, Laplace's equation for a scalar field, for example, takes the form:

$$r^2 \frac{\partial^2 V}{\partial r^2} + r \frac{\partial V}{\partial r} + \frac{\partial^2 V}{\partial \theta^2} = 0. \tag{1.74}$$

See Chapters 4 and 9 for discussion of two- and three-dimensional and vector fields.

Note that all of the above equations have identical left-hand sides with the operator ∇^2, known as the Laplacian, operating in the appropriate way on the chosen field function, depending upon whether this is scalar or vector. The solutions themselves naturally increase in complexity as the number of space coordinates increase from one to three, with vector as compared with scalar field functions, and with finite right-hand sides, particularly involving time variation. This is reflected in the progressive development of the analytical solutions in Chapter 4.

1.9 SUMMARY

The material of this chapter provides the basis from which to compute most electromagnetic fields when properly defined, it is important to have an understanding of some of the basic concepts and equations for simple field geometries. Some of these solutions are listed in an appendix: the electric field of a point and line charge, the use of flux and potential functions in two dimensions to produce field maps, the calculation of capacitance and inductance, the expression of the field of line currents and charges for use in conformal transformation and the use of conjugate functions. In addition the fundamental equations of the Biot–Savart law are presented in another appendix.

The chapter has the aim of providing, without undue explanation, the concepts and equations required for an understanding of later chapters.

REFERENCES

[1] J. Simkin and C. W. Trowbridge, On the use of the total scalar potential in the numerical solution of field problems in electromagnets, *IJNME*, **14**, 423 (1979).

[2] T. W. Preston and A. B. J. Reece, Solution of 3-Dimensional eddy current problems: T-Ω method, *IEEE Transactions on Magnetics*, **18** (March), 486 (1982).

[3] R. Albanese, R. Martone, G. Miano, and G. Rubinacci, A T formulation for 3D finite element eddy current computation, *IEEE Transactions on Magnetics*, **21** (November), 2299 (1985).

[4] T. Nakata, N. Takahashi, K. Fujiwara, and Y. Okada, Improvements of the T-Ω method for 3-D eddy current analysis, *IEEE Transactions on Magnetics*, **24** (January), 94 (1988).

[5] C. W. Trowbridge, *Applications of Integral Equation Methods for the Numerical Solution of Magnetostatic and Eddy Current Problems*. Chichester: Wiley (1979).

[6] R. F. Harrington, *Field Computation by Moment Methods*. New York: The Macmillan Company (1968).

[7] J. Moore *et al.*, *Moment Methods in Electromagnetics*. Chichester, UK: Research Studies Press (1983).

[8] M. A. Jaswon, Integral equation methods in potential theory, *Proc. Roy. Soc. A*, **275**, 23 (1963).

[9] J. Simkin and C. W. Trowbridge, Magnetostatic fields computed using an integral equation derived from Green's theorem, in *Compumag Conference on the computation of magnetic fields*, (Chilton, Didcot, Oxon, UK), p. 5, Rutherford Appleton Laboratory (1976).

[10] G. Liebmann, Electrical analogues, *Br. J. Appl. Phys.*, **4**, 193 (1953).

[11] E. Billig, The calculation of the magnetic field of rectangular conductors in a closed slot and its application to the reactance of transformer windings, *Proc. Instn. Elect. Engrs.*

[12] C. J. Carpenter, Surface-integral methods of calculating forces in magnetized iron parts, *Proc. Instn. Elect. Engrs.*, **107C**, 19 (1960).

[13] G. W. Carter, Distribution of mechanical forces in magnetized material, *Proc. Instn. Elect. Engrs.*, **112**, 1771 (1965).

[14] P. Hammond, Forces in electric and magnetic fields, *Bull. Elect. Engng. Educ.*, **17** (1960).

[15] B. Aldefeld, Forces in electromagnetic devices, *Conf. Proc.*, COMPUMAG-78, paper 8.1, Grenoble. J. Penman and M. D. Grieve, 'Efficient calculation of Force in Electromagnetic Devices', *Proc. IEE*, **133B**(4), 212–216.

[16] J. L. Coulomb, A methodology for the determination of global electromechanical quantities from a finite element analysis and its application to the evaluation of magnetic forces, torques and stiffness, *IEEE Trans. on Magnetics*, **MAG-19**(6), 2514–2519 (Nov. 1983).

[17] J. L. Coulomb and G. Meunier, Finite element implementation of virtual work principle for magnetic or electric force and torque computation, *IEEE Trans. on Magnetics*, **MAG-20**(5), 1894–1896 (Sept. 1984).

[18] A. O. Mohammed, A virtual work—magnetic vector potential method for the calculation of three dimensional forces on magnetized ferrous cores, *Conf. Proc.*, IEEE SOUTHEASTCON-85, pp. 64–68, Rayleigh.

[19] D. A. Lowther and P. P. Silvester, *Computer-Aided Design in Magnetics*, pp. 194–195, Springer-Verlag, New York (1985).

[20] J. Simkin, Recent developments in field and force computation, *J. Phys. Colloq.* (France), **45**(C-1), 851–860 (Jan. 1984).

[21] S. McFee and D. A. Lowther, Towards accurate and consistent force calculation in finite element based computational magnetostatics, accepted for publication, *IEEE Trans. on Magnetics* (1987).

2

IMAGES

2.1 INTRODUCTION

The method of images can be used to give solutions to some important problems involving straight-line or circular boundaries and in a particularly simple manner; for it offers certain ready-made solutions which eliminate the need for formal solutions of Laplace's and Poisson's equations. The idea of images for field problems is due to Lord Kelvin, but Maxwell, Lodge [1], and Searle [2] extended the scope of the method.

The essence of the method consists in replacing the effects of a boundary on an applied field by simple distributions of currents or charges *behind* the boundary line (called images), the desired field being given by the *sum* of the *applied* and the *image* fields. A different system of images is required for the field on each side of a boundary, but a knowledge of one group of images quickly leads to the other, since the solutions for the two regions are connected by the boundary conditions.

In the following discussion, the magnitudes and positions of images for the single straightline and circular boundaries are established, for convenience, in terms of the electric field of a line charge using a method first indicated by Hammond [3]; the distribution of surface charge (or polarity) representing the influence of the boundary is first found and is then replaced by a simple equivalent array (the images). (As shown in Appendix 1, the electric field of a line charge and the magnetic field of a line current are analogous, and so, from a knowledge of the images of the electric one, the images of the magnetic one may be deduced directly.) There is no general method of deriving the images for any given problem with multiple boundaries, though a method of successive approximation (see Maxwell, article 315), gives correct results in certain cases.

It is possible to derive image solutions from a solution to Poisson's equation obtained by use of the complex Fourier transform. This method is described in a paper by Mullineux and Reed [4]. However, the mathematical manipulation involved in transforming the field equation is considerable compared with the elementary trigonometry needed in the method to be described. Also some image systems can be derived using theorems developed originally for the solution of problems in hydrodynamics [5]. Thus the images for a circular boundary can be deduced from Milne-Thomson's circle theorem (see Milne-Thomson, p. 154). It is possible to check all solutions obtained (by the above or other means) by considering the boundary conditions. If a set of images is proposed for each region of a field, their validity can be tested directly. If a set is proposed for one region only, it can be verified by consideration of surface charge distribution. A simpler check is possible when the boundaries are equipotential or flux line, for then the boundary conditions are easily seen to be satisfied by the system of images which is symmetrically disposed about all the boundary lines.

It is convenient here to consider images under two headings: firstly those due to plane boundaries, and, secondly, those due to circular boundaries. For both of these, consideration is given to the field of line charges (or currents), but the effect of a single circular boundary on an applied uniform field (images of doublets) is also discussed. The image representation of the fields of distributed currents is discussed in section 2.4.

Images are also very useful for solving *dimensional* fields, involving plane boundaries with *any shape of conductor*, for example, in problems involving windings near iron boundaries [3, 6, 13, 14], but a discussion of these does not fall within the scope of the book.

2.2 PLANE BOUNDARIES

The images for a line charge near an infinite plane boundary are first established, and then the results for this boundary are extended to give solutions for various combinations of plane boundaries. These include two parallel and up to four intersecting plane boundaries, the angles of intersection being submultiples of π. For the single boundary the solution for finite permittivity is given but, for multiple-boundary problems the discussion is restricted to equipotential and flux-line boundaries[†]. In the cases considered the images are symmetrically disposed about all of the boundaries. Their positions are those of the optical images, in reflecting surfaces coincident with the boundaries, of an object coincident with the charge.

2.2.1 Single plane boundary

Consider the derivation of the image charges which give the solution for the field of a line charge q near to an infinite plane boundary (Fig. 2.1). The charge lies in region 1 of permittivity $\varepsilon_0\varepsilon_1$, region 2 having permittivity $\varepsilon_0\varepsilon_2$. Let the two regions be in the (x, y) plane, the boundary coinciding with the x-axis and the charge being at the point $y = a$.

At any point P on the boundary line the normal component of applied field E_n is given by

$$E_n = \frac{q\cos\alpha}{2\pi\varepsilon_0\varepsilon_1\sqrt{x^2 + a^2}}, \tag{2.1}$$

where α is as shown in Fig. 2.1. Substituting for α in terms of a and x gives

$$E_n = \frac{qa}{2\pi\varepsilon_0\varepsilon_1(x^2 + a^2)}, \tag{2.2}$$

[†]Maxwell, article 317, has given solutions for fields due to a point charge involving parallel boundaries of different conductivities, and these solutions can be extended by superposition to apply to line charges.

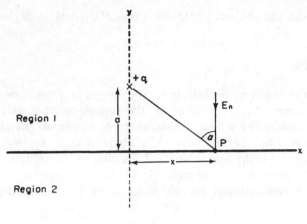

Fig. 2.1

and this component of the applied field can be considered to induce charge along the boundary. As shown in section 1.5, the normal component of field strength E'_n due to this equivalent distribution of surface charge is simply related to the applied normal field by the equation

$$E'_n = E_n \left(\frac{\varepsilon_2 - \varepsilon_1}{\varepsilon_2 + \varepsilon_1} \right). \tag{2.3}$$

Thus the effect of the boundary on the field in region 1 can be completely specified by the component E'_n along the boundary, and this is given, by substitution from eqn (2.2) in eqn (2.3), as

$$E'_n = \frac{q}{2\pi\varepsilon_0\varepsilon_1} \left(\frac{\varepsilon_2 - \varepsilon_1}{\varepsilon_2 + \varepsilon_1} \right) \frac{a}{x^2 + a^2}. \tag{2.4}$$

This is the same field distribution, however, as would result either from a charge $-q(\varepsilon_2 - \varepsilon_1)/(\varepsilon_2 + \varepsilon_1)$ at a distance a behind the boundary line, or, equally, from a charge $+q(\varepsilon_2 - \varepsilon_1)/(\varepsilon_2 + \varepsilon_1)$ at a distance a from the boundary in region 1. Hence either of these charges can be used to account for the effect of the boundary. However, only one of them can be used in the representation of the field in each of the two regions. Thus, considering first the field in region 1, it is seen that the second of the charges would cause a change in the total flux entering or leaving any curve enclosing the charge q in region 1 [that is eqn (1.1) for the divergence of flux would not be obeyed], and so it cannot be used to represent the field in region 1. The other charge, however, lies *outside* region 1, and so it does not affect the divergence of flux there. This charge, called an image charge, gives, in conjunction with the actual charge q, the field solution in region 1. (It should be emphasized that this image charge does not apply to the field in region 2.)

In order to derive the image charge for region 2, it is simplest to consider again E'_n. In region 2 this opposes the applied field, and so the equivalent point charges, giving the correct distribution of the normal component of field in region 2, have opposite signs to those above. Of these two, the image charge must again be outside the field region, and so the required one is that of strength $-q(\varepsilon_2 - \varepsilon_1)/(\varepsilon_2 + \varepsilon_1)$ lying at the point $y = +a.$[†]

Image of line currents

In a similar way to that above, the images of a line current, of strength i, in a medium of permeability μ_1, near an infinite straight-line boundary behind which is a medium of permeability μ_2, can be shown to be similar to the above images. For the field in region 1, the required image is of strength $i(\mu_2 - \mu_1)/(\mu_2 + \mu_1)$ and it lies at $y = -a$ in region 2; for the field in region 2 the image is of strength $-i(\mu_2 - \mu_1)/(\mu_2 + \mu_1)$ and lies at the point $y = +a$, in region 1. A field map is shown in Fig. 2.2 for the case $\mu_2/\mu_1 = 5$, and it demonstrates the 'refraction' of the flux lines at the boundary.

General remarks

Certain general points emerging from the above discussion should be emphasized. First, the fields are calculated as the resultant of the applied field and the image fields.

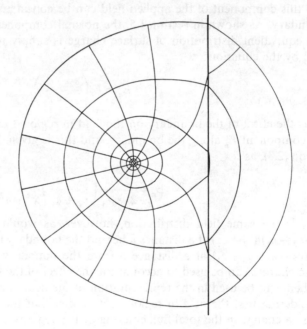

Fig. 2.2

[†]The above solutions can be confirmed by showing that the tangential component of field strength is continuous across the boundary. This is easily seen, since the tangential components due to both images are equal, being given by

$$\left(\frac{\varepsilon_2 - \varepsilon_1}{\varepsilon_2 + \varepsilon_1}\right) \frac{q}{2\pi\varepsilon_0} \frac{x}{(x_2 + a^2)}$$

and the applied field is obviously continuous.

Secondly, the images for the two regions are of equal magnitude, but whilst they are of the same sign for charge they are of opposite sign for currents. Finally, for the region where the actual charge or current is, the image is in the same position as an optical image of the applied source in the boundary; whilst, for the region not containing the influence, the image is in the same position as the influence.

Complex variable form

The solutions obtained above are used later in the book in complex variable form, which is generally the most suitable for calculation and is given below. The field of a line charge or a line current may be expressed by

$$w = \frac{1}{2\pi} \log(t - t_0),$$ (2.5)

where $t = t_0$ gives the position of the current in the t-plane. Hence the field of a charge and its images is simply expressed by the sum of terms of the above form. Taking, for instance, the case of a charge at a distance a from an infinite plane, with the charge at the point $t = ja$ and the plane coinciding with the real axis, the field of the charge and its image is

$$w = \frac{1}{2\pi} \left[\log(t - ja) - \left(\frac{\varepsilon_2 - \varepsilon_1}{\varepsilon_1 + \varepsilon_2} \right) \log(t + ja) \right].$$ (2.6)

If the plane is conducting ($\varepsilon_2 = \infty$), this solution reduces to

$$w = \frac{1}{2\pi} \log\left(\frac{t - ja}{t + ja} \right).$$

Similarly, it can be shown that the field of a line current near an infinitely permeable plane is expressed by

$$w = \frac{1}{2\pi} \log(t - ja)(t + ja)$$

$$= \frac{1}{2\pi} \log(t^2 + a^2),$$ (2.7)

and that of a line current near an impermeable plane by

$$w = \frac{1}{2\pi} \log\left(\frac{t - ja}{t + ja} \right).$$ (2.8)

2.2.2 Parallel plane boundaries

The solution for the field in the region between two parallel boundaries, due to a charge or current in that region, has been given by Kunz and Bayley [7] and Hague, p. 173. For either case, two solutions arise depending upon whether the boundaries are equipotentials or flux-lines.

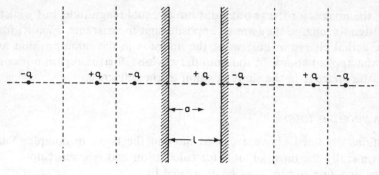

Fig. 2.3

For the case of the electric field due to a line charge between conducting (equipotential) boundaries, the images are as shown in Fig. 2.3 and the field map as shown in Fig. 2.4. It is seen that there is an infinite number of images, which occur as equally spaced pairs of equal and opposite charges and which are symmetrically disposed about each boundary line. It is to be noted that (because of this last feature) the flux crosses the boundary lines at right angles, indicating that the boundaries are equipotential, and so confirming that the solution is a valid one.

By summing the complex potential functions, for the positive and negative charges separately, it is simply shown, see [8], that the solution for the region between the boundaries is

$$w = \frac{1}{2\pi} \log \frac{\sin \pi[(t + a)/2l]}{\sin \pi[(t - a)/2l]}, \tag{2.9}$$

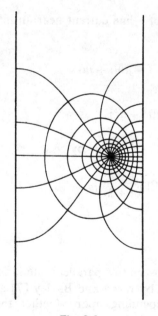

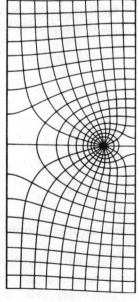

Fig. 2.4   **Fig. 2.5**

where the field is in the t-plane and the distances a and l and the origin are as shown in Fig. 2.3. This solution obviously applies also to the magnetic field of a line current between parallel impermeable boundaries.

For infinitely permeable boundaries the images of a line current have the same positions as those above, but they are all of the same sign. A map of the field is shown in Fig. 2.5, and it is seen that the images are such as to make the boundary lines magnetic equipotentials. Since the images are all of the same sign as the line current, the field is expressed by

$$w = \frac{1}{2\pi} \log\left[\sin \pi\left(\frac{t+a}{2l}\right) \right]\left[\sin \pi\left(\frac{t-a}{2l}\right) \right].$$ (2.10)

These solutions have been applied to a variety of problems, e.g. by Kunz and Bayley [7] to the calculation of the capacitance of a wire between conducting boundaries by Walker [8] to determining the characteristics of a triode valve, by Hague to the calculation of force on a conductor in a machine air gap, and by Frankel [9] to the determination of the impedance of conductors near parallel boundaries.

2.2.3 Intersecting plane boundaries

Solutions have been obtained, in the first instance by Lodge [1], for a number of problems involving interesecting boundaries. The range of these solutions is, however, limited: a maximum of four boundaries can be handled, and the angles of intersection in the field region must in all cases be submultiples of π. (These limitations do not apply to conformal transformation methods, see section 7.2). The discussion is to be restricted to the magnetic field since the analogy with the electric field is now obvious and since flux-line boundaries occur only in the magnetic case.

Two intersecting boundaries

Consider two straight boundaries of zero permeability intersecting at an angle π/n, where n is an integer, and enclosing a region containing a line current, i. The images which can be used to give the field *inside* the region are shown in Fig. 2.6(a), (b), and (c) for the values $n = 2, 3$ and 4. (With the value $n = 1$ the boundary becomes as infinite

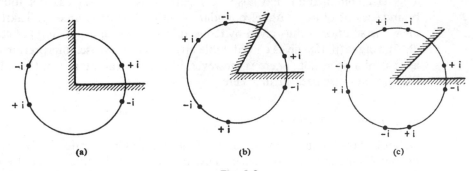

(a)   (b)   (c)

Fig. 2.6

straight line, discussed earlier.) For all values of n, the image currents lie on a circle, its centre at the point of intersection of the boundaries and passing through the current, i. The currents in the complete system (the images and the actual currents) are alternately positive and negative and are symmetrically disposed about the boundaries. Consequently, this system gives rise to flux lines along the boundaries, so confirming the validity of the solution.

It can be shown that in general the number of images is $(2n-1)$, and that the angles subtended by the images at the intersection of the boundaries form the series

$$\left(\frac{2\pi}{n}-\theta\right),\left(\frac{2\pi}{n}+\theta\right),\left(\frac{4\pi}{n}-\theta\right),\left(\frac{4\pi}{n}+\theta\right),\dots,\left(\frac{2\pi}{n}(n-1)+\theta\right),(2\pi-\theta),$$

where θ is the angular displacement of the current from one of the boundary lines.

When the boundaries are infinitely permeable, the images are in the same positions as the above, but they all have the same sign as the actual current.

There have been many applications of these solutions to practical problems, a recent example occurring in the calculation of force on the end-windings of turbogenerators [10].

Three intersecting boundaries

There are only four combinations of three interesecting boundaries which have interior angles which are submultiples of π. These combinations are:

$$\frac{\pi}{3},\frac{\pi}{3},\quad\text{and}\quad\frac{\pi}{3};$$

$$\frac{\pi}{2},\frac{\pi}{4},\quad\text{and}\quad\frac{\pi}{4};$$

$$\frac{\pi}{2},\frac{\pi}{3},\quad\text{and}\quad\frac{\pi}{6};$$

and

$$\frac{\pi}{2},\frac{\pi}{2},\quad\text{and}\quad 0.$$

This last combination may also be regarded as a special case of four intersecting boundaries or of two parallel boundaries with a line of symmetry. Taking the former view, the solution is thus given by reduced forms (m or $n=\pm 1$ only) of eqns (2.11) and (2.12). Hague (p. 188) gives considerable attention to this case, interpreting the boundary as that of a deep slot. There is, however, little practical interest in the other cases and they are not discussed further here.

Four intersecting boundaries

Because of the limitation in the values of the angles of intersection, only one combination of four boundaries is possible, namely, that with all the interior angles equal to $\pi/2$. The positions of the images of a line current for the field with flux-line boundaries are

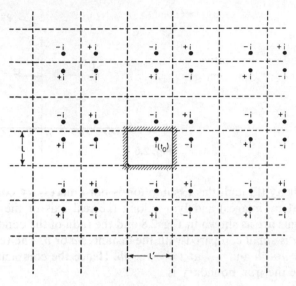

Fig. 2.7

shown in Fig. 2.7. The distribution of the images is doubly periodic, and they are symmetrical about each of the boundary lines, so that, as required, these are flux lines.

Letting the field be in the t-plane, the origin and dimensions being as shown in Fig. 2.7, the field solution is

$$w = \frac{1}{2\pi} \log \prod_{\substack{m=-\infty \\ n=-\infty}}^{\infty} \frac{(t + \bar{t}_0 + a)(t - \bar{t}_0 + a)}{(t + t_0 + a)(t - t_0 + a)}, \tag{2.11}$$

where m and n are integers and where

$$a = 2ml' + 2jnl.$$

Equation (2.11) has been used, for example, to determine the capacitance of a wire in a rectangular cylinder [7].

The images of a line current in the infinitely permeable boundary occupy the same positions as those for a flux-line boundary, but they all have the same sign as the current itself, and so the solution is

$$w = \frac{1}{2\pi} \log \prod_{\substack{m=-\infty \\ n=-\infty}}^{\infty} (t + \bar{t}_0 + a)(t - \bar{t}_0 + a)(t + t_0 + a)(t - t_0 + a). \tag{2.12}$$

In section 4.3.3. this image distribution is considered in connection with the field of distributed currents.

2.2.4 Inductance of parallel bus-bars near an iron surface

To show the usefulness and simplicity of the image method, it is now applied to the calculation of the inductance of two bus-bars carrying equal and opposite currents $\pm I$ near to a highly permeable surface. The surface is assumed infinitely permeable for

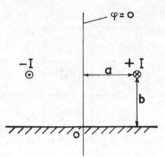

Fig. 2.8

simplicity (though this assumption is not a necessary condition for the problem to be solved by images) and lying along the real axis of the t-plane. The positions of the currents are as shown in Fig. 2.8 and the radii of the conductors are r, it being assumed that r is small compared with the distances a or b. The required image currents are $+I$ at $t = a - jb$ and $-I$ at $t = -a - jb$. Hence the complete field solution for the region above the iron boundary is

$$\omega = \frac{I}{2\pi} \log \frac{[t - (a + jb)][t - (a - jb)]}{[t - (-a + jb)][t - (-a - jb)]}$$

which can be simplified to

$$\omega = \frac{I}{2\pi} \log \frac{(t - a)^2 + b^2}{(t + a)^2 + b^2}. \tag{2.13}$$

The imaginary axis is a line of symmetry and along it lies the flux line $\phi = 0$ as may be confirmed by showing that the real part of eqn (2.13) is zero for all purely imaginary values of t. If the small quantity of flux which passes within the conductors is neglected, the total flux passing between the two bus-bars is twice that passing between the line $\phi = 0$ and a point $t = (a + jb + r)$ on the surface of the conductor at $t = a + jb$. (Since r is small the surface of the conductor is very nearly a flux line.)

Hence the total flux linking the bus-bars is given by

$$2Rl \left[\frac{\mu_0 I}{2\pi} \log \frac{|(r + jb)^2 + b^2|}{(2a + jb + r)^2 + b^2} \right]$$

and so the inductance per unit length of the bus-bar circuit is

$$\frac{\mu_0}{\pi} Rl \log \left| \frac{(r + jb) + b}{(2a + jb + r) + b} \right|. \tag{2.14}$$

In a similar way it is possible to calculate from the electric field the capacitance of parallel transmission lines (see Bewley, p. 47, and [11]).

2.3 CIRCULAR BOUNDARIES

Whereas for the single straight-line boundary there are only two images to be considered, for the single circular boundary there are four sets of images—the two for the charge inside the boundary and the two for the charge outside. Solutions are developed in detail for the latter case using the concept of equivalent surface charge. The results for the case of a charge interior to the boundary are given for both regions. Consideration is also given to the field of a circular cylinder in a uniform applied field. The applied field is due to charges at infinity, but the boundary effects can still be accounted for by images, which in this case form a doublet.

2.3.1 Charge or current near a circular boundary

Line charge near an isolated cylinder

Consider first the field of a line charge in a medium of permittivity $\varepsilon_1\varepsilon_0$ near a circular cylinder of permittivity $\varepsilon_2\varepsilon_0$ (Fig. 2.9). The normal component of the applied field strength E_n at any point P on the circular boundary is given by

$$E_n = \frac{q\cos\alpha}{2\pi\varepsilon_0\varepsilon_1 b},$$ (2.15)

the angle α and distance b being as shown in Fig. 2.9. The normal component of field due to the equivalent surface charge distribution is then, from eqn (2.3),

$$E_n' = \frac{q}{2\pi\varepsilon_0\varepsilon_1}\left(\frac{\varepsilon_2-\varepsilon_1}{\varepsilon_2+\varepsilon_1}\right)\frac{\cos\alpha}{b}.$$ (2.16)

However, by consideration of triangles PAB and PBC, where B is the inverse of A in the circle, it can be shown that

$$a\cos\alpha + b\cos\beta = \frac{b\sin(\alpha+\beta)}{\sin\alpha},$$

and

$$a\sin\alpha = r\sin(\alpha-\beta),$$

which may be combined to give

$$\frac{\cos\alpha}{b} = \frac{\cos\beta}{a} - \frac{1}{r}.$$ (2.17)

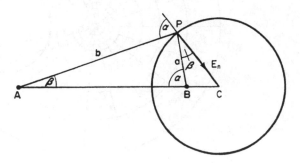

Fig. 2.9

Hence eqn (2.16) can be rewritten as

$$E'_n = \frac{q}{2\pi\varepsilon_0}\left(\frac{\varepsilon_2 - \varepsilon_1}{\varepsilon_2 + \varepsilon_1}\right)\frac{\cos\beta}{a} - \frac{q}{2\pi\varepsilon_0}\left(\frac{\varepsilon_2 - \varepsilon_1}{\varepsilon_2 + \varepsilon_1}\right)\frac{1}{r}, \tag{2.18}$$

and this gives the normal component of field strength along the perimeter of the circle due to the distributed charge. But this is the normal component of field due to a charge $-q(\varepsilon_2 - \varepsilon_1)/(\varepsilon_2 + \varepsilon_1)$ at inverse point B together with a charge $+q(\varepsilon_2 - \varepsilon_1)/(\varepsilon_2 + \varepsilon_1)$ at the centre of the circle. Therefore, since these charges are inside the boundary they are the required image charges which, together with the applied charge, give the field outside the circular boundary.

For the region inside the boundary, the images giving rise to E'_n must be outside the boundary, and must give rise to a normal component of field

$$E'_n = -\frac{q}{2\pi\varepsilon_0}\left(\frac{\varepsilon_2 - \varepsilon_1}{\varepsilon_2 + \varepsilon_1}\right)\frac{\cos\alpha}{b}. \tag{2.19}$$

This obviously requires the image charge to be of magnitude $-q(\varepsilon_2 - \varepsilon_1)/(\varepsilon_2 + \varepsilon_1)$ and to lie at the same point as the charge q.

In a similar way to the above it is possible to establish the magnitudes and positions of the images of a charge inside the circular boundary. The field inside the circle is given by an image of magnitude $q(\varepsilon_2 - \varepsilon_1)/(\varepsilon_2 + \varepsilon_1)$ at the exterior inverse point, and the field outside is given by an image of magnitude $-q(\varepsilon_2 - \varepsilon_1)/(\varepsilon_2 + \varepsilon_1)$ at the same point as the charge q together with an image of magnitude $q(\varepsilon_2 - \varepsilon_1)/(\varepsilon_2 + \varepsilon_1)$ at the centre. Figure 2.10 shows the field of a line charge outside a boundary of circular section for the case $\varepsilon_2/\varepsilon_1 = 5$. Images solutions for a charge near a circular boundary have been used in the determination of the field in an electronic valve (see Bewley and References [3] and [7]).

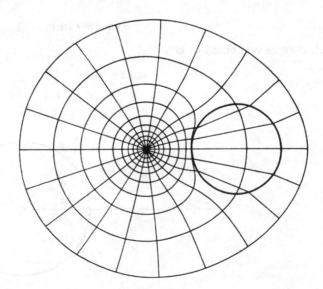

Fig. 2.10

Line current near a permeable cylinder

The images of a line current near a circular boundary are, of course, analogous to the above image charges. When the current i is outside the boundary, the field outside is given by an image current $i(\mu_2 - \mu_1)/(\mu_2 + \mu_1)$ at the inverse point, together with a current $-i(\mu_2 - \mu_1)/(\mu_2 + \mu_1)$ at the centre, and, of course, the actual current i; the field inside is given by a current i and an image $-i(\mu_2 - \mu_1)/(\mu_2 + \mu_1)$ at the same position as the current i. When the current is inside the boundary, the field outside requires an image current $-i(\mu_2 - \mu_1)/(\mu_2 + \mu_1)$ at the same point as the current i together with an image of magnitude $i(\mu_2 - \mu_1)/(\mu_2 + \mu_1)$ at the centre; and the field inside is that of the actual current and image current $-i(\mu_2 - \mu_1)/(\mu_2 + \mu_1)$ at the inverse point.

Line charge near an earthed conducting cylinder

When an isolated conducting cylinder is near to a line charge, the field is obtained from the general solution for a circular boundary by making the substitution $\varepsilon_2 \to \infty$. Hence, the images are of magnitude $-q$ at the inverse point and $+q$ at the centre. The conducting cylinder, being isolated, acquires a potential but has no net charge on it, since all the flux from the charge at A passes to infinity (some of it through the cylinder).

If the cylinder is earthed its potential changes to zero, which means that the image charge at its centre must change. Also, all of the flux from the charge at A now passes into the cylinder, which acquires a net charge of $-q$. For this to be possible, the image charge at the centre must change to zero, leaving the other image charge of $-q$ at B, and the potential of the cylinder changes to zero as a result of the removal of image from the centre.

Finally, it is interesting to note that the image solutions for charges or currents near circular boundaries reduce, as the radius tends to infinity, to those for an infinite straight-line boundary. In the next chapter, section 3.2.2, the field of a line current exterior to a permeable cylinder is analysed by the direct solution of Laplace's equation. Also, in section 5.4.3, the solution for infinite permeability is obtained by conformal transformation. In both cases, the equivalence of the solutions with that obtained by images is demonstrated.

2.3.2 Doublets: circular cylinder in a uniform field

Doublets

The influence of a circular cylinder on an applied uniform field can be accounted for by a doublet, i.e. by a combination of an infinite positive charge (or current) and an infinite negative charge (or current), an infinitesimal distance apart, so arranged that the product of charge and distance is finite. Before examining its influence on the uniform field, the field of a doublet is first developed from a consideration of that of line charges, and then the images of doublets for plane and circular boundaries are obtained. All field functions are expressed in complex variable form because of its convenience in calculation.

The complex potential function of a line charge of strength q at the point $t = a$ of the complex t-plane is given by

$$w = \frac{q}{2\pi} \log(t - a).$$
(2.20)

Therefore, the field due to equal positive and negative line charges at $t = a + jb$ and $t = -a - jb$ respectively is expressed by

$$w = \frac{q}{2\pi} \log[t - (a + jb)] - \frac{q}{2\pi} \log[t + (a + jb)].$$
(2.31)

This can be written as

$$w = \frac{q}{2\pi} \log\left(1 - \frac{a + jb}{t}\right) - \frac{q}{2\pi} \log\left(1 + \frac{a + jb}{t}\right),$$

or, expanding in series form,

$$w = -\frac{2q}{2\pi}\left[\frac{1}{t}(a + jb) + \frac{1}{3t^3}(a + jb)^3 + \frac{1}{5t^5}(a + jb)^5 + \cdots\right].$$
(2.22)

Consider now the formation of a doublet from the two discrete charges; the distance $2|a + jb|$ between the charges tends to zero and the magnitude of the charge q tends to infinity in such a way that $(q/\pi)|a + jb|$, known as the strength, d, of the doublet, remains finite. Under these conditions all terms but the first in eqn (2.22) disappear, and the equation of the doublet (at the origin) becomes

$$w = -\frac{q}{\pi t}(a + jb).$$

Writing

$$d = -\frac{q}{\pi}|a + jb|,$$

this reduces to

$$w = \frac{d}{t}\varrho^{j\alpha},$$
(2.23)

where α, which equals $\tan^{-1}(b/a)$, is the inclination of the axis of the doublet to the real axis. The field of a doublet, for which $\alpha = 0$, is shown in Fig. 2.11; the flux and potential lines form orthogonal families of circles having as common tangent the axes of the doublet.

The equation of a doublet situated not at the origin but at $t = t_1$, and with its axis parallel with the x-axis, is

$$w = \frac{d}{t - t_1},$$
(2.24)

and this form is used in section 7.2.2 to obtain solutions by transformation for field exterior to boundaries of complicated shape. The relationship between w and t is that of complex inversion which is discussed in section 5.3.1.

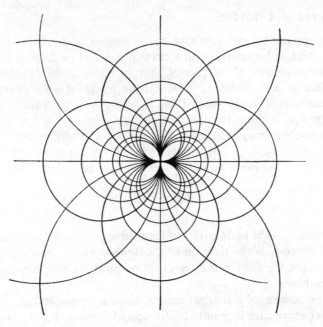

Fig. 2.11

Doublet at infinity

In a similar way to the above, it is also possible to develop the field of a doublet at the point at infinity. Consider again the two unlike charges, the field of which is given by eqn (2.31). That equation may be rewritten as

$$w = \frac{q}{2\pi} \log\left(1 - \frac{t}{a + jb}\right) - \frac{q}{2\pi} \log\left(1 + \frac{t}{a + jb}\right) + \frac{q}{2\pi} \log(-1),$$

or expanding the first two terms on the right in a series, as

$$w = \frac{q}{\pi}\left[\frac{t}{a + jb} + \frac{t^3}{3(a + jb)^3} + \cdots \right]. \tag{2.25}$$

Now the doublet at infinity is formed by letting both charges move away to infinity (where they converge on the single point, $t = \infty$) whilst, at the same time, $q \to \infty$ in such a way that d, the strength of the doublet, given by

$$-\frac{2q}{(a + jb)} = d\,e^{j\alpha},$$

remains finite. As $|a + jb|$ and q both approach infinity, eqn (2.25) reduces to

$$w = d\,e^{j\alpha}t, \tag{2.26}$$

which is thus the equation of the field of a doublet at infinity. It is seen that this equation describes a uniform field of strength d inclined at an angle α to the real axis, and such a field, therefore, is produced by a doublet charge at infinity. Equation (2.26) could be derived directly from eqn (2.23) by inversion and this point is discussed in section 5.3.

Images of a doublet

It is apparent from the above developments of doublets that the images of a doublet in a straight boundary or in a circular one can be derived from the combination of the separate images of the pair of charges forming the doublet. The result for the plane surface is particularly easy to see; the image of each charge has the position of the optical image in the surface and $-(\varepsilon_2 - \varepsilon_1)/(\varepsilon_2 + \varepsilon_1)$ times the strength of the charge, where $\varepsilon_1\varepsilon_0$ is the permittivity in the field region, and $\varepsilon_2\varepsilon_0$ is that behind the boundary. The doublet image, therefore, has the position of the optical image of the doublet and $-(\varepsilon_2 - \varepsilon_1)/(\varepsilon_2 + \varepsilon_1)$ times its strength. For example, the field in the region containing a doublet at the point t_1 and influenced by a boundary coinciding with the real axis is

$$w = \frac{d\,e^{j\alpha}}{(t - t_1)} - \left(\frac{\varepsilon_2 - \varepsilon_1}{\varepsilon_2 + \varepsilon_1}\right)\frac{d\,e^{j\alpha}}{(t - \bar{t}_1)}, \tag{2.27}$$

where $\varepsilon_1\varepsilon_0$ is the permittivity in the field region, and $\varepsilon_2\varepsilon_0$ is that inside the second region, and where t_1 is the position of the doublet and $\bar{t}_1$ is the complex conjugate of t_1. The image for the field in the region of permittivity $\varepsilon_2\varepsilon_0$ can easily be found, but it is not given here.

The images of a doublet near a circular boundary can be found similarly, though rather more care is required. The case of the doublet exterior to the boundary and the solution for the region containing the doublet is examined. Consider first the images of two equal finite charges of opposite sign, at A and B, outside the boundary as shown in Fig. 2.12. The positive charge, $+q$, at A has an image $-q(\varepsilon_2 - \varepsilon_1)/(\varepsilon_2 + \varepsilon_1)$ at its inverse point C with respect to the circle, and an image $q(\varepsilon_2 - \varepsilon_1)/(\varepsilon_2 + \varepsilon_1)$ at the centre O of the circle. The images of the negative charge at B are $+q(\varepsilon_2 - \varepsilon_1)/(\varepsilon_2 + \varepsilon_1)$ at the corresponding inverse point D and $-q(\varepsilon_2 - \varepsilon_1)/(\varepsilon_2 + \varepsilon_1)$ at the centre. The two charges at the centre of the circle thus cancel one another, leaving the pair of equal and opposite image charges at the inverse points. Figure 2.12 shows the position of these charges. As the charges at A and B approach one another, the product $q(AB)$ remaining finite, they form a doublet, the image of which is formed by the two image charges at C and D. The strength of the doublet causing the applied field is given by

$$d = q(AB),$$

and the strength of the image doublet by

$$d_i = -d \, \underset{A \to B}{\text{Lt}} \left(\frac{CD}{AB}\right)\left(\frac{\varepsilon_2 - \varepsilon_1}{\varepsilon_2 + \varepsilon_1}\right). \tag{2.28}$$

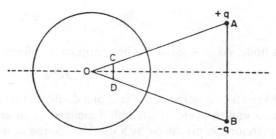

Fig. 2.12

But, since the triangles OCD and OBA are similar,

$$\frac{CD}{AB} = \frac{OC}{OA};$$

also if, as the pairs of charges come together,

$$OA = OB = a$$

(say), then

$$OC = OD = \frac{r^2}{a}, \tag{2.29}$$

r being the radius of the circle. Therefore, substituting for (CD/AB) in eqn (2.28) gives the strength of the image doublet as

$$d_i = -d\left(\frac{r^2}{a^2}\right)\left(\frac{\varepsilon_2 - \varepsilon_1}{\varepsilon_2 + \varepsilon_1}\right). \tag{2.30}$$

Also the position of the image is that of the inverse point of the position of the doublet with respect to the circle. The inclination of the doublet is seen, from a consideration of the similar triangles, as $AB \to 0$, to be equal and opposite to that of the doublet (with respect to the line joining them).

The image of a doublet inside the circular boundary can be found in a similar way to the above.

Cylindrical boundary in applied uniform field

The influence of a circular boundary on a uniform applied field can be accounted for simply by using a doublet at infinity to give the uniform field and then by forming

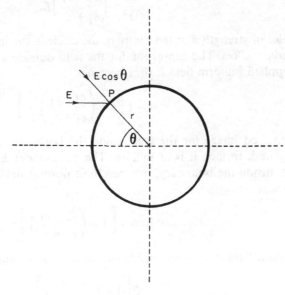

Fig. 2.13

images, of this doublet, in the boundary. Since the centre of the circular boundary is the inverse of the point at infinity, it is apparent that the resultant field is given by the sum of the actual doublet at infinity and the image doublet at the centre. However, it is interesting to develop this solution from a consideration of the surface charge distribution.

In a uniform electric field of strength E, the normal component at any point P on the circular boundary is $E\cos\theta$, where θ is the angle between the direction of the field, taken parallel to the real axis and a radius to the point P (Fig. 2.13). The normal field E'_n due to the charging of the boundary is, therefore,

$$E'_n = \left(\frac{\varepsilon_2 - \varepsilon_1}{\varepsilon_2 + \varepsilon_1}\right) E\cos\theta.$$

Consider now the field component normal to the circular boundary, due to a doublet, strength d, with its axis parallel to the field and placed at the centre of the circle. Its field is given by

$$w = \frac{d}{t},$$

and the field strength (see section 5.4) by

$$\frac{d}{r^2}\exp\left[j(\pi - 2\theta)\right].$$

This means that the field strength, due to the doublet, is of magnitude d/r^2, and is inclined to the real axis by an angle 2θ and so to the radius P by an angle θ. Hence, the normal component of field at the boundary due to the doublet is $(d/r^2)\cos\theta$ and so, by making

$$\frac{d}{r^2} = \left(\frac{\varepsilon_2 - \varepsilon_1}{\varepsilon_2 + \varepsilon_1}\right) E,$$

the doublet of strength d at the centre of the circle is the image required to account for the boundary effect. The expression for the field outside a circular boundary of radius r in an applied uniform field E thus becomes

$$w = E\left[t - \left(\frac{\varepsilon_2 - \varepsilon_1}{\varepsilon_2 + \varepsilon_1}\right)\frac{r^2}{t} \right]. \tag{2.31}$$

The required image for the field inside the boundary must, of course, outside the boundary and, in fact, it is at infinity. The component E'_n opposing the applied field gives rise, inside the boundary, to a resultant normal field having along the boundary a component

$$E\cos\theta\left[1 - \left(\frac{\varepsilon_2 - \varepsilon_1}{\varepsilon_2 + \varepsilon_1}\right) \right].$$

The resultant field everywhere inside the boundary is clearly uniform and of strength

$$E\left[1 - \left(\frac{\varepsilon_2 - \varepsilon_1}{\varepsilon_2 + \varepsilon_1}\right) \right],$$

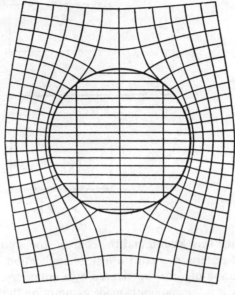

Fig. 2.14

which equals

$$\frac{2\varepsilon_1 E}{\varepsilon_2 + \varepsilon_1}.$$

Figure 2.14 is a field plot for $\varepsilon_2/\varepsilon_1 = 5$.

2.4 GENERAL CONSIDERATIONS

It has been shown that the method of images can be used to analyse fields due to line currents or charges (including those at infinity which give uniform applied fields) influenced by a variety of straight-line or circular boundaries. A combination of a straight-line and a circular boundary has also been treated in several papers, e.g. by Milne-Thomson and in Reference [12].

The above treatment is restricted to the fields of line influences, but it should be noted that the method also applies to influences with finite areas of cross-section. This is because any such influence can be treated as the aggregate of line influences, each of which forms its own image in the usual way. Thus, for example, the field in air of a rectangular conductor carrying current of uniform density J, and near an infinite plane of relative permeability μ, is given by introducing a rectangular image current with uniform density $J(\mu - 1)/(\mu + 1)$ (Fig. 2.15). (Further discussion of the images of rectangular conductors is given in Chapter 4.) However, the application of the method is restricted to straight-line boundaries, since the image in a circular boundary has an awkward shape and a non-uniform current density and this leads to excessive difficulty in the analysis.

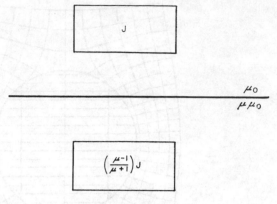

Fig. 2.15

To conclude this section and the chapter, the image method is used to demonstrate a number of general points to which its application is very suitable. The first of these is the calculation of force experienced by a boundary; the second is an indication of the error involved in the frequently made assumption that boundaries of finite permeability can be treated as being infinitely permeable; and the third is an approximate method for the treatment of field regions with boundaries in which eddy-currents are induced by an alternating field.

The total force on a boundary

The total force acting on a boundary can be calculated very conveniently using the image method because it is equal to the total force on the images due to the applied field or to the force on the field sources due to the field of the images.

As an example, consider the determination of the force between a line current in air and an infinite plane boundary of relative permeability μ (Fig. 2.16). The field strength

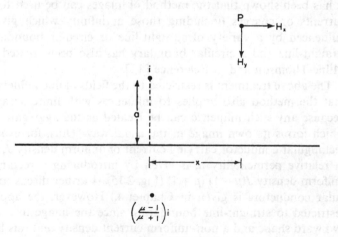

Fig. 2.16

H_x, at the position of the current, due to the image, is given by

$$H_x = \frac{i}{2\pi} \left(\frac{\mu - 1}{\mu + 1} \right) \frac{1}{2a}, \tag{2.32}$$

there being no component in the y-direction. Therefore, the force on the current due to this field (and that on the boundary due to the current) is in the y-direction, and it is given by

$$F_y = \frac{\mu_0}{2\pi} i^2 \left(\frac{\mu - 1}{\mu + 1} \right) \frac{1}{2a}. \tag{2.33}$$

This is an attractive force when the boundary is permeable, and a repulsive one when the effective permeability is less than unity.

The influence of finite permeability

The simplifying assumption, commonly made in magnetic problems, that a boundary is of infinite permeability, clearly involves some error. The magnitude of this error varies with position in a different way for each problem but, in order to demonstrate its nature, in a simple case, consideration is again given to the line current in air near an infinite plane boundary of relative permeability μ. In the air, the field is that of the current i and its image of magnitude $i(\mu - 1)/(\mu + 1)$ (Fig. 2.16). The component of field strength parallel to the plane is given by

$$H_x = \frac{i}{2\pi} \frac{a - y}{[(a - y)^2 + x^2]} - \frac{i}{2\pi} \left(\frac{\mu - 1}{\mu + 1} \right) \frac{a + y}{[(a + y)^2 + x^2]}, \tag{2.34}$$

and that normal to the plane by

$$H_y = \frac{i}{2\pi} \frac{x}{[(a - y)^2 + x^2]} + \frac{i}{2\pi} \left(\frac{\mu - 1}{\mu + 1} \right) \frac{x}{[(a + y)^2 + x^2]}. \tag{2.35}$$

The error involved in assuming the permeability infinite is thus given by substituting $\mu = \infty$ and subtracting the resulting values from the above equations. The error in the x-component is

$$\frac{i}{2\pi} \frac{2}{\mu + 1} \frac{a + y}{[(a + y)^2 + x^2]}, \tag{2.36}$$

and that in the y-component is

$$\frac{i}{2\pi} \frac{2}{\mu + 1} \frac{x}{[(a + y)^2 + x^2]}. \tag{2.37}$$

At any point along the boundary (given by $y = 0$) the ratio of the error to the applied field, for each component, is $2/(\mu + 1)$ (which becomes less than 5% for $\mu > 39$, and less than 0.5% for $\mu > 399$), and this is negligibly small in practice. Further, the error decreases with movement away from the boundary, and so this value $[2/(\mu + 1)]$ gives the upper bound for the error at any point in the field. For most problems which can be solved for finite permeability it is found that the assumption of $\mu = \infty$ causes little error in practice, and this assumption is made for many problems considered in this book.

Eddy-current problems

All problems treated in this book are essentially steady-state ones, but an approximate method, using steady-state results, can be used for certain transient problems involving boundaries carrying eddy-currents induced by the applied field. The effect of the eddy-current is to reduce the quantity of flux crossing the boundary and, if the assumption is made that no flux penetrates (that is the boundary is a flux line), solutions may be obtained by taking the boundary permeability to be zero. The field of a line-current near a semi-finite block of impermeable material is given by eqn (2.8), and this equation is used later in conjunction with conformal transformation (see sections 7.2.1 and 7.2.2), for the analysis of problems involving alternating currents near complicated conducting boundaries.

REFERENCES

[1] O. J. Lodge, On some problems connected with the flow of electricity in a plane, *Phil. Mag.*, 5th series, **1**, 373–89 and **2**, 37–47 (1876).

[2] G. F. C. Searle, On the magnetic field due to a current in a wire placed parallel to the axis of a cylinder of iron, *The Electrician*, 453 and 510 (1898).

[3] P. Hammond, Electric and magnetic images, *Proc. Instn. Elect. Engrs.*, **107** C, 306 (1960).

[4] N. Mullineux and J. R. Reed, Images of line charges and currents, *Proc. Instn. Elect. Engrs.*, **111**, 1343.

[5] F. Chorlton, Circle and sphere theorems in potential theory and the determination of image systems, *Birm. Coll. Adv. Tech. Technical Report*, No. **7** (1965).

[6] C. J. Carpenter, The application of the method of images to machine end-winding fields, *Proc. Instn. Elect. Engrs.*, **107** A, 487 (1960).

[7] J. Kunz and P. L. Bayley, Some applications of the method of images, *Phys. Rev.*, **17**, 147–156 (1921).

[8] G. B. Walker, Electric field of a single grid radio valve, *Proc. Instn. Elect. Engrs.*, Part III, **98**, 57 (1951).

[9] S. Frankel, Characteristic impedance of parallel wires in rectangular troughs, *Proc. Inst. Rad. Engs.*, 182 (1942).

[10] D. Mayer, Mechanicke sily pusobici na cela vinuti statoru stridavych stroju (Forces acting on the faces of stator windings in a.c. machines), *Electrotechnicky Obzor*, **44** (8), 395 (1955).

[11] J. H. Gridley, The shielding of overhead lines against lightning, *Proc. Instn. Elect. Engrs.*, **107** A, 325 (1960).

[12] B. V. Jayawant, Flux distribution in a permeable sheet with a hole near an edge, *Proc. Instn. Elect. Engrs.*, **107** C, 238 (1960).

[13] J. F. H. Douglas, Reactance of end-connections, *Trans. Am. Inst. Elect. Engrs.*, **56**, 257 (1937).

[14] P. J. Lawrenson, The magnetic field of the end-windings of turbo-generators, *Proc. Instn. Elect. Engrs.*, **108** A, 538 (1961).

Additional reference

Mack, C., The field due to an infinite dielectric cylinder between two parallel conducting plates, *Br. J. Appl. Phys.*, **6**, 59 (1955).

THE SOLUTION OF LAPLACE'S EQUATION BY SEPARATION OF THE VARIABLES

3.1 INTRODUCTION

Whilst for certain simpler problems the method of images is of considerable practical value, direct solution of the field equations is necessary or at least more convenient for problems which involve multiple boundaries or, particularly, specified distributions of potential or potential gradient. In this chapter consideration is given to such direct solutions of Laplace's equation, and in the next to direct solutions of Poisson's equation, see section 1.8.

Emphasis is placed on two-dimensional fields and, particularly, on those with field regions which are either rectangular or circular and are describable by field functions which are periodic (see below). The principal reason for these emphases is practical utility, as brought out in section 3.4. In addition the treatment is kept as simple as possible to bring out as clearly as possible the relationship between the physical nature of the fields and their corresponding mathematic descriptions and also the relationship to other methods of solution. This should be helpful later in the book when more abstract or numerical formulations are in question. (Non-linear fields, for example allowing for saturation of permeable regions, cannot be treated by the purely analytical methods of this chapter. Also of course, fields described by Laplace's equation are by their nature independent of the time variable.)

Essentially, the direct-solution method involves the determination of a field function which satisfies the field equation (Laplace's) and also satisfies imposed boundary and other field conditions for a particular region. This field function, which may be for example the field strength, the (scalar) potential function, or the flux function, is, in general, the sum of several parts (each of which separately is a solution). One part, usually in the form of a series, describes the effect of the boundary influences, and the others describe the effect of field sources such as currents and charges. However, field functions cannot be determined for any given problem—not because of the difficulty of finding solutions which satisfy the equation, but because of the difficulty choosing solutions appropriate to particular boundary and field conditions. There is an infinite number of solutions of Laplace's equation, e.g. nx, $x^2 - y^2$, $e^{nx} \sin ny$, and any linear combination of these—but finding combinations of them which satisfy field conditions on boundary lines of particular shape can be difficult and not infrequently is impossible.

However, when a coordinate system exists for which constant values of one (or more) coordinates express the boundary shape(s), it is always possible to obtain a solution by a routine method. The two coordinate systems which are of most use to the engineer or physicist are the cartesian one for the solution of fields with rectangular boundaries, and the circular-cylinder one for the solution of those with concentric circular and/or radial line boundaries. The method of solution is discussed for these two coordinate systems and it is shown to reduce merely to the determination of constants in a general form of field function. Discussion of other coordinate systems, e.g. parabolic and elliptic cylinder, or bipolar, may be found for example in the books by Morse and Feshbach, Stratton, Weber, Batcman, and Moon and Spencer. Particular attention is drawn to the Moon and Spencer book, the whole of which is devoted to separation-of-the-variables techniques, and which includes not only a wide variety of coordinate systems but also 3-dimensional fields and the diffusion and wave equations.

When a direct solution is possible, the part of the field function satisfying the boundary conditions can always be expressed as the product of two terms, each term being a function of one coordinate only. As a result, the partial differential equation can be converted to a pair of ordinary differential equations, related by a constant known as the 'separation' constant, and so solved. A general solution is then taken which consists of a suitable linear combination of the products of the pairs of particular solutions from the ordinary equations. In practice it is not necessary to develop the general solution for each problem; instead, the known general solution for the given boundary shape is fitted to the boundary and field conditions.

In the sections dealing with rectilinear boundaries, consideration is restricted to fields arising from specified flux density and potential distributions at the boundary, since the method of images gives a simpler treatment of fields due to line sources. In the sections dealing with circular boundaries, emphasis is placed on fields developed by currents and poles, but the treatment of fields arising from, or giving rise to, specified distributions of potential or potential gradient is also discussed briefly. Attention is given first to circular boundaries and the equivalence of the solutions to those obtained by the method of images is demonstrated.

3.2 CIRCULAR BOUNDARIES

3.2.1 The solution of Laplace's equation in circular-cylinder coordinates

Consider first Laplace's equation in circular-cylinder coordinates in terms of the potential function ψ

$$r^2 \frac{\partial^2 \psi}{\partial r^2} + r \frac{\partial \psi}{\partial r} + \frac{\partial^2 \psi}{\partial \theta^2} = 0, \tag{3.1}$$

and *assume* as a solution the potential function

$$\psi(r, \theta) = R(r) S(\theta), \tag{3.2}$$

where R is a function of r only, and S is a function of θ only. Partial differentiation of eqn (3.2) leads to expressions for $\partial\psi/\partial r$, $\partial\psi^2/\partial r^2$, and $\partial^2\psi/\partial\theta^2$, and substitution of these quantities in eqn (3.1) gives

$$\frac{1}{R}\left(r^2\frac{d^2R}{dr^2}+r\frac{dR}{dr}\right)=-\frac{1}{S}\frac{d^2S}{d\theta^2}. \tag{3.3}$$

The right-hand side of this equation is independent of r and the left-hand side is independent of θ, so that both sides must be equal to a constant, known as the *separation constant*, and the assumption concerning the form of the potential function is justified. For any field in which the effect of currents is ignored, the potential at (r,θ') is the same as that at $(r,\theta'+2n\pi)$, where n is an integer, so that the solution sought for must be periodic in θ. To fulfil this condition, the constant is chosen as a positive integer, and taking it to be m^2 to avoid root signs later, gives

$$r^2\frac{d^2R}{dr^2}+r\frac{dR}{dr}-m^2R=0 \tag{3.4}$$

and

$$\frac{d^2S}{d\theta^2}=-m^2S. \tag{3.5}$$

Well-known solutions of these equations are

$$R=cr^m+dr^{-m} \tag{3.6}$$

and

$$S=g\cos m\theta+h\sin m\theta, \tag{3.7}$$

where c, d, g, and h are constants. Hence, substituting for R and S in eqn (3.2) gives a particular solution of the original equation as

$$\psi(r,\theta)=(cr^m+dr^{-m})(g\cos m\theta+h\sin m\theta). \tag{3.8}$$

This function is known as a circular harmonic of order m.

No limitation has been imposed upon the values of the constants in this solution, so that there is an infinite number of particular solutions of the form (3.8) capable of satisfying eqn (3.1). Since a linear combination of these particular solutions also satisfies eqn (3.1) (see, for example, Churchill, p. 3), a more general solution can be written in the convenient form

$$\psi(r,\theta)=\sum_{m=1}^{\infty}(c_mr^m+d_mr^{-m})(g_m\cos m\theta+h_m\sin m\theta), \tag{3.9}$$

where m takes all integral values between 1 and ∞. For any fixed value of r in the range $0\leqslant r\leqslant\infty$, the right-hand side of this equation is a Fourier series in θ (which can be used to represent any single-valued and periodic function, such as ψ) and, therefore, it is capable of describing the field due to any physically realizable boundary influences. The values of the constants (in the absence of currents and poles) depend only on the boundary conditions.

The right-hand side of eqn (3.9) cannot account for the effect of line currents and poles. To account for these influences, additional terms are required in the potential function, and these are found to correspond to the particular solutions of the ordinary differential eqns (3.4) and (3.5) for the case $m = 0$. For a line current i situated at any point, the potential function (particular solution) is $(i/2\pi)\alpha$, where the angle α, at the position of the current, measures the angular position of any point; and for a line pole p the potential function is $(p/2\pi) \log R$, where R is the distance between the pole and a point in the field (see Appendix 1). The complete solution, including the influence of q currents and s poles, is, therefore,

$$\psi(r, \theta) = \sum_{m=1}^{\infty} (c_m r^m + d_m r^{-m})(g_m \cos m\theta + h_m \sin m\theta)$$

$$+ \sum_q \frac{i_q}{2\pi} \alpha_q + \sum_s \frac{p_s}{2\pi} \log R_s. \tag{3.10}$$

For purposes of manipulation, the angles α_q and the distances R_s are expressed, in terms of r and θ, in series of the same form as those occurring in the first part of the function. To express the angles α_q in series form, consider Fig. 3.1(a); a current i is situated at the point A a distance b from the origin of coordinates, and α is the angle subtended by any point P, (r, θ) at the current. From the geometry of the figure it is seen that

$$\alpha = \tan^{-1} \frac{r \sin \theta}{r \cos \theta - b}$$

$$= \text{Im} \log (re^{j\theta} - b).$$

This function can be expanded in terms of $(b/r)e^{-j\theta}$ for $r > b$, or in terms of $(r/b)e^{j\theta}$ for $b > r$, to give two distinct convergent series, valid in the above regions, for the angle α: thus, where $r > b$,

$$\alpha = \theta + \sum_{m=1}^{\infty} \frac{1}{m} \left(\frac{b}{r} \right)^m \sin m\theta, \tag{3.11}$$

and where $r < b$,

$$\alpha = \pi - \sum_{m=1}^{\infty} \frac{1}{m} \left(\frac{r}{b} \right)^m \sin m\theta. \tag{3.12}$$

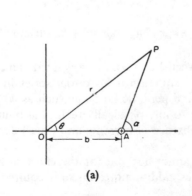

(a)

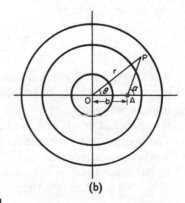

(b)

Fig. 3.1

Further, where $r = b$, it is immediately obvious from the geometry that

$$\alpha = \frac{\theta + \pi}{2}. \tag{3.13}$$

It is to be noted that in using eqns (3.11) and (3.12) for regions not actually containing the current, their series parts are not needed explicitly, being in effect included in the general Fourier series describing boundary effects. Thus the only term to be added to the general solutions is $(i/2\pi)\theta$ for a region enclosing that containing the current, and zero for a region enclosed by the one containing the current. This is explained with the aid of Fig. 3.1(b) in which are shown three annular regions of different permeability, the middle one containing a current i at A. For the inner region, $r < b$ everywhere, so that eqn (3.12) is relevant for the expansion of α; whilst, for the outer region, $r > b$ everywhere, and eqn (3.11) is relevant. Thus, combining the series parts of these equations with the general series, the terms to be added to the general solutions to account for the effect of the current in these regions can be reduced to 0 and $(i/2\pi)\theta$ respectively. In the middle region, a change of 2π in α is associated with changes in θ of 2π or 0, depending upon whether r is greater or less than b, and so eqns (3.11) and (3.12) must be employed separately as appropriate.

The physical interpretation of the above is that the term $(i/2\pi)\theta$ accounts for the (non-conservative) property of the field of the current; it is therefore required for any region within which closed paths can be drawn to contain the current but not for regions in which no path contains the current. The series terms account for the remainder of the effect of the current, the boundary magnetization, and so are naturally described by the general series of eqn (3.10).

In a similar manner to that above (using the cosine rule) it can be shown (see Smythe, p. 65) that the expansions in terms of r and θ of the function $\log R$ for a line pole (or charge), situated at the point A, Fig. 3.1(a), are as follows:
where $r > b$,

$$\log R = \log r - \sum_{m=1}^{\infty} \frac{1}{m} \left(\frac{b}{r} \right)^m \cos m\theta; \tag{3.14}$$

where $r < b$,

$$\log R = \log b - \sum_{m=1}^{\infty} \frac{1}{m} \left(\frac{r}{b} \right)^m \cos m\theta; \tag{3.15}$$

and where $r = b$ (θ not a multiple of π),

$$\log R = \log b - \sum_{m=1}^{\infty} \frac{1}{m} \cos m\theta. \tag{3.16}$$

Use of all three of these equations is necessary only in the potential function for the region containing the pole. In all other regions the effect of the pole can be attributed to boundary magnetization and so included in the general series expression.

In employing the method of separation of the variables it is simplest to use separate functions for each region of a field, these functions being connected by the boundary conditions between the regions. The boundary conditions in terms of flux density and field strength connect the *gradients* of potential functions (see section 1.5.1) so that, in general, the resulting values of potential are discontinuous across the boundaries. If it is required to derive a solution in which values of potential are continuous everywhere, appropriate constants, determined from additional equations expressing the equality of the potential functions at a boundary, can be added to the potential functions. However, when the field strength rather than the absolute potential values are required, this offers no advantages and, as in the examples below, it is not done.

3.2.2   Iron cylinder influenced by a current

As a simple example of the determination of the constants in the general solution, and to show the relationship of the present method to that of images, consider the field of a current outside an iron cylinder (Fig. 3.2). The development of the solution involves the determination of two potential functions—one for the air region ψ_A and one for the iron ψ_I, the two being connected by the boundary conditions at the surface of the cylinder. The constants of ψ_A are chosen so that the boundary requirements at infinity and at the cylindrical surface are satisfied, those of ψ_I so that the requirements are satisfied at the cylindrical surface and at the centre of the cylinder, which is taken to be the origin of coordinates.

First note that the whole field is symmetrical about the line $\theta = 0$ through the current and the centre of the circle. Thus, choosing the origin of the potential functions ψ_A and ψ_I to be the line $\theta = 0$, and, since the field due to the current is odd, the potential functions of the whole field are odd and therefore contain no cosine terms. Hence the general solution describing boundary influences in both regions, eqn (3.9), reduces to

$$\psi(r, \theta) = \sum_{m=1}^{\infty} (c_m r^m + d_m r^{-m}) \sin m\theta. \tag{3.17}$$

In the air region, r can be infinite, but, as the potential at infinity must be finite (equipotential lines stretch from infinity to the current and $(\partial\psi/\partial r)_{r=\infty} = 0$) all values of c_{Am} must be zero. Also the air space contains a current, so it is necessary to include a term $(i/2\pi)\alpha$ in the potential function, giving

$$\psi_A = \frac{i}{2\pi}\alpha + \sum_{m=1}^{\infty} d_{Am} r^{-m} \sin m\theta. \tag{3.18}$$

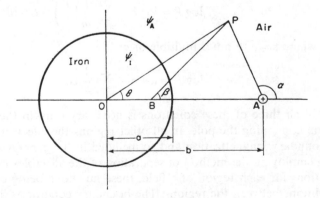

Fig. 3.2

In the iron, r can be zero so that, as the potential at the origin must be finite, d_{Am} must be zero. This region does not contain a current or pole, so there are no additional terms. Hence

$$\psi_I = \sum_{m=1}^{\infty} c_{Im} r^m \sin m\theta, \tag{3.19}$$

and it remains to determine d_{Am} and c_{Im} from the continuity of the two functions ψ_A and ψ_I on the cylinder surface. The boundary conditions, see section 1.5, when expressed in cylindrical coordinates are, for the tangential component of field strength,

$$-\left(\frac{\partial \psi_A}{r \partial \theta}\right)_{r=a} = -\left(\frac{\partial \psi_I}{r \partial \theta}\right)_{r=a}; \tag{3.20}$$

and for the radial component of flux density,

$$-\left(\frac{\partial \psi_A}{\partial r}\right)_{r=a} = -\mu\left(\frac{\partial \psi_I}{\partial r}\right)_{r=a}; \tag{3.21}$$

where μ is the relative permeability of the iron. Substituting in ψ_A for α from eqn (3.12), then differentiating ψ_A and ψ_I and substituting in eqns (3.20) and (3.21), gives

$$d_{Am} - a^{2m} c_{Im} = \frac{i}{2\pi} \frac{a^{2m}}{mb^m}, \tag{3.22}$$

and

$$d_{Am} + \mu a^{2m} c_{Im} = -\frac{i}{2\pi} \frac{a^{2m}}{mb^m}, \tag{3.23}$$

which, when solved for c_{Im} and d_{Am}, lead to the following complete solutions:

$$\psi_A = \frac{i}{2\pi}\left(\alpha + \sum_{m=1}^{\infty} \frac{\mu-1}{\mu+1} \frac{1}{m} \frac{a^{2m}}{b^m r^m} \sin m\theta\right), \tag{3.24}$$

and

$$\psi_I = -\frac{i}{(\mu+1)\pi} \sum_{m=1}^{\infty} \frac{1}{m} \frac{r^m}{b^m} \sin m\theta. \tag{3.25}$$

The radial and tangential components of field strength are given by $-(\partial \psi/\partial r)$ and $-(\partial \psi/r \partial \theta)$ respectively. If forming them, α is expressed in terms of θ as described in the previous section.

The equivalence of the solution obtained by the method of images to that obtained above may be shown by summing the potential functions of the source and image currents, and by representing them in series form using eqns (3.11) and (3.12). With current i at A the field outside the cylinder is represented by image currents of $i(\mu-1)/(\mu+1)$ at B, the inverse point of A, and $-i(\mu-1)/(\mu+1)$ at 0. The potential function for these is

$$\psi_A = \frac{i}{2\pi}\left(A + \frac{\mu-1}{\mu+1}\beta - \frac{\mu-1}{\mu+1}\theta\right), \tag{3.26}$$

and substituting for β using eqn (3.11) ($r_A > OB$), and remembering that $OB = a^2/b$, this expression becomes that obtained above (3.24). The equivalence of the solutions relevant to the iron region may be demonstrated similarly.

A number of similar harmonic function solutions involving a single cylindrical boundary and a single current are given by Hague and Moullin, but in practice it is quicker to solve such simple problems using the image method. Consideration is now given to some problems in which the method of separation of the variables gives a simpler, or in some cases, the only practicable method of solution.

3.2.3   The screening effect of a permeable cylinder

Consider the effect of a permeable cylinder used, for example, to screen an instrument or circuit from an external magnetic field. The cylinder, of relative permeability μ and having the dimensions shown in Fig. 3.3, is placed in a (say) uniform magnetic field of strength H. The complete solution for the field in all the regions involves three potential functions: ψ_A for the outer air space, ψ_I for the iron, and ψ_B for the inner air space, the region now of particular interest. Since there are no currents or poles in the field and since, from symmetry, $\psi(r, \theta') = \psi(r, -\theta')$, so that there are no sine terms in the solution, the general form for each of these potential functions can be reduced to

$$\psi = \sum_{m=1}^{\infty} (c_m r^m + d_m r^{-m}) \cos m\theta. \tag{3.27}$$

Examining the outer air space first, it is seen that the boundary condition at infinity is not, as in the last example, one of potential value but is of potential gradient. At infinity the field strength is uniform and its radial component is

$$H \cos \theta = -\frac{\partial \psi_A}{\partial r}; \tag{3.28}$$

that is

$$H \cos \theta = -\sum_{m=1}^{\infty} (m c_{Am} r^{m-1} - m d_{Am} r^{-(m+1)}) \cos m\theta, \tag{3.29}$$

from which it is seen that m can have the value unity only and hence, substituting $m = 1$

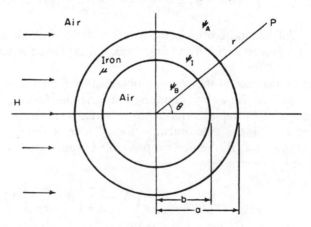

Fig. 3.3

and $r = \infty$, that $c_{A1} = -H$. Thus, for the outer air space, eqn (3.27) reduces to

$$\psi_A = -\left(Hr - \frac{d_{A1}}{r}\right)\cos\theta. \tag{3.30}$$

Expressing the equality of the radial components of flux density and of the tangential components of field strength on opposite sides of each iron–air interface gives

$$\left(\frac{\partial\psi_A}{\partial r}\right)_{r=a} = \mu\left(\frac{\partial\psi_I}{\partial r}\right)_{r=a}, \tag{3.31}$$

$$\frac{1}{a}\left(\frac{\partial\psi_A}{\partial\theta}\right)_{r=a} = \frac{1}{a}\left(\frac{\partial\psi_I}{\partial\theta}\right)_{r=a}, \tag{3.32}$$

$$\mu\left(\frac{\partial\psi_I}{\partial r}\right)_{r=b} = \left(\frac{\partial\psi_B}{\partial r}\right)_{r=b}, \tag{3.33}$$

and

$$\frac{1}{b}\left(\frac{\partial\psi_I}{\partial\theta}\right)_{r=b} = \frac{1}{b}\left(\frac{\partial\psi_B}{\partial\theta}\right)_{r=b}. \tag{3.34}$$

Then, since from eqn (3.30)

$$\left(\frac{\partial\psi_A}{\partial r}\right)_{r=a} = -\left(H + \frac{d_{A1}}{a^2}\right)\cos\theta, \tag{3.35}$$

it is clear from this equation and eqns (3.32) and (3.33) that the only value which m may take in ψ_I and ψ_B is also unity. So

$$\psi_I = -\left(c_{I1}r - \frac{d_{I1}}{r}\right)\cos\theta, \tag{3.36}$$

and

$$\psi_B = -\left(c_{B1}r - \frac{d_{B1}}{r}\right)\cos\theta. \tag{3.37}$$

The remaining boundary condition requires that the field strength remains finite, even when $r = 0$, so that $d_{B1} = 0$ and

$$\psi_B = -c_{B1}r\cos\theta. \tag{3.38}$$

Thus the field in the central air space has the direction of the line $\theta = 0$, and its strength, $-\partial\psi_B/\partial r$, is c_{B1} (a constant) and so uniform. Differentiating ψ_A, ψ_I and ψ_B, and substituting in the interface boundary conditions, (3.31) to (3.34), gives

$$\left.\begin{aligned}
H + \frac{d_{A1}}{a^2} &= \mu\left(c_{I1} + \frac{d_{I1}}{a^2}\right), \\
H - \frac{d_{A1}}{a^2} &= \left(c_{I1} + \frac{d_{I1}}{a^2}\right), \\
\mu\left(c_{I1} + \frac{d_{I1}}{b^2}\right) &= c_{B1}, \\
c_{I1} - \frac{d_{I1}}{b^2} &= c_{B1}.
\end{aligned}\right\} \tag{3.39}$$

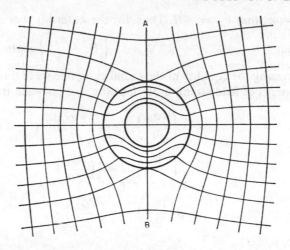

Fig. 3.4

From these equations the constants are determined. The constant c_{B1}, which from eqn (3.38) is seen to be the field strength inside the cylinder, is

$$c_{B1} = \frac{4\mu a^2 H}{a^2(\mu + 1)^2 - (\mu - 1)^2 b^2},$$ (3.40)

and, in practice, since $\mu \gg 1$, this expression can be simplified to give

$$c_{B1} \doteqdot \frac{4}{\mu}\frac{H}{1 - (b/a)^2}.$$ (3.41)

It is seen, therefore, that the strength of the field inside the cylinder varies inversely with the permeability for given proportions of the cylinder. For example, with mumetal at field strengths up to 1 oersted for which the permeability is 10 000 or more, it would be reduced to 0.04% of the outside value. A map of the whole field is shown in Fig. 3.4.

A discussion of this problem in terms of surface polarity has been given by Hammond [1].

Doublet representation of the cylinder

An interesting result demonstrable from the above is that the effect of a cylinder of finite thickness on a uniform field can be accounted for by the use of a doublet current. (It is shown in section 2.3 that a doublet can also be used to account for the effect of a *solid* cylinder on a uniform field.) By substituting $t = r\,e^{j\theta}$ in eqn (2.23) it is readily seen that the potential function in circular cylinder coordinates of a doublet at the origin, with its axis on the line $\theta = 0$ and of strength d, is given by

$$\psi = \mathrm{R}\left(\frac{d}{r}e^{-j\theta}\right)$$

$$= \frac{d}{r}\cos\theta.$$ (3.42)

But the potential function ψ_A for the region outside the cylinder is given by the right-hand side of eqn (3.30), the first term of which is the potential function of the applied uniform field and the second of which is recognizable from the above as the potential function of a doublet. Thus the effect of the hollow cylinder on the uniform field can be accounted for by a doublet, placed at the centre of the cylinder, of strength $d = d_{A1}$, when it can be simply shown that

$$d_{A1} = \frac{(\mu - 1)[1 - (b/a)^2]a^2H}{(\mu + 1)[1 - (\mu - 1)^2 b^2/(\mu + 1)^2 a^2]}. \tag{3.43}$$

3.2.4 The force between rotor and stator conductors in a cylindrical electrical machine

The magnetic circuit of a cylindrical electrical machine can for some purposes usefully be assumed to consist essentially of a thick outer iron shell (the stator) separated by an air space (the air gap) from a solid iron cylinder (the rotor) coaxial with it (Fig. 3.5). The tangential force between typical rotor and stator conductors is considered here, the presence of slots being neglected. The field at the position of the rotor conductor due to the stator current is evaluated first.

Let the potential functions in the four regions be: ψ_O in the outer air space, ψ_S in the stator, ψ_G in the air gap, and ψ_R in the rotor. Noting that the stator region contains a current, that in all regions the field is symmetrical and the potential is odd about the line $\theta = 0$, and that r can become infinite in the outer air region and zero in the rotor, then the potential functions for the field (due to the stator current alone) may be written:

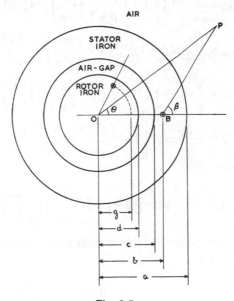

Fig. 3.5

for the outer air space,

$$\psi_O = \frac{i_S}{2\pi}\theta + \sum_{m=1}^{\infty} D_{Om}r^{-m}\sin m\theta;$$

(3.44)

for the stator,

$$\psi_S = \frac{i_S}{2\pi}\beta + \sum_{m=1}^{\infty} (C_{Sm}r^m + D_{Sm}r^{-m})\sin m\theta,$$

(3.45)

where β is the angle subtended at the stator conductor;
for the air gap,

$$\psi_G = \sum_{m=1}^{\infty} (G_{Gm}r^m + D_{Gm}r^{-m})\sin m\theta;$$

(3.46)

and for the rotor,

$$\psi_R = \sum_{m=1}^{\infty} C_{Rm}r^m \sin m\theta.$$

(3.47)

Capital letters are used for the constants to avoid confusion with the dimensions.

There are three interfaces in the field region, and substituting the above four equations in the boundary conditions at the interfaces, six equations arise from which the constants of the potential functions may be evaluated. Using the notation of Fig. 3.5, and for iron parts of relative permeability μ, the values of these constants are:

$$D_{Om} = \frac{i_S\mu}{\pi(1+\mu)}\left\{\begin{array}{l} b^{2m}[(1+\mu)^2c^{2m}-(\mu-1)^2d^{2m}] \\ -c^{2m}[(\mu^2-1)(c^{2m}-d^{2m})] \end{array}\right\}\frac{1}{Q},$$

(3.48)

$$C_{Sm} = \frac{i_S}{2\pi}\left(\frac{\mu-1}{\mu+1}\right)\frac{1}{a^{2m}}\left\{\begin{array}{l} b^{2m}[(\mu+1)^2c^{2m}-(\mu-1)^2d^{2m}] \\ -c^{2m}[(\mu^2-1)(c^{2m}-d^{2m})] \end{array}\right\}\frac{1}{Q}$$

(3.49)

$$D_{Sm} = \frac{i_S}{2\pi}c^{2m}\left\{\left[1-\left(\frac{\mu-1}{\mu+1}\right)\left(\frac{b}{a}\right)^{2m}\right][(\mu^2-1)(c^{2m}-d^{2m})]\right\}\frac{1}{Q},$$

(3.50)

$$C_{Gm} = -\frac{i_S}{\pi}\mu(\mu+1)c^{2m}\left\{1-\left(\frac{\mu-1}{\mu+1}\right)\left(\frac{b}{a}\right)^{2m}\right\}\frac{1}{Q},$$

(3.51)

$$D_{Gm} = \frac{i_S}{\pi}\mu(\mu-1)c^{2m}d^{2m}\left\{1-\left(\frac{\mu-1}{\mu+1}\right)\left(\frac{b}{a}\right)^{2m}\right\}\frac{1}{Q},$$

(3.52)

$$C_{Rm} = -\frac{2i_S}{\pi}\mu c^{2m}\left[1-\left(\frac{\mu-1}{\mu+1}\right)\left(\frac{b}{a}\right)^{2m}\right]\frac{1}{Q},$$

(3.53)

where

$$Q = \left\{[(\mu+1)^2c^{2m}-(\mu-1)^2d^{2m}]-\left(\frac{\mu-1}{\mu+1}\right)\left(\frac{c}{a}\right)^{2m}[(\mu^2-1)(c^{2m}-d^{2m})]\right\}mb^m.$$

(3.54)

The radial component of flux density due to the stator current at the position of the rotor conductor is

$$B_r = -\mu\left(\frac{\partial\psi_R}{\partial r}\right)_{r=g},$$

(3.55)

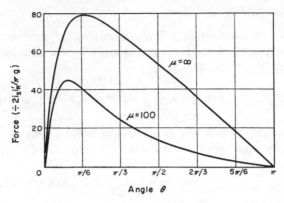

Fig. 3.6

and, if the rotor conductor carries a current i_R, the tangential force F_t causing the rotor to turn is

$$F_t = i_R B_r = - i_R \mu \sum_{m=1}^{\infty} m C_m g^{m-1} \sin m\theta$$

$$= - \frac{2 i_S i_R}{\pi} \sum_{m=1}^{\infty} \frac{\dfrac{g^{m-1}}{b^m} \mu^2 \left[1 - \dfrac{\mu-1}{\mu+1} \left(\dfrac{b}{a} \right)^{2m} \right] \sin m\theta}{\left[(\mu+1)^2 - (\mu-1)^2 \left(\dfrac{d}{c} \right)^{2m} \right] - (\mu-1)^2 \dfrac{1}{a^{2m}} [c^{2m} - d^{2m}]}, \qquad (3.56)$$

its sense being dependent upon the directions of i_S and i_R.

Using values of $a = 1.4$, $b = 1.12$, $c = 1.02$, $d = 1.0$, and $g = 0.9$, evaluation of this expression for $\mu = \infty$ and 100 gives the results plotted in Fig. 3.6. The influence of permeability is seen to be considerable, and this is because, in the model, the currents are embedded in the iron. In an actual machine the conductors are housed in slots so that the above representation of them is unsatisfactory. However, the representation of an actual machine, achieved by placing the currents in the model on the *surfaces* of the rotor and stator, is entirely satisfactory for the calculation of the force causing rotation. As is the case in many solutions of this type, the series is rather slowly convergent: 25 terms are necessary to reduce the error to less than 1%.

3.2.5 Specified distributions of potential or potential gradient on the perimeter of a circular boundary

The preceding discussion is restricted mainly to cases in which the field sources are lines (currents), and attention is turned now to the solution of fields due to specified distributions of potential or potential gradient on the boundaries of circular regions and in particular to the general solution—known as the Poisson integral—for the field inside a circle, due to specified values of potential on the perimeter.

The use of Fourier series

The general solution for a field, containing no line sources, in a region with a circular boundary centred on the origin, is given by the potential function of eqn (3.9) with the addition of a constant, i.e. by

$$\psi(r, \theta) = k + \sum_{m=1}^{\infty} (c_m r^m + d_m r^{-m})(g_m \cos m\theta + h_m \sin m\theta). \tag{3.57}$$

Taking first the case of a single region *inside* a circular boundary, $r = r'$, it is seen that, since ψ remains finite when $r = 0$, d_m is zero, and the solution becomes

$$\psi(r, \theta) = k + \sum_{m=1}^{\infty} r^m (G_m \cos m\theta + H_m \sin m\theta), \tag{3.58}$$

where $G_m = c_m g_m$ and $H_m = c_m h_m$. If the potential distribution $\psi'(r', \theta')$ is specified on the boundary, then the complete solution for the field due to this distribution is obtained by choosing the constants $k, r'^m G_m$, and $r'^m H_m$, so that the right-hand side of eqn (3.58) is equal to $\psi'(r', \theta')$. That is, the constants must be chosen to give the Fourier series representing $\psi'(r', \theta')$ in the range 0 to 2π; hence,

$$k = \frac{1}{2\pi} \int_0^{2\pi} \psi'(r', \theta') \, d\theta', \tag{3.59}$$

$$r'^m G_m = \frac{1}{\pi} \int_0^{2\pi} \psi'(r', \theta') \cos m\theta' \, d\theta' \tag{3.60}$$

and

$$r'^m H_m = \frac{1}{\pi} \int_0^{2\pi} \psi'(r', \theta') \sin m\theta' \, d\theta'. \tag{3.61}$$

For a single region *exterior* to the boundary $r = r'$, the solution is identical with the above except that r^m is replaced by $(1/r^m)$ in eqn (3.58) and r'^m is replaced by $(1/r'^m)$ in eqns (3.59), (3.60), and (3.61). When potential gradient is specified, the constants of eqn (3.58) are chosen in a similar way to give the Fourier series for $\partial\psi'/\partial r$ or $\partial\psi'/r\partial\theta$ on the boundary.

No applications to this class of problem are given here, but some are described in the papers by Rudenberg [2,3] in which the flux distribution inside machine stators due to an impressed sinusoidal field is examined.

The Poisson integral

The general solution of the Dirichlet problem (specified boundary potentials) for the interior of a circle is known as the Poisson integral, and it is of considerable importance, particularly in connection with transformation methods (see sections 6.4 and 7.7). Its development from the above equations is briefly indicated here, taking, as is usual, the radius of the bounding circle to be unity. Substituting for the Fourier coefficients (with $r' = 1$) in the potential function, eqn (3.58), gives

$$\psi(r, \theta) = \frac{1}{2\pi} \int_0^{2\pi} \psi'(\theta') \, d\theta' \left[1 + 2 \sum r^m \cos m(\theta - \theta') \right], \tag{3.62}$$

and the expression is square brackets from this equation may be reduced using the following identities:

$$1 + 2\sum r^m \cos m(\theta - \theta') = R\left[1 + 2\sum (re^{j(\theta - \theta')})^m\right]$$

$$= R\left[\frac{1 + re^{j(\theta - \theta')}}{1 - re^{j(\theta - \theta')}}\right]$$

$$= \frac{1 - r^2}{1 + r^2 - 2r\cos(\theta - \theta')}. \qquad (3.63)$$

Therefore eqn (3.62) can be reduced to

$$\psi(r, \theta) = \frac{1}{2\pi}\int_0^{2\pi} \frac{1 - r^2}{1 + r^2 - 2r\cos(\theta - \theta')}\psi'(\theta')\,d\theta', \qquad (3.64)$$

which is the Poisson integral, yielding directly the solution for the inside of the unit circle with any potential $\psi'(\theta')$ impressed at the circumference. (See also the discussion in section 7.7.)

3.3 RECTANGULAR BOUNDARIES

3.3.1 Solution of Laplace's equation in Cartesian coordinates

In cartesian coordinates Laplace's equation using the potential function ψ is

$$\frac{\partial^2 \psi}{\partial x^2} + \frac{\partial^2 \psi}{\partial y^2} = 0, \qquad (3.65)$$

and a solution for it can be obtained using the same method as for the cylindrical polar form of the equation. Thus, assuming a solution

$$\psi(x, y) = X(x)Y(y), \qquad (3.66)$$

where X is a function of x only and Y is a function of y only, the partial differential equation reduces to the two ordinary differential equations

$$\frac{d^2 X}{dx^2} + m^2 X = 0, \qquad (3.67)$$

and

$$\frac{d^2 Y}{dy^2} - m^2 Y = 0, \qquad (3.68)$$

in which m is a constant. Solutions to these equations lead to the particular solution

$$\psi(x, y) = (c\sin mx + d\cos mx)(g\sinh my + h\cosh my), \qquad (3.69)$$

and the general potential function (see section 3.2.1)

$$\psi(x, y) = \sum_{m=1}^{\infty} (c_m \sin mx + d_m \cos mx)(g_m \sinh my + h_m \cosh my). \qquad (3.70)$$

For any value of y in the range $0 \leqslant y \leqslant \infty$, this function is a Fourier series periodic in x and it is, therefore, capable of representing the field due to boundary effects. For the value of $m = 0$, the ordinary differential equations (3.67) and (3.68) give rise to the additional terms

$$k_1 + k_2 x + k_3 y, \tag{3.71}$$

which are also solutions of the field equation. The constant k_1 defines a reference potential, and k_2 and k_3 uniform fields in the x- and y-directions respectively. Thus, a more general solution is

$$\psi(x, y) = \sum_{m=1}^{\infty} (c_m \sin mx + d_m \cos mx)(g_m \sinh my + h_m \cosh my) + k_1 + k_2 x + k_3 y. \tag{3.72}$$

This solution does not account for the influence of line sources in the field; to enable it to do so, the potential functions of these sources have to be added to it. These functions, which are familar in the more convenient polar forms (see section 3.2.1), are, for the current i, at (x_0, y_0),

$$\psi = \frac{i}{2\pi} \tan^{-1}\left(\frac{y - y_0}{x - x_0}\right), \tag{3.73}$$

and for the pole, p, at (x_0, y_0),

$$\psi = \frac{p}{2\pi} \log \sqrt{[(x - x_0)^2 + (y - y_0)^2]}. \tag{3.74}$$

However, a general solution is not written here since, as stated in the introduction, and as is now evident from the form of eqns (3.73) and (3.74), the solution of problems with rectangular boundaries and line sources is more conveniently achieved using the image method.

Equation (3.70) gives, for positive integral values of m, a solution periodic in x; that is, a solution found for a region $a \leqslant x \leqslant a + \lambda$ applies to a succession of similar problems in regions $a + n\lambda \leqslant x \leqslant a + (n + 1)\lambda$, where n is an integer taking values between $-\infty$ and ∞. By choosing the constant m^2 with the opposite sign, a solution periodic in y is derived.

Consideration is now given to the two (similar) boundary shapes for which Laplace's equation is cartesian coordinates may be solved by the function (3.72), and to the ways in which the solutions may be extended.

3.3.2   The semi-infinite strip and the rectangle

Solutions, based upon eqn (3.72), are possible for the interior region of (i) a rectangle, and (ii) a semi-infinite strip, or rectangle with one side at infinity. They are not possible for the exterior regions for reasons discussed at the end of this subsection. The reference potential k_1 of the field under investigation may be taken for convenience as zero, and also the presence of uniform fields parallel to the axes is disregarded, so that $k_2 = k_3 = 0$.

The semi-infinite strip

Consider first the evaluation of the constants to give a solution for the inside of the semi-infinite strip $0 \leqslant x \leqslant a$, $0 \leqslant y \leqslant \infty$ (Fig. 3.7), the edges $x = 0$, $x = a$ and $y = \infty$ being maintained at zero potential whilst along $y = 0$ is impressed a potential distribution $f(x)$. (Physically there is only one such strip, but the mathematical treatment is the same as if there were a succession of strips side-by-side, as shown dotted in the figure.)

The boundary conditions for the potential in the sheet are:

$$\psi(0, y) = 0, \tag{3.75}$$

$$\psi(a, y) = 0, \tag{3.76}$$

$$\psi(x, \infty) = 0, \tag{3.77}$$

and

$$\psi(x, 0) = f(x). \tag{3.78}$$

The first of them requires that

$$d_m = 0, \tag{3.79}$$

and the second that

$$m = \frac{\alpha \pi}{a}, \tag{3.80}$$

where α is an integer. From the third condition, eqn (3.77),

$$g_m \tanh \infty + h_m = 0,$$

and, therefore,

$$g_m = -h_m. \tag{3.81}$$

Substituting these values for m, d_m, and g_m in eqn (3.70), and combining $\sinh my$ and $\cosh my$ in exponential form, gives

$$\psi(x, y) = \sum_{\alpha=1}^{\infty} C_\alpha \sin \frac{\alpha \pi}{a} x \exp [-(\alpha \pi/a)y], \tag{3.82}$$

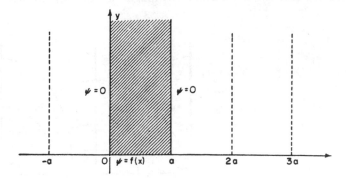

Fig. 3.7

where C_α is written for $c_\alpha h_\alpha$. The fourth condition, eqn (3.78), requires that

$$f(x) = \sum_{\alpha=1}^{\infty} C_\alpha \sin \frac{\alpha\pi}{a} x, \qquad (3.83)$$

or that C_α be chosen as the coefficients of the Fourier sine series for $f(x)$. Now it is well known (see, for example, Churchill) that the coefficients C_α of the Fourier sine series representing the function $f(x)$ in the range 0 to a are given by

$$C_m = \frac{2}{a} \int_0^a f(x) \sin mx \, dx. \qquad (3.84)$$

So, for example, if the potential along $y = 0$ is unity at all points, then m is odd only with

$$C_{(2\alpha-1)} = \frac{4}{\pi(2\alpha - 1)}, \qquad (3.85)$$

and the potential at any point in the strip is given by

$$\psi(x, y) = \frac{4}{\pi} \sum_{\alpha=1}^{\infty} \frac{1}{2\alpha - 1} \exp\{-[(2\alpha-1)/a]\pi y\} \sin\left[\frac{(2\alpha-1)\pi}{a}\right]x. \qquad (3.86)$$

From a family of curves of the variation of ψ with x for a range of values of y, the lines of constant potential can be found by interpolation. In the same manner, lines of flux may be derived from a family of curves for the flux function ϕ, this being obtained from the potential function by the Cauchy–Riemann relationships, Appendix 1. Curves of flux and potential are shown in Fig. 3.8.

This solution is probably visualized most easily in 'potential', say electrostatic, terms, but it is instructive to note that it refers also to certain arrays of currents. First, the field is identical with that which would be set up (in the positive half-plane) by an infinite array of currents of alternate sign placed on the x-axis at $\ldots, -a, 0, a, 2a, \ldots$, all the currents being of magnitude 4 amperes. (A current of n amperes produces a potential difference $n/4$ between two lines meeting at right-angles in the current.) Secondly, the field is also that which would be set up by currents at 0 and a of magnitude 2 amperes and of opposite sign, where the boundaries $x = 0$ and $x = a$ are infinitely permeable. Finally, a further equivalent field array is derivable from each of the above by considering the x-axis as the edge of an infinitely permeable boundary in the negative half-plane, and the currents to have half the magnitude of those above.

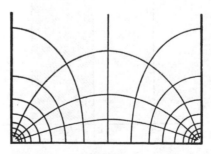

Fig. 3.8

The rectangle

The solution for the field inside a rectangular region is also of considerable interest. For instance, it has been used to investigate the field in a plane section of an electron multiplier, where different potentials exist on the pairs of sides diagonally opposite (see Zworykin *et al.*, p. 369); the treatment involved the superposition of the separate solutions for the different potential distributions on each of the sides of the rectangle (with the potentials on the remaining sides, in each case, assumed zero). Consideration is given below to typical basic solutions for the rectangular region.

Figure 3.9(a) and (b) shows such a region $0 \leqslant x \leqslant a$, $0 \leqslant y \leqslant b$, and the field in this region is to be solved for two different sets of boundary conditions. The difference between the two sets of conditions is that in the first the edge $y = b$ is a line of zero potential, whilst in the second it is impervious to flux. The remaining conditions along $x = 0$, $x = a$, and $y = 0$, in both cases, are identical with those, eqns (3.75), (3.76), and (3.78), in the example of the strip. Consequently they result in the same values for the constants d_m, m, and C_α, eqns (3.79), (3.80), and (3.83). The remaining condition on $y = b$ requires in the first case ($\psi = 0$) that

$$g_m \sinh mb + h_m \cosh mb = 0$$

or

$$\frac{g_m}{h_m} = -\coth mb; \tag{3.87}$$

and in the second case, the condition $(\partial \psi / \partial y = 0)$ on $y = b$ requires that

$$mg_m \cosh mb + mh_m \sinh mb = 0$$

or

$$\frac{g_m}{h_m} = -\tanh mb. \tag{3.88}$$

Thus, the solutions are; for the case with the line $y = b$ at zero potential,

$$\psi(x, y) = \sum_{\alpha=1}^{\infty} C_\alpha \sin\left(\frac{\alpha\pi}{a}x\right) \frac{\sinh\left[\alpha\pi(b - y)/a\right]}{\sinh\left[\alpha\pi b/a\right]}, \tag{3.89}$$

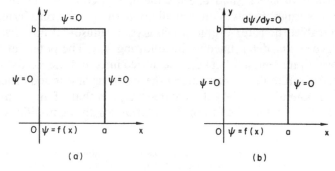

(a) (b)

Fig. 3.9

and for the case with the line $y = b$ impervious to flux,

$$\psi(x, y) = \sum_{\alpha=1}^{\infty} C_{\alpha} \sin\left(\frac{\alpha\pi}{a} x\right) \frac{\cosh\left[\alpha\pi(b - y)/a\right]}{\cosh\left[\alpha\pi b/a\right]}, \tag{3.90}$$

in both of which the values C_{α} are the coefficients of the Fourier sine series representing $f(x)$ [see eqn (3.84)].

Interconnected regions

The results of this sub-section may be extended to interconnected rectangular regions provided that these have the same width (and so m), see section 4.5, by using the equations expressing, at the boundaries, the continuity of the normal component of flux, and the tangential component of field strength, but this technique is adequately discussed in the treatment of circular boundaries.

The reason that it is not feasible to derive solutions for the (infinite) region exterior to a rectangular boundary is coupled with the use of such inter-regional continuity equations. In order to treat the whole exterior region, this must be divided into several sub-regions (formed by extending the sides of the rectangle to infinity), and then the separate potential functions for each sub-region must be connected to those in adjacent regions by the continuity equations. But the potential functions for the regions infinite in both the x- and y-directions, being integrals (see the next section), are very difficult to handle, and, because of this and because of the large number of continuity equations, the resulting formal solution is generally unmanageable.

A final example demonstrates how a solution for fields with basically simple rectangular boundaries can be used to solve a physical problem having boundaries the *shape* of which is described by a complicated equation.

3.3.3   Pole profile in the inductor alternator for a sinusoidal flux distribution

The inductor alternator offers the prospect in theory, of providing an output having a sinusoidal wave-shape. The pole shape necessary to achieve this has been determined by F. W. Carter (published in a paper by Walker [4]) and by Hancock [5], their methods being rather different from the one employed here. The shape is represented approximately in Fig. 3.10, where the armature surface is shown as a straight line and the rotor surface is shaded. Since the pole pitch t is small, the representation of the air gap as straight involves only small approximation. Considering the iron to be infinitely permeable, the rotor profile which gives a sinusoidal flux distribution at the armature surface can be determined in the following way. The potential distribution in the typical semi-infinite region — between the dotted lines and the x-axis — which gives the required distribution of flux density along the x-axis, is first found. Then, since the rotor surface is equipotential ($\mu = \infty$), its required shape is that of any convenient equipotential line of the field, determined from the solution for the semi-infinite space.[†]

[†]This device of representing a physical boundary of non-simple shape by a field equipotential line is of value in extending simpler solutions, and it has found various applications. See, for example, section 6.2.3 and Reference [9], Chapter 6.

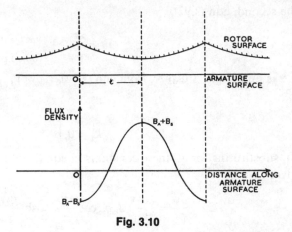

Fig. 3.10

The field is naturally periodic in x and may be analysed by considering the typical vertical strip bounded by the lines $x = 0$, $x = t$, and $y = 0$. Now the lines $x = 0$ and $x = t$ obviously coincide with flux lines, and it is required that the flux density on $y = 0$ should vary harmonically with period $2t$, so that the solution must satisfy the boundary conditions

$$\left(\frac{\partial \psi}{\partial x}\right)_{x=0} = 0, \tag{3.91}$$

$$\left(\frac{\partial \psi}{\partial x}\right)_{x=t} = 0, \tag{3.92}$$

and

$$\left(\frac{\partial \psi}{\partial y}\right)_{y=0} = B_A - B_S \cos\left(\frac{\pi}{t} x\right), \tag{3.93}$$

where B_A is the steady and B_S is the peak periodic component of flux density on the armature surface. In addition, the armature surface can be assumed, like the rotor, to be infinitely permeable, so that $(\partial \psi / \partial x)_{y=0} = 0$. This condition is fulfilled and the solution is given its simplest form by taking the potential at the armature surface to be zero. Then the remaining boundary condition is

$$\psi(x, 0) = 0. \tag{3.94}$$

The general solution for the region, eqn (3.72), is

$$\psi = \sum_{m=1}^{\infty} (c_m \sin mx + d_m \cos mx)(g_m \sinh my + h_m \cosh my) + k_1 + k_2 x + k_3 y, \tag{3.95}$$

and substituting from this in the boundary conditions gives the values of the constants. The first condition, eqn (3.91), gives

and

$$\left.\begin{array}{l} c_m = 0 \\[2mm] k_2 = 0; \end{array}\right\} \tag{3.96}$$

and the second, eqn (3.92),

$$m = \frac{\alpha\pi}{t},$$

(3.97)

where α is an integer. Further, from eqn (3.94),

$$\left.\begin{array}{l} h_m = 0 \\ \\ k_1 = 0, \end{array}\right\}$$

and

(3.98)

so that, substituting for all these constants in eqn (3.95) and differentiating with respect to y gives

$$\frac{\partial\psi}{\partial y} = \sum_\alpha \frac{\alpha\pi}{t} d_m g_m \cos\frac{\alpha\pi x}{t} \cosh\frac{\alpha\pi y}{t} + k_3.$$

(3.99)

Thus, using eqn (3.93), it is seen that

$$\left.\begin{array}{l} \alpha = 1 \quad \text{(only)}, \\ k_3 = B_A, \\ \\ d_m g_m = -\frac{t}{\pi} B_S, \end{array}\right\}$$

and

(3.100)

so that finally the solution for the potential is

$$\psi = B_A y - \frac{t}{\pi} B_S \sinh\frac{\pi}{t} y \cos\frac{\pi}{t} x.$$

(3.101)

When $\psi = 0$, this reduces to the equation $y = 0$ of the armature surface, and for any other value gives an equation for a possible rotor shape.

The profile best suited to a given application depends upon the ampere turns available to set up the magnetic potential difference ψ between rotor and stator, upon the ratio of mean flux density to the amplitude of its variation, and upon a consideration of the shapes which are physically realizable and which may be manufactured conveniently. Having decided suitable values for ψ and B_A/B_S, eqn (3.101) must be evaluated for pairs of values of x and y giving points lying on the rotor surface. Examination of the equation shows that x must be evaluated for a range of values of y between y_0 and y_t, the end points, which have to be determined graphically or numerically. Figure 3.11 shows a field map of some physically realizable profiles.

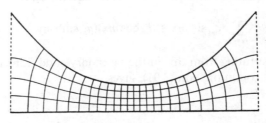

Fig. 3.11

It is evident that the profile for a flux wave of any shape can be found, for if the waveform is $f(x)$ it may be represented as a Fourier cosine series $\sum B_m \cos m\theta$. Then the required equipotentials have the form

$$\psi = B_A y - \sum_{m=1}^{\infty} \frac{B_m}{m} \sinh my \cos mx. \tag{3.102}$$

3.4 CONCLUSIONS

Integral solutions

In the preceding sections the general solutions developed are based upon the treatment of the fields as periodic, either inherently so as when polar coordinates are used, or artificially so as when cartesian coordinates are used. Because of this it is possible to express the single-valued potential function of the field region in a general form applicable to any physically realizable boundary conditions by taking only *integral* values of the separation constant m and utilizing Fourier series. An alternative form of expression for the potential function, which is necessary when the field cannot be treated as periodic in the coordinate system used (e.g. when it is infinite in both directions and cartesian coordinates are used), is as an integral over the whole range of values of m; that is, for the coordinate systems studied, as

$$\psi(r, \theta) = \int_{-\infty}^{\infty} f(m)RS \, dm, \tag{3.103}$$

and as

$$\psi(x, y) = \int_{-\infty}^{\infty} f(m)XY \, dm, \tag{3.104}$$

where R, S, X and Y are respectively functions of r, θ, x and y only. These integrals, which can be regarded as the limiting forms of the Fourier series (see Byerly, p. 52), are more difficult to handle analytically than the series and they are of limited application particularly in view of the powerful developments in numerical techniques, see later. For an example however, see Smythe (p. 70) who analyses the field, due to a line charge, within, and exterior to, a dielectric wedge.

In section 7.7.1 an alternative form of solution for the field in a region bounded by an infinite, straight boundary on which is impressed a potential distribution, is obtained from the Schwarz complex potential function. Note, too, the relationship of this function to the Poisson integral for the unit circle.

Scope of the method of separation of variables

For the coordinate systems considered it is seen that the scopes of the method of separation of variables and the method of images overlap. However, the types of problem to which each is best suited differ: thus, for two-dimensional problems involving line or distributed sources and relatively few boundaries and (with the exception of other cases treated in Chapter 2) giving rise to a finite number of images (see also section 4.2.5), the image method is simpler; but for other two-dimensional problems having such sources and multiple boundaries, separation of the variable is to be preferred; and for all problems involving specified distributions of potential or gradient it is essential.

Throughout this chapter currents when considered have been in the form of line currents, but it is to be noted that the fields of *current sheets* can be simply derived by integration (over the range of the sheet) of the solution for a line current. See, for example, section 4.2.2 and Hague or, for a different approach, Ferraro.

Sinusoidally distributed current sheets on cylindrical surfaces have been extensively employed in the study of rotating electrical machine field problems, including the three-dimensional fields in the end region, and examples are included in the Additional References. Other closely related applications of the separation-of-variables technique to end-field problems are also included for completeness.

Concerning application to three-dimensional fields it has to be said that, because of the boundary shape constraints which emerge above, the increased complexity and computational demands from the series form of the solutions in going from two to three dimensions, and the difficulty arising in the simultaneous treatment of field sources (such as electrical circuits) of practical shape, the practical use of the separation of the variables method is largely restricted to boundaries and field sources of the types considered. Outside such cases, and indeed for some within, the numerical procedures discussed in the second half of this book are more likely to be fruitful.

The question of treatment of multi-region problems has been touched upon at a number of points, and in summary it may be noted that any number of regions can be treated provided that the functions for adjoint regions involve the same constant of separation (m). With concentric circular boundaries, the constant is automatically the same, and several examples are given in this chapter. Other examples in connection with circular boundaries and many related problems of more general interest, are given by Ralph [6]. With rectangular boundaries, the constant is the same provided that all the regions are adjoint and have the same width (periodicity). No examples have been included in this chapter, but the procedure should be immediately apparent by comparison with the circular-boundary cases. It is, anyway, made clear through two problems considered in sections 4.5.2 and 4.5.3—these problems differ from those of this chapter only in that the field in one of the regions satisfies the slightly more complex Poisson equation rather than the Laplace equation. See also References [23], [24], [25] and [26] in Chapter 4. The use of boundary matching techniques (using a finite number of series terms) is discussed by Alwash [7] and Midgley and Smethurst [8].

In the next chapter, solutions are given in terms of the vector potential function A for Poisson's equation, and it will be evident that the solutions described in this chapter could also be obtained in terms of this function. However, by so doing no advantages would accrue and, indeed, for a solution requiring values of scalar potential, the use of A is inconvenient.

REFERENCES

[1] P. Hammond, The magnetic screening effects of iron tubes, *Proc. Instn. Elect. Engrs.*, **103** C, 112 (1956).

[2] R. Rudenberg, Über die Verteilung der magnetischen Induktion in Dynamoankern und die Berechnung von Hysterese und Wirbelstromverlusten, *Electrotechn. Z.*, **27** (6), 109 (1906).

[3] R. Rudenberg, Energic der Wirbelströme in elektrischen Bremsen und Dynamomaschinen, *Samml. Electrotech. Vortr.*, **10**, 269 (1907).

[4] J. H. Walker, The theory of the inductor alternator, *Proc. Instn. Elect. Engrs.*, **89** II, 227 (1942).

[5] N. N. Hancock, The production of a sinusoidal flux wave with particular reference to the inductor alternator, *Proc. Instn. Elect. Engrs.*, **104** C (new series), 167 (1957).

[6] M. C. Ralph, Eddy currents in rotating electrical machines, Ph.D. thesis, University of Leeds, 1967.

[7] S. R. Alwash, Theoretical and experimental determination of the electrical parameters of superconducting a.c. generators, Ph.D. thesis, University of Leeds (1980).

[8] D. Midgley and S. W. Smethurst, Magnetic field problems with axial symmetry, *Proc. Instn. Elect. Engrs.*, **110**, 1465–72 (1963).

Additional references

Apanasewicz, S., Method of electromagnetic field calculation in transformer of high power using integral equations, *Rozpr, Electrotech.*, **32**, 1, 141–71 (1985).

Ashworth, D. S. and Hammond, P., The calculation of the magnetic field of rotating machines: Part 2, The field of turbo-generator end-windings, *Proc. Instn. Elec. Engrs.*, **108A**, 527 (1961).

Bayburin, V. B., Three-dimensional solution of the potential problem of electron bunches in crossed fields, *Radiotekh and Electron.*, **29**, 4, 751–756 (April 1984).

Bol'shakov, A. E. and Gol'din, L. L., A method of calculating the magnetic field from the configuration of the poles of the magnet and of establishing the field of configuration from the results of magnetic measurements, *Prib. Tekh. Eksp.*, **25**, 1, 192–195 (Jan.–Feb. 1982).

Elmoursi, A. A. and Castle, G. S. P., Mapping of field lines and equipotential contours in electric field problems using the charge-simulation technique, *J. Electrostat*, **19**, 3, 221–234 (1987).

Elmoursi, A. A. and Castle, G. S. P., Mapping of field lines and equipotential contours in electric field problems using the charge-simulation technique, *IEEE Industry Applications Society Annual Meeting*, 1467–1473 (1985).

Ferrell, T. L. *et al.*, A solution to Laplace's equation for hyperboloidal electrodes with applications to dielectric testing in nonuniformn electric dields, *Gaseous Dielectrics II. Proceedings of the Second International Symposium on Gaseous Dielectrics*, pp. 383–388 (1980).

Gribovskiy, A. V., Rigorous calculation of spatially periodic electrostatic and magnetostatic fields, *Radiotekh. and Elektron.*, **29**, 3, 566–572 (March 1984).

Honsinger, V. B., Theory of end-winding leakage reactance, *Trans. Am. Instn. Elect. Engrs.*, **78** III, 417 (1959).

Ignatov, V. A., Investigation of magnetic-field distribution in the active volume of disk-rotor electric machines with a built-up magnetic circuit, *Elektrotekhnika*, **54**, 8, 27–30 (1983).

Leuer, J. A., Electromagnetic modeling of complex railgun geometries, *IEEE Trans. Magn.*, **22**, 6, 1584–1590 (1986).

Lewis, A. M. *et al.*, Thin-skin electromagnetic fields around surface-breaking cracks in metals, *J. Appl. Phys.*, **64**, 8, 3777–3784 (1988).

Marcuse, D., Electrostatic field of coplanar lines computed with the point matching method, *IEEE J. Quantum Electror*, **25**, 5, 1, 939–947 (1989).

Morisue, T., A new formulation of the magnetic vector potential method for three dimensional magnetostatic field problems, *IEEE Trans. Magn.*, **21**, 6, 2192–2195 (Nov. 1985).

Ramakrishna, K. *et al.*, Two-dimensional analysis of electrical breakdown in a nonuniform gap between a wire and a plane, *J. Appl. Phys.*, **65**, 1, 41–50 (1989).

Reece, A. B. J. and Pramanik, A., Calculation of the end region field of AC machines, *Proc. Instn. Elect. Engrs.*, **112**, 1355 (1965).

Ryan, P. M. *et al.*, Determination of fields near an ICRH antenna using a 3D magnetostatic Laplace formulation, *AIP Conf. Proc. (USA)*, **190**, 322–325 (1989).

Segal, A. M., The electrical field and resistance of a parallelepiped with small contact surface, *Izv. Akad. Nauk SSSR Energ. & Transp.*, **24**, 1, 100–103 (1986).

Smith, R. T., End component of armature leakage reactance of round rotor generators, *Trans. Am. Instn. Elect. Engrs.*, **74** III, 636 (1958).

Suzuki, M. and Asano, K., A mathematical model of droplet charging in ink-jet printers, *J. Phys. D.*, **12**, 4, 529–537 (April 1979).

Tegopulos, J. A., Determination of the magnetic field in the end zone of turbine generators, *Trans. Am. Instn. Elec. Elect. Engrs.*, **82** III, 562 (1963).

Tsaknakis, H. J. and Kriezis, E. E., Field distribution due to a circular current loop placed in an arbitrary position above a conducting place, *IEEE Trans. Geosci. & Remote Sensing*, **GE-23**, 6, 834–840 (Nov. 1985).

Uflyand, Y. A. S., Electrostatic field of a torus and spherical sector, *ZH. Tekh. Fiz.*, **48**, 8, 1741–1744 (Aug. 1978).

Vazhnov, A. I. and Grinbaum, I. N., Analytic calculations for magnetic field of excitation winding of turbogenerator with low-magnetic rotor, *Electrotechnika*, **47**, 7, 18–20 (1976).

FIELDS WITH DISTRIBUTED CURRENTS: POISSON AND DIFFUSION EQUATIONS

4.1 INTRODUCTION

Attention is turned in this chapter to fields which include regions of distributed current. Whereas in the last chapter currents were restricted to lines or sheets, the methods are now extended to deal with currents (as field sources) distributed uniformly over conducting circuits of finite cross-section and to deal with currents induced in continuous media—under both steady-state and transient conditions.

In terms of the field equation to be solved this means working with Poisson's equation, eqn (1.28), or the diffusion equation, eqn (1.33), and see section 1.8. As may be expected with both of these equations having the same left-hand side as the Laplace equation (∇^2, the Laplacian), similarities with the earlier solution forms exist in many cases. The equations for the fields *within* current-carrying regions have to be described in terms of the vector potential, A, see Chapter 1; and where both current-carrying and current-free regions exist, it is generally convenient to work in terms of A in both types of region at least for analytical solutions (as here). For solutions involving only fields exterior to current-carrying regions, however, it generally remains simpler to work with a scalar potential.

Principal attention is again given to two-dimensional field situations and for reasons very much the same as given in Chapter 3 for Laplace's equation. However, some consideration is given to three-dimensional fields also. This includes completely general solutions for the diffusion equation in both rectangular and cylindrical polar coordinates. In addition some useful approaches are described to certain types of field which could equally properly have been classified in the preceding discussion of Laplace's equation. These are, first, the fields (and forces or inductances) due to filamentary current sources of arbitrary three-dimensional shape and, secondly, fields excited by filamentary or sheet currents and involving multiple regions.

We begin by studying the fields of current-carrying conductors in free space, then move in section 4.3 to two-dimensional, time-invariant fields combining conductors of finite cross-section with multiple boundaries, in section 4.4 to time-varying fields and, finally, in section 4.5 to some consideration of three-dimensional fields. For two-dimensional fields, as discussed in section 1.2.1, the vector potential reduces to a single component, normally taken to be the z component, and for convenience here the subscript is omitted.

4.2 CURRENT-CARRYING CONDUCTORS IN FREE SPACE

This section begins with a general analytical approach to the determination of the two-dimensional fields (using A) of 'long' conductors of finite cross-section and to the calculation of force due to such conductors. It then presents a powerful semi-analytical method to determine the three-dimensional fields due to, the forces on and the inductances, self and mutual, of arbitrarily shaped conductors of negligible cross-section; and finally the use of images to extend both of the above methods is noted.

Consideration of permeable conductors is restricted to the case with rectangular cross-section (Appendix 2) because, though a variety are known, see Strutt and Hague, saturation effects would be likely in practice to be significant and numerical solutions would be more reliable.

4.2.1 Basic method: vector potential of a line current

The field of a current distributed over the cross-section of a conductor can be found by summing (integrating) the fields of an infinite number of line currents which can be deemed to constitute the distributed current. Assuming, as is likely, that the permeability of the conductor material is the same as that of free space μ_0 there is quite a variety of conductor cross-sections which can be handled analytically. These include circular, rectangular, triangular, linear, elliptical and, of course, any combination (by super-position) of these.

The starting point is the expansion of the vector potential of a line current and this can be arrived at in various ways. For example, beginning with the complex potential function (see Appendix 1), the vector potential function is identical (in the two-dimensional situation considered) with the real part or flux function component of the complex function, i.e. for the line current i

$$A = \frac{\mu_0}{2\pi} i \log r, \tag{4.1}$$

where r is the distance from the conductor to the point at which A exists. An alternative would be to note that A for the line current is identical with A outside a conductor of finite circular cross-section. The latter is

$$A = \tfrac{1}{2}\mu_0 J a^2 \log r, \tag{4.2}$$

where J is the current density and a is the radius of the conductor.

As an example of the use of eqn (4.1), the case of a conductor of rectangular section is examined.

4.2.2 The field of a rectangular bus-bar

Figure 4.1 shows a bus-bar of rectangular cross-section with sides of length $2a$ and $2b$. For a current I in the bar, the current density is $I/4\,ab$, and the current i carried by a filament of cross-section $dx'dy'$, where x' and y' are coordinates of any filament in the bar, is $(I/4\,ab)\,dx'\,dy'$. The field of all the elements of the bar is given by substituting for

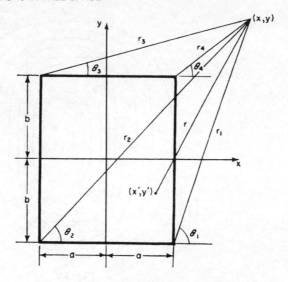

Fig. 4.1

i in eqn (4.1) and integrating over the section of the rectangle. The vector potential at any point (x, y), distance r from a typical filament with coordinates (x', y') is

$$A = \frac{I\mu_0}{8\pi ab} \int_{-a}^{a} \int_{-b}^{b} \log r \, dx' \, dy'.$$

Expressing r in terms of the coordinates gives

$$A = \frac{I\mu_0}{16\pi ab} \int_{-a}^{a} \int_{-b}^{b} \log[(x' - x)^2 + (y' - y)^2] \, dx' \, dy',$$

and this can be integrated in terms of simple functions. The result, in the form given by Strutt [1], is

$$A = \frac{I\mu_0}{16\pi ab} \left\{ \begin{aligned} &(a - x)(b - y)\log[(a - x)^2 + (b - y)^2] \\ &+ (a + x)(b - y)\log[(a + x)^2 + (b - y)^2] \\ &+ (a - x)(b + y)\log[(a - x)^2 + (b + y)^2] \\ &+ (a + x)(b + y)\log[(a + x)^2 + (b + y)^2] \\ &+ (a - x)^2 \left[\tan^{-1}\frac{b - y}{a - x} + \tan^{-1}\frac{b + y}{a - x} \right] \\ &+ (a + x)^2 \left[\tan^{-1}\frac{b - y}{a + x} + \tan^{-1}\frac{b + y}{a + x} \right] \\ &+ (b - y)^2 \left[\tan^{-1}\frac{a - x}{b - y} + \tan^{-1}\frac{a + x}{b - y} \right] \\ &+ (b + y)^2 \left[\tan^{-1}\frac{a - x}{b - y} + \tan^{-1}\frac{a + x}{b + y} \right] \end{aligned} \right\}. \tag{4.3}$$

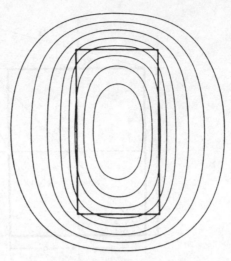

Fig. 4.2

This expression for the vector potential equates, as noted, to the flux function of the field, so that curves corresponding to constant values of A are flux lines. A set of flux lines for a conductor having $b = 2a$ is shown in Fig. 4.2.

The field components H_x and H_y are given by

$$H_x = \frac{1}{\mu_0} \frac{\partial A}{\partial y} \quad \text{and} \quad H_y = -\frac{1}{\mu_0} \frac{\partial A}{\partial x}.$$

The resulting expressions are rather long, but can be simplified by expressing them in terms of the distances r_1, r_2, r_3, and r_4 from a point in the field to the corners of the rectangle, and the angles θ_1, θ_2, θ_3, and θ_4 which the lines to the corners make with lines parallel to the x-axis. This gives

$$H_x = \frac{I}{8\pi ab} \left[\begin{array}{l} (y+b)(\theta_1 - \theta_2) - (y-b)(\theta_4 - \theta_3) \\[2mm] + (x+a)\log\frac{r_2}{r_3} - (x-a)\log\frac{r_1}{r_4} \end{array} \right], \tag{4.4}$$

$$H_y = \frac{I}{8\pi ab} \left[\begin{array}{l} (x+a)(\theta_2 - \theta_3) - (x-a)(\theta_1 - \theta_4) \\[2mm] + (y+b)\log\frac{r_2}{r_1} - (y-b)\log\frac{r_3}{r_4} \end{array} \right]. \tag{4.5}$$

As the length of one side of the rectangle tends to zero the field approaches that of a current sheet.(the current per unit area is infinite but the line density is finite). Such sheets are useful for the representation of thin, distributed windings (see also section 3.4), and as an approximation for thin, rectangular conductors. The field of one is now determined.

Consider a sheet of width $2a$, Fig. 4.3(a) carrying a total current I. In an elemental strip of width dx' the current i is $(I/2a)dx'$ and, substituting this value in eqn (4.1), the

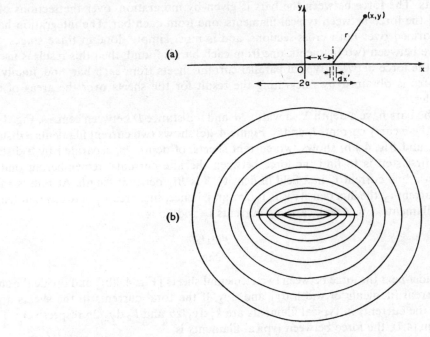

Fig. 4.3

vector potential for the sheet is

$$A = \frac{I\mu_0}{4\pi a} \int_{-a}^{a} \log r \, dx'$$

$$= \frac{I\mu_0}{8\pi a} \int_{-a}^{a} \log[(x'-x)^2 + y^2] \, dx'$$

$$= \frac{I\mu_0}{8\pi a} \left\{ \begin{array}{l} (a+x)\log[(a+x)^2 + y^2] + (a+x)\log[(a-x)^2 + y^2] \\ \\ + 2y\left[\tan^{-1}\frac{a+x}{y} + \tan^{-1}\frac{a-x}{y} \right] - 4a \end{array} \right\}. \qquad (4.6)$$

Flux lines determined from this equation are shown in Fig. 4.3(b). The components of field strength can be found from the expressions

$$H_x = \frac{1}{\mu_0}\frac{\partial A}{\partial y} \quad \text{and} \quad H_y = -\frac{1}{\mu_0}\frac{\partial A}{\partial x}.$$

4.2.3 The force between parallel rectangular bus-bars

The forces between adjacent conductors carrying large currents are a matter of considerable practical importance; see, for example, References [2] and [3]. The force between two parallel, rectangular bus-bars [4] is analysed here and, as in the previous section, the approach is to treat the actual currents as being the aggregates of many current filaments. The force between the bars is given by integration, over the sections of the bars, of the force between typical filaments, one from each bar. The integration has to be performed over both cross-sections, and is most simply done in three stages: first, the force between two filaments, one from each bar, is found; then this result is used to obtain the force between typical parallel current sheets from each bar; and, finally, the total force is obtained by integrating the result for the sheets over the areas of both rectangles.

Let the bars have a depth $2b$, a width $2a$, and a distance D between centres, Fig. 4.4(a), and let them carry currents i_1 and i_2. Figure 4.4(c) shows two current filaments a distance r apart, and Fig. 4.4(b) shows two current sheets, of depth $2b$, separated by a distance d. The first step is to find the force between the line currents, remembering that the force on a line current i_1 in a field of density B is Bi_1 per unit length. At radius r from a line current i_2, the flux density is $\mu_0 i_2/2\pi r$ and, hence, the force F_f between unit lengths of the filaments, Fig. 4.4(c), carrying currents i_1 and i_2, is

$$F_f = \frac{\mu_0 i_1 i_2}{2\pi r}. \tag{4.7}$$

Consider next the force between two elemental sheets [Fig. 4.4(b)] and divide the sheets into current filaments of width dy_1 and dy_2. If the total currents in the sheets are k_1 and k_2, the currents in typical filaments are $k_1\,dy_1/2b$ and $k_2\,dy_2/2b$ respectively. Thus, from eqn (4.7), the force between typical filaments is

$$F_f = \frac{\mu_0 k_1 k_2}{8\pi b^2} \frac{dy_1\,dy_2}{[d^2 + (y_1 - y_2)^2]^{1/2}}. \tag{4.8}$$

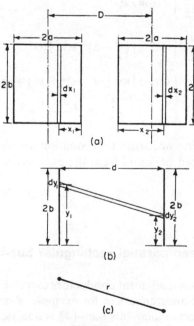

Fig. 4.4

This force has components parallel to both axes but, from considerations of symmetry, it can be seen that the resultant force on the sheets is *normal* to them. Therefore the resultant force F_s between sheets is given by integration over the two sheets of the normal component of F_f as

$$F_s = \frac{\mu_0 k_1 k_2 d}{8\pi b^2} \int_{-b}^{b} \int_{-b}^{b} \frac{dy_1 dy_2}{[d^2 + (y_1 - y_2)^2]}$$

$$= \frac{\mu_0 k_1 k_2}{2\pi b} \left[\tan^{-1} \frac{2b}{d} - \frac{d}{4b} \log\left(1 - \frac{4b^2}{d^2}\right) \right].$$

This equation can now be used to find the force between the bus-bars by expressing k_1 and k_2 in terms of the actual currents I_1 and I_2, expressing d in terms of D, a, x_1, and x_2, and by integrating the resultant expression over the widths of the rectangular sections. Now

$$k_1 = \frac{I_1 dx_1}{2a}, \qquad k_2 = \frac{I dx_2}{2a},$$

and

$$d = D - 2a + x_1 + x_2;$$

and, therefore, the force, F_b, between bars is

$$F_b = \frac{\mu_0 I_1 I_2}{8\pi a^2 b} \int_0^{2a} \int_0^{2a} \left\{ \tan^{-1} \left[\frac{2b}{D - 2a + x_1 + x_2} \right] \right.$$

$$\left. - \left[\frac{D - 2a + x_1 + x_2}{4b} \right] \log\left[1 + \frac{4b^2}{(D - 2a + x_1 + x_2)^2} \right] \right\} dx_1 dx_2$$

$$= \frac{\mu_0 I_1 I_2}{32\pi a^2 b^2} \left\{ 2b \left[(D + 2a)^2 - \frac{4b^2}{3} \right] \tan^{-1} \frac{2b}{D + 2a} \right.$$

$$+ 2b \left[(D - 2a)^2 - \frac{4b^2}{3} \right] \tan^{-1} \frac{2b}{D - 2a}$$

$$- 4b \left(D^2 - \frac{4b^2}{3} \right) \tan^{-1} \frac{2b}{D} - D\left(4b^2 - \frac{d^2}{3} \right) \log \frac{(D^2 + 4b^2)}{D^2}$$

$$+ \frac{1}{2}(D + 2a) \left[4b^2 - \frac{(D + 2a)^2}{3} \log\left(\frac{(D + 2a)^2 + 4b^2}{D^2} \right) \right]$$

$$+ \frac{1}{2}(D - 2a) \left[4b^2 - \frac{(D - 2a)^2}{3} \log\left(\frac{(D - 2a)^2 + 4b^2}{D^2} \right) \right]$$

$$+ \frac{1}{3}(D + 2a)^3 \log\left(\frac{D + 2a}{D} \right) + \frac{1}{3}(D - 2a)^3 \log\left(\frac{D - 2a}{D} \right) \right\}. \tag{4.9}$$

The variation of this force with the distance D for a range of values of the ratio a/b is shown in Fig. 4.5: the ratio F_b/F_f, where F_p is calculated as the force between line currents I_1 and I_2 a distance D apart, is plotted against the ratio $(D - 2a)/(2a + 2b)$, i.e. against the ratio of the distance between adjacent conductor faces to the conductor perimeter.

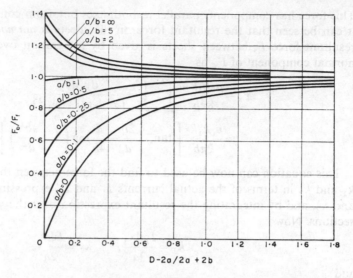

Fig. 4.5

A useful general conclusion which can be drawn from these results is that the force between two rectangular conductors having their longer sides adjacent *is always less* than the corresponding force between the central filaments, whilst the force between the rectangular conductors when their shorter sides are adjacent is *higher* than that on the central filaments. This fact has been used, for example, in determining upper bounds for the forces in turbogenerator end windings [7].

4.2.4 Filamentary conductors of arbitrary three-dimensional shape: Biot–Savart law

Field and forces

There has always been extensive use of the Biot–Savart law (see Carter or Hammond) to determine the fields of conductor configuration of various forms but, for a long time, with an inevitable interest in achieving purely analytical results. The point about the method to be described here is that it recognizes that the basic form of the law makes it perfect for application partly numerically—allowing conductor shapes to be modelled as accurately (or approximately) as desired, and involving negligible software effort. In short the approach, developed by Lawrenson [5, 6, 7] in connection with the complex fields of the end-windings of rotating machines involves an analytical integration to obtain the field of a 'short' straight current segment, following by simple numerical summation of the result for each of whatever number of segments is chosen to model the winding.

Consider Fig. 4.6 which shows a segment of a current filament RS carrying a current i_q and with end-points (x_q, y_q, z_q) and $(x_{q+1}, y_{q+1}, z_{q+1})$. Consider particularly the general point T (X, Y, Z) at which the field strength **H**, or the flux density **B**, due to i_q is to be

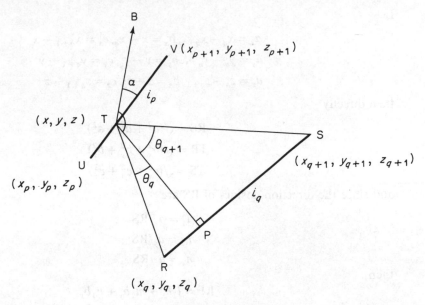

Fig. 4.6

evaluated. TP is the normal from T to RS and the direction of $\mathbf{B}$ (or $\mathbf{H}$) is normal to the plane RTS. The magnitude of the field and its direction need to be determined and this can be readily done wholly in terms of the segment coordinates.

First, however, consider also a segment of a second current filament UV with its end coordinates (x_p, y_p, z_p) and $(x_{p+1}, y_{p+1}, z_{p+1})$. From a known value of $\mathbf{B}$ and the current $\mathbf{i}_p$ the force density $\mathbf{F}$ due to RS on UV can be calculated as

$$\mathbf{F} = \mathbf{B} \times \mathbf{i}_q \qquad (4.10)$$

where T is taken most reasonably to be the mid-point of UV.

To calculate $\mathbf{B}$, the Biot–Savart law is integrated over the line RS and the result is

$$|\mathbf{B}| = \frac{\mu_0 |i_q|}{4\pi \cdot \text{TP}} (\sin \theta_q + \sin \theta_{q+1})$$

$$= \frac{\mu_0 |i_q|}{4\pi \cdot \text{TP}} \left\{ \frac{\text{RP}}{\text{TR}} + \frac{\text{PS}}{\text{TS}} \right\} \sin \alpha \qquad (4.11)$$

the direction being that of the normal to the plane RTS.

Thence, from eqn (4.11) the force density $|\mathbf{F}|$ is given by

$$|\mathbf{F}| = \frac{\mu_0 |i_q| |i_p|}{4\pi \cdot \text{TP}} \left\{ \frac{\text{RP}}{\text{TR}} + \frac{\text{PS}}{\text{TS}} \right\} \sin \alpha \qquad (4.12)$$

with the direction being normal to the plane containing UV and $\mathbf{B}$.

To use eqns (4.11) and (4.12) it is now only necessary to express the various dimensions and $\sin \alpha$ in terms of the segment coordinates, and similarly to determine the direction cosines of $\mathbf{B}$ and of $\mathbf{F}$. This involves simple results in coordinate geometry as follows.

Let

$$a_x = x_q - x_{q+1}, b_x = x - x_q, c_x = x_{q+1} - x$$
$$a_y = y_q - y_{q+1}, b_y = y - y_q, c_y = y_{q+1} - y$$
$$a_z = z_q - z_{q+1}, b_z = z - z_q, c_z = z_{q+1} - z$$

then directly

$$\text{RS} = \sqrt{(a_x^2 + a_y^2 + a_z^2)}$$
$$\text{TR} = \sqrt{(b_x^2 + b_y^2 + b_z^2)}$$
$$\text{TS} = \sqrt{(c_x^2 + c_y^2 + c_z^2)}$$

and since the direction cosines of RS are

$$l_1 = a_x/\text{RS}$$
$$m_1 = a_y/\text{RS}$$
$$n_1 = a_z/\text{RS}$$

then

$$\text{RP} = l_1 b_x + m_1 b_y + n_1 b_z$$

being positive or negative according to whether P is on the opposite or the same side of R as is S.

Also

$$\text{TP} = \sqrt{[\text{TR}^2 - \text{RP}^2]}$$

so that all lengths are now determined.

The direction cosines (l', m', n') of the vector **B** (i.e. of the normal to the plane TRS), using a well known result of coordinate geometry are:

$$l' = (a_z c_y - a_y c_z)/G$$
$$m' = (a_x c_z - a_z c_x)/G$$
$$n' = (a_y c_x - a_x c_y)/G$$

where

$$G = \sqrt{[(a_z c_y - a_y c_z)^2 + (a_x c_z - a_z c_x)^2 + (a_y c_x - a_x c_y)^2]}$$

Also, the direction cosines (l_2, m_2, n_2) of UV are

$$\left. \begin{array}{l} l_2 = (x_p - x_{p+1})/\text{UV} \\ m_2 = (y_p - y_{p+1})/\text{UV} \\ n_2 = (z_p - z_{p+1})/\text{UV} \end{array} \right\}$$

where

$$\text{UV} = \sqrt{[(x_p - x_{p-1})^2 + (y_p - y_{p+1})^2 + (z_p - z_{p+1})^2]}$$

l_2, m_2 and n_2 can be used to give the components of **B** as

$$B_x = l_2 |\mathbf{B}|, B_y = m_2 |\mathbf{B}| \quad \text{and} \quad B_z = n_2 |\mathbf{B}| \tag{4.13}$$

They also lead, using $\sin \alpha = \sqrt{1 - \cos^2 \alpha}$, to the value of $\sin \alpha$ since $\cos \alpha = l' l_2 + m' m_2 + n' n_2$ and so $|\mathbf{F}|$ can be evaluated from eqn (4.12).

Finally, the direction cosines of **F** (which has the direction of the common perpendicular to **B** and UV) are required. If they are (L, M, N), then

and

$$\left. \begin{array}{c} Ll' + Mm' + Nn' = 0 \\ \\ Ll_2 + Mm_2 + Nn_2 = 0 \end{array} \right\}$$

whence

$$\frac{L}{m_2 n' - m' n_2} = \frac{M}{l' n_2 - l_2 n'} = \frac{N}{l_2 m' - l' m_2} = \frac{1}{\sin \alpha}$$

and the components of **F** can be found as

$$F_x = L|\mathbf{F}|, F_y = M|\mathbf{F}| \quad \text{and} \quad F_z = N|\mathbf{F}|. \tag{4.14}$$

Finally therefore to obtain either the field strength or flux density at any point due to the current i_q flowing in the whole circuit of which RS is a part, it is only necessary to use eqn (4.13) and to sum over all the values of q for the segments (and their coordinates) chosen to define the whole circuit.

Similarly the force density at any point on a second conductor due to the first conductor carrying i_q is given by eqn (4.14) and, again, summing over all values of q. Note that if the force on a conductor due to itself is needed, whilst the procedure is identical, it is necessary to omit from the summation the contribution to **F** from the segment at the point at which F is being evaluated (to avoid infinities). The error in so doing is negligible for any 'reasonable' length of RS or UV.

Various aspects of the practical application of the methods, choice of segments, accuracy of solution and allowance for the finite cross-section of the actual conductors are assessed in References [5] and [6].

Inductance

To apply this method to the evaluation of inductance it is appropriate to make use of a vector potential form of the Biot–Savart law. This is because the line integral of **A** along any contour gives immediately the total flux contained by that contour, and hence the inductance. The inductance is the mutual or the self-inductance depending upon whether the contour is chosen respectively to define the appropriate contour of a separate conductor 'linking' with the flux of the exciting conductor or to define an appropriate contour adjacent to the conductor itself exciting the flux, see below.

As given in Reference [7] the expression for the magnitude of the vector potential at T due to i_q in RS is

$$|\mathbf{A}| = \frac{i_q \mu_0}{4\pi} \left(\sinh^{-1} \frac{PR}{PT} - \sinh^{-1} \frac{PS}{PT} \right) \tag{4.15}$$

Since the direction of **A** is parallel to RS the direction cosines are l_1, m_1, n_1 (above). Regarding UV in this case as being a segment of the appropriate integration contour, the component of **A** along UV is needed, being $|\mathbf{A}| \cos \alpha$ and

$$\cos \alpha = l_1 l_2 + m_1 m_2 + n_1 n_2 \tag{4.16}$$

where l_2, m_2 and n_2 are the direction cosines of UV as above.

As hinted at above, some thought needs to be given in each case as to what contour of integration of **A** is 'appropriate'. In the case of mutual inductances, as will be fairly apparent, it will generally be best to take the central filament of the (linking) conductor. In the case of self-inductance, matters are not quite so self-evident but it will generally be best to take the contour as the central track on the 'inner' surface of the conductor. In addition, it will be necessary to include a component of inductance due to fluxes inside the (exciting) conductor. This can be done using results such as those for example in eqns (4.2) and (4.3) covering the cases of circular and rectangular cross-section and including fluxes within and up to the appropriate contour. This and other aspects of the implementation of the method are discussed in Reference [7].

4.2.5   Boundary effects: use of the image method

All of the above results can be extended in varying but useful ways by making use of the image methods discussed in Chapter 2. This is apparent for two-dimensional cases because, regarding the finite cross-section as replaced by an infinite set of filament currents, images can be found filament by filament. With plane boundaries this can be seen to lead in fact to images which are of the same shape and disposition as the existing conductor and results can be economically computed even if infinite sets of images have to be limited to a modest number. Circular boundaries are not so easy, however, and conceiving of the images filament by filament will be seen to lead to images with non-uniform densities of current. This is obviously not impossible to handle but may or may not be worthwhile.

Turning to the case of filamentary circuits of the type treated in section 4.2.4 (which includes, of course, simple geometrical shapes parallel or inclined to the boundaries), images the equivalent of optical ones are seen to apply so far as plane surfaces are concerned, and for many practical situations yield wholly adequate solutions.

Attempts to use images combining circular cylindrical boundaries with three-dimensional circuit geometries throw up not only more complicated but apparently formally intractable questions because of the absence of a basis for imaging circumferential components of the circuit. Approximate approaches placing image points on a contour intermediate between the inverse and equidistance points may be useful in some situations. See Reference [5].

4.3   SOLUTION OF POISSON'S EQUATION IN SINGLE AND DOUBLE SERIES

4.3.1   Introduction

The solutions in the preceding section are all basically by integration over the field source (current)—subsequently augmented where appropriate to handle certain boundary effects by the use of the image method. Another approach which can be more powerful, particularly with two-dimensional fields, is by extension of the separation of variables method described for Laplace's equation. Mathematically the solution to the

Poisson equation is obtained by using a field function transformation such that the Poisson equation is reduced to the Laplace equation—effectively that a particular integral is added to the Laplace solution, see Stratton or Moon and Spencer for example.

The Poisson equation expressed in the vector potential, A, appropriate to the fields of distributed (steady) current density J is $\nabla^2 \mathbf{A} = -\mu\mu_0 J$, and for the two-dimensional cases to be considered here this reduces to

$$\frac{\partial^2 A}{\partial x^2} + \frac{\partial^2 A}{\partial y^2} = -\mu\mu_0 J \qquad (4.17)$$

A being to a single component everywhere parallel to J.

The coincidence between practical interest and mathematical possibility is greatest for fields due to rectangular cross-section conductors near magnetic boundaries and attention is restricted to these. Two approaches are illustrated, one making use of single Fourier series as in Chapter 3, the other using double Fourier series and providing an analytically concise result. In the former, due to Rogowski [8], the field is subdivided into a number of sub-regions bounded by lines parallel to the axis of the coordinates in which the series is expressed. In the latter, due to Roth [10–16], the whole field region is considered together, the solution being the product of two series each in terms of one coordinate alone.

In the next sub-section the single series approach is taken up for a field configuration like that of a rotating machine with an air gap winding and, so far as current distribution is concerned, a generalization on that treated in subsection 3.2.4 (with cylindrical boundaries).

4.3.2   Single series solutions: Rogowski method

Rectangular conductors in an infinite, parallel air gap

Figure 4.7(a) shows part of an infinitely long air gap of width g bounded by two parallel plane surfaces of permeability $\mu\mu_0$ in the lower of which lies a series of equal rectangular conductors carrying currents of magnitude I, alternatively positive and negative. Let the conductors have a depth b, a width $2(l-a)$, and let them be equally spaced with a distance $2l$ between their centres.

To analyse the field, the whole region is divided into several parallel subregions so chosen that, for each, a simple expression can be found to describe the current density (including $J = 0$) at all points. Separate vector potential functions are used for each region and they are chosen to satisfy, firstly, Laplace's or Poisson's equation, eqn (4.17) (depending upon whether $J = 0$ or not), and, secondly, the boundary conditions between the regions.

It is convenient to subdivide the field, by lines parallel to the permeable surfaces, into four regions: I comprises the lower permeable block, II the region containing the currents between the lines $y = 0$ and b, III the region above the currents but below the upper surface, and, IV the upper permeable block.

In the regions I, III, and IV, J is everywhere zero, and so the respective vector potentials A_{I}, A_{III}, and A_{IV} must be chosen to satisfy Laplace's equation. In the region II, A_{II} must be chosen to satisfy Poisson's equation (4.17) in which J is an expression

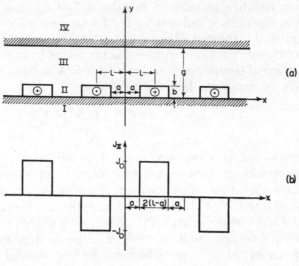

Fig. 4.7

describing the current density distribution throughout the region. Since J_{II} is independent of y and varies periodically with x, in the manner shown in Fig. 4.7(b), it may be expressed in terms of the Fourier series,

$$J_{II} = \frac{4J_0}{\pi} \sum_{m=1,3,5}^{\infty} \frac{1}{m} \cos(mka) \sin(mkx), \qquad (4.18)$$

where J_0 (the amplitude of the waveform of J) is the current density within the conductors, and $k = \pi/2l$.

The next step is the determination of the general form which the vector potentials can take. So far as the regions, I, III, and IV are concerned, the form of the necessary functions can be written down immediately from the general solution of Laplace's equation (3.72) in which A replaces ψ. The form is

$$A = \sum_{m=1}^{\infty} (c_m \sin mx + d_m \cos mx)(g_m \sinh my + h_m \cosh my) + k_1 + k_2 x + k_3 y, \qquad (4.19)$$

but this can be simplified. Since the origin of A is arbitrary, it is convenient to take $k_1 = 0$; and since the solution involves no uniform fields in either the x- or y-directions, $k_2 = k_3 = 0$. Further, with the origin of coordinates chosen as indicated in Fig. 4.7 the function of A must be odd so that $d_m = 0$. Therefore, expressing the right-hand bracketed term of eqn (4.19) in simple exponential form, a sufficiently general form of the field valid for the three Laplacian regions is

$$A = \sum_{m=1}^{\infty} (G_m e^{mky} + H_m e^{-mky}) \sin mkx. \qquad (4.20)$$

A suitable form for the solution of Poisson's equation in region II is obtained by adding the solution for the Laplace equation to a convenient particular integral for the Poisson equation, the right-hand side of which is equal to $\mu_0 J_{II}$, where J_{II} is given by eqn (4.18). The simplest particular integral A_p is clearly

$$A_p = \frac{4J_0 \mu_0}{\pi \ k^2} \sum_{m=1,3,5,...}^{\infty} \frac{1}{m^3} \cos{(mka)} \sin{(mkx)} \tag{4.21}$$

Hence the form of solution for region II may be written, combining eqns (4.20) and (4.21), and noting that only odd values of m can arise, as

$$A_{\text{II}} = \sum_{m=1,3,5,...}^{\infty} \left(G_m e^{mky} + H_m e^{-mky} + \frac{4\mu_0 J_0}{\pi m^3 k^2} \cos{mka} \right) \sin{mkx}. \tag{4.22}$$

It remains to find the values of the constants G and H in equations having the forms of (4.20) and (4.22) which satisfy the boundary conditions between the four regions. The constants F and G are expressed in terms of C, D, E, and M (primed for G) for the functions A_{I}, A_{II}, A_{III}, and A_{IV} respectively. It would be possible to use all the boundary conditions immediately and to solve them simultaneously to find the constants, but the amount of work involved can be reduced by first using some of them, chosen from experience, to effect simplification of the functions A_{I}, A_{II}, A_{III}, and A_{IV}.

In region I at $y = -\infty$, B_x is zero; thus, $(\partial A_{\text{I}}/\partial y)_{y=-\infty} = 0$, and so G_m must be zero. Hence from eqn (4.20) and noting from eqn (4.21) that only odd values of m can arise,

$$A_{\text{I}} = \sum_{m=1,3,...}^{\infty} C_m e^{mky} \sin{mkx}. \tag{4.23}$$

No simplification is possible in A_{II} which in terms of the constants D and D' becomes

$$A_{\text{II}} = \sum_{m=1,3,...}^{\infty} \left(D_m e^{mky} + D'_m e^{-mky} + \frac{4\mu_0 J_0}{\pi m^3 k^2} \cos{mka} \right) \sin{mkx}. \tag{4.24}$$

The only simplification in A_{III} results from dropping the even values of m and so, from eqn (4.20),

$$A_{\text{III}} = \sum_{m=1,3,...}^{\infty} (E_m e^{mky} + E'_m e^{-mky}) \sin{mkx}. \tag{4.25}$$

In the region IV, $B_x = 0$ at $y = \infty$ and so $F_m = 0$. Thus, from eqn (4.20), again dropping even values of m,

$$A_{\text{IV}} = \sum_{m=1,3,...}^{\infty} M'_m e^{-mky} \sin{mkx}. \tag{4.26}$$

The remaining boundary conditions relating the above simplified functions, and expressing the equality of the normal components of flux and the tangential components of field at the interfaces are:

$$\left(\frac{\partial A_{\text{I}}}{\partial x} \right)_{y=0} = \left(\frac{\partial A_{\text{II}}}{\partial x} \right)_{y=0} \quad \text{and} \quad \frac{1}{\mu} \left(\frac{\partial A_{\text{I}}}{\partial y} \right)_{y=0} = \left(\frac{\partial A_{\text{II}}}{\partial y} \right)_{y=0} ;$$

on the surface $y = b$,

$$\left(\frac{\partial A_{\text{II}}}{\partial x} \right)_{y=b} = \left(\frac{\partial A_{\text{III}}}{\partial x} \right)_{y=b} \quad \text{and} \quad \left(\frac{\partial A_{\text{II}}}{\partial y} \right)_{y=b} = \left(\frac{\partial A_{\text{III}}}{\partial y} \right)_{y=b} ; \tag{4.27}$$

and, finally, on the surface $y = g$,

$$\left(\frac{\partial A_{\text{III}}}{\partial x}\right)_{y=g} = \left(\frac{\partial A_{\text{IV}}}{\partial x}\right)_{y=g} \quad \text{and} \quad \left(\frac{\partial A_{\text{III}}}{\partial y}\right)_{y=g} = \frac{1}{\mu}\left(\frac{\partial A_{\text{IV}}}{\partial y}\right)_{y=g}.$$

Substituting in these equations from eqns (4.23)–(4.26) yields the six simultaneous equations

$$
\begin{aligned}
- C_m + D_m - D'_m &= 0, \\
- \mu C_m + D_m + D'_m &= - j, \\
\alpha^2 D_m - D'_m - \alpha^2 E_m + E'_m &= 0, \\
\alpha^2 D_m + D'_m - \alpha^2 E_m - E'_m &= - \alpha j, \\
\beta E_m - E'_m + M'_m &= 0, \\
\beta E_m + E'_m - \mu M'_m &= 0,
\end{aligned}
$$

in which $\alpha = \exp(mkb)$, $\beta = \exp(2mkg)$ and $j = (4J_0\mu_0/\pi m^3 k^2)\cos mka$. Solving these equations gives the values of the constants as

$$C_m = 4[1 - e^{-mkb}]\left[\frac{e^{2mkg}(\mu + 1) + e^{mkb}(\mu - 1)}{e^{2mkg}(\mu + 1)^2 + (\mu - 1)^2}\right]\frac{J_0\mu_0}{\pi m^3 k^2}\cos mka, \tag{4.28}$$

$$D_m = 2e^{-mkb}\left[\frac{(\mu - 1)e^{mkb}\{(\mu + 1)e^{mkb} - 2\} - e^{2mkg}(\mu + 1)^2}{e^{2mkg}(\mu + 1)^2 - (\mu - 1)^2}\right]\frac{J_0\mu_0}{\pi m^3 k^2}\cos mka, \tag{4.29}$$

$$D'_m = 2e^{-mkb}\left[\frac{(\mu - 1)^2 e^{2mkb} - (\mu + 1)e^{2mkg}\{(\mu - 1) + 2e^{mkb}\}}{e^{2mkg}(\mu + 1)^2 - (\mu - 1)^2}\right]\frac{J_0\mu_0}{\pi m^3 k^2}\cos mka, \tag{4.30}$$

$$E_m = 2(\mu - 1)[1 - e^{-mkb}]\left[\frac{e^{mkb}(\mu + 1) - (\mu - 1)}{e^{2mkg}(\mu + 1)^2 - (\mu - 1)^2}\right]\frac{J_0\mu_0}{\pi m^3 k^2}\cos mka, \tag{4.31}$$

$$E'_m = 2(\mu + 1)e^{2mkg}[1 - e^{-mkb}]\left[\frac{e^{mkb}(\mu + 1) + (\mu - 1)}{e^{2mkg}(\mu + 1)^2 - (\mu - 1)^2}\right]\frac{J_0\mu_0}{\pi m^3 k^2}\cos mka, \tag{4.32}$$

$$M'_m = 4e^{2mkg}[1 - e^{-mkb}]\left[\frac{e^{mkb}(\mu + 1) + (\mu - 1)}{e^{2mkg}(\mu + 1)^2 - (\mu - 1)^2}\right]\frac{J_0\mu_0}{\pi m^3 k^2}\cos mka, \tag{4.33}$$

where $k = \pi/2l$, and m is an odd integer. The term m^{-3} in all of these expressions results in rapid convergence of the series and, for each value of A, two or three terms only

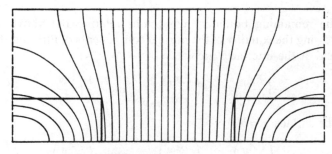

Fig. 4.8

need be taken to achieve an accuracy of 1%. The map shown in Fig. 4.8 is for the case $\mu = \infty$.

The example was chosen in the simplest form to demonstrate Rogowski's method of analysis, but by making only slight modification to the value for J_{II}, as given by eqn (4.18), to allow for more general periodic arrays of currents, it is possible to obtain solutions of a form used by several writers in the analysis of machine problems.

Rectangular conductor in a slot

Numerous problems concerned with the leakage inductance of single and multiple conductors in the slots of electrical machines are of importance and have been studied using single Fourier series solutions, for example by Robertson and Terry [9]. There is no detailed examination of these here but the basis on which the preceding type of solution, presuming upon infinite periodicity in the exciting current, is noted.

Consider the conductor slot configuration shown in Fig. 4.9(a), and assume that the bottom and sides of the slot are infinitely permeable. It can be seen by inspection of the lines of symmetry in the field, or by reference to the image solutions in section 2.2.3, that the boundary conditions on the slot sides are correctly represented by the infinite series of currents shown in Fig. 4.9(b). This being so the solution can proceed by essentially the same route as above. Thus the field is subdivided into three sub-regions

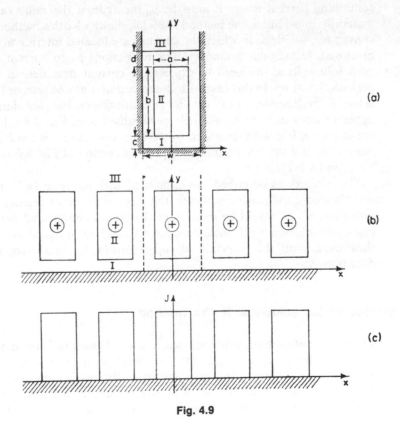

Fig. 4.9

I, II and III interfaced on the lines $y = c$ and $y = c + b$ and the boundary conditions for continuity of A (or normal flux) and $\partial A / \partial y$ (tangential field) are applied.

At the lower boundary of $I(y = 0)$, $\partial A_I / \partial y = 0$ since μ is infinite; and at 'the' upper boundary of III either, for the case of a deep slot, at $y = \infty$ $(\partial A_{III}) / \partial y = I/w$ or, for a finite depth of slot represented by a closing boundary taken to be a straight flux line (a reasonable approximation), $A_{III} = 0$ and $\partial A_{III} / (\partial y) = 0$.

Another difference to note is that in J_{II}, because of the choice of the coordinate origin, cosine terms appear (instead of sines as above) and also there is a constant term because all the current are of the same sign, thus

$$J_{II} = J_0 \left\{ \frac{a}{w} + \frac{2}{\pi} \sum_{m=1}^{\infty} \frac{1}{m} \sin\left(\frac{mka}{2}\right) \cos mkx \right\}.$$

Also constant terms and terms in y are needed in both A_{II} and A_{III}. All of these conditions are then applied in the same way as in the previous example to derive the solution.

Scope of the method

Rogowski's method is valuable in the solution of fields with distributions of current which are periodic or can be treated as such, which are rectangular or can be synthesized from rectangles, and which have sub-regions and boundaries all of which are parallel. The permeability of the sub-regions can have any (constant) value except in those containing current where it must be μ_0 (or at least the same value as the conductor material). In addition, the method is useful, though Roth's method below may be more convenient, for fields in which the currents are located interior to an open (three-sided) or closed, rectangular boundary. The restrictions as to current and boundary shape both follow from the need to express the current densities, in all regions, as singly periodic functions. In this case with closed boundaries one side, or two (opposite parallel) sides, of the boundary can be of finite permeability, but, for simplicity, the other two opposite sides must be (i) infinitely permeable [as in Fig. 4.9(a)]; or (ii) flux lines (e.g. the section of Fig. 4.7(a) bounded by the lines $x = 0$ and $x = 2 = 2l$]; or (iii) one infinitely permeable and the other a flux aline [e.g. the section of Fig. 4.7(a) bounded by the lines $x = 0$ and $x = l$].

The method, as described, could be used with many or (with the above restriction) complicated conductors. However, the large number of regions and separate vector potentials which would be required make this a complicated matter. Instead, in such cases, the solution for a single conductor should be obtained in general terms and this then used, with the principle of superposition, to synthesize the required current distribution.

4.3.3   Double series solutions: Roth's method

The method of solution now discussed, and which was originally described and developed in an electrical engineering context by Roth in a long series of papers [10–16], is applicable to a similar class of problems as Rogowski's method. However, it is much easier to use algebraically than Rogowski's method, the solution being obtained as a single function, a double Fourier series, which defines the field in the whole region of the conductors and the air. In addition it has been developed to deal with a number of problems for which Rogowski's method is not suited.

Multiple conductors in a rectangular window

To demonstrate the general method, consider an infinitely permeable, rectangular boundary containing an arrangement of p rectangular conductors, as shown by the continuous lines in Fig. 4.10. Let the lower left-hand corner of the boundary be the origin of coordinates, and let the dimensions be as shown. The order of procedure in establishing the solution is the opposite of that used in Chapter 3 or earlier in this chapter: here a function is first found which satisfies the boundary conditions of the problem, and the constants of the function are then chosen to satisfy the field equations.

The necessary form of the vector potential function can be seen from a consideration of the method of images (see section 2.2.3) and, as discussed below, this will also serve to reveal certain limitations of the method. By this method the boundaries (Fig. 4.10) could be replaced by a doubly infinite set of images, shown dotted, and so the desired function defining the resultant field inside the rectangle must be periodic with both

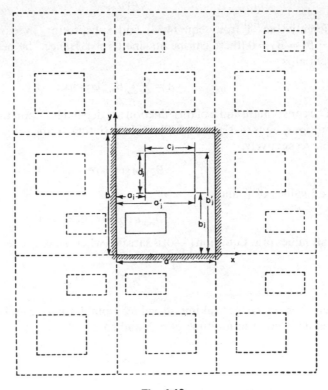

Fig. 4.10

coordinates. Its general form is given, therefore, by the product of two single Fourier series, one in x and the other in y, as

$$A = \sum_m \sum_n B_1 \cos mx \cos ny \sum_m \sum_n B_2 \cos mx \sin ny$$
$$+ \sum_m \sum_n B_3 \sin mx \cos ny + \sum_m \sum_n B_4 \sin mx \sin ny, \tag{4.34}$$

in which B, m, and n are constants dependent upon the field and boundary conditions.

As the boundaries are infinitely permeable, there are no tangential components of flux at their surfaces, and so the boundary conditions are

$$\left(\frac{\partial A}{\partial x}\right)_{x=0} = 0, \tag{4.35}$$

$$\left(\frac{\partial A}{\partial y}\right)_{y=0} = 0, \tag{4.36}$$

$$\left(\frac{\partial A}{\partial x}\right)_{x=a} = 0, \tag{4.37}$$

and

$$\left(\frac{\partial A}{\partial y}\right)_{y=b} = 0. \tag{4.38}$$

Differentiating A from eqn (4.34) and substituting in eqns (4.35) and (4.36) gives $B_2 = B_3 = B_4 = 0$ (there can be no sine terms). Hence, the necessary form of the vector potential is

$$A = \sum_m \sum_n B_{m,n} \cos mx \cos ny. \tag{4.39}$$

The constants m and n of this function are dependent upon the two remaining boundary conditions, eqns (4.37) and (4.38). Differentiating A and substituting in these equations gives, respectively,

$$B_{m,n} m \sin ma \cos ny = 0, \tag{4.40}$$

for all values of y, and

$$B_{m,n} n \cos mx \sin nb = 0, \tag{4.41}$$

for all values of x. Equation (4.40) is satisfied when ma is an even multiple of $\pi/2$, and so

$$m_h = 2(h-1)\frac{\pi}{2a}, \tag{4.42}$$

where h is an integer taking all values from 1 to ∞. Similarly, eqn (4.41) is satisfied when nb is an even multiple of $\pi/2$, and so

$$n_k = 2(k-1)\frac{\pi}{2b}, \tag{4.43}$$

where k is an integer taking all values from 1 to ∞. Hence, substituting for m_h and n_k

in eqn (4.39), the final form of the vector potential function is

$$A = \sum_{h=1}^{\infty} \sum_{k=1}^{\infty} B_{h,k} \cos(h-1)\frac{\pi x}{a} \cos(k-1)\frac{\pi y}{b}. \tag{4.44}$$

For this function to be the required field solution it must also obey Poisson's equation over the cross-section of the conductors, and Laplace's equation elsewhere in the air region. Hence, differentiating eqn (4.39) and substituting in the field eqn (4.17), this requires that

$$\sum_{h} \sum_{k} (m_h^2 + n_k^2) B_{h,k} \cos m_h x \cos n_k y = \mu_0 J, \tag{4.45}$$

where the current density, J, is equal to J_j (a constant) over the jth conductor, and to 0 over all the air regions. Equation (4.45) is satisfied provided that $B_{h,k}$ is chosen as the general coefficient of the double Fourier series of cosine terms, the sum of which is equal to $\mu_0 J_j$ over the area of each of the conductors, and 0 elsewhere. The values of $B_{h,k}$ can be determined by a method exactly analogous to that used in finding the coefficients of a single Fourier series, namely by multiplying both sides of eqn (4.45) by $\cos m_h x \cos n_k y$ and by integrating over the area of the field space (one period of the doubly infinite array, Fig. 5.10). Since all other integrated terms vanish, this gives, for the left-hand side of the equation,

$$\int_0^a \int_0^b (m_h^2 + n_k^2) B_{h,k} \cos^2 m_h x \cos^2 n_k y \, dx \, dy = \frac{ab}{4}(m_h^2 + n_k^2) B_{h,k}; \tag{4.46}$$

and for the right-hand side,

$$\int_0^a \int_0^b \mu_0 J \cos m_h x \cos n_k y \, dx \, dy,$$

which, since J is zero except over the area of the p conductors, is equal to

$$\mu_0 \sum_{j=1}^{p} J_j \int_{a_j}^{a'_j} \int_{b_j}^{b'_j} \cos m_h x \cos n_k y \, dx \, dy$$

$$= \mu_0 \sum_{j=1}^{p} J_j \left[\frac{(\sin m_h a'_j - \sin m_h a_j)}{m_h} \frac{(\sin n_k b'_j - \sin n_k b_j)}{n_k} \right]. \tag{4.47}$$

Hence, equating the right-hand sides of eqns (4.46) and (4.47) gives

$$B_{h,k} = \frac{4\mu_0}{ab} \frac{1}{(m_h^2 + n_k^2)} \sum_{j=1}^{p} J_j \left[\frac{(\sin m_h a'_j - \sin m_h a_j)}{m_k} \frac{(\sin n_k b'_j - \sin n_k b_j)}{n_k} \right]. \tag{4.48}$$

This equation gives the value of $B_{h,k}$ provided that h or k are not equal to 1, but, when either h or k, or both, are equal to 1 (i.e., when m_h or n_k, or both, are zero) eqn (4.48) takes special forms and it is necessary to return to eqn (4.45) to evaluate $B_{1,k}$, $B_{h,1}$, and $B_{1,1}$. The three cases are now taken in order.

(a) For $h = 1$ and $k > 1$ (i.e. $m_h = 0$ and $n_k > 0$) the left-hand side of eqn (4.45) reduces (after integration) to $B_{1,k} n_k^2 ab/2$. On the right-hand side the term

$$\frac{\sin m_1 a'_j - \sin m_1 a_j}{m_1}$$

is indeterminate, as $m_1 = 0$, but differentiating top and bottom with respect to m_1 and again putting $m_1 = 0$ gives, for its limiting value,

$$a'_j - a_j = c_j$$

(see Fig. 4.10). Therefore

$$B_{1,k} = \frac{2\mu_0}{n_k^2 ab} \sum_{j=1}^{p} J_j c_j \left[\frac{\sin n_k b'_j - \sin n_k b_j}{n_k} \right]. \tag{4.49}$$

(b) For $h > 1$ and $k = 1$ it is easily shown in a similar manner that

$$B_{h,1} = \frac{2\mu_0}{m_h^2 ab} \sum_{j=1}^{p} J_j d_j \left[\frac{\sin m_h a'_j - \sin m_h a_j}{m_h} \right]. \tag{4.50}$$

(c) Similarly for $h = 1$ and $k = 1$ it is easily shown that $B_{1,1}$ is a constant which, because the origin of A is arbitrary, can be ignored.

Finally, therefore, the solution for the field, inside and outside the conductors, is given by the single vector potential function A, eqn (4.44), in which the value of $B_{h,k}$ is obtained from eqn (4.48) in the general case, but from eqns (4.49) and (4.50) when h or k equals 1.

This general solution is employed directly in the next section to obtain the field in a transformer winding. It applies, of course, to any number and arrangement of conductors within the boundary and, in practice, to adapt it to suit other problems only a change in the value of m or n is involved.

The forces on, and the inductance of, a transformer winding

The forces experienced by the windings of transformers are very large and, to ensure that no damage results from them, it is important to be able to estimate their magnitudes. It is also important to be able to calculate the inductances of the windings. Analysis of forces and inductances has been made by Roth in papers referred to earlier—Reference [12] is devoted to a problem similar to that examined below, and Reference [15] to the case of a transformer with cylindrical windings—and also by Billing [17] (who has summarized some of Roth's work in English), Rabins [18], and Vein [19].

Consider the arrangement of the windings in the 'window' of a simple core-type transformer, the cross-section of which is shown in Fig. 4.11; it is a special case of that represented in Fig. 4.10. Also, it is symmetrical about the vertical line through the centre of the window and so, in performing the analysis, it is sufficient to consider one half of the window area. The line of symmetry is an equipotential line of the field and so, by choosing the coordinate axes as shown, the left-hand half corresponds directly with the field space in Fig. 4.10. Consequently, the solution given in eqns (4.44) and (4.48)–(4.50) is, with the number of conductors p, equal to 2, the solution for the field in the present problem. It has been used to prepare the field map given in Fig. 4.12.

The force F on a conductor is calculated from the components of flux density acting within the conductor. These are simply obtained in terms of the gradients of the vector potential and are used to obtain expressions for the forces on a filament of cross-section $dx\, dy$ which are then integrated over the conductor sections. Thus the x- and y-components of F, per unit lingth in the z-direction, acting on the jth conductor, are,

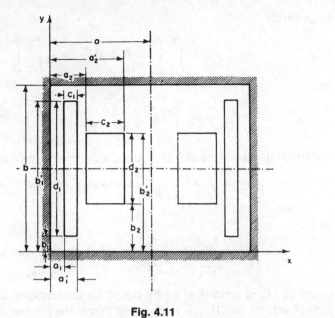

Fig. 4.11

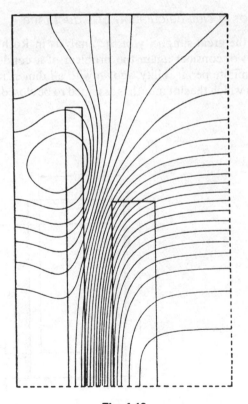

Fig. 4.12

respectively,

$$F_{xj} = \int_{a_j}^{a_j'} \int_{b_j}^{b_j'} J_j B_y \, \mathrm{d}x \, \mathrm{d}y = -J_j \int_{a_j}^{a_j'} \int_{b_j}^{b_j'} \frac{\partial A}{\partial x} \, \mathrm{d}x \, \mathrm{d}y$$

and

$$F_{yj} = \int_{a_j}^{a_j'} \int_{b_j}^{b_j'} J_j B_x \, \mathrm{d}x \, \mathrm{d}y = J_j \int_{a_j}^{a_j'} \int_{b_j}^{b_j'} \frac{\partial A}{\partial y} \, \mathrm{d}x \, \mathrm{d}y.$$

Substituting the values of $\partial A/\partial x$ and $\partial A/\partial y$ from eqn (4.44) gives, respectively,

$$F_{xj} = -J_j \sum_{h=1}^{\infty} \sum_{k=1}^{\infty} B_{h,k} \left[\frac{(\cos m_h a_j' - \cos m_h a_j)(\sin n_k b_j' - \sin n_k b_j)}{n_k} \right] \tag{4.51}$$

and

$$F_{yj} = J_j \sum_{h=1}^{\infty} \sum_{k=1}^{\infty} B_{h,k} \left[\frac{(\cos n_k b_j' - \cos n_k b_j)(\sin m_h a_j' - \sin m_h a_j)}{m_h} \right], \tag{4.52}$$

where $B_{h,k}$ is, in general, given by eqn (4.48) but has special values when h or $k=1$ [see eqns (4.49) and (4.50)]. It is to be noted that in the present case, because of the symmetry about the x-axis, $F_{yj}=0$ for both conductors.

Conductor in slot: calculation of inductance

To show the great simplicity of the analysis in Roth's method as compared with that of Rogowski, consider again the problem of a conductor in a parallel-sided slot in a block of infinite permeability, shown with all dimensions in Fig. 4.13. Further, consider the case in which the slot mouth is assumed to be closed by a straight flux line at $y=b$.

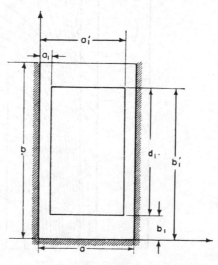

Fig. 4.13

The boundary conditions on the slot sides and bottom, those of zero tangential component of flux density on the lines $x = 0$, $y = 0$, and $x = a$, are identical with those expressed, for the detailed analysis, in eqns (4.35)–(4.37). Therefore the form of the vector potential for the field in the slot is expressed by eqn (4.39) and the value of m_h is given by eqn (4.42). The remaining boundary condition differs from those considered previously in this section. At the line $y = b$ the field is bounded by a flux line, i.e. by a line of constant vector potential to which it is most convenient to attribute the value zero. Thus

$$(A)_{y=b} = 0, \tag{4.53}$$

and, substituting for A from eqn (4.39), this requires that

$$\cos mx \cos nb = 0$$

for all values of x. Therefore nb must be an odd multiple of $\pi/2$, so that

$$n_k = (2k - 1)\frac{\pi}{2b}. \tag{4.54}$$

Substituting for m_h from eqn (4.42) and for n_k from eqn (4.54) in eqn (4.39) then gives the vector potential function

$$A = \sum_{h=1}^{\infty} \sum_{k=1}^{\infty} B_{h,k} \cos\left[(h-1)\frac{\pi}{a}\right] \cos\left[(2k-1)\frac{\pi}{2b}\right]. \tag{4.55}$$

This function must also satisfy the field equations, and this is achieved, exactly as in section 4.3.3, by choosing the coefficients $B_{h,k}$ to be the coefficients of a double Fourier series the values of which are given by eqn (4.48) with $p = 1$. Note, from eqn (4.54), that when $k = 1$, $n_k \neq 0$, and so it is only necessary to consider the special values of $B_{h,k}$ covered by eqn (4.49).

This example is a particularly suitable one with which to demonstrate the analytical determination of the inductance L of a conductor, section 1.6.2,

$$L = \frac{1}{SI} \int \int_S (A_0 - A) \, dx \, dy,$$

to this end is described below. The physical quantity of most interest is the slot leakage inductance, i.e. the inductance due to flux crossing the slot from tooth to tooth and circulating within the conductor itself. It is usual to consider this flux to be bounded by the straight flux line crossing the slot mouth, and in this case the value of vector potential function A_0 associated with this line is 0 [see eqn (4.53)]. Further, the form of the function A describing the field is given by eqn (4.55), where $B_{h,k}$ is calculated for the total current I in the conductor. Therefore, using the above notation for the conductor and slot dimensions and substituting for A, gives the expression for inductance as

$$L = \frac{-1}{c_j d_j I} \int_{a_j}^{a'_j} \int_{b_j}^{b'_j} \sum_h \sum_k B_{h,k} \cos(h-1)\frac{\pi x}{a} \cos(2k-1)\frac{\pi y}{2b} \, dx \, dy$$

$$= \frac{1 - 2ab}{\pi^2 c_j d_j I} \sum_h \sum_k \frac{B_{h,k}}{(h-1)(2k-1)} \left\{ \left[\sin(h-1)\frac{\pi a'_j}{a} - \sin(h-1)\frac{\pi a_j}{a} \right] \right.$$

$$\left. \times \left[\sin(2k-1)\frac{\pi b'_j}{2b} - \sin(2k-1)\frac{\pi b_j}{2b} \right] \right\}. \tag{4.56}$$

4.3.4 Scope of the methods

Roth's method, its scope and extensions, was extensively discussed by Hammond [21] and by Pramanik [22,23]. The general scope of the method may be deduced from the form of the solution, eqn (4.34), and is basically defined by the requirement that the boundaries must effectively be equivalent to lines of symmetry within some doubly periodic field pattern. As an example, Fig. 4.10 shows a small section of such a pattern (in terms of actual and image currents), and it is immediately apparent that a basically rectangular shape of boundary is implied. Also, it can be seen, in this particular case, that the permeability of *all* of the 'dividing' lines of the field can be infinite (when actual and image currents have the same sign) or zero (when the signs of the currents alternate between adjacent rectangular regions). Further consideration reveals, however, that, more generally, the boundaries can involve any combination of $\mu = \infty$ or 0. Thus, for example, Fig. 4.13 applies to a combination of three boundaries having $\mu = \infty$ and one having $\mu = 0$; and Fig. 4.14 could apply for a combination with two boundaries (*HA* and *AE*) having $\mu = \infty$, and two opposite boundaries (*EI* and *IH*) having $\mu = 0$, this being regarded, if convenient, as one quarter of a symmetrical field with four boundaries (*AB*, *BC*, *CD*, and *DA*) having $\mu = \infty$.

It is to be emphasized, however, that the word boundary in the sense just used need only be taken to be an 'outer' boundary. It is possible, as Pramanik has shown, to subdivide a field bounded by a rectangle such as *ABCD* in Fig. 4.14 into separate regions which may have different *and finite* permeabilities. For example, provided *ABCD* has $\mu = \infty$ or 0, then the medium of *ABFH* could be μ and that of *HFCD* could be μ_2. It has also been shown by Pramanik that, with similar restrictions on the 'outer' boundary, L-shaped regions (e.g. *ABCGIHA*) can be treated. However, with both of these extensions it is necessary to work with the complete form of eqn (4.34) for each separate region, and this aggravates difficulties of computation which, as discussed below, can exist even with solutions involving only one double series. (In Roth's original work, solutions to problems involving regions of finite permeability inside an 'outer' boundary were developed by introducing current sheets, of appropriate strength and distribution, along the surface of the region [13]. This approach is of some theoretical interest but adds to be conceptual difficulties of the solution. Additional comments on it have been given by Hammond [21].

With regard to conductor cross-sectional shapes, although most applications of Roth's method have been concerned with rectangles placed parallel to the boundaries, there is

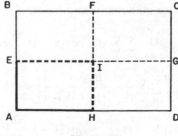

Fig. 4.14

little real restriction on the shapes or their disposition. In the first place, any shape synthesizable from rectangular elements can be treated, and there is no requirement (as there is with Rogowski's method) for the elements to be parallel to the boundaries. In fact, any shape of conductor bounded by straight lines or mathematically definable curves can be treated simply by integrating the right-hand side of eqn (4.45) appropriately over the actual area of the conductors. An example of this procedure for the case of conductors of traingular section is given elsewhere [25].

Perhaps the principal attraction of Roth's method, apart from its elegance, is its concise algebraic form, with a single function, which can be virtually written down by inspection, providing the solution for relatively complex combinations of Poissonian and Laplacian field regions. Unfortunately, a price has to be paid for this convenience and it is, not surprisingly, that the computational effort involved in evaluating solutions can be considerable. This follows from rather slow convergence of some of the double series. To combat this difficulty, Roth and Kouskoff [20] developed methods for the reduction of the double series to single series, and these methods have been appraised very fully by De Kuijper [24, 25]. As pointed out by Hammond, when such reductions are possible they must imply the availability of single series solutions by another method possibly Rogowski's.)

4.4 TIME-DEPENDENT FIELDS: EDDY CURRENTS

4.4.1 Introduction

Attention is turned now to the important practical case of fields which involve time variation and induced currents, commonly termed eddy currents, within conducting regions. The fields within such regions satisfy the diffusion equation which whether expressed in terms of flux density, **B**, field strength, **H**, current density, **J**, or vector potential, **A**, has the same form expressible as

$$\nabla^2 \mathbf{A} = \sigma \mu \mu_0 \frac{\partial \mathbf{A}}{\partial t}. \tag{4.57}$$

The actual variable is chosen as most convenient in any particular situation to suit boundary conditions, desired information etc. As part of the progressive development through the book, this sub-section whilst introducing the time variable, limits consideration of space dimensions to one or two and continues to concentrate on boundary shapes which are rectangular, including plane, or cylindrical. The Laplacian ∇^2 has the identical forms occurring in the Laplace and Poisson equations and so, in terms of flux density for example, the cartesian form of the diffusion equation is

$$\frac{\partial^2 B}{\partial x^2} + \frac{\partial^2 B}{\partial y^2} = \sigma \mu \mu_0 \frac{\partial B}{\partial t} \tag{4.58}$$

and the circular cylindrical form is

$$\frac{\partial^2 B}{\partial r^2} + \frac{1}{r} \frac{\partial B}{\partial r} + \frac{1}{r^2} \frac{\partial^2 B}{\partial \theta^2} = \sigma \mu \mu_0 \frac{\partial B}{\partial t}. \tag{4.59}$$

In view of this it is not surprising that the methods of separation of the variables can be used in much the same way as for the Laplace and Poisson equation and, indeed, can be extended to include the time-dependent portion of the equation also. This extension is taken up in the next sub-section. Subsequently a number of examples illustrative of the types of field and of the coordinate and boundary shapes are discussed, the approach being to draw as much as possible on discussion of similar points treated earlier, e.g. boundary conditions, and to introduce as wide a range of solutions of practical interest as reasonable.

4.4.2   Treatment of time variation

The treatment of time variation poses little difficulty and can be handled mathematically by treating the time variable very much as an additional space variable (coordinate). Indeed, it can be seen that since only the right-hand term in eqns (4.57) (or (4.58) or (4.59)) contains t, that term can always be decoupled (separated) from the remainder of the equation. Thus a solution, $\mathbf{A}$, to eqn (4.57) in coordinates u_1, u_2, u_3, can be assumed in the form

$$\mathbf{A}(u_1, u_2, u_3, t) = \mathbf{S}(u_1, u_2, u_3)\, T(t) \qquad (4.60)$$

substituting into eqn (4.57) yields

$$\frac{1}{\mathbf{S}}\nabla^2\mathbf{S} = \sigma\mu\mu_0\frac{1}{T}\frac{\partial T}{\partial t} = -k^2, \qquad (4.61)$$

where $\gamma^2 = (\sigma\mu\mu_0)^{-1}$ and where k^2 is the separation constant (constant because it must be independent of variations in both t and u, as in x and y or r and θ earlier, see Chapter 3). Thus, T can be found as the solution of

$$\frac{\partial T}{\partial t} + k^2\gamma^2 T = 0 \qquad (4.62)$$

and in its general form is

$$T = K\,\mathrm{e}^{-k^2\gamma^2 t}, \qquad (4.63)$$

where K is a constant depending upon the particular field.

The part of the solution in the space coordinates (u_1, u_2, u_3) becomes similarly the solution of

$$\nabla^2\mathbf{S} + k^2\mathbf{S} = 0 \qquad (4.64)$$

which is the Helmholtz equation—forms of which arose and were solved earlier for particular cases in sections 3.2.1 and 3.3.1 and in previous sections of this chapter.

Returning to eqn (4.63), this solution has been obtained independently of the particular choice of space coordinates and shows that all 'diffusion' type fields are characterized by exponential variation of the field components in time. The handling of the general case is discussed further in sub-section 4.4.4 and attention is focused first on the practically very important case of harmonic time variation (often referred to as the periodic case) arising from the widespread use of a.c. systems and with implications for power losses, effective circuit impedances and various industrial and scientific processes.

For this case both mathematical and physical experience indicate that, for the linear media to which attention has to be restricted for analytical solutions, all field quantities in the steady-state must have amplitudes which are constant, and time variations which are at the same frequency as that of the 'exciting' field component. Taking that frequency to be ω, all time variations can be expressed as $e^{j\omega t}$, the full solution (as above) can be written

$$\mathbf{A}(u_1, u_2, u_3, t) = \mathbf{S}(u_1, u_2, u_3)e^{j\omega t} \tag{4.65}$$

and substitution in eqn (4.57) yields

$$e^{j\omega t} \nabla^2 \mathbf{S} = j\omega\sigma\mu\mu_0 \mathbf{S}e^{j\omega t}. \tag{4.66}$$

This shows that the explicit appearance of time is eliminated and that the solution reduces to the solution of eqn (4.64) in which

$$k^2 = -j\omega\sigma\mu\mu_0 \tag{4.66a}$$

and S becomes complex with its modulus being the magnitude and its argument being the phase (with respect for some suitable reference) of the field vector.

4.4.3 Harmonic time variation with spatially fixed fields

A variety of boundary shapes and excitation functions with cases involving one or two space coordinates are now studied briefly to illustrate first the use of the above results, particularly eqn (4.64), and secondly something of the range of problems that can be handled.

Flux and current in a plane-faced block: skin effect

The simplest of all situations, but one which provides easy insight and also quantitative results useful in a number of cases is shown in Fig. 4.15. This shows a plane-faced block of conducting material with its surface in the xz-plane, subject to a uniform applied (harmonically varying) magnetic field, B, everywhere directed in the x direction and the associated distribution of current density, J, flowing within the block in the z direction,

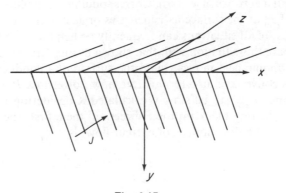

Fig. 4.15

with magnitude independent of x but decaying in density with depth (increasingly y). This field is one (space) dimensional and is described in terms of J (scalar having only the z component) by the reduced form of eqn (4.58) as

$$\frac{\partial^2 J}{\partial y^2} = \sigma\mu\mu_0 \frac{\partial J}{\partial t}. \tag{4.67}$$

The solution, being the complex phasor $\bar{J}$ defining the magnitude and phase of J, is the solution of the reduced form of eqn (4.64)

$$\frac{\partial^2 \bar{J}}{\partial y^2} - j\omega\sigma\mu\mu_0 \bar{J} = 0 \tag{4.68}$$

that is

$$\bar{J} = a_1 e^{ky} + a_2 e^{-ky},$$

where $k = \sqrt{(-j\omega\sigma\mu\mu_0)}$ and a_1 and a_2 are constants.

The boundary conditions on $\bar{J}$ can be expressed as,

$$\bar{J} = J_0, y = 0,$$

where J_0 is the magnitude of the current at the surface of the block and may be chosen to be real and to define the origin of phase; and,

$$\bar{J} = 0, y = \infty.$$

Hence $a_1 = 0$, $a_2 = J_0$ and the solution becomes

$$\bar{J} = J_0 e^{-ky}. \tag{4.69}$$

This result can be simply re-expressed to bring out a number of points. Hence introducing δ the *skin depth*, where $\delta = \sqrt{(2/\omega\sigma\mu\mu_0)}$ and separating the modulus and argument of $\bar{J}$ in eqn (4.69) gives

$$\bar{J} = J_0 e^{-y/\delta} e^{-jy/\delta}. \tag{4.70}$$

It is now easy to determine and visualize the way in which the induced current varies with position (specifically depth y below the surface) and also with ω, σ and μ—since increases (for example) in each of these lead equally to decreases in the value of δ. Thus the magnitude of current density decreases on an exponential basis from the maximum value J_0 at the surface both with increasing depth y and with decreases in δ (finally becoming zero when $y = \infty$). Correspondingly the phase of $\bar{J}$ lags progressively behind that of J_0 at the surface as y increases or as δ decreases. This physical picture from this simplest of all situations can frequently be helpful in understanding complex situations. Compare this situation with that shown in Fig. 4.19 which shows the actual flux lines as an harmonic field moves round the surface of a conducting cylinder.

It is convenient also to illustrate how power loss P due to J, current I and effective resistance R_{eff} can be simply calculated. Considering in what follows a unit width of the conducting block (in the z direction) then, first, the power loss in a length l of the block (in the direction, x, of current flow) is

$$P = l \int_0^\infty \frac{|\bar{J}|^2}{\sigma} \, dy = \frac{J_0^2 l}{\sigma} \int_0^\infty e^{-2y/\delta} \, dy$$

$$= l J_0^2 \delta/2\sigma. \tag{4.71}$$

Secondly the total current is

$$I = \int_0^\infty J_0 \, dy = J_0 \int_0^\infty e^{-(1+j)y/\delta} \, dx$$

$$= \frac{J_0 \delta}{\sqrt{2}} \angle -45° \tag{4.72}$$

and thirdly the effective resistance is

$$R_{eff} = \frac{P}{|I|^2} = l/\delta\sigma. \tag{4.73}$$

Straight conductor of circular cross-section

The above results can be used to approximate more complex situations provided that the skin depth is small compared with the physical dimensions involved. A very simple example is the case of a solid circular conductor, radius r, with its axis in the direction of the field. Provided $\delta \ll r$ the above can be used, but otherwise results have to be derived for the field which correctly represents the circular cylinder nature of the conducting region. This means working with the field equation of the form of eqn (4.59) reduced to reflect the (still) 1-dimensional nature of the field which is independent of θ. The corresponding form of the Helmholtz equation is

$$\frac{\partial^2 \bar{J}}{\partial r^2} + \frac{1}{r}\frac{\partial \bar{J}}{\partial r} - \tau^2 \bar{J} = 0 \tag{4.74}$$

and even in this simple case leads to a solution involving Bessel functions (eqn (4.74) being Bessel's equation). The details are not pursued here but the reader can find them for example in McLachlan, Stoll, or Moon and Spencer (who plot current density as a function of frequency and radius), or they can be deduced from the two- and three-dimensional results presented later in this chapter. Other examples of problems involving one space dimension such as this, but in some cases multiple field regions, can be found in Stoll, Silvester, and Tegopoulos and Kriezis, and include particularly magnetic screening and power losses in cylindrical shells.

Straight conductor of rectangular cross-section

The field in such a conductor again in a time varying magnetic field which is everywhere uniform (with peak flux density B_0) and is coaxial with the axis of the conductor, provides a simple example of a two-dimensional situation. The basic field equation is (4.58) and for harmonic excitation this leads to the corresponding Helmholtz equation

$$\frac{\partial^2 B}{\partial x^2} + \frac{\partial^2 B}{\partial y^2} - \tau^2 B = 0. \tag{4.75}$$

The solution to this is obtained using the method of separation of the variables as discussed in the preceding chapter though with a small generalization. Assuming the solution to be in the form

$$B = X(x) \cdot Y(y) \tag{4.76}$$

eqn (4.75) becomes

$$\frac{1}{X}\frac{d^2 X}{dx^2} + \frac{1}{Y}\frac{d^2 Y}{dy^2} = \tau^2. \tag{4.77}$$

As before, and since τ is a constant, the two terms to the left must each be constant. Letting

$$\frac{1}{X}\frac{d^2 X}{dx^2} = -q_m^2, \tag{4.78}$$

where q_m is a constant, X has a particular solution

$$X = c \sin q_m x + d \cos q_m x, \tag{4.79}$$

where c and d are constants. Also, however, substituting for the X term in eqn (4.77) gives

$$\frac{1}{Y}\frac{d^2 Y}{dy^2} = \tau^2 + q_m^2 \tag{4.80}$$

which in turn has the solution

$$Y = g \sinh\left(\sqrt{\tau^2 + q_m^2}\right) y + h \cosh\left(\sqrt{\tau^2 + q_m^2}\right) y. \tag{4.81}$$

Further solutions arise of course by starting the above process by setting the term in Y in eqn (4.77) to a constant, say, $-p_m^2$, leading to a constant $\tau^2 + p_m^2$ in respect of the X term in eqn (4.77). Substituting back into eqn (4.76) and allowing m to range over all positive integer values leads finally to the general solution.

It remains to determine the various constants in this solution and this is done in ways similar to those illustrated earlier in the book: first, accounting for the even symmetry of the field about both axes (choosing for simplicity that the central axis of the conductor coincides with the coordinate axis); secondly setting the flux density at all surfaces of the conductor to be equal to B_0; and thirdly determining the remaining constants in the resulting Fourier series in the same way as was done in section 4.3.3. Further details can be found in Stoll (p. 38) or in Silvester (p. 221) who interprets the same solution as applying to the magnetic core of a relay.

Conductors embedded in permeable material

This situation arises for example with solid conducting circuits in electric motors (particularly cage rotors and damper circuits). It serves to illustrate a basic difference in the type of field configuration as compared with those first considered because the directions of the currents and those of the fluxes are not parallel. The currents flow wholly axially in the direction of the conductor but the fluxes (being cross-slot) are orthogonal to the current. They are also two-dimensional and the overall solution, whilst following the same essential route illustrated above, is considerably more complex algebraically. Consideration of boundary conditions also raises some new questions.

As an example brief consideration is given to the case of a circular conductor in a semi-closed slot first studied by Swann and Salmon [26] and shown in Fig. 4.16). Assuming the permeability of the embedding material to be infinite, the solution needs only to address the single circular region occupied by the conductor. The relevant Helmholtz equation following from eqn (4.59) working in terms of current density J

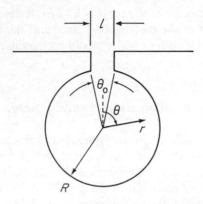

Fig. 4.16

is

$$\frac{\partial^2 J}{\partial r^2} + \frac{1}{r}\frac{\partial J}{\partial r} + \frac{1}{r^2}\frac{\partial^2 J}{\partial \theta} + k^2 J = 0. \tag{4.82}$$

The term in θ can be decoupled and leads to a solution which must be an even periodic function of the form $\sum_{m=0}^{\infty} a_n \cos n\theta$, where n is the separation constant. The remainder of eqn (4.82) becomes

$$\frac{d^2 J}{dr^2} + \frac{1}{r}\frac{dJ}{dr} + \left(k^2 + \frac{n^2}{r^2} \right) J = 0 \tag{4.83}$$

for which the solution is of the form

$$\sum_{n=0}^{\infty} b_n I_n(kr) + d_n K_n(kr), \tag{4.84}$$

where I_n and K_n are modified Bessel functions of the first and second kind respectively. Since J remains finite where $r \to 0$, the d_m must vanish and the complete solution becomes

$$J = \sum_{n=0}^{\infty} g_n I_n(kr) \cos n\theta, \tag{4.85}$$

where $g_n = b_n d_n$, and $n = 0$ must be included to account for current components.

The boundary conditions to be applied to determine the constants are, at the surface of the (infinitely) permeable material

$$H_\theta = B_\theta = 0 \tag{4.86}$$

and, at the air/conductor interface, since the whole magnetic potential drop due to the current I in the conductor occurs across this interface which is (reasonably) assumed to be straight, is

$$H_0 = \frac{I}{R\theta_0} = \frac{I}{l}. \tag{4.87}$$

Hence the values of c_n can be found by equating eqn (4.85) to the Fourier series expansion for H_θ round the boundary $r = R$. See Reference [26] for more details and the determination of the effective resistance and the reactance of the conductor.

Similar problems for conductors of rectangular and T-cross-section are dealt with in Stoll.

4.4.4 Harmonic time variation with travelling fields

The fields considered in the preceding subsection are all characterized by having distributions of the flux (and current) which are fixed in space, even though time-dependent. A variety of problems with such fields received attention from a number of writers relatively early but in the last 20 to 25 years there has been strong interest in situations in which the distribution of the field 'moves', e.g. Bondi and Mukherji [27], Stoll and Hammond [28], Lawrenson *et al.* [29], Ralph [30], Greig and Freeman [31], Freeman and Smith [32], West and Hesmondhalgh [33], Preston and Reece [34], Miller and Lawrenson [35], Eastham and Alwash [37], and others. Particularly prominent has been the interest in fields such as those occurring in electrical machines in which, due to the actual movement of permanent magnets or excited poles or due to the aggregate effect of currents of different time phases, flux waves move at the synchronous or actual rotational speeds of the machine. Commonly only a fundamental component is considered and this is a sufficient basis anyway since non-sinosoidal shapes can easily be handled by combining the appropriate harmonic components. A related simplified but illuminating situation, dealing with a circular current near to a conducting surface, was treated by Hammond [47].

Solution approach

Some writers have tended to discuss the solutions for this type of field as though in some way they involved a new start to the solution process and with the implication that the nature of the harmonic travelling field gives rise of itself to increased difficulties. In fact, as is shown below, the harmonic nature of the space distribution leads to a simplification—though, of course, the overall form of the associated fields does mean that a minimum of two space coordinates is always involved—and the whole solution can be arrived at easily using the results developed above.

Thus, reviewing the whole approach: the equation to be solved here remains, depending upon the coordinate system, as (4.58) or (4.59); the component of the solution involving time remains, as always, in the form of eqn (4.63); and the components of the solution involving the space coordinates are the solution of the arising form of eqn (4.64) and can be inferred from the solutions for Laplace's equation as in Chapter 3. In the present case, the time variation of the field at any (fixed) point in space remains harmonic, as in the preceding sub-section, and reduces to $e^{j\omega t}$. So far as the space component solution is concerned, the situation as rehearsed above is also simplified in a way similar to that for the time component, because of the harmonic space variation in one of the space coordinate directions. The solution sub-component for the relevant coordinate (following from the separation of the Helmholtz equation) reduces to a single harmonic term, effectively $c \cos mx$ for example in rectangular coordinates (see eqn (3.70)) or $c \cos m\theta$ for example in cylindrical coordinates. The other solution sub-component in the other space coordinate retains its general form from the Helmholtz separation e.g. $(g_m \sinh my + h_m \cosh my)$ for rectangular coordinates (see eqn (3.70)) or $(C_m r^m + d_m r^{-m})$ for cylindrical coordinates (see eqns (3.1) to (3.9)).

The implementation of this approach is illustrated below for a simple but useful situation, and includes the first example with the diffusion equation for which fields have to be determined in more than one region.

Eddy-current losses and flux pulsations in solid poles

The stator slots in large electrical machines can lead to significant flux pulsations in the air gap, and hence eddy-currents and troublesome power losses in the solid pole faces which are often used. The eddy-currents in turn influence the magnitude of the resultant flux density in the air gap and this effect also needs to be included. A solution to this problem was given by Lawrenson *et al.* [29] and is reviewed here. For simplicity the air gap surfaces are modelled as flat rather than cylindrical and the field source is treated as being an harmonic current sheet, see Fig. 4.17, though the characterizing parameter is, of course, the tooth-ripple flux density.

It is instructive to note the relationship of this field to the one shown in Fig. 4.16. First, whereas the earlier case involved only one interface and a single field region for which a field function was needed, the present case involves two iron/air boundaries plus the current–sheet interface, leading to the need for four field functions—for the solid pole, 1, air gap, 2, the armature between gap and current, 3, and the armature between current and ∞, 3'. In addition, because the fields move in the direction of the y-axis, two space coordinates y and z (rather than one) have to be considered (though the fields remain invariant with x).

Dealing with the conducting region, 1, we can proceed to the solution following the approach described, first by noting that the time dependence is expressible in the form $K_1 e^{j\omega t}$. Secondly, the separated component of the solution in the coordinate the direction of which coincides with the direction of the field movement, y, must be periodic in a way identical to that of the travelling field. Its form is obvious from this and it corresponds with the periodic component of the Helmholtz solution inferable from eqn (3.70) reduced to a single fundamental term. For algebraic simplicity it is convenient to express it not in trigonometrical form, as earlier, but in the exponential form $K_2 e^{j2\pi y/\lambda}$, where λ is the wavelength, i.e. the spatial period (and equal to two pole pitches). The final component

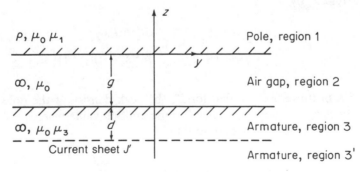

Fig. 4.17

of the solution is the remaining term arising from the solution of the Helmholtz equation (also inferable from eqn (3.70)), but chosen again to be expressed in exponential form as $L e^{mz} + M e^{-mz}$. Hence the complete solution for the flux density in region 1 becomes

$$B_1 = (L e^{mz} + M e^{-mz}) e^{j2\pi(y-vt)/\lambda}, \tag{4.88}$$

where the two periodic terms are combined introducing the speed of travel v, K_1 and K_2 are subsumed in L and M, and m, from the solution to the Helmholtz equation, is given by

$$m^2 = \left(\frac{2\pi}{\lambda}\right)^2 - \frac{j 2\pi\sigma\mu\mu_0 v}{\lambda} = \left(\frac{2\pi}{\lambda}\right)^2 - \frac{j2}{\delta^2}. \tag{4.89}$$

Similarly, it may be shown that the corresponding component of the fields in the air gap and the armature take the form

$$B_z = (L e^{2\pi z/\lambda} + M e^{-2\pi z/\lambda}) e^{j2\pi(y-vt)/\lambda}. \tag{4.90}$$

In all the regions the component of field in the direction of motion is given by

$$B_y = \frac{\lambda}{2\pi j} \frac{\partial B_z}{\partial z}. \tag{4.91}$$

The boundary conditions to be satisfied by these fields are:

(i) $\displaystyle \lim_{z \to \pm\infty} B = 0$

(ii) B_z is continuous at all interfaces

(iii) H_z is continuous at the interfaces at $z = 0$ and $-g$

(iv) at the current sheet

$$[H_{3y}]_{z=-(g+d)} - [H_{3'y}]_{z=-(g+d)} = J' e^{j2\pi(y-vt)/\lambda}.$$

Applying these (eight) boundary conditions yields the z component field in the pole

$$B_{1z} = B_0 e^{\{-\gamma z + j2\pi(y-vt)/\lambda\}},$$

where B_0 is the peak normal field at the pole surface, and may be related to the exciting current J' by

$$B_0 = \frac{2 j\mu_0\mu_1\mu_3 J' e^{-2\pi d/\lambda}}{(\mu_3+1)\left(\mu_1 + \frac{\lambda\gamma}{2\pi}\right) e^{2\pi g/\lambda} - (\mu_3-1)\left(\mu_1 - \frac{\lambda\gamma}{2\pi}\right) e^{-2\pi g/\lambda}} \tag{4.92}$$

From this equation, that for B', the peak density at the pole face in the absence of eddy currents, may be obtained as the limiting value when $v \to 0$. Under this condition, $(\lambda m/2\pi) = 1$ (see eqn (4.89)), so that

$$B' = -\frac{2 j \mu_0\mu_1\mu_3 J' e^{-2\pi d/\lambda}}{(\mu_3+1)(\mu_1+1)e^{2\pi g/\lambda} - (\mu_3-1)(\mu_1-1)e^{-2\pi g/\lambda}}. \tag{4.93}$$

Hence B_0 may be expressed in terms of B', an interesting result directly reflecting the reaction of the eddy-currents on the field, as follows,

$$B_0 = B' \frac{(\mu_1 + 1)}{\mu_1 + (1 - j\mu_1 K)^{1/2}} \cdot \frac{1 - \dfrac{(\mu_3 - 1)(\mu_1 - 1)}{(\mu_3 + 1)(\mu_1 + 1)} e^{-4\pi g/\lambda}}{1 - \dfrac{(\mu_3 - 1)\{\mu_1 - (1 - j\mu_1 K)^{1/2}\}}{(\mu_3 + 1)\{\mu_1 + (1 - j\mu_1 K)^{1/2}\}} e^{-4\pi g/\lambda}}, \qquad (4.94)$$

where

$$K = \sigma \mu_0 f \lambda^2 / 2\pi \qquad (4.95)$$

Using eqn (4.94), Fig. 4.18 shows as an example the ratio of flux density at the pole face with and without eddy-currents as a function of K (proportional to frequency) for permeability and conductivity values typical of a turbo-generator rotor. It also shows the phase-lag angles at each end of the ranges of K. Figure 4.19 below shows current flow maps for situations similar to this one discussed below.

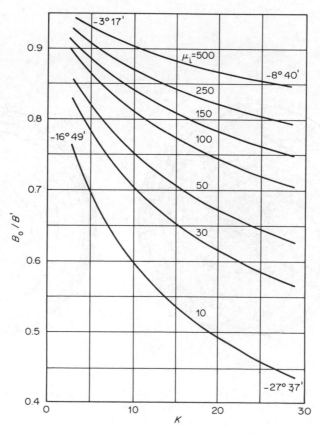

Fig. 4.18 (Reproduced by permission of the Institution of Electrical Engineers.)

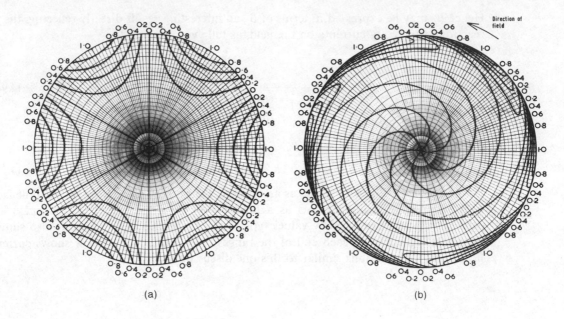

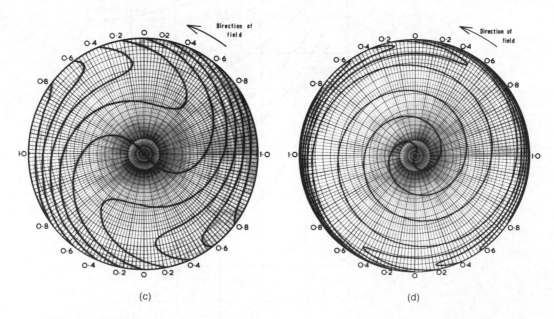

Fig. 4.19   (a) 6-pole field, penetration depth = ∞ (field stationary), (b) 6-pole field, penetration depth = 0.05 × radius, (c) 2-pole field, penetration depth = 0.2 × radius, (d) 2-pole field, penetration depth = 0.05 × radius. (Reproduced from Reference [30], with thanks to the author and the University of Leeds.)

General treatments of multiple regions

Because of the underlying similarity between problems considered in this section, and various others, generalized formulations have been developed making solutions as routine as possible and minimizing algebraic manipulation as much as possible. Greig and Freeman [31] deal with the case of *N plane-faced* adjacent regions each with its own permeability, conductivity and speed of movement (in the same direction as the travelling field). They use the idea of a transfer matrix introduced by Pipes [38] to obtain elegant and efficient formulations and briefly illustrate applications of it to a linear motor problem and to the problem of pole-face losses above.

The general problem in which harmonic fields travel along the surfaces of *cylinders* in an axial direction (as opposed to the circumferential direction that corresponds most simply with the above case with plane interfaces) is dealt with by Freeman and Smith [32]. They again deal with *N* regions (concentric cylinders) using the transfer matrix approach but, additionally, introduce *surface impedance* concepts to minimize algebraic manipulation. Fields with axially moving excitation of this type are pertinent to a variety of surface heating problems.

A particularly thorough and wide-varying study of multi-region cylindrical structures subject to *circumferentially* travelling harmonic fields was made by Ralph [30]. His thesis, which should be studied by any serious reader, provides a particularly good review and correlation of the pertinent literature and, so far as two-dimensional models are concerned, presents substantial studies of pole-face loss, composite rotor-induction motors, hollow-rotor motors, laminated rotors and shaft effects. He includes a number of field maps showing the flux paths in solid rotors with 2, 4 and 6 poles and with varying depths of penetration. A group of these are shown in Fig. 4.19(a)–(d), and are very useful in visualizing the physical situation (including for conducting regions of other shapes), and in illustrating the general nature of eddy-current fields including phase changes and the effects of skin depth and pole pitch. Note that the flux map for the 2-pole field with zero relative velocity (infinite skin depth) consists of equally-spaced parallel lines. The thesis also deals extensively with three-dimensional fields as discussed in section 4.5.

4.4.5 General time dependence: transients

The literature on situations and solutions in which the field variation is non-periodic seems to be very limited and this appears to be due partly to relatively limited interest and partly to the increased difficulty and effort in obtaining solutions by analytical means. Examples include the transient diffusion of fields in rods, tubes and flat surfaces by Miller [39], transient skin-effect in embedded rectangular conductors by Rolicz [40] and transient screening effects in superconducting a.c. generators by Lawrenson and Miller [36]. In view of this, discussion here is brief.

Basically the solution requires a return to eqns (4.57) and (4.60) and to the solution eqn (4.63) for the decoupled time-dependent component of the solution. Whereas in the preceding discussion it has been sufficient to consider only one term in the time component, it now becomes necessary to include an infinite series (or continuum) of terms. This purely analytical approach is discussed by Smythe, Carslaw and Jaeger and Tegopoulos and Kriezis but is not pursued here. Instead a very brief introduction is provided to an effective approach based on the Fourier transformation using a discrete formulation—see Papoulis, Bergland [41] and Cooley and Tukey [42] for general background.

The Fourier transformation has long been very familiar in solutions for the wave equation—involving second time derivatives of the field variable (e.g. $\partial^2 B/\partial t^2$ instead of $\partial B/\partial t$ as here)—but has been largely ignored in relation to diffusion type fields. It is, however, similarly effective for these fields. The nature of the transformation can be thought of in two ways: similar to the use of Fourier series but with a summation which is continuous (an integral) rather than one restricted to specific components (albeit infinite in number); or, like the possibly more familiar Laplace transformation with ordinary differential equations, as a transformation of variable enabling solutions to be obtained in a simpler and more routine way via (in the space of) the 'introduced' variable. In the familiar Laplace transformation, time derivatives are replaced by a complex number (to an appropriate power). In the present case the 'natural' expressions in terms of the time variable are transformed to corresponding ones in terms of frequency, and the solution, derived in terms of frequency, is then (inversely) transformed to yield the required solution in the time variable.

The classical Fourier transform pair defining these operations can be written

$$h(t) = F^{-1}\{H(f)\} = \int_{-\infty}^{\infty} H(f)e^{j2\pi ft}\,df \qquad t \sim (-\infty, \infty)$$

and

$$H(f) = F\{h(t)\} = \int_{-\infty}^{\infty} h(t)e^{-j2\pi ft}\,dt \qquad f \sim (-\infty, \infty),$$

(4.90)

where $h(t)$ is the sought, time-dependent field solution due to some excitation $h_0(t)$, and $H(f)$ is the Fourier transform (in the frequency variable, f, of $h(t)$, denoted by F. $h(t)$ is retrieved as the inverse Fourier transform, denoted F^{-1}, from $H(f)$.

The relationship between the Fourier transform approach and that of section 4.4 is not hard to see. In effect the transform replaces the operator $\partial/\partial t$ by $j\,2\pi f$ and, whilst a solution is required for the full range of values of f, the general form of it is already familiar as the solution for the harmonic excitation case discussed in section 4.4.2. Thus, it can be seen that the solution in the frequency domain is expressible (in phasor terms) as

$$H(f) = S(f)\hat{H}_0\,e^{j2\pi ft} \qquad f \sim (-\infty, \infty),$$

(4.91)

where $\hat{H}_0\,e^{j2\pi ft}$ is the excitation phasor, $H(f)$ is the phasor value of any component of $H(f)$ at any point, and $S(f)$ can be regarded as a complex frequency response function characterizing the field at each point. $S(f)$ depends only on the space terms in the field equation and, in fact, corresponds with the solution of the Helmholtz eqn (4.64) as discussed above (though expressed in terms of f). The desired solution of the time-varying field is then available as

$$h(t) = F^{-1}\{S(f)H_0(f)\}.$$

(4.92)

In these terms the general solution emerges as a straightforward extension of the harmonic solution considered initially. Moreover, two further important points are worth emphasis. First, because $S(f)$ depends only on geometry and material properties, it can be used for the given problem without further effort in conjunction with any excitation function $h_0(t)$. Secondly, whilst the discussion has naturally presumed an analytical derivation of $S(f)$, either a numerical or an experimental derivation could be employed. The transform method can thus be used to obtain general transient results from numerical or experimental results derived over a range of fixed (DFT) frequencies.

Turning now to the practical evaluation of eqns (4.90), the classical approach has of course been by analytical means, but powerful numerical procedures are now available and under continuing development. As with all numerical procedures these involve the 'approximation' of continuous functions in infinite domains and eqn (4.90) is replaced by the so-called discrete Fourier transform (DFT) pair

$$h_d(k) = h_d(k\Delta t) = \frac{1}{T} \sum_{l=0}^{N-1} H_d(l)e^{j2\pi kl/N} \qquad k = 0, 1, 2, \ldots, N-1$$

and

$$H_d(l) = H_d(l\Delta f) = \sum_{k=0}^{N-1} h_d(k)e^{j2\pi kl/N} \qquad l = 0, 1, 2, \ldots, N-1 \qquad (4.93)$$

in which $h(t)$ and $H(f)$ are each represented by the N samples of $h_d(k)$ and $H_d(l)$ respectively. The inverse DFT of $H_d(l)$ is $h_d(k)$ which approximates $h(t)$ for $t < T/2$, where $T = N\Delta t$ is the 'length' or duration of the sampled time response and Δt is the time interval between samples. Similarly $H_d(l)$ approximates $H(f)$ for $f < F/2$ where $F = N\Delta f$ is the 'length' or bandwidth of the sampled frequency response and Δf is the spacing between samples in the frequency domain.

Detailed discussion of the numerical procedures is not possible here but poses no special problems and the fast Fourier transform (FFT) provides an efficient and flexible approach. The main considerations are discussed by Miller and Lawrenson [35] who also present practical results for the frequency response of, and rate of flux penetration through, cylindrical screens in superconducting a.c. generation. Greater detail is available in the texts referenced above.

4.5 THREE-DIMENSIONAL FIELDS

4.5.1 Introduction

Progression through this chapter has inevitably highlighted the way in which the algebraic complexity of solution builds up very rapidly as we add to the effects of field sources those of simple boundary influences, then more complex including multiple boundary effects, time dependence—first harmonic then general, and not least boundary shape and associated coordinate system effects. Also it has been seen that purely closed-form analytical solutions need to be augmented, not unnaturally, by numerical procedures because of their greater generality or efficiency—first, in dealing with current sources of arbitrary shape (even in the absence of boundaries) and, immediately above, in the effective implementation of the Fourier transform (and it could be added in the efficient evaluation of various functions such as those of Bessel in section 4.4.3). All of these influcneces and trends become most marked when three-space dimensions need to be addressed. A variety of such problems are amenable to analytic solution but, generally, the need to explore purely numerical procedures (as comprehensively discussed in the later chapters of the book) tends to arise quite quickly.

In the light of all of this, not least the overriding effects of algebraic complexity, which has already led to a progressive reduction in detail throughout the chapter, the discussion in this section is brief. Treatment of particular applications is impracticable in the space available; and solutions for Laplace's or Poisson's equations are left to be extracted as reduced forms of the more general solutions presented for the diffusion equation. Rectangular and cylindrical polar coordinate systems are mentioned but little more than the powerful general solutions are given.

4.5.2 Diffusion equation in rectangular coordinates

As discussed in general terms in section 4.4.2, the solution for the time component of the field decouples from that for the space component(s) and takes the form given by eqn (4.63) or, for harmonic variation, simply $e^{j\omega t}$. Separation of the space variables for the Helmholtz (or Laplace) equations is entirely straightforward and the three-dimensional case is a simple extension of the two-dimensional case. See sections 3.3.1, 4.3.2 and 4.4.3, eqns (4.76)–(4.82). Thus, it is readily seen that the general solution for the three-dimensional solution of the diffusion equation in rectangular coordinates may be written conveniently recognizing interchangeability in the x, y, z terms, in the form

$$A = (a_m \cos mx + b_m \sin mx)(c_n e^{ny} + d_n e^{-ny})\, e^{(j\omega t + gz)}. \tag{4.94}$$

One accessible and informative application of this solution was given by Preston and Reece [34] in their study of transverse edge effects in linear induction motors. Other useful studies are given by Wood and Concordia [43], Angst [44] and Vilnitis [45].

4.5.3 Diffusion equation in cylindrical polar coordinates

With the possible exception of spherical polar coordinates, which are of limited practical importance, the next easiest to handle coordinate system after the rectangular one is the cylindrical polar one. For two-dimensional Laplacian fields solutions are of similar complexity to those of the rectangular case but, if time dependence is added, the complexity is increased, and with three-space dimensions there is a further major increase in complexity to a level probably the maximum which it is reasonable to contemplate. This arises because of the increased difficulty in the separation of the space variables in conjunction with the effects of time on the Helmholtz equation.

An outstanding study covering the general solution and a variety of applications of practical significance was provided by Ralph [30]. See also Lawrenson and Ralph [46]. His detailed approach was in terms of the flux density B and he achieved separation by obtaining equations in terms of rB_r and rB_θ (rather than B_r and B_θ). By this means he was able to obtain a general solution for each component, B_r, B_θ, B_z, of the field as follows. For $p \neq \pm 1$ or 0 and $\mathbf{p} \neq 0$.

$$B_r = \frac{1}{r}\sum_\omega \left[\sum_p \left\{ \left(\sum_m [\{\mathbf{g}_\alpha \mathbf{J}_p(\alpha r) + \mathbf{h}_\alpha \mathbf{Y}_p(\alpha r)\} \{\mathbf{L}_m \exp(\mathbf{m}z) + \mathbf{M}_m \exp(-\mathbf{m}z)\}] \right. \right. $$
$$\left. + \left[\mathbf{g}_\omega J_p \left\{ j^{3/2}\left(\frac{r}{\mathbf{d}}\right)\right\} + \mathbf{h}_\omega Y_p \left\{ j^{3/2}\left(\frac{r}{\mathbf{d}}\right)\right\} \right] z + (\mathbf{g}_p r^p + \mathbf{h}_p r^{-p})\{\mathbf{L}_\omega \exp(j^{1/2}z/\mathbf{d})$$

$$+ \mathbf{M}_\omega \exp(-j^{1/2}z/\mathbf{d})\} \Big)(\mathbf{C_p}\sin \mathbf{p}\theta + \mathbf{D_p}\cos \mathbf{p}\theta)\Big\}\mathbf{l}_\omega \exp(j\,\omega t)\Bigg]$$

$$- \sum_\omega \Bigg[\sum_p \Bigg\{\Bigg(\sum_m \Bigg[\frac{m}{\alpha}\{g_\alpha J_{p+1}(\alpha r) + h_\alpha Y_{p+1}(\alpha r)\}\{L_m\exp(mz) - M_m\exp(-mz)\}\Bigg]$$

$$+ \frac{d}{j^{3/2}}\Bigg[g_\omega J_{p+1}\Big\{j^{3/2}\Big(\frac{r}{d}\Big)\Big\} + h_\omega Y_{p+1}\Big\{j^{3/2}\Big(\frac{r}{d}\Big)\Big\}\Bigg]$$

$$+ \frac{j^{1/2}r}{2d}\Big(\frac{g_p}{p+1}r^p - \frac{h_p}{p+1}r^{-p}\Big)\{L_\omega\exp(j^{1/2}z/d)$$

$$+ M_\omega \exp(-j^{1/2}z/d)\}\Big)(C_p\sin p\theta + D_p\cos p\theta)\Bigg\}\mathbf{l}_\omega \exp(j\,\omega t)\Bigg]. \tag{4.95}$$

when $p = -1$, $g_p/(p+1)$ is replaced by $2g_p\ln r$; when $p=0$, $g_p r^p$ is replaced by $g_p(\ln r - 1)$; when $p = 1$, $h_p(p-1)$ is replaced by $-2h_p\ln r$; and when $\mathbf{p}=0$, $\mathbf{g_p}r^{\mathbf{p}}$ is replaced by $\mathbf{g_p}\ln r$. Otherwise the above expression remains the same for $p = \pm 1$ or 0 and for $\mathbf{p}=0$.

$$B_\theta = \sum_\omega \Big(\sum_{m'}[\{g'_\alpha \mathbf{J}_1(\alpha' r) + \mathbf{h}'_{\alpha'}\mathbf{Y}_1(\alpha' r)\}\{\mathbf{L}_{m'}\exp(\mathbf{m}'z) + \mathbf{M}_{m'}\exp(-\mathbf{m}'z)\}]\mathbf{l}'_{\omega'}\exp(j\omega' t)\Big)$$

$$+ \frac{1}{r}\sum_\omega \Bigg\{\sum_p \Bigg(\Bigg\{\sum_m \Bigg[g_\alpha\Big\{J_p(\alpha r) - \frac{\alpha r}{\mathbf{p}}J_{p+1}(\alpha r)\Big\} + \mathbf{h}_\alpha\Big\{Y_p(\alpha r)$$

$$- \frac{\alpha r}{\mathbf{p}}Y_{p+1}(\alpha r)\Big\}\Bigg]\{L_m\exp(mz) + M_m\exp(-mz)\}\Bigg)$$

$$+ \Bigg[g_\omega\Big\{J_p\Big(j^{3/2}\frac{r}{\mathbf{d}}\Big) - \frac{j^{3/2}r}{\mathbf{p}\mathbf{d}}J_{p+1}\Big(j^{3/2}\frac{r}{\mathbf{d}}\Big)\Big\} + \mathbf{h}_\omega\Big\{Y_p\Big(j^{3/2}\frac{r}{\mathbf{d}}\Big)$$

$$- \frac{j^{3/2}r}{\mathbf{p}\mathbf{d}}Y_{p+1}\Big(j^{3/2}\frac{r}{\mathbf{d}}\Big)\Big\}\Bigg]z + (\mathbf{g_p}r^{\mathbf{p}} - \mathbf{h_p}r^{-\mathbf{p}})\{\mathbf{L}_\omega\exp(j^{1/2}z/\mathbf{d})$$

$$+ \mathbf{M}_\omega\exp(-j^{1/2}z/\mathbf{d})\}(\mathbf{C_p}\cos \mathbf{p}\theta - \mathbf{D_p}\sin \mathbf{p}\theta)\Big)\mathbf{l}_\omega\exp(j\,\omega t)\Bigg\}$$

$$+ \sum_\omega \Bigg[\sum_p \Bigg\{\Bigg(\sum_m \Bigg[\frac{m}{\alpha}\{g_\omega J_{p+1}(\alpha r) + h_\omega Y_{p+1}(\alpha r)\}\{L_m\exp(mz) - M_m\exp(-mz)\}\Bigg]$$

$$+ \frac{d}{j^{3/2}}\Big\{g_\omega J_{p+1}\Big(j^{3/2}\frac{r}{d}\Big) + h_\omega Y_{p+1}\Big(j^{3/2}\frac{r}{d}\Big)\Big\}$$

$$+ \frac{j^{1/2}r}{2d}\Big(\frac{g_p}{p+1}r^p + \frac{h_p}{p-1}r^{-p}\Big)\{L_\omega\exp(j^{1/2}z/d)$$

$$- M_\omega\exp(-j^{1/2}z/d)\}\Big)(C_p\cos p\theta + D_p\sin p\theta)\Big\}\mathbf{l}_\omega\exp(j\,\omega t)\Bigg]. \tag{4.96}$$

for $p \neq \pm 1$ or 0 and $\mathbf{p} \neq 0$. When $p = -1$, $g_p/(p+1)$ is replaced by $-2(g_p/p)\ln r$; when $p=0$, $g_p/(p+1)$ is replaced by g_p/p; when $p=1$, $h_p/(p-1)$ is replaced by $-2h_p/p\ln r$ and when $\mathbf{p}=0$ $\mathbf{g_p}$ is replaced by $\mathbf{g_p}/\mathbf{p}$. Otherwise, the above expression remains the same in all cases. (It should be noted in the above that dividing by p or $\mathbf{p}$ when they are zero does not necessarily make the solution infinite, since $\mathbf{g}_p$ or g_p may be proportional to $\mathbf{p}$ or p, respectively.)

$$B_z = \sum_\omega \left[\sum_p \left\{ \left(\sum_m \left[\{ g_\alpha J_p(\alpha r) + h_\alpha Y_p(\alpha r) \} \{ L_m \exp(mz) + M_m \exp(-mz) \} \right] \right. \right. \right.$$

$$+ \left\{ g_\omega J_p \left(j^{3/2} \frac{r}{d} \right) + h_\omega Y_p \left(j^{3/2} \frac{r}{d} \right) \right\} z + (g_p r^p + h_p r^{-p}) \{ L_\omega \exp(j^{1/2} z/d)$$

$$+ M_\omega \exp(-j^{1/2} z/d) \} \bigg) (C_p \sin p\theta + D_p \cos p\theta) \bigg\} l_\omega \exp(j\omega t) \right] \qquad (4.97)$$

for $p \neq 0$.

For $p = 0$, g_p is replaced by $g_p \ln r$, the remainder of the expression being unchanged.

These are powerful solutions and several general points should be made in respect of them. First, the terms proportional to z, which had not previously been known to exist, arise from a consideration of the separated form of the equation for B_z; terms in the solution dependent upon z only yield the same effect as zero variation in z, and so exist in the final solution in the same form as the solution terms which are independent of z; i.e. the part involving L_m and M_m with $m = 0$. Secondly, exponential terms similar to those in eqn (4.97) had been given previously by Moon and Spencer for fields with negligible conduction current (and significant displacement current), but they omitted further terms from the solutions for B_r and B_θ. Finally, it should be noted that, since m may be complex, the exponential terms in the solutions *allow for harmonic variation with z* (as well as in θ).

These solutions embrace, and reduce to, all of those for cylindrical boundaries discussed earlier in the book and, indeed, to all previously known ones. They obviously also permit the derivation of a wide variety of solutions not previously attempted and apart from those in two dimensions mentioned at the end of section 4.4.4, Ralph [30] applied them successfully to problems involving *short* cylinders and various arrays of cylinders with periodicity in both the axial and circumferential directions. The complexity of the solutions, however, probably leaves the reader sharing the view expressed above that contemplation of anything more algebraically complex is unlikely to be worthwhile and purely numerical approaches must be taken up.

REFERENCES

[1] M. Strutt, Das magnetische Feld eines rechteckigen, von Gleichstrom durchflossenen Leiters, *Arch. Elektrotech.*, **17**, 533, and **18**, 282 (1927).
[2] O. R. Schurig and M. F. Sayre, Mechanical stresses in busbar supports during short-circuits, *J. Amer. Instn. Elect. Engrs.*, **44**, 365 (1925).
[3] T. J. Higgins, Formulas for calculating short-circuit forces between conductors of structural shape, *Trans. Am. Instn. Elect. Engrs.*, **62 III**, 10, 659 (1943).
[4] H. B. Dwight, Repulsion between strap conductors, *Elect. World*, **70**, 522 (1917).
[5] P. J. Lawrenson, The magnetic field of the end-windings of turbo generators, *Proc. Instn. Elect. Engrs.*, **108A**, 538–549 and 552–553 (1961).
[6] P. J. Lawrenson, Forces in turbo-generator end-windings, *Proc. Inst. Elect. Engrs.*, **112**, 1144–1158 (1965).

[7] P. J. Lawrenson, Calculation of machine end-winding inductances with special reference to turbo-generators, *Proc. Instn. Elect. Engrs.*, **117**, 1129–1134 (1970).

[8] W. Rogowski, Über das Streufeld und der Streuinduktionskoeffizierten eines Transformators mit Schiebenwicklung und geteilten Endspulen, *Mitt. Forsch Arb. VDI*, **71** (1909).

[9] B. L. Robertson and J. A. Terry, Analytical determination of magnetic fields, *Trans. Am. Instn. Elect. Engrs.*, **48**, 4, 1242–1262 (1929).

[10] E. Roth, Introduction à l'étude analytique de l'échauffement des machines électriques, *Bull. Soc. Franc. Élect.*, **7**, 840 (1927).

[11] E. Roth, Étude analytique due champ propre d'une encoche, *Rev. Gén. Elect.*, **25**, 417 (1927).

[12] E. Roth, Étude analytique due champ de fuites des transformateurs et des efforts mécaniques exercés sur les enroulements, *Rev. Gén. Elect.*, **23**, 773 (1928).

[13] E. Roth, Étude analytique des champs thermique et magnétique lorsque la conductibilité thermique ou la perméabilité n'est pas la même dans toute l'étendue due domaine considéré, *Rev. Gén., Elect.*, **24**, 137–148, and 179–187 (1928).

[14] E. Roth, Étude analytique due champ résultant d'une encoche de machine électrique, *Rev. Gén. Elect.*, **32**, 761 (1932).

[15] E. Roth, Inductance due aux fuites magnétiques dans les transformateurs à bobines cylindriques et efforts exercés sur les enroulements, *Rev. Gén. Elect.*, **40**, 259, 291 and 323 (1936).

[16] E. Roth, Champ magnétique et inductance d'un systéme des barres rectangulaires parralléles, *Rev. Gén. Elect.*, **44**, 275 (1938).

[17] E. Billig, The calculation of the magnetic fields of rectangular conductors in a closed slot and its application to the reactance of transformer windings, *Proc. Instn. Elect. Engrs.*, **98**(4), 55 (1951).

[18] L. Rabins, Transformer reactance calculations with a digital computer, *Trans. Am. Instn. Elect. Engrs.*, **25** I (1956).

[19] P. R. Vein, A method based on Maxwell's equations for calculating the short-circuit forces on the concentric windings of an idealized transformer, *Elec. Res. Assoc.*, Report Q/T151 (1960).

[20] E. Roth and G. Kouskoff, Sur une méthode de sommation de certaines séries due Fourier, *Rev. Gén. Elect.*, **23**, 1061 (1928).

[21] P. Hammond, Roth's method for the solution of boundary-value problems in electrical engineering, *Proc. Instn. Elect. Engrs.*, **114**(12), 1969 (1967).

[22] A. Pramanik, Magnetic field and eddy current distributions in the core-end region of a.c. machines, Ph.D. thesis, University of Birmingham (1967).

[23] A. Pramanik, Extension of Roth's method to two-dimensional rectangular regions containing conductors of any cross-section, *Proc. Instn. Elect. Engrs.*, **116**(7), 1286 (1969).

[24] C. E. M. De Kuijper, Bijdrage tot de berekening van de spreidings-reactantie van transformatoren en van de krachten, welke op de wikkelingen van transformatoren werken, Delftsche Uitgevers Maatschappij, based on Ph.D. thesis at Technische Hogeschool, Delft, (1949).

[25] N. Mullineux, J. R. Reed and C. E. M. De Kuijper, Roth's method for the solution of boundary-value, *Proc. Instn. Elect. Engrs.*, **116**(2), 291 (1969).

[26] S. A. Swann and J. W. Salmon, Effective resistance and reactance of a solid cylindrical conductor placed in a semi-closed slot, *Proc. Instn. Elect. Engrs.*, **109**, 611–615 (1962).

[27] H. Bondi and K. C. Mukherji, An analysis of tooth-ripple phenonema in smooth laminated pole shoes, *Proc. Instn. Elect. Engrs.*, **104C**, 349–356 (1957).

[28] R. L. Stoll and P. Hammond, Calculation of the magnetic field of rotating machines, Pt. 4, Approximate determination of the field and losses associated with eddy currents in conducting surfaces, *Proc. Instn. Elect. Engrs.*, **112**, 2083–94 (1965).

[29] P. J. Lawrenson, P. Reece and M. C. Ralph, Tooth ripple losses in solid poles, *Proc. Instn. Elect. Engrs.*, **113**, 657–662 (1966).

[30] M. C. Ralph, Eddy-current effects in the cylindrical members of rotating electrical machinery, Ph.D. Thesis, University of Leeds (1968).

[31] J. Greig and E. M. Freeman, Travelling wave problem in electrical machines, *Proc. Instn. Elect. Engrs.*, **114**, 1681–1683 (1967).

[32] E. M. Freeman and B. E. Smith, Surface impedance method applied to multilayer cylindrical induction devices with circumferential exciting currents, *Proc. Instn. Elect. Engrs.*, **117**, 2012–2013 (1970).

[33] J. C. West and D. E. Hesmondhalgh, The analysis of thick cylinder induction machines, *Proc. Instn. Elect. Engrs.*, **109C**, 171–181 (1962).

[34] Preston and A. B. J. Reece, Transverse edge effects in linear induction motors, *Proc. Instn. Elect. Engrs.*, **116**, 973–979 (1969).

[35] T. J. E. Miller and P. J. Lawrenson, Penetration of transient magnetic fields through conducting cylindrical structure with particular reference to superconducting a.c. machines, *Proc. Instn. Elect. Engrs.*, **123**, 437–443 (1976).

[36] P. J. Lawrenson and T. E. J. Miller, Transient solution of the diffusion equation by discrete Fourier transformation, Chapter 8 of *Finite Elements in Electrical and Magnetic Field Problems*, Ed. M. Chari and P. Silvester, John Wiley (1980).

[37] J. F. Eastham and J. H. Alwash, Transverse flux tubular motors, *Proc. Instn. Elect. Engrs.*, **119**, 1709–1718 (1972).

[38] L. A. Pipes, Matrix theory of skin effect in laminations, *Jl. Franklin Inst.*, **262**, 127–138 (1956).

[39] K. W. Miller, Diffusion of electric current into rods, tubes and flat surfaces, *Trans. Am. Instn. Elect. Engrs.*, **66**, 1496–1502 (1947).

[40] P. Rolicz, Transient skin-effect in a rectangular conductor placed in a semi-closed slot, *Archiv für Electrotechnik*, **57**, 329–338 (1976).

[41] G. B. Bergland, A guided tour of the fat Fourier transform, *Inst. Elect. Electr. Engrs. Spectrum*, **6**, 41–52 (1969).

[42] J. Cooley and J. Tukey, An algorithim for the machine calculation of complex Fourier series, *Math Comp.*, **19**, 297–301 (1965).

[43] A. J. Wood and C. Concordia, Analysis of solid rotor induction machines: Pt III—finite length effects, *Trans. Am. Instn. Elect. Engrs.*, **79**, 21–26 (1960).

[44] G. Angst, Polyphase induction motor with solid rotor: effects of saturation and finite length, *Trans. Am. Instn. Elect. Engrs.*, **80**, 902–909 (1962).

[45] A. Vilnitis, Transverse edge effect in flat induction magnetodynamic machines in Movement of conducting bodies in magnetic fields, *Instn. of Physics of Latvian Academy of Sciences Monograph*, pp. 63–94 (1966).

[46] P. J. Lawrenson and M. C. Ralph, The general 3-dimensional solution of eddy-current and Laplacian fields in cylindrical structures, *Proc. Instn. Elect. Engrs.*, **117**, 469–472 (1970).

[47] Hammond P., The calculation of the magnetic field of rotating machines, Pt. 3, Eddy currents induced in a solid slab by a circular current loop, *Proc. Instn. Elect. Engrs.*, **109C**, 508–15 (1962).

Additional references

Apanasewicz, S., Method of electromagnetic field calculation in transformer of high power using integral equations, *Rozpr, Electrotech.*, **32**, 1, 141–171 (1985).

Bayburin, V. B., Three-dimensional solution of the potential problem of electron bunches in crossed fields, *Radiotekh and Electron.*, **29**, 4, 751–756 (April 1984).

Carpenter, K. H. and Yeh, H. T., Eddy current calculations for thin cylinders of finite length with driving fields of ramp time dependence, *IEEE Trans. Magn.*, **Mag-12**, 6, 1059–1061 (Nov. 1976).

Demetrios, T. and Petros, D., Eddy currents and forces in a three phase gas insulated cable with steel enclosure, *IEEE Trans. Magn.*, **Mag-19**, 5, 2210–2212 (Sept. 1983).

Dokopoulos, P. and Tampakis, D., Analysis of field and losses in a three phase gas cable with thick walls, *IEEE Trans. Power Appar. and Syst.*, **Pas-103**, 9, 2728–2734 (Sept. 1984).

Fabricatore, G. *et al.*, A one-dimensional solution of the homogeneous diffusion equation, *IEEE Trans. Educ.*, **32**, 4, 454–456 (Nov. 1989).

Findlay, R. D. *et al.*, Application of the harmonic analysis technique to determining eddy currents in conducting plates, *Electromagnetic Fields in Electrical Engineering*, 263–268 (1988).

Hammond, P., The magnetic screening effects of iron tubes, *Proc. Instn. Elect. Engrs.*, **103C**, 112–120 (1955).

Hatzikonstantinou, P. and Moyssides, P. G., Explanation of the ball bearing motor and exact solutions of the related Maxwell equations, *J. Phys. A, Math. Gen.*, **23**, 14, 3183–3197 (July 1990).

Higgins, T. J. and Messinger, H. P., Equations for the inductance of three-phase coaxial busses comprised of square tubular conductors, *J. Appl. Phys.*, **18**, 1009 (1947).

Higgins, T. J., Formula for the geometrical mean distances of rectangular areas and of line segments, *J. Math. Phys.*, **14**(4), 188 (1943).

Higgins, T. J., New formulas for calculating short-circuit stresses in bus supports for rectangular tubular conductors, *J. Math. Phys.*, **14**(3), 151 (1943).

Higgins, T. J., Inductance of hollow rectangular conductors, *J. Franklin Inst.*, **230**(3), 375 (1940).

Higgins, T. J., A comprehensive review of Saint-Venant's torsion problem, *Am. J. Phys.*, **10**, 248 (1942).

Jack, A. G. and Harris, M. R., Complex permeability approach to the one-dimensional diffusion equation, including hysteresis, *Colloquium on 'Problems of Modelling Non-linear Material Properties in Electromagnetics'*, pp 13/1–4 (1983).

Jones, D. E., Mullineux, N., Reed, J. R. and Stoll, R. L., Solid rectangular and T-shaped conductors in semi-closed slots, *Jl. Eng. Math.*, **3**, 123–35 (1969).

Kocketkov, V. M. *et al.*, The theory of electrodynamic levitation, fundamental results and further problems, *IZV. Akad. Nauk SSSR Energ. and Transp.*, **19**, 1, 72–91 (1981).

Krakowski, M., Boundary-value problem for eddy currents induced in thin metallic plates by external magnetic field, *Bull. Pol. Acad. Sci. Tech.* Sci., **36**, 7–9, 551–556 (1988).

Krakowski, M., Eddy-current distribution in a thin metallic plate after applying a step function of an external magnetic field, *Bull. Acad. Pol. Sci. Ser. Sci. Tech.*, **27**, 7, 611–671 (1979).

Kwok, S. C., Solution of Poisson equation in p-Al/sub y/Ga/sub 1-x/AS/p-Al/sub 0.55/As/n-GaAs structures, *Electron. Lett.*, **26**, 13, 894–896 (June 1990).

Lewis, A. M. *et al.*, Thin-skin electromagnetic fields around surface-breaking cracks in metals, *J. Appl. Phys.*, **64**, 8, 3777–3784 (1988).

Martinelli, G., Screening of transient electromagnetic fields in superconducting turbogenerators, *Proceedings of the International Conference on Electrical Machines*, **1**, S1/7/1–12 (1978).

Morisue, T., A new formulation of the magnetic vector potential method for three dimensional magnetostatic field problems, *IEEE Trans. Magn.*, **21**, 6, 2192–2195 (Nov. 1985).

Morjaria, M. *et al.*, A boundary integral method for eddy current flow around cracks in thin plates, *IEEE Trans. Magn.*, **Mag-18**, 2, 467–472 (March 1982).

Nevins, R. J., Eddy current loss distribution in a conducting plate near polyphase filaments, *IEEE Power Engineering Society Summer Meeting (Text of Papers)*, pp A76 307–9/1–9 (1976).

Peterson, W., Fixed point solution of saturable steady-state eddy current problem, *Model. Simul. Control A*, **20**, 3, 53–64 (1989).

Pratap, S. B., Transient eddy current distribution in the shield of the passively compensated, compensated pulsed alternator: iron-core machines. *IEEE Trans. Magn.*, **26**, 4, 1256–1269 (July 1990).

Ramakrishna, K. *et al.*, Two-dimensional analysis of electrical breakdown in a nonuniform gap between a wire and a plane, *J. Appl. Phys.*, **65**, 1, 41–50 (1989).

Reece, A. B. J. and Pramanik, A., Calculation of the end-field region of AC machines, *Proc. Instn. Elect. Engrs.*, **112**, 1355–1368 (1965).

Roger, D. and Eastham, J. F., Dynamic behaviour of linear induction machines in the heave mode, *Trans. Instn. Elect. Electr. Engrs.*, **VT-31**, 100–116 (1982).

Russell, R. L. and Norsworthy, K. H., Eddy currents and wall losses in screened-rotor induction motors, *Proc. Instn. Elect. Engrs.*, **105A**, 163–175 (1958).

Ryan, P. M. *et al.*, Determination of fields near an ICRH antenna using a 3D magnetostatic Laplace formulation, *AIP Conf. Proc. (USA)*, **190**, 322–325 (1989).

Segal, A. M., The electrical field and resistance of a parallelepiped with small contact surface, *Izv. Akad. Nauk SSSR Energ. & Transp.*, **24**, 1, pp 100–103 (1986).

Such, V. *et al.*, The local dynamic magnetic characteristic in the solution of the electromagnetic diffusion equation, *Workshop on Electromagnetic Field Computation. Proceedings*, **11**, 6–10 (1987).

Szabados, B. *et al.*, A new approach to determine eddy current losses in the tank walls of a power transformer, *IEEE Trans. Power Delivery*, **PWRD-2**, 3, 810–816 (1987).

Tsaknakis, H. J. and Kriezis, E. E., Field distribution due to a circular current loop placed in an arbitrary position above a conducting place, *IEEE Trans. Geosci. & Remote Sensing*, **GE-23**, 6, 834–840 (Nov. 1985).

Weigelt, K., Expressions for solution of the Laplace and Helmholtz equations in three-dimensional eddy-current problems, *Arch. Elektrotech.*, **72**, 3, 195–203 (1989).

5

CONFORMAL TRANSFORMATION: BASIC IDEAS

Conformal transformation is a particularly powerful method for the analytical solution of Laplacian fields with boundaries of complicated shape. It can be used to analyse the fields, for instance, between non-concentric circular cables, in waveguides or high-frequency transmission lines of many different cross-sections, exterior to charged conductors of polygonal section, and in the slotted air gap of a rotating machine. Also, many solutions take simple analytical forms and readily yield expressions for flux density and permeance in magnetic fields (or potential gradient and capacitance in electrostatic fields), and frequently allow the direct calculation of field maps. The chief limitation in applying transformation techniques, is that, for most problems, boundaries have to be assumed to be equipotentials (infinitely permeable or infinitely conducting), or coincident with flux lines, or combinations of these two types. More general problems are reviewed briefly in Chapter 7.

Because of the value of conformal transformation and the absence of any full up-to-date treatment, it is discussed in some detail with basic background material reviewed in Appendix 1. This chapter treats the basis of the method and discusses some simpler applications with particular emphasis on curved boundaries. The next chapter deals thoroughly with polygonal boundaries and after that powerful numerical methods and various more difficult applications and extensions are discussed (Chapter 7). An extensive and easy to use tabulation of boundary shapes and transformation equations is given in Appendix 3.

5.1 CONFORMAL TRANSFORMATION AND CONJUGATE FUNCTIONS

5.1.1 Conformal transformation

Consider the properties of a regular function of the type

$$z = f(t) = x(u, v) + j y(u, v), \tag{5.1}$$

which defines a complex variable $z = x + j y$ as some function of another complex variable $t = u + jv$. A particular value, t', can be represented by a point in the complex plane of

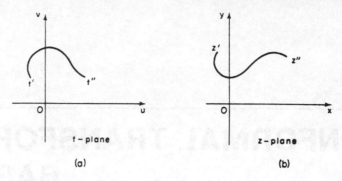

Fig. 5.1

t, Fig. 5.1(a). Through eqn (5.1) some particular value (or values) z' corresponds to t', and it may be represented by a point in the complex plane of z, Fig. 5.1(b). Further, there is a similar correspondence for a succession of pairs of points in the t- and z-planes so that to some curve $t't''$ there corresponds a curve $z'z''$ which is said to have been *transformed or mapped* from $t't''$ by eqn (5.1).

As an example, consider the curves in the z-plane which correspond to straight lines in the t-plane through the equation

$$z = \sin t. \tag{5.2}$$

To facilitate this, it is necessary to form equations, connecting x and y, from which either u or v is absent. Expanding $\sin(u + jv)$ gives

$$z = \sin u \cos jv + \cos u \sin jv$$
$$= \sin u \cosh v + j \cos u \sinh v,$$

so that, equating the real and imaginary parts,

$$x = \sin u \cosh v, \tag{5.3}$$
and
$$y = \cos u \sinh v. \tag{5.4}$$

Squaring and adding these equations eliminates u to give

$$\frac{x^2}{\cosh^2 v} + \frac{y^2}{\sinh^2 v} = 1, \tag{5.5}$$

and squaring and subtracting them eliminates v, so that

$$\frac{x^2}{\sin^2 u} - \frac{y^2}{\cos^2 u} = 1. \tag{5.6}$$

A straight line parallel to the u-axis in Fig. 5.2(a) is described by $v = $ constant. However, for a constant value of v, eqn (5.5) represents an ellipse in the z-plane, so that any straight line parallel to the u-axis is transformed by the equation $z = \sin t$ into an ellipse in the z-plane, Fig. 5.2(b). Any straight line parallel to the v-axis has the equation $u = $ constant, and from eqn (5.6) it is seen that such a line is transformed into a hyperbola in the z-plane.

In this simple example it is possible to obtain recognizable equations for the

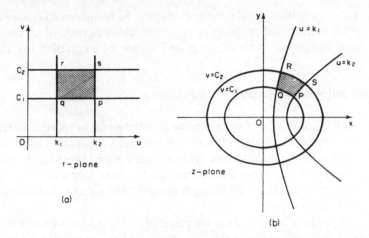

Fig. 5.2

transformed curves. However, this is not usually possible, and to obtain the curves it is often necessary that a series of particular values of t be substituted in the equation. Doing this in the example for the real axis between 0 and ∞ (i.e. $v = 0$), it is seen, from eqns (5.3) and (5.4), that

(a) $y = 0$, for all values of u,
(b) as u varies from 0 through $\pi/2$, π and $3\pi/2$ to 2π, x varies from 0 through $1, 0$ and -1 to 0.

This shows [as is apparent from eqn (5.3)] that x is periodic and that consequently an infinite number of values of u between $-\infty$ and ∞ can give rise to one value of x. Similarly, y also is periodic with u and, in addition, it may be noted that in an equivalent way one point (u, v) in the t-plane can give rise to several in the z-plane. This correspondence of a single point in one plane with more than one point in another plane occurs frequently and, in some cases, requires that care be exercised when interpreting results.

The correspondence between *curves* in the two planes has been indicated, and it is evident also that particular *regions* in the two planes correspond. So, for example, the region between the lines $v = 0$ and $v = c_1$ corresponds in the z-plane to the interior of the ellipse for which $v = c_1$ (the ellipse for $v = 0$ degenerates, as shown above, into the straight line between $x = 1$ and -1); and the rectangle $pqrs$ between the lines $v = c_1$, $v = c_2$, $u = k_1$, and $u = k_2$ corresponds to the area PQRS (and to images of this in the axes) in the z-plane.[†]

Transformations of the general type of eqn (5.1) are termed *conformal*. The means they are such that if two curves cross at a given angle in one plane, the transformed curves in another plane cross at the same angle and in such a way that the senses of the two angles are the same.[‡] So, in Fig. 5.2, $p\hat{q}r = P\hat{Q}R$. In particular, if two curves

---

[†]Note that any strip of width 2π in the upper half t-plane corresponds with the whole of the z-plane.
[‡]The general proof of the property is simple and can be found in any elementary discussion of functions of a complex variable.

cross at right-angles (as in the example), the transformed curves also cross at right-angles. This orthogonal property is, of course, characteristic of all conjugate functions (the real and imaginary parts of a regular function of a complex variable), see Appendix 1.

5.1.2 The solution of Laplace's equation

In addition to possessing the above orthogonal property, conjugate functions (as shown in Appendix 1) are solutions of Laplace's equation. Thus it can be seen that any conformal transformation provides a simple relationship between two Laplacian fields. The use of transformation techniques to derive field solutions turns on the determination of a suitable equation relating a given field to another field for which a solution is known or is easily found.

Consider first the complex plane of t, Fig. 5.3, in which is represented a uniform field parallel to the axes, so that the lines $u = $ constant are flux lines and the lines $v = $ constant are equipotential lines. Let the gradient of the potential function (proportional to flux density) be k at all points.[†] Then, arbitrarily taking the u-axis as the zero potential line, the potential function, ψ (though not necessarily the absolute value of potential), at any point (u, v) is given by $\psi = kv$; and the flux function ϕ, taking the line $u = 0$ as the line $\phi = 0$, is given by $\phi = ku$. Thus, combining these two equations, the expression for the complex potential w of any point in the t-plane is

$$w = \phi + j\psi = k(u + jv)$$

or

$$w = kt. \tag{5.7}$$

The uniform field (represented here in the t-plane) is the most simple of all Laplacian fields and is the basic one to which all other solutions are related. It is convenient to regard the complex potential, $w = \phi + j\psi$, as being represented in the plane of w, so that eqn (5.7) defines a simple form of transformation, involving only a change in scale

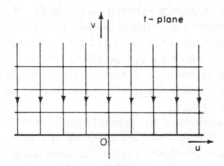

Fig. 5.3

[†]The relative numerical values of ϕ and ψ are chosen for convenience to give the simplest form to w.

proportional to k. Hence for any given field, the process of solution by transformation reduces ultimately to the derivation of an equation,

$$w = f(z), \qquad (5.8)$$

which relates points in the given field in the z-plane to points in the w-plane. More specifically this means the boundary shape and conditions of the two planes are connected through eqn (5.8). It will be seen that, in general, derivation of eqn (5.8) involves the introduction of intermediate planes and variables.

As a simple example, consider the influence of the transformation

$$z = t^{1/2} \qquad (5.9)$$

on the uniform field with lines parallel to the axes in the t-plane. It is seen that the positive real axis in the t-plane transforms into the positive real axis in the z-plane, but that the negative half of the real axis, because of the root of a negative sign, becomes the imaginary axis between 0 and $j\infty$ in the z-plane. Thus, the equation transforms the upper half of the t-plane into the first quadrant of the z-plane. The shapes of the parallel lines when transformed can be found by eliminating u or v from eqn (5.9). Squaring this equation and equating the real and imaginary parts gives

$$x^2 - y^2 = u \qquad (5.10)$$

and

$$2xy = v, \qquad (5.11)$$

from which it is seen that lines parallel to the v-axis transform into rectangular hyperbolae, as also do those parallel to the u-axis, Fig. 5.4

The physical picture of the influence of this transformation is that straight flux lines which enter an equipotential boundary coinciding with the real axis of the t-plane, become hyperbolic in shape in the z-plane and enter equipotential boundaries coinciding with the positive real and imaginary axes. The boundary conditions in the two planes are identical because of the conformal property of the transformation. Equally, the

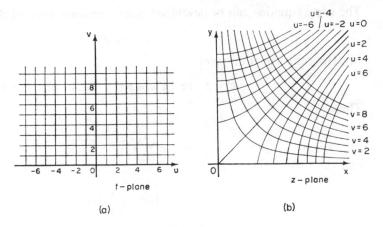

(a) (b)

Fig. 5.4

boundary in the z-plane is a limiting line ($\psi = 0$) in a set of hyperbolically shaped equipotential lines.

The equations of the flux and potential functions in the z-plane are given by elimination of t from eqns (5.7) and (5.9) as

$$kz^2 = w, \tag{5.12}$$

or, for the flux function,

$$\phi = k(x^2 - y^2), \tag{5.13}$$

and, for the potential function,

$$\psi = 2kxy. \tag{5.14}$$

For a given value of ϕ or ψ these equations map out in the z-plane the shape of the corresponding flux or equipotential line (including the boundary). Equally, substitution of any point (x, y) in eqn (5.12) gives the complex potential w with respect to an appropriate origin, of the point (x, y).

5.1.3   The logarithmic function

The logarithmic transformation is of great importance, being used in the majority of all solutions by transformation techniques. Its mapping properties and the three basic fields which it can be used to describe are considered below.

Field of a line current

Consider again the field of a line current. Appendix 1 shows that for a line current, i, placed at the origin of the z-plane, the complex potential function of the field is given by the function $(i/2\pi) \log z$; that is,

$$\phi + j\psi = \frac{i}{2\pi} \log z. \tag{5.15}$$

The same equation can be developed from a consideration of the transformation

$$t = \log z. \tag{5.16}$$

Inverting this equation gives

$$e^t = z \quad \text{or} \quad e^u \cos v = x \quad \text{and} \quad e^u \sin v = y.$$

Thus

$$\sqrt{x^2 + y^2} = e^u \tag{5.17}$$

and

$$\tan^{-1}\left(\frac{y}{x}\right) = v. \tag{5.18}$$

Equation (5.17) shows that a line parallel to the v-axis in the t-plane transforms into a circle, centre the origin, in the z-plane; and eqn (5.18) shows that a line parallel to the u-axis in the t-plane transforms into a straight line, passing through the origin and making an angle v radians with the x-axis of the z-plane (Fig. 5.5).

Now if a uniform field described by

$$w = kt \qquad (5.19)$$

is put in the t-plane, lines corresponding to constant values of u and v are flux and equipotential lines respectively. However, the lines in the z-plane to which these correspond are the same as the flux and equipotential lines in the field of a line current. Hence, combining eqns (5.19) and (5.16) gives the solution for the field of a line current as

$$w = k \log z, \qquad (5.20)$$

where the value of k is dependent on the magnitude of the current.

To evaluate k it is most convenient to make use of the equation for the potential function. Equating the imaginary parts of eqn (5.20) gives

$$\psi = k \arg z. \qquad (5.21)$$

Now, in tracing a path once round the current, there is a potential change of i, so that, expressing this condition in eqn (5.21) gives $i = k2\pi$.

Thus substitution for k in eqn (5.20) yields the complex potential function of the field due to the current, eqn (5.15) in absolute terms.

Field of a line charge

Equally the solution (5.20) applies, with interchange of the flux and potential functions, to the field of a line charge. But the value of the constant k is dependent in this case on the magnitude of the charge, q units per unit length of the line. The equation corresponding to (5.21) is

$$\phi = k \arg z, \qquad (5.22)$$

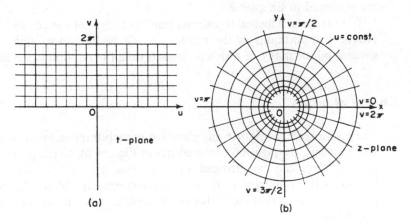

Fig. 5.5

and, expressing the condition that a path traced round the charge cuts all the lines of flux emanating from it, leads to

$$k = \frac{q}{2\pi}.$$

Field of two semi-infinite equipotential planes

The field in the upper half plane of Fig. 5.5(b) is identical with that due to two semi-infinite equipotential planes, one lying between 0 and ∞, the other between 0 and −∞, with a difference in potential of $i/2$ between them. In general, with a difference in potential of ψ_1 maintained between the two halves (dividing at the origin) of the real axis, the field in the positive half plane is described by the complex potential function

$$\phi + j\psi = \frac{\psi_1}{\pi} \log z. \tag{5.23}$$

This equation is frequently used in the analysis of fields in which two (different) equipotential sections form the boundary.

5.2 APPROACHING THE SOLUTION

In order to obtain a field solution by conformal transformation, two things are necessary: a transformation equation must be found relating the given field to a simpler one, and a solution must be found for this simple field. The possibility of doing the latter depends upon whether or not the transformed boundary conditions can be identified and used. The general problem of identifying and using these conditions can be very difficult or even impossible, but there are many important types of problem for which it is relatively simple. These include those in which, at the boundaries, values of ϕ or ψ are known or, though unknown, are constant so that the field reduces effectively to a single region. Attention is concentrated on such fields but the more general capabilities of the method are reviewed in Chapter 7.

The basis of conformal transformation have been set out above but, before considering practical applications of the method, it will be helpful to elaborate on a number of general points many of which arise each time the method is applied.

5.2.1 Choice of origin

The position of the origin in any plane is quite arbitrary and may be chosen as convenient. For example, the position of the corner in Fig. 5.4(b), or the position of the line current in Fig. 5.5(b) can be changed to any point $z = x_1 + jy_1$ by replacement of z by $[z - (x_1 + jy_1)]$ in eqn (5.9) or (5.16) respectively. Similarly, replacement of t by $[t - (u_1 + jv_1)]$ makes the vertex in the z-plane correspond with the point $t = u_1 + jv_1$, Fig. 5.5(a).

5.2.2 Multiple transformations

In applying transformation methods it is necessary to find equations which convert a simple boundary into the more complicated boundary containing the field which it is required to solve. It is not generally possible to find, directly, a single equation which does this and, usually, the w- and z-planes have to be connected through intermediate variables and boundary shapes. So, for example, the field pattern of Fig. 5.6 can be derived by combining transformations of the types (5.9) and (5.15). The pattern shown in Fig. 5.5(b) is mapped into the p-plane from the w-plane by the equation

$$w = \frac{i}{2\pi} \log p, \tag{5.24}$$

and, in turn, the upper half of the p-plane is mapped into the z-plane, Fig. 5.6, by

$$z = k(p - a)^{1/2}, \tag{5.25}$$

where a is a real constant defining the point, on the real axis of the p-plane, which corresponds to the vertex in the z-plane. Eliminating p from eqns (5.24) and (5.25) gives the solution in the z-plane as

$$\phi + j\psi = \frac{i}{2\pi} \log\left(\frac{z^2}{k^2} + a\right). \tag{5.26}$$

This may be regarded as describing either the field of a line current placed on the surface of an infinitely permeable corner, or one half of the field of a line current parallel to an infinite, infinitely permeable plane.

5.2.3 Field maps

The solution for a field in the z-plane is often derived as an equation of the form

$$w = f(z) \tag{5.27}$$

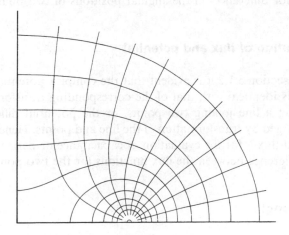

Fig. 5.6

If this equation is such that it can be written

$$z = g(w)$$
$$= g(\phi + j\psi), \qquad (5.28)$$

as, for instance, eqn (5.26) can be, then a field map can be calculated directly. Substituting a value ψ_1 for ψ gives the equation connecting points (x, y) lying on the equipotential line of value ψ_1, and this line may be traced by further substitution of a succession of values of ϕ. Similarly, a flux line may be traced by selecting a fixed value of ϕ and substituting a series of values of ψ. To produce a map of curvilinear *squares* the increments in ϕ and ψ must be equal. The squares can be chosen to have a convenient size by using eqn (5.27) to find the desired increment in flux or potential.

5.2.4 Scale relationship between planes

The equation

$$z = k f(t)$$

directly relates one *number* in the t-plane to another *number* in the z-plane. Consequently it relates a distance between two points (numbers) in the t-plane to a distance between the corresponding points in the z-plane; for example, eqn (5.9) makes a length of 4 units measured from the origin of the t-plane correspond with 2 units of length measured from the origin in the z-plane. When plotting a field map, any convenient unit and scale may be selected, because the *shape* of a map is not dependent upon its scale. However, when numerical information, such as flux density, is required, the unit chosen must be the same for both planes; it is clear, for example, that the flux density at a point, for given boundary shapes and field excitation, is inversely proportional to the size of the field. The scale constant k is then evaluated so as to give direct correspondence between points and field strengths in the z- and t-planes (see section 5.1.3). In certain cases, to allow for differences in the angular positions of two boundaries, k may be complex.

5.2.5 Conservation of flux and potential

From section 5.1.2 it is evident that the complex potential associated with a particular point is identical with that of the corresponding transformed point. Therefore the flux crossing a line joining two points, or the potential difference between the points, is unchanged by transformation of the line and points. Hence, for example, the calculation of total flux—for the evaluation of capacitance or permeance—merely involves taking the difference between the flux functions for the two points.

5.2.6 Field strength

Since flux density and field strength are simply related by a multiplicative constant (see section 5.1.2), it is necessary to consider only one of them. The general solution for a field in the z-plane is

$$w = \phi + j\psi = f(z),$$

and the field strength in the x-direction E_x is

$$E_x = -\frac{\partial \psi}{\partial x},$$

and the field strength in the y-direction E_y is

$$E_y = -\frac{\partial \psi}{\partial y}.$$

Consider the quantity $E_x - jE_y$. This can be written

$$E_x - jE_y = -\frac{\partial \psi}{\partial x} + j\frac{\partial \psi}{\partial y},$$

which, substituting for $\partial \psi / \partial y$ from the Cauchy–Riemann equations (Appendix 1), gives

$$E_x - jE_y = -\frac{\partial \psi}{\partial x} + j\frac{\partial \phi}{\partial x}$$

$$= j\left(\frac{\partial \phi}{\partial x} + j\frac{\partial \psi}{\partial x}\right)$$

$$= j\frac{\partial w}{\partial x}$$

$$= j\frac{dw}{dz}. \tag{5.29}$$

Now the quantity $E_x - jE_y$ has the same modulus but minus the argument of the quantity $E_x + jE_y$, which is equal to the field strength $\mathbf{E}$. Therefore from eqn (5.29) the magnitude of the field strength is given simply by

$$|\mathbf{E}| = \left|\frac{dw}{dz}\right| \tag{5.30}$$

and the argument of the field strength by

$$\arg \mathbf{E} = -j\arg\left(\frac{dw}{dz}\right). \tag{5.31}$$

These are important results, and eqn (5.30) particularly is frequently required.

5.3 THE BILINEAR TRANSFORMATION

Attention is turned now, in the light of the preceding introduction of the basic ideas of conformal transformation, to a more detailed study of a small number of particular transformations. These are chosen partly because of their relative simplicity and partly because of their usefulness both directly in relation to particular (curved) boundary shapes, and in conjunction with other general transformation equations. In addition they are convenient in bringing out some basic transformation properties and in illustrating more fully the ways in which actual solutions are constructed.

This section focuses particularly on the *bilinear transformation*. This provides for the treatment of boundaries which are circular but non-concentric or intersecting, which combine circular and straight lines and which convert circles into straight lines—a property used in the next chapter to develop the general transformation connecting a circular boundary with a polygonal one. These have direct practical application in relation to such matters as the capacitance between cables or waveguides of circular cross-section and the unbalanced magnetic pull on an eccentrically mounted machine rotor (none of which are amenable to the methods of Chapter 4 these being restricted to concentric boundaries only). The bilinear transformation has also been useful in handling problems which have to be treated as having infinite regions: Silvester [3] has studied eddy-current effects in isolated conductors (in conjunction with both direct and analogue methods); and Boothroyd *et al.* [2] made ingenious use of a circular electrolytic tank to represent the whole complex plane in connection with network problems. Such approaches could also be fruitful in conjunction with the numerical methods discussed in the later chapters.

Before proceeding to use the transformation to solve particular problems, its general mapping properties are explored.

5.3.1 Mapping properties

The bilinear transformation is of the form

$$z = \frac{at + b}{ct + d}, \tag{5.32}$$

where a, b, c, and d are constants which may be real or complex. It is the general transformation for which one, and only one, value of one variable corresponds to one, and only one, value of the other. It can be regarded as being built up from three successive, simple transformations:

$$t_1 = ct + d, \tag{5.33}$$

$$t_2 = \frac{1}{t_1}, \tag{5.34}$$

and

$$z = \frac{a}{c} + (bc - ad)\frac{t_2}{c}. \tag{5.35}$$

The first and third of these introduce magnification, translation, and rotation of the field map depending upon the values of the constants, but do not change its *shape*. The expression $(bc - ad)$ is called the *determinant* of the transformation (5.32) and clearly, from eqn (5.35), must not become zero. Only eqn (5.34), which is called the *inversion* transformation (since one variable is the inverse of the other), changes the shape of the map. Therefore the complete bilinear transformation produces, for a given map, the same change of shape as does the inversion transformation. The bilinear and inversion transformations are most useful when applied to circles or straight lines, and the influence of the inversion transformation on these curves is now described.

The transformation of straight lines, and the doublet

Consider the inversion transformation

$$t = \frac{1}{z}.$$

(5.36)

Rationalizing this equation and equating real and imaginary parts gives

$$u = \frac{x}{x^2 + y^2} \quad \text{and} \quad v = \frac{y}{x^2 + y^2},$$

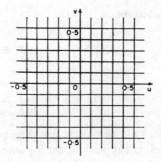

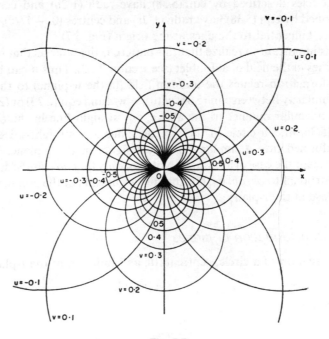

Fig. 5.7

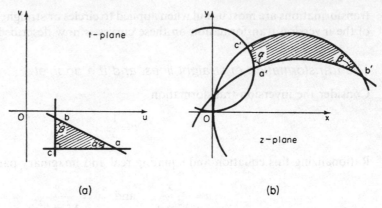

Fig. 5.8

which may be rewritten as

$$\left(x - \frac{1}{2u}\right)^2 + y^2 = \left(\frac{1}{2u}\right)^2 \tag{5.37}$$

and

$$x^2 + \left(y + \frac{1}{2v}\right)^2 = \left(\frac{1}{2v}\right)^2. \tag{5.38}$$

For constant values of u and v these are the equations of circles in the z-plane; that is, straight lines parallel to the axes of the t-plane are mapped into circles in the z-plane. The circles described by eqn (5.38) have radii $(1/2u)$ and centres $(1/2u, 0)$, and those described by eqn (5.38) have radii $(1/2v)$ and centres $(0, -1/2v)$, so that circles from each set are tangential to the axes at the origin (Fig. 5.7).

It is important to realize that this pattern is the same as that of the flux and equipotential lines in the field of a doublet (see section 2.3.2). Thus it can be seen that the inversion transformation relates the uniform field (in the w-plane) to the field of a doublet, and the similarity between eqn (5.36) with $z = w/d$ and eqn (2.23) or (2.24) should be noted.

In a similar manner to the above, any straight line in the t-plane is transformed by eqn (5.36) into a circle in the z-plane. Thus any straight-sided figure in the t-plane is transformed into a similar curvilinear figure in the z-plane. This is demonstrated in Fig. 5.8 for the case of a triangle. Note not only that ao', ob', oc' have magnitudes inversely proportional to oa, ob, oc, but that the arguments of the z-plane vectors are the negative of those in the t-plane.

The transformation of circles

The equation of a circle, centre $(a, 0)$, and radius r, in the t-plane, is

$$t = |t - a|, \tag{5.39}$$

but, since

$$t - a = \frac{1}{x + jy} - a,$$

the expression for the radius can be written

$$r = |t - a|$$

$$= \left| \frac{(1 - ax) - jay}{x + jy} \right|.$$

Therefore

$$r^2 = \frac{(1 - ax)^2 + a^2 y^2}{x^2 + y^2},$$

which may be rearranged as

$$\left(x - \frac{a}{a^2 - r^2} \right)^2 + y^2 = \left(\frac{r}{a^2 - r^2} \right)^2. \tag{5.40}$$

This is the equation of a circle, radius $r/(a^2 - r^2)$, centre $(a/(a^2 - r^2), 0)$ in the z-plane, mapped by eqn (5.36) from the circle, radius r, centre a, in the t-plane. For $a > r$ it follows that

$$\frac{a}{a^2 - r^2} > \frac{r}{a^2 - r^2} > 0, \tag{5.41}$$

so that the transformed circle lies entirely in the positive real half of the z-plane. For $a = r$, i.e. for a circle passing through the origin in the t-plane, the transformed curve is a straight line,

$$2ax = 1, \tag{5.42}$$

and, for $a < r$,

$$\frac{a}{a^2 - r^2} < \frac{r}{a^2 - r^2} < 0, \tag{5.43}$$

so that the transformed circle extends into the positive real half of the plane, though its centre is on the negative real axis. These three cases are shown in Fig. 5.9 where a set of concentric circles, and the corresponding transformed curves, are shown.

Alternatively, by suitable positioning of the origin in the z-plane, it is possible to transform any two circles (the straight line is a circle of infinite radius) which are non-concentric into two concentric circles in the t-plane. There are three different arrays in the z-plane:

   (i) a circle within another circle;
  (ii) a circle near a straight line; and
 (iii) a circle exterior to another circle;

and a field having boundaries falling into any one of these classes can be analysed by first determining the solution of the field for the corresponding concentric boundaries with appropriate boundary conditions.

Geometrical and complex inversion

Geometrical and complex inversion are not identical: a point having an argument θ inverts, in the geometrical case, into a point with the same argument whilst, in the

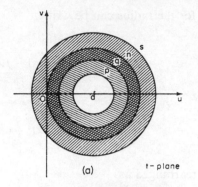

(a)   t-plane

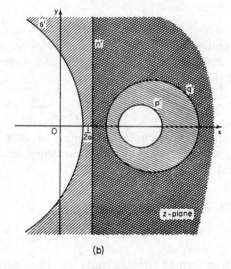

(b)

Fig. 5.9

complex case, the inverted point has argument $-\theta$. This is evident when z [eqn (5.36)] is written in polar form,

$$t = \frac{1}{r\,e^{j\theta}} = \frac{1}{r}e^{-j\theta}. \tag{5.44}$$

As mentioned above, this feature is demonstrated in Fig. 5.8.

5.3.2 The cross-ratio

Consider again the general transformation (5.32) and let z_1, z_2, z_3, z_4 and t_1, t_2, t_3, t_4, be two sets of corresponding points. Then, by substitution in eqn (5.32),

$$z_1 - z_4 = \frac{at_1 + b}{ct_1 + d} - \frac{at_4 + b}{ct_4 + d}$$

$$= \frac{ad - bc}{(ct_1 + d)(ct_4 + d)}(t_1 - t_4). \tag{5.45}$$

Similar expressions for $z_3 - z_2$, $z_1 - z_2$ and $z_3 - z_4$ can be derived and combined with the above to eliminate all terms involving the constants a, b, c, and d to give

$$\frac{(z_1 - z_4)(z_3 - z_2)}{(z_1 - z_2)(z_3 - z_4)} = \frac{(t_1 - t_4)(t_3 - t_2)}{(t_1 - t_2)(t_3 - t_4)}. \tag{5.46}$$

The right-hand side of eqn (5.46) is called the *cross-ratio* of the four points t_1, t_2, t_3, and t_4. It is a constant for the transformation (5.32), and the relationship

$$\frac{(z_1 - z)(z_3 - z_2)}{(z_1 - z_2)(z_3 - z)} = \frac{(t_1 - t)(t_3 - t_2)}{(t_1 - t_2)(t_3 - t)}. \tag{5.47}$$

defines the unique bilinear transformation connecting the curves passing through the particular corresponding sets of point t_1, t_2, t_3 and z_1, z_2, z_3.

As an example of the use of eqn (5.37), the transformation, used in the next chapter, which converts the real axis of one plane into the unit circle in another, is developed here. The points $t_1 = 0$, $t_2 = 1$, and $t_3 = \infty$ define the real axis of the t-plane. The unit circle in the z-plane may be defined by the corresponding points $z_1 = 1$, $z_2 = j$, $z_3 = -1$. Substitution of all these values in eqn (5.37) yields the equation transforming the real axis of the t-plane into the unit circle in the z-plane in such a way that the perimeter of the circle in the upper half of the plane corresponds with the positive real axis. This equation is

$$t = j \frac{(1 - z)}{(1 + z)}. \tag{5.48}$$

It is easiest to see which regions in the two planes correspond by expressing z in polar form, $z = \varrho\, e^{j\theta}$. This gives for eqn (5.48)

$$t = j \frac{1 - \varrho\, e^{j\theta}}{1 + \varrho\, e^{j\theta}}$$

$$= \frac{2\varrho \sin\theta}{1 + 2\varrho \cos\theta + \varrho^2} + j \frac{(1 - \varrho^2)}{1 + 2\varrho \cos\theta + \varrho^2}. \tag{5.49}$$

Since the quantity $(1 + 2\varrho \cos\theta + \varrho^2)$ cannot be negative, this shows that the interior of the unit circle ($\varrho < 1$) corresponds to the upper half of the t-plane, and that the exterior of the unit circle ($\varrho > 1$) corresponds to the lower half of the t-plane.

5.3.3 The magnetic field of currents inside an infinitely permeable tube

Equation (5.48) can be used, for instance, in the analysis of the magnetic field of currents flowing inside an infinitely permeable tube of circular cross-section, taken for convenience to have unit inside radius. (The equivalent current-flow and electrostatic fields will be apparent.) First consider the field of a single current, of strength i, placed at the origin on the edge of an infinitely permeable sheet in the lower half of the t-plane, see Fig. 5.10(a),

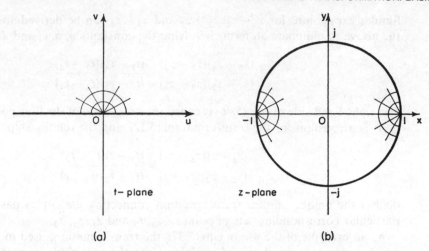

(a)   (b)

Fig. 5.10

and described by

$$w = \psi + j\phi = \frac{i}{\pi} \log t. \tag{5.50}$$

Magnetic equipotential lines from the point $t = 0$ terminate at infinity, so that at $t = \infty$ there is effectively a 'return' current $-i$. When the field in the upper half plane is transformed by eqn (5.48) into the interior of the circular tube represented in the z-plane by the unit circle, centre $z = 0$, the point $z = 1$ is made to correspond with $t = 0$, and the point $z = -1$ with $t = \infty$. Thus the field interior to the tube is that of two currents of opposite signs at the opposite ends of a diameter, Fig. 5.10(b). Elimination of t from eqns (5.48) and (5.50) gives the solution for this field as

$$w = \psi + j\phi = \frac{i}{\pi} \log j \frac{(1-z)}{(1+z)}. \tag{5.51}$$

To calculate the field map in the z-plane, z is expressed in terms of $\psi + j\phi$ [by inverting eqn (5.51)] as

$$z = \frac{1 + j\,e^{w\pi/i}}{1 - j\,e^{w\pi/i}}, \tag{5.52}$$

and values of w are substituted as described earlier (see section 5.2.3).

The solution for the field in the w-plane with the currents at the points $z = -1$ and $z = z_s$ (on the surface of the tube) is obtained by noting that t in eqn (5.50) must be replaced by $(t - t_s)$ where t_s is the point corresponding to z_s. The result is

$$w = \frac{i}{\pi} \log j \left\{ \frac{(1-z)}{(1-z)} - \frac{(1-z_s)}{(1+z_s)} \right\}. \tag{5.53}$$

Treatment of the more general problem in which the two (or more) currents are inside the tube away from the boundary involves the use of field solutions [in place of eqn (5.50)] which are discussed in section 7.2.

5.3.4 Capacitance of and voltage gradient between two cylindrical conductors

As a demonstration of the use of the special form of the bilinear transformation, inversion, the field between two charged cylindrical conductors, situated as shown in Fig. 5.11 is considered. The solution of this field is of value, for example, in determining the breakdown voltage of air gaps and the necessary insulation for and the capacitance of parallel cables: in the former case the space between the cylinders is filled with air, and in the latter with air or other insulating material. It should be noted that for any problem the analysis given requires the assumption that a single uniform dielectric be present between the cylinders.

Let the non-concentric boundaries of Fig. 5.11 be placed in the z-plane, see Fig. 5.12(a), and let the origin of inversion be placed d units to the left of the smaller circle, radius R_2. d must be so chosen that these circles are transformed into concentric circles in the t-plane, see Fig. 5.12(b). From eqn (5.40) it is seen that the smaller circle, centre $(d, 0)$, in the z-plane, transforms into a circle in the t-plane with centre coordinates

$$\left(\frac{d}{d^2 - R_2^2}, 0\right)$$

and radius

$$r_2 = \frac{R_2}{d^2 - R_2^2}.$$

Similarly, the large circle, with radius R_1 and centre $(d - D, 0)$ transforms into a circle, centre

$$\left(\frac{d - D}{(d - D)^2 - R_1^2}, 0\right)$$

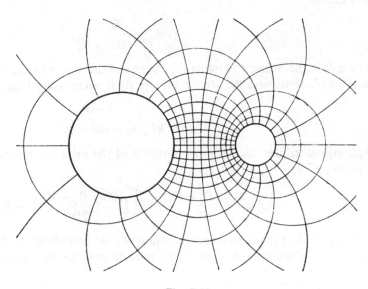

Fig. 5.11

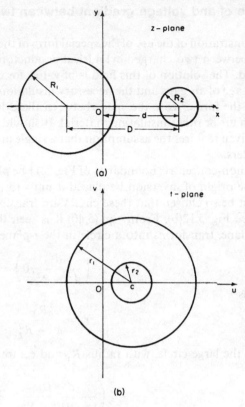

Fig. 5.12

and radius

$$r_1 = \frac{R_1}{(d-D)^2 - R_1^2},$$

where D is the distance between the centres of the circles in the z-plane. For the circles in the t-plane to have the same centre, $(c, 0)$, it is necessary that

$$\frac{d}{d^2 - R_2^2} = \frac{d-D}{(d-D)^2 - R_1^2}.$$

This equation fixes the necessary position of the origin of inversion in the z-plane by defining $d^\dagger$ as

$$d = \frac{D^2 + R_2^2 - R_1^2}{2D} \pm \sqrt{\left[\left(\frac{D^2 + R_2^2 - R_1^2}{2D}\right)^2 - R_2^2\right]}. \qquad (5.54)$$

The cylinders in the z-plane have equipotential boundaries at different potentials and so, in the t-plane, the concentric cylindrical boundaries must be equipotential lines with

---

†The value of d making r positive must be taken.

the same difference in potential between then. The field between two concentric cylindrical conductors, centred at the point $(c, 0)$, separated by a medium of relative permittivity ε and carrying a charge of q units per unit length is, see Appendix 1

$$w = \psi + j\phi = \frac{q}{2\pi\varepsilon\varepsilon_0} \log(t - c). \tag{5.55}$$

By considering two points, one on each of these cylinders, most simply, $t_1 = c + r_1$ and $t_2 = c + r_2$, with the same flux function, the potential difference between them may thus be expressed as

$$\psi_1 - \psi_2 = \frac{q}{2\pi\varepsilon\varepsilon_0} \log \frac{r_1}{r_2}. \tag{5.56}$$

Therefore the capacitance per unit length C between these cylinders, i.e. the total flux per unit length (which is equal to q) divided by the potential difference, is

$$C = \frac{2\pi\varepsilon\varepsilon_0}{\log(r_1/r_2)}. \tag{5.57}$$

Further, since corresponding points and boundaries in the t- and z-planes have the same flux and potential functions, eqn (5.57) also gives the capacitance of the non-concentric cylinders.

The potential gradient $\mathbf{E}$, at any point in the z-plane has, as is described earlier (see section 5.2.6), a magnitude

$$|\mathbf{E}| = \left| \frac{dw}{dz} \right|$$

$$= \left| \frac{dw}{dt} \right| \left| \frac{dt}{dz} \right|. \tag{5.58}$$

Differentiating eqn (5.55) gives

$$\frac{dw}{dt} = \frac{q}{2\pi\varepsilon\varepsilon_0} \frac{1}{t - c},$$

and substituting for q from eqn (5.56) gives

$$\frac{dw}{dt} = \frac{\psi_1 - \psi_2}{\log(r_1/r_2)} \frac{1}{t - c}.$$

Also, from eqn (5.36)

$$\frac{dt}{dz} = -\frac{1}{z^2},$$

so eqn (5.58) can be written

$$|\mathbf{E}| = \left| \frac{\psi_1 - \psi_2}{\log(r_1/r_2)} \frac{1}{(t - c)} \frac{1}{z^2} \right|$$

$$= \left| \frac{\psi_1 - \psi_2}{\log(r_1/r_2)} \frac{1}{(1 - cz)z} \right|. \tag{5.59}$$

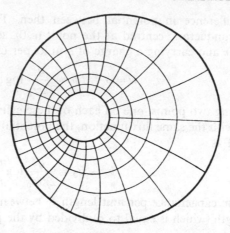

Fig. 5.13

The maximum value of potential gradient is important in the consideration of the breakdown voltage between the cylinders. It occurs at the point on the shortest line between the cylinders at the boundary with the greater curvature. This point is $z = d - R_2$ to which corresponds the point $t = c + r_2$ and, therefore, substitution of these values in eqn (5.59) gives the expression for maximum gradient as

$$|\mathbf{E}|_{max} = \left| \frac{\psi_1 - \psi_2}{\log(r_1/r_2)} \frac{1}{r_2(d - R_2)^2} \right|. \tag{5.60}$$

The field map in the z-plane (Fig. 5.11) is transformed from that in the t-plane, which consists of concentric circles, and radial lines centre $(c, 0)$. It is calculated using the equation

$$z = \frac{1}{e^W + c}, \tag{5.61}$$

where

$$W = \frac{w \log(r_1/r_2)}{\psi_1 - \psi_2}, \tag{5.62}$$

derived by eliminating t from eqns (5.36) and (5.55) and using eqn (5.56). As will be clear from the subsection *The transformation of circles* of section 5.3.1 the same equations can also be used to plot the field map for two cylindrical boundaries one within the other (Fig. 5.13), or, after first determining the appropriate values of d, r_1, and r_2, for one cylinder near a plane surface.

For the maps, Figs 5.11 and 5.13, the boundary potentials and dimensions are such as to give the same quantity of total flux in each.

5.4   THE SIMPLE JOUKOWSKI TRANSFORMATION

5.4.1   The transformation

The equation

$$Kt = z + \frac{a^2}{z} \tag{5.63}$$

where K and a are constants, and which may also be written,

$$z = \frac{K}{2}\left[t \pm \sqrt{ t^2 - \left(\frac{2a}{K} \right)^2 } \right],$$ (5.64)

transforms a circle in the z-plane into an ellipse in the t-plane, in such a way that the regions exterior to both curves, and the regions interior to them, correspond. It also has a second useful mapping property, that it can be used to transform the real axis in the t-plane into the real axis of the z-plane for $|t| \geqslant 2a/K$, and into a semicircle, radius a, centre $z = 0$, for $|t| \leqslant 2a/K$. More generally, with the equation,

$$z = \frac{K}{2}\left[t \pm \lambda \sqrt{ t^2 - \left(\frac{2a}{K} \right)^2 } \right],$$ (5.65)

where λ is a real constant ($\neq 1$), the curve in the z-plane for $|t| < 2a/K$ is an ellipse (Fig. 5.14).

This may be seen from the following considerations. When t is real and $|t| \geqslant 2a/K$, the expression under the root sign is positive, so that, provided the root is given the sign of t, z is wholly real with the sign of t. When t is real and $|t| < 2a/K$, the quantity under the root sign is negative, and separation of the real and imaginary parts of eqn (5.65) gives

$$x = \frac{K}{2}u \quad \text{and} \quad y = \frac{K\lambda}{2}\sqrt{ \left(\frac{2a}{K} \right)^2 - u^2 },$$

from which u can be eliminated to give

$$\frac{x^2}{a^2} + \frac{y^2}{(a\lambda)^2} = 1.$$ (5.66)

This is the equation of an ellipse describing the shape of the curve in the z-plane corresponding with the real axis of the t-plane between $\pm 2a/K$. The ellipse cuts the y-axis at $y = \pm a\lambda$. For $\lambda = 1$ the curve degenerates into the circle of radius a. It should be noted that to transform the upper half t-plane into the region above the real axis (and curve) in the z-plane, the imaginary part of the square root must be taken as positive throughout, whilst the real part takes the same sign as the real part of t.

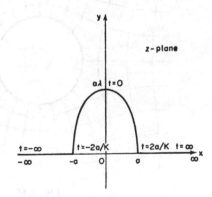

Fig. 5.14

(This transformation is also considered in section 7.3 as a particular example of a transformation for curved boundaries with vertices.)

5.4.2 Flow round a circular hole

As an example of the use of eqn (5.63) [eqn (5.65) in which $\lambda = 1$], the very simple example of the flux distribution round a circular hole in an infinite sheet is considered (Fig. 5.15). The field is symmetrical, and so one half of it can be represented in the upper half of the z-plane with the boundary (flux line of zero flow) shown in Fig. 5.16(a). This boundary, assuming the hole to have a radius a, is related to the real axis in the t-plane, Fig. 5.16(b), by eqn (5.63) with $K = 2a$, i.e. by

$$t = \frac{1}{2a}\left(z + \frac{a^2}{z}\right). \tag{5.67}$$

Now the field in the z-plane is such that the whole boundary is a flux line and, therefore, the real axis in the t-plane must be a flux line. This condition is satisfied by a uniform field (in the t-plane) parallel to the real axis described by

$$w = \psi + j\phi = kt, \tag{5.68}$$

where k defines the density of the flow. Thus, eliminating t between eqns (5.67) and (5.68), the solution for the field in the z-plane may be written as

$$w = \frac{k}{2a}\left(z + \frac{a^2}{z}\right). \tag{5.69}$$

Inverting this gives the form from which, by substitution of values of ϕ and ψ, the field map (there is only one pattern) is calculated.

It is important to note that the field described by eqn (5.69) is formed by the superposition of two simpler fields, one uniform (z), the other that of a doublet (a^2/z). Thus,

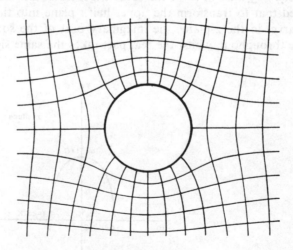

Fig. 5.15

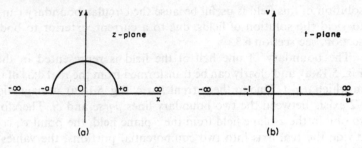

Fig. 5.16

as demonstrated in sections 2.3.2 and 3.2.3, the influence of a circular hole, or cylinder, in a uniform applied field, is described by a doublet at the circle centre.

5.4.3 Permeable cylinder influenced by a line current

Equation (5.63) is applied now to the analysis of the field of a permeable cylinder of unit radius influenced by a line current (Fig. 5.17), a problem previously treated by images, section 2.3, and the direct solution of Laplace's equation, section 3.2.2. The

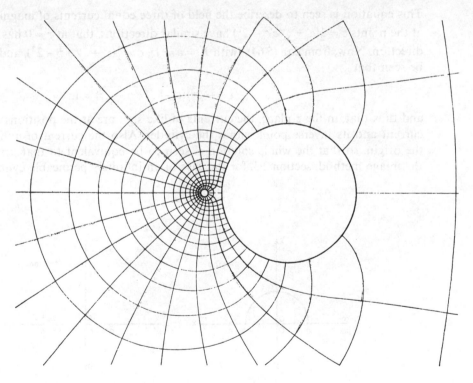

Fig. 5.17

solution of this field is useful because the circular boundary can be further transformed to yield the solution of fields, due to a current, exterior to bodies of polygonal cross-section, see section 6.3.

The boundary of one half of the field is represented in the z-plane as shown in Fig. 5.18(a), and clearly can be transformed from the real axis of the t-plane by eqn (5.63) in which $a = 1$. Due to the current i at c, Fig. 5.18(a), a magnetic potential difference of $i/2$ exists between the two boundary lines $pqrsc$ and cl. Therefore, for it to be possible to obtain the z-plane field from the t-plane field, the point c', corresponding to c, must divide the real axis into two equipotential portions, the values of which differ by $i/2$. That is, the field in the (whole) t-plane must be that of a line current i at c', described by

$$w = \phi + j\psi = \frac{i}{2\pi} \log(t - c'). \tag{5.70}$$

This solution is transformed by eqn (5.63) (remembering $a = 1$ and taking $K = 1$, so that $s' = 1$ and $q' = -1$) into that for the z-plane,

$$w = \frac{i}{2\pi} \log\left(\frac{z^2 + 1}{z} - c'\right). \tag{5.71}$$

By factorizing the term in brackets, eqn (5.71) can be rewritten as follows:

$$w = \frac{i}{2\pi} \left\{ \log\left[z - \tfrac{1}{2}(c' + \sqrt{c'^2 - 2^2})\right] + \log\left[z - \tfrac{1}{2}(c' - \sqrt{c'^2 - 2^2})\right] - \log z \right\}. \tag{5.72}$$

This equation is seen to describe the field of three equal currents of magnitude i: those at the points $z = \tfrac{1}{2}(c' \pm \sqrt{c'^2 - 2^2})$ have similar directions, that at $z = 0$ has the opposite direction. Now, from eqn (5.64), (with $K = a = 1$), $c = \tfrac{1}{2}(c' + \sqrt{c'^2 - 2^2})$, and it can easily be seen that

$$\tfrac{1}{4}(c' + \sqrt{c'^2 - 2^2})(c' - \sqrt{c'^2 - 2^2}) = 1,$$

and thus that, in the z-plane, the currents of like sign are at the positions of the actual current and its inverse point within the cylinder. Also the current of unlike sign is at the origin, so that the whole solution is seen to be equivalent to that first derived by the image method, section 2.3, for the case of an infinitely permeable cylinder.

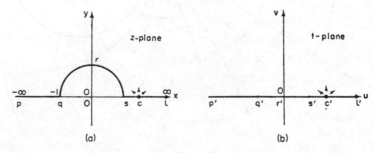

Fig. 5.18

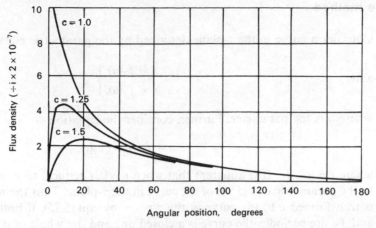

Fig. 5.19

The field map (Fig. 5.17) is calculated in the usual manner after inverting eqn (5.71). Also, the flux density B_z at any point in the z-plane can be evaluated from the equation

$$B_z = \left| \frac{dw}{dz} \right|$$

$$= \left| \frac{dw}{dt} \frac{dt}{dz} \right|$$

$$= \left| \frac{i}{2\pi} \frac{1}{(t-c')} \left(1 - \frac{1}{z^2} \right) \right|. \tag{5.73}$$

This equation can be evaluated in two ways by expressing it wholly in terms either of z or of t. If it is expressed in terms of t it is also necessary to evaluate the points in the z-plane corresponding with the values of t for which B_z is evaluated, but, even so, this can be the simpler method when t is real. In Fig. 5.19 are shown curves of flux density over the cylinder surface. $c = 1$ gives the case in which the current touches the cylinder.

5.5 CURVES EXPRESSIBLE PARAMETRICALLY: GENERAL SERIES TRANSFORMATIONS

The equation, transforming any curve, the shape of which is expressible in terms of a parameter, into a straight line, can easily be obtained. This is demonstrated in the first sub-section, and then, after consideration of a simple example, the result is used to develop general series transformations for both open and closed curved boundaries. Employing these series with curve-fitting techniques, it is possible to derive solutions for problems involving a wide range of boundary shapes.

5.5.1 The method

Consider a curve in the z-plane described by the parametric equations

and
$$\left. \begin{array}{c} x = f_1(u) \\ y = f_2(u), \end{array} \right\} \tag{5.74}$$

where u is the parameter. Further, consider the equation

$$z = f_1(t) + jf_2(t), \tag{5.75}$$

where $t = u + jv$. It is apparent that when $v = 0$, t reduces to u, and eqn (5.75) reduces to the parametric eqn (5.74) of the curve in the z-plane. Thus the real axis of the t-plane is transformed into the curve in the z-plane by eqn (5.75). If both of the functions, f_1 and f_2, are periodic, the curve is a closed one and the whole of it corresponds with the portion of the real axis in the t-plane between 0 and 2π. Also, the region exterior to the curve in the z-plane corresponds with the vertical strip of width 2π in the upper half of the t-plane. Figure 5.20 shows the corresponding regions (marked E and E') and the boundaries in the two planes. If the curve in the z-plane is an open one, not more than one of the functions, f_1 or f_2, is periodic, the whole of the real axis of the t-plane corresponds with the whole of the curve, and the whole of the upper half of the t-plane is transformed into the region exterior to the boundary in the z-plane.

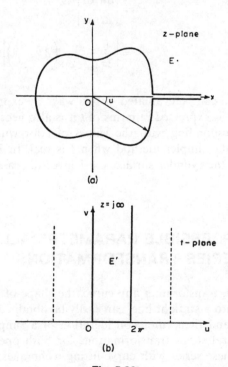

(a)

(b)

Fig. 5.20

5.5.2 The field outside a charged, conducting boundary of elliptical shape

As a simple example of the above method, consider the analysis of the field exterior to a charged conducting boundary with the shape of the ellipse described by

$$\frac{x^2}{a^2} + \frac{y^2}{b^2} = 1.$$

The coordinates of any point on this boundary can be expressed in terms of the parameter u by

$$x = a\cos u \quad \text{and} \quad y = b\sin u,$$

and thus, from eqn (5.75), the equation transforming the boundary into the real axis of the t-plane, between 0 and 2π, is

$$z = a\cos t + jb\sin t. \tag{5.76}$$

The boundary in the z-plane is equipotential, and so the real axis in the t-plane must also be equipotential. Thus the required field in the t-plane is uniform, described, most simply, by the equation

$$w = t,$$

and the solution in the z-plane, substituting for t in eqn (5.76), is

$$z = a\cos w + jb\sin w. \tag{5.77}$$

This can be simplified by substituting $a = k\cosh\alpha$ and $b = k\sinh\alpha$, where $k^2 = a^2 - b^2$, to give

$$z = k\cos(w + j\alpha). \tag{5.78}$$

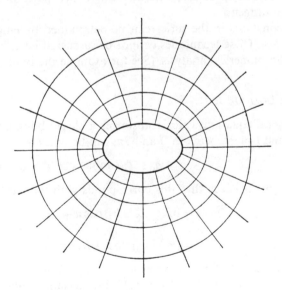

Fig. 5.21

Shifting the origin in the w-plane by $(\pi/2 - j\alpha)$ reduces this to

$$z = k \sin w, \tag{5.79}$$

which is the transformation equation considered earlier (section 5.1.1), generalized by the scale constant k. The field equipotential lines are ellipses described by

$$\frac{x^2}{k^2 \cosh^2 \psi} + \frac{y^2}{k^2 \sinh^2 \psi} = 1. \tag{5.80}$$

the flux lines are the hyperbolae described by

$$\frac{x^2}{k^2 \sin^2 \phi} - \frac{y^2}{k^2 \cos^2 \phi} = 1, \tag{5.81}$$

and both sets are shown in Fig. 5.21. By writing ψ for ϕ the solution becomes that for a field in which the elliptical boundary is a flux line.

5.5.3 General series transformations

As described in section 5.5.1, it is possible, for any curve expressible in terms of parametric equations, to derive simply a transformation equation. Parametric equations are well known for a rather limited number of curves, but a most important point is that they can be determined approximately in the form of power series for open boundaries or Fourier series for closed ones. One limitation to be noted, however, is that since a vertex in a curve requires a discontinuity in the gradient of the curve, it is not possible to treat curves having vertices by employing a power series or a finite number of terms of a Fourier series. An exception to this is the representation, with closed curves (expressed in Fourier series), of cusps (points where two parts of a curve intersect and have a common tangent).

The constants in the series can be determined by employing standard curve-fitting techniques. These techniques cannot be discussed here, but they are described in most books on numerical analysis. (See for example the book by Lanczos.)

Closed boundaries

The general equation which transforms a closed boundary in the z-plane into the real axis of the t-plane between 0 and 2π, can be written

$$z = a_0 e^{-jt} + b_0 + b_1 e^{jt} + b_2 e^{2jt} + \cdots, \tag{5.82}$$

and it is readily seen that this corresponds with the parametric equations

and
$$\left.\begin{aligned} x &= b_0 + (a_0 + b_1)\cos u + b_2 \cos 2u + \cdots \\ y &= (b_1 - a_0)\sin u + b_2 \sin 2u + \cdots \end{aligned}\right\} \tag{5.83}$$

(which are simply Fourier series for x and y in terms of the angle u subtended at the origin, see section 5.5.1 and Fig. 5.20. For certain simple forms of eqn (5.82) the closed boundary takes the form of well-known geometrical curves. This can be seen by substituting $v = 0$, when the transformation equation reduces to recognizable parametric equations.

(a) When $b_n = 0$ for all n,

$$z = a_0 e^{-jt}$$
$$= a_0(\cos t - j \sin t). \tag{5.84}$$

and this reduces to the parametric equation of a circle of radius a_0 centred on the origin of the z-plane. [It is easily seen that eqn (5.84) also transforms the *interior* of the circle into the strip of width 2π in the *lower* half t-plane.]

(b) When $a_0 = (a + b)/2$, $b_1 = (a - b)/2$ and all other values of b are zero,

$$z = \frac{a+b}{2} e^{-jt} + \frac{a-b}{2} e^{jt}$$
$$= a \cos t - jb \sin t, \tag{5.85}$$

and this reduces to the equation of an ellipse with semi-axes a and b and centred on the point $z = 0$. It is interesting to note that the Joukowski transformation, eqn (5.63), can be derived by eliminating e^{jt} between eqns (5.84) and (5.85).

(c) When terms up to and including b_2 are present, the curves range from the hypocycloid with three cusps to symmetrical aerofoils, and they have been considered in detail by Wrinch [1]. More complicated forms of eqn (5.82) do not appear to have been studied.

Open boundaries

A general form of transformation for open boundaries without vertices in the z-plane can be written

$$z = jt + b_0 + b_1 t + b_2 t^2 + \cdots \tag{5.86}$$

This is seen to transform a curve having the equation

$$x = b_0 + b_1 y + b_2 y^2 + \cdots, \tag{5.87}$$

which is equivalent to the parametric equation

and
$$\left.\begin{array}{l} x = b_0 + b_1 u + b_2 u^2 + \cdots \\ \\ y = u. \end{array}\right\} \tag{5.88}$$

As with eqn (5.82), certain simple forms of eqn (5.86) can be recognized immediately. For example, when all constants in eqn (5.87) except b_0 and b_2 are zero, the points (x, y) lie on a parabola, for which the transformation equation is thus

$$z = b_0 + jt + b_2 t^2. \tag{5.89}$$

5.5.4 Field solutions

The solutions for fields with equipotential or flux-line boundaries are particularly simple to obtain; since the required field in the t-plane is uniform, these solutions are given, by writing $w = t$ in eqn (5.75), as

$$z = f_1(w) + jf_2(w). \tag{5.90}$$

[See eqn (5.77).]

It should also be noted that the fields of line sources can be derived by transformation of the appropriate image solutions: for open curved boundaries the image solution is that for an infinite plane and, for closed boundaries, it is that for three intersecting plane boundaries, two of which are parallel (see section 2.2.3). This technique could be used, for example, as an alternative to that described in section 5.5.3 for the field of a current near an infinitely permeable circular cylinder.

For a general discussion of the transformation of field sources the reader is referred to Chapter 7.

REFERENCES

[1] D. Wrinch, Some problems of two dimensional hydrodynamics, *Phil. Mag.*, **48**, 1089 (1924).
[2] A. R. Boothroyd, E. C. Cherry and R. Makar, An electrolytic tank for measurement of steady-state response, transient response and allied properties of networks, *Proc. Instn. Elect. Engrs.*, **96II**, 176 (1949).
[3] P. Silvester, Network analog solution of skin and proximity effect problem, *Trans. Inst. Elect. Electr. Engrs.*, PAS-86, 241 (1967).

Additional references

Adams, E. P., Electrical distributions on circular cylinders, *Proc. Am. Phil. Soc.*, **75**, 11 (1935).

Adams, E. P., Split cylindrical condenser, *Proc. Am. Soc.*, **76**, 251 (1936).

Fry, T. C., Two problems in potential theory, *Am. Math. Mon.*, **39**, 199 (1932).

Hodgkinson, J., A note on a two-dimensional problem in electrostatics, *Q. J. Math.*, Oxford Series **9**, 5 (1938)

Jayawant, B. V., Flux distribution in a permeable sheet with a hole near an edge, *Proc. Instn. Elect. Engrs.*, **107**(C), 238 (1960).

Richmond, W. H., Notes on the use of the Schwarz–Christoffel transformation in electrostatics (and hydrodynamics), *Proc. Lond. Math. Soc.*, **22**, 483 (1923).

Snow, C., Electric field of a charged wire and a slotted cylindrical conductor, *Bureau of Standards*, Sci. Papers **542**, 631 (1926).

Swann, S. A., Effect of rotor eccentricity on the magnetic field in the air gap of a non-salient pole machine, *Proc. Instn. Elect. Engrs.*, **110**, 903 (1963).

Wrinch, D., On the electric capacity of certain solids or revolution, *Phil. Mag.*, **50**, 60 (1925).

POLYGONAL BOUNDARIES

6.1 INTRODUCTION

In practice many fields are encountered with boundaries which are or can be treated as being made up of straight-line segments—for instance, the field of a plate condenser or microwave strip line or the field in the slotted air gap of a machine. Boundaries of this type are referred to here as polygonal though, as will emerge the vertices of the polygons to be considered are not infrequently at infinity in the complex plane in which they are located, see Fig. 6.1.

This chapter describes the routine methods which are available for the derivation of equations transforming a polygonal boundary into either an infinite straight line or the perimeter of a circle. These methods involve the determination, merely by inspection of the boundaries to be related, of a differential equation which is integrated to yield the transformation equation. This differential equation can be written down for *any shape of polygon*, but the integration of it varies from being very simple to being impossible by analytical means. The various types of integral which occur and the ways in which they are dependent upon the form of the boundary are discussed at the end of this chapter. However, so that attention is concentrated on the function of the transformation and not on the evaluation of integrals, the examples considered involve simple functions only. Examples of different boundary types are classified according to the number of vertex angles which define the boundary, see section 6.2.1.

Also included in this chapter are examples of the use of transformation methods in the calculation of flux densities (e.g. sections 6.2.3 and 6.2.6), capacitance (section 6.2.3), and force on a magnetized surface (section 6.2.8). The calculation of inductance (or capacitance) of a conductor of small, circular cross-section is obtained from the field solution exactly as was demonstrated in section 2.2.4 (but see also Reference [14]).

6.2 TRANSFORMATION OF THE UPPER HALF PLANE INTO THE INTERIOR OF A POLYGON

6.2.1 The transformation

The transformation equation which connects the real axis in one plane with the boundary of a polygon in another plane, in such a way that the upper half of the first plane trans-

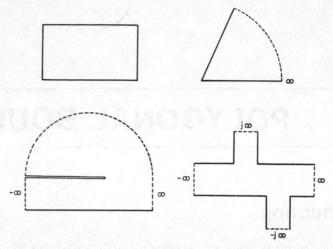

Fig. 6.1

forms into the *interior* of the polygon, was first given, independently, by Schwarz [1] and Christoffel [2].

Consider the two planes, shown in Fig. 6.2, in which the regions to be transformed lie on the unshaded sides of the boundary lines and corresponding points are similarly lettered. Then the transformation from the real axis of the *t*-plane to the polygon boundary in the *z*-plane is obtained by integrating the equation

$$\frac{dz}{dt} = S(t-a)^{(\alpha/\pi)-1}(t-b)^{(\beta/\pi)-1}(t-c)^{(\gamma/\pi)-1}(t-d)^{(\delta/\pi)-1}\cdots \qquad (6.1)$$

This is the Schwarz–Christoffel differential equation, and in it S is a constant of scale and rotation, $a, b, c, d, \ldots$, are points on the real axis of the *t*-plane corresponding to the vertices of the polygon in the *z*-plane, and $\alpha, \beta, \lambda, \delta, \ldots$, are the interior angles of the

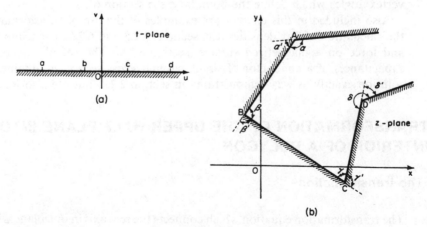

Fig. 6.2

polygon at $A, B, C, D, \ldots$, respectively. The number of products in eqn (6.1) is equal to the number of vertices required to define the polygonal boundary.

It is most simply demonstrated that the Schwarz–Christoffel differential equation introduces a succession of vertices in the z-plane boundary, as described, by expressing eqn (6.1) in terms of the angles $\alpha', \beta', \gamma', \delta', \ldots$, where each primed variable is equal to π minus the respective unprimed one and represents the *change in direction* of the z-plane boundary at the related vertex, Fig. 6.2(b). Equation (6.1) then takes the simpler form[†]

$$\frac{\mathrm{d}z}{\mathrm{d}t} = S(t-a)^{-(\alpha'/\pi)}(t-b)^{-\beta'/\pi}(t-c)^{-(\gamma'/\pi)}(t-d)^{-\delta'/\pi}\cdots,$$

and, like eqn (6.1), describes the length and inclination of a small element $\mathrm{d}z$ in terms of the length and inclination of the corresponding element $\mathrm{d}t$. Further, since the direction of all the elements $\mathrm{d}t$ constituting the real axis in the t-plane is constant, the argument of this equation defines directly, for continuously varying real values of t between $-\infty$ and ∞, the inclination of elements $\mathrm{d}z$ tracing out the boundary of the polygon. Now it is seen that

$$\arg\left(\frac{\mathrm{d}z}{\mathrm{d}t}\right) = \arg S - \frac{\alpha'}{\pi}\arg(t-a) - \frac{\beta'}{\pi}\arg(t-b) - \frac{\lambda'}{\pi}\arg(t-c) - \frac{\delta'}{\pi}\arg(t-d) - \cdots,$$

and the right-hand side expression can be simply evaluated for real values of t. For all values of $t < a$, $\arg(\mathrm{d}z/\mathrm{d}t)$ remains constant because all the terms in brackets are real and negative; and, therefore, the corresponding points on the boundary in the z-plane trace out a straight line. However, as the point t passes through a, $(t-a)$ becomes positive, so that $\arg(t-a)$ changes by $-\pi$ and the argument of $\mathrm{d}z$ changes by α'. For values of t between a and b, $\arg(\mathrm{d}z/\mathrm{d}t)$, and the inclination of elements in the z-plane, remain constant at the new value, and so the corresponding portion of the z-plane boundary is a further straight line, inclined by α' radians to the first section. Again, when the point t passes through b, the term $(t-b)$ is the only one to change sign and, therefore, the slope in the z-plane changes by β' radians on passing through the point corresponding with $t = b$, thereafter remaining constant until the next vertex is reached. Similarly, as t increases further and passes through the points $c, d, \ldots$, the direction of the z-plane boundary changes by $\gamma', \delta', \ldots$, respectively, until the whole polygon is traced out.

In this way, *all* the angles of the polygon are completely defined by those at vertices corresponding with finite points in the t-plane; if the point $t = \pm \infty$ corresponds with a vertex, the angle at that vertex is fixed by the remaining angles since the sum of the interior angles of a polygon is $\pi(N-2)$, where N is the number of vertices. Thus the number of factors in eqn (6.1) is $(N-1)$ when the point $t = \infty$ corresponds with a vertex, and is N when $t = \infty$ corresponds with a finite point on the boundary of the polygon.

A wise choice of corresponding points can often considerably simplify the form of solution derived from eqn (6.1). The various possibilities for a simple boundary shape are discussed in section 6.2.3, but some general guidance is given here. For polygons which are infinite, the simplest form of solution is usually obtained by choosing the limits of the real axis to correspond with the limits of a pair of adjacent sides of the

---

[†]This form of the equation is not normally used in practice beause it is necessary to deterine the appropriate signs to be attributed to the angles. This is avoided with eqn (6.1).

polygon which go to infinity. For polygons which are finite and also symmetrical, the point $t = \infty$ is usually chosen to lie where the line of symmetry cuts the boundary. In cases for which this line passes through a vertex and the centre of one side, the point is chosen to be at the vertex.

So far consideration of eqn (6.1) has been restricted to its argument. However, it is clear that the modulus of this equation determines the length of an element dz in terms of the length of the corresponding element dt, and thus, in transforming the real axis of the t-plane into a polygon with given dimensions, it is necessary to choose the constants $a, b, c, \ldots$, to give the desired vertex positions in the z-plane. The ways in which this can be accomplished are demonstrated in the examples, and the general problem of scale relationship between planes is discussed in section 6.2.5.

6.2.2 Polygons with two vertices

Quadrant bounded by the real and imaginary axes

As a first example of the use of the Schwarz–Christoffel differential equation, consider the transformation of the real axis of the t-plane, Fig. 6.3(a), into the polygon consisting of two semi-infinite, straight lines meeting in a right-angle at the point $z = 0$, Fig. 6.3(b), so that the region of the upper half of the t-plane becomes the region of the first quadrant in the z-plane. Let the points $t = -\infty$, $z = j\infty$, and $t = \infty$, $z = \infty$ correspond, and let the point $t = a$ correspond with the vertex at $z = 0$. The interior angle of the polygon at $z = 0$ is $\pi/2$, and so the Schwarz–Christoffel equation (6.1), gives

$$\frac{dz}{dt} = S(t - a)^{-1/2}. \tag{6.2}$$

This, when integrated, yields

$$z = S'(t - a)^{1/2} + k. \tag{6.3}$$

Now the vertex in the z-plane may be made to correspond with any real value of t since this is equivalent merely to shifting the t-plane origin (see section 5.2.1), and so, taking the simplest case, $a = 0$, eqn (6.3) reduces to

$$z = S't^{1/2} + k.$$

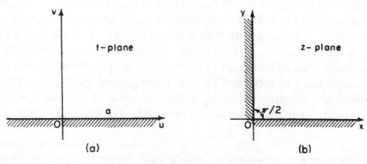

(a)   (b)

Fig. 6.3

The constants S' and k depend upon the points which, in the two planes, are made to correspond. The problem has been defined in such a way that the points $z = 0$ and $t = a$ correspond and, therefore, substitution of these values in eqn (6.3) gives $k = 0$. A more general view in connection with the evaluation of k is that by leaving the origin in one plane free to take up any necessary value, the constant k can always be made equal to zero. The scale constant S' can be given any convenient value but, by defining a further pair of corresponding points in the two planes, the scale relationship is fixed. If, for example, the points $z = 2$ and $t = 4$ are made to correspond, then $S' = 1$, and the transformation equation reduces finally to the form

$$z = t^{1/2},$$

which was used earlier (see section 5.1.2) to describe the field of flow near a perfectly conducting corner. In practice the choice of corresponding points in the two planes is not arbitrary but depends upon the nature of the field to be solved. This will become clear in the examples which follow.

Polygon with two parallel sides

As a second simple example consider the transformation of the real axis of the t-plane, Fig. 6.4(a), into the polygon formed by two infinite, parallel lines, Fig. 6.4(b). For a reason which will become evident, let the polygon be placed in the w-plane and let the sides be a distance $j\psi_0$ apart. Let the points $t = -\infty$ and $t = \infty$ be chosen to correspond with the points $w = \infty + j\psi_0$ and $w = \infty$ respectively. Further, let the point $t = 0$ correspond with the vertex of the polygon at $w = -\infty$ where there is an interior angle of 0. Then the Schwarz–Christoffel equation gives

$$\frac{dw}{dt} = S(t - 0)^{-1}, \tag{6.4}$$

which, when integrated, is

$$w = S \log t + k. \tag{6.5}$$

The points $w = -\infty$ and $t = 0$ have been made to correspond, so $k = 0$ and

$$w = S \log t, \tag{6.6}$$

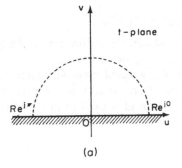
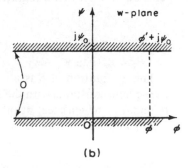

(a)   (b)

Fig. 6.4

which will be recognized as the important equation describing, in terms of the complex potential w, the field of a line current or the field between two semi-infinite equipotential lines meeting at the origin of the t-plane (see section 5.1.3). [Note that the points $t = -\infty$ and ∞ cannot be chosen to correspond with the points $w = -\infty$ and $-\infty + j\psi_0$. This requires a vertex at $w = \infty$ to correspond with a finite value of t, which is not possible for eqn (6.6).]

The constant S may be evaluated, as in the previous example, by the substitution of corresponding values of t and w in the transformation eqn (6.6). To determine corresponding points between the planes, it is helpful to recall that a vertical flux line in the w-plane transforms into a semi-circular one in the t-plane.[†] For, considering a flux line with radius R in the t-plane, it is seen that its end points, Re^{jo} and $Re^{j\pi}$, correspond with points $w = \phi'$ and $w = \phi' + j\psi_0$ respectively where ϕ' is the flux function at the radius R. Substituting these points in eqn (6.6) gives

$$\phi = S \log Re^{j\pi} = S \log R$$

and

$$\phi' + j\psi_0 = S \log R\, e^{j\pi} = S \log R + S j\pi.$$

Subtracting these equations leads to

$$j\psi_0 = S j\pi \quad \text{or} \quad S = \frac{\psi_0}{\pi},$$

and the equation of transformation can be written

$$w = \frac{\psi_0}{\pi} \log t.$$

A method equivalent to, but neater than, the above, can be used to evaluate the constant S. Let $t = Re^{j\theta}$, then, by differentiation,

$$dt = jRe^{j\theta}\, d\theta,$$

and eqn (6.4) can be rewritten

$$dw = \frac{SjRe^{j\theta}\, d\theta}{Re^{j\theta}} = Sj\, d\theta.$$

This equation when integrated yields

$$[w]_{w_1}^{w_2} = j\,S\,[d\theta]_{\theta_1}^{\theta_2}, \tag{6.7}$$

where w_1, $R\,e^{j\theta_1}$ and w_2, $R\,e^{j\theta_2}$ are pairs of corresponding points. In particular, if $\theta_1 = 0$ and $\theta_2 = \pi$, then $w_2 - w_1 = j\psi_0$ and eqn (6.7) yields $S = \psi_0/\pi$.

This evaluation of S is an example of a general method which can be usd to relate the t-plane constants with the dimensions of a polygon whenever the polygon has parallel boundaries which meet at infinity. The method is used in many of the examples which follow and is discussed in section 6.2.5.

[†]The shape of the flux lines is, of course, a consequence, not a cause, of the correspondence of points between the planes.

6.2.3 Parallel plate capacitor: Rogowski electrode

As an example of the transformation for a polygon with three vertices, the effect of flux fringing on the capacitance of a parallel plate capacitor, Fig. 6.5(a), is considered. The plates of the capacitor, a distance $2d$ apart, are assumed, for simplicity, to be of negligible thickness; they are also assumed to be charged to potentials of ψ_1 and $-\psi_1$ respectively. The field is symmetrical and the boundaries of one half of it can be represented in the z-plane as shown in Fig. 6.5(b), where the real axis corresponds with the line of symmetry.

The Schwarz–Christoffel equation is used to transform the real axis of the t-plane, Fig. 6.5(c), into the boundary of the z-plane. The upper surface of the plate is represented by the portion of the real axis of the t-plane between $-\infty$ and a, and the lower surface of the plate by the portion between a and b. Corresponding points in the two planes are then:

$$
\begin{aligned}
t &\to -\infty, & z &\to -\infty + jd; \\
t &= a, & z &= 0 + jd; \\
t &= b, & z &= -\infty \\
t &\to +\infty, & z &\to +\infty + j0;
\end{aligned}
$$

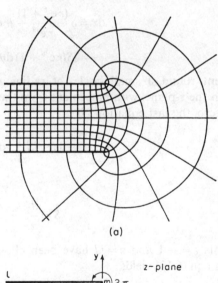

(a)

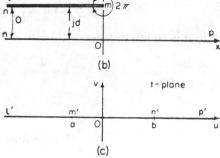

(b)

(c)

Fig. 6.5

and, since the interior angles of the polygon are 0 at $z = -\infty$ and 2π at $z = 0 + jd$, eqn (6.1) gives

$$\frac{\mathrm{d}z}{\mathrm{d}t} = S(t - a)(t - b)^{-1}. \tag{6.8}$$

The constants S, a, and b must be chosen so that the transformation equation, derived by integration of eqn (6.8), gives the required shape and size of the z-plane boundary. However, the *shape* of the z-plane configuration is independent of the single dimension d, so that any two of these constants may be given convenient values, the third being determined to give the required value for d. By taking $a = -1$ and $b = 0$, eqn (6.8) is simplified to the form

$$\frac{\mathrm{d}z}{\mathrm{d}t} = \frac{S(t + 1)}{t}, \tag{6.9}$$

which gives, when integrated,

$$z = S(t + \log t) + k. \tag{6.10}$$

To evaluate the constants S and k, consider first the expression of eqn (6.9) in polar coordinates

$$\mathrm{d}z = S \frac{(r\,\mathrm{e}^{j\theta} + 1)}{r\,\mathrm{e}^{j\theta}} jr\,\mathrm{e}^{j\theta}\,\mathrm{d}\theta$$

$$= jS(r\,\mathrm{e}^{j\theta} + 1)\,\mathrm{d}\theta.$$

Now movement round a small circle, of radius r, and centre at the point $t = 0$, corresponds, in the z-plane, with movement from the real axis to the line $z = jd$ and so, as r tends to zero, the last equation gives

$$\int_{-\infty}^{-\infty + jd} \mathrm{d}z = jS \int_{0}^{\pi} \mathrm{d}\theta \tag{6.11}$$

or

$$S = \frac{d}{\pi}.$$

Since the points $t = -1$ and $z = jd$ have been chosen to correspond, substitution of these values in eqn (6.10) yields

$$jd = S(-1 + j\pi) + k.$$

Putting $S = d/\pi$ in this equation gives

$$k = \frac{d}{\pi}.$$

and

$$z = \frac{d}{\pi}(1 + t + \log t) \tag{6.12}$$

as the equation of transformation.

The boundaries of the z-plane have a potential difference ψ_1 of half that between the capacitor plates. Hence the required field in the t-plane is expressed, from eqn (5.23), by

$$w = \frac{\psi_1}{\pi} \log t, \tag{6.13}$$

and the solution for the z-plane field is, eliminating t between eqns (6.12) and (6.13),

$$z = \frac{d}{\pi}\left(1 + e^{w\pi/\psi_1} + \frac{w\pi}{\psi_1}\right). \tag{6.14}$$

This gives as corresponding values: $z = -\infty$, $w = -\infty$ and $z = jd$, $w = j\psi_1$.

Because the constants a and b have been chosen as -1 and 0 rather than, for instance, 0 and 1, the potential division in the t-plane is at the point $t = 0$ and the solution for the field in the t-plane has its simplest form, eqn (6.13). Also, as pointed out in section (6.2.2), the constant k may be given the value zero and the transformation eqn (6.12) is then

$$z = \frac{d}{\pi}(t + \log t), \tag{6.15}$$

the origin in the z-plane being displaced through a distance d/π to the right of its first position.

Capacitance

The capacitance of a parallel plate capacitor is often calculated on the assumption that the flux density everywhere between the plates has the value that would exist if the plates were infinite. However, the value so derived is low, because the charge density on the inner faces increases towards the edges of the plates, and also because the charge on the outer faces of the plate is ignored. At any point on the surface of the plates the charge density, ϱ is equal to the flux density there; that is,

$$\varrho = \left(\frac{\partial \psi}{\partial x}\right)\varepsilon_0$$

$$= \left|\frac{\partial w}{\partial z}\right|\varepsilon_0, \tag{6.16}$$

where ε is the relative permittivity of the medium surrounding the plates. From eqn (6.14) by differentiation

$$\left|\frac{dw}{dz}\right| = \frac{\psi_1}{d}\frac{1}{|e^{w\pi/\psi_1} + 1|},$$

and, hence, substitution in eqn (6.16) gives the charge density at any point as

$$\varrho = \frac{\varrho_0}{|e^{w\pi/\psi_1} + 1|}, \tag{6.17}$$

where $\varrho_0 = \psi_1\varepsilon\varepsilon_0/d$ is the density of charge calculated on the assumption that the plates are infinite. Equation (6.17) shows that the charge density increases towards, and becomes infinite at, the end of the plate ($z = jd$, $t = -1$, $w = j\psi_1$).

The additional charge on the inner surface of the plates, due to the fringing, is the difference between the charge actually present and that calculated assuming the field uniform everywhere. It is equal to

$$\int_{-\infty + jd}^{jd} (\varrho - \varrho_0) \, dz,$$

which can be expressed in terms of w as

$$\int_{-\infty}^{j\psi_1} \varrho_0 \left(\frac{1}{1 + e^{w\pi/\psi_1}} - 1 \right) \frac{d}{\psi_1} (e^{w\pi/\psi_1} + 1) \, dw,$$

and simplified and integrated, to yield

$$\frac{\varrho_0 d}{\pi}.$$

Thus this additional charge is equivalent to a lengthening of the uniform field by the amount d/π.

The charge on the outer surface of the plates is

$$\int_{jd}^{-\infty} \varrho \, dz = \frac{\varrho_0 d}{\psi_1} \int_{j\psi_1}^{-\infty} dw.$$

$$= \infty.$$

However, for a capacitor of finite width, it is possible to estimate the additional charge fairly closely by assuming the charge distribution near both edges to be identical with that near the edge of the infinite plate. This is a reasonable assumption since the density of charge decreases very rapidly away from an edge—at a distance from the edge equal to $1.5d$ the charge density is less than $\varrho_0/10$—and the resulting value for the charge on the outer surface is calculable as

$$2 \int_{jd}^{-l + jd} \varrho \, dz,$$

where $2l$ is the width of the plate.

Rogowski electrode

The solution for the field of the boundary shown in Fig. 6.5(b) was used in an interesting and valuable way by Rogowski [3] to establish electrode shapes suitable for the measurement of the breakdown strength of gases and liquids. For these measurements it is important that breakdown always occurs in the uniform portion of the field. This may be ensured by choosing an electrode with the shape of an equipotential line along which the gradient is nowhere greater than ψ_1/d. For other uses of equipotential lines in representing boundary shapes, see sections 3.3.3 and 7.3.

6.2.4 The choice of corresponding points

For a given polygon, the choice of t-plane constants can be made in many ways, all of which lead, with varying degrees of difficulty, to the solution. To demonstrate the main considerations involved in making a choice, the various possibilities are examined for the following simple problem.

Current between two infinite, parallel permeable surfaces

A general form of this problem is treated earlier in the book by the method of images (see section 2.2.2), and it is sufficient for the present purpose to consider the particular case with the current midway between the surfaces. Since the field is symmetrical about the line through the current normal to the parallel surfaces, it is necessary to consider only the field to one side of this line. (Equally, since this field region is also symmetrical, it would be sufficient to consider one half of it.) The boundary of the field is shown in Fig. 6.6, where the current acts at the point $z = 0$, the finite side of the polygon corresponds to the line of symmetry normal to the surfaces, and the semi-infinite lines, a distance $2d$ apart, correspond with the permeable surfaces. There are three vertices, at $l\,(z = j\infty)$, $m\,(z = -d)$, and $n\,(z = d)$, and the angles are respectively 0, $\pi/2$, and $\pi/2$.

It is possible to choose the constants on the real axis of the t-plane in three reasonable ways and, because the polygon is defined by a single dimension, $2d$, two (or more when there is symmetry) of each set of constants can be given convenient values. The choices of pairs of corresponding points are as follows:

Choice 1. The corresponding points are:

$$
\begin{aligned}
z &= -d + j\infty, & t &= -\infty; \\
z &= d + j\infty, & t &= \infty; \\
z &= -d, & t &= -1; \\
z &= d, & t &= 1.
\end{aligned}
$$

Equation (6.1) gives

$$\frac{dz}{dt} = \frac{S}{\sqrt{t^2 - 1}},$$

which, when integrated, yields the transformation equation

$$z = S\cosh^{-1} t + k. \tag{6.18}$$

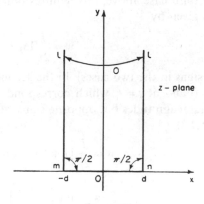

Fig. 6.6

Choice 2. The corresponding points are:

$$z = -d, \qquad t = \pm\infty;$$
$$z = \pm d + j\infty, \qquad z = 0;$$
$$z = d, \qquad t = 1.$$

Equation (6.1) gives

$$\frac{dz}{dt} = \frac{S}{t\sqrt{t-1}},$$

which yields the transformation equation

$$z = 2S\tan^{-1}\sqrt{t-1} + k. \tag{6.19}$$

Choice 3. The corresponding points are:

$$z = 0, \qquad t = \pm\infty;$$
$$z = \pm d + j\infty, \qquad t = 0;$$
$$z = -d, \qquad t = -1;$$
$$z = d, \qquad t = +1.$$

Equations (6.1) gives

$$\frac{dz}{dt} = \frac{S}{t\sqrt{t^2-1}}.$$

which yields the transformation equation

$$z = S\log\frac{\sqrt{t^2-1}-1}{t} + k. \tag{6.20}$$

Consider now the field in the z-plane and the forms of t-plane fields appropriate to each of the above transformation equations. In the z-plane, the current I sets up a potential difference of $I/2$ between the two halves of the boundary meeting at the point, $z = 0$, and so each of the t-planes must have the same potential difference, $I/2$, between the sections of the real axis corresponding with the two halves of the z-plane boundary. In the first and third case above, this requires potential divisions at $t = 0$ and $t = \pm\infty$, and the field is given by

$$w = \frac{I}{2\pi}\log t \tag{6.21}$$

(with opposite signs in the two cases). In the second case above, potential divisions are required at $t = 0$ and at $t = t_c$ which corresponds with $z = 0$: the field is that due to currents, of equal magnitudes but opposite signs, at the points 0 and t_c, and the solution is (see section 5.1.3)

$$w = \frac{I}{2\pi}[\log t - \log(t - t_c)]$$

$$= \frac{I}{2\pi}\log\left(\frac{t}{t - t_c}\right). \tag{6.22}$$

A comparison of the three cases shows, firstly that of the three transformation equations, eqns (6.18) and (6.19) are (marginally) simpler than (6.20) and, secondly that, of the two field equations, eqn (6.21) is simpler. Therefore the solution of this problem is best obtained using the first choice of constants.

Substitution for the pairs of corresponding points, $z = -d$, $t = -1$, and $z = d$, $t = 1$ in eqn (6.18) gives $k = d$ and $S = 2jd/\pi$. Then, eliminating t between eqns (6.18) and (6.22), the solution for the field in the z-plane is

$$w = \frac{I}{2\pi} \log \left[\cosh \frac{(z-d)\pi}{2jd} \right]. \tag{6.23}$$

[From a consideration of the image solution to this problem it is apparent that eqn (6.23) describes also the field due to an infinite array of equally spaced line currents.]

In the above example there is little variation in difficulty associated with the possible choices of constants, but, for more complicated problems, a wise choice can be very important. As has been seen, the two features to be considered are the difficulty of integrating and using the transformation equation, and the determination and complexity of the equation of the t-plane field. In practice, the latter is usually the more important and first consideration should be given to it.

6.2.5 Scale relationship between planes

General considerations

Use of the Schwarz–Christoffel equation automatically constructs a polygon with the required *angles*. The *dimensions* of a polygon are obtained by a suitable choice of the constants in the transformation equation. For a consideration of scale, the positions of the origins are of no concern, and so the equation can be written

$$z = Sf(t, a, b, c \cdots), \tag{6.24}$$

where f is a function not only of the complex variable t but also of the real constants $a, b, c \cdots$.

It is shown in the previous section that when only two constants a and b (in addition to the scale constant S) occur in eqn (6.24), both can be given arbitrary values, and it is generally true that, for any number of constants $a, b, c, \ldots$, two of them can be given convenient values. This is so because, by fixing the values of two constants, a distance between the corresponding points in the z-plane is defined, and this can be given the correct value by a suitable choice of S. The remaining constants in the transformation equation are then defined by the other dimensions of the polygon. It should be noted that the number of constants to be determined is equal to the number of *ratios* of the dimensions which define the proportions of the polygon: the polygons of the above examples are defined by no more than a single *dimension* and so no constants, apart from S, require evaluation; the polygon of the succeeding section is defined by two dimensions, or one ratio, and requires the evaluation of one constant in addition to S. The two 'free' constants are chosen to simplify the form of the transformation eqn (6.24) or the solution of the field in the t-plane, and the values most commonly used are 0 and ± 1.

Evaluation of constants

Two methods are available for the evaluation of the constants which, in eqns (6.1) and (6.24), define the proportions of the polygon. The first method merely involves substitution in the transformation equation, (6.24), for pairs of corresponding values of z and t; each substitution yields one equation connecting the constants with the polygon dimensions. In many cases the equation gives the value of a constant directly [see eqn (6.18)], but in others a graphical or numerical method may be necessary [see eqn (6.43)] and, in the most difficult, it is necessary to invoke an iterative numerical scheme (see section 7.6).

The second method, *the method of residues*,[†] can be applied when the polygon has parallel sides meeting at infinity and, using it, one relationship can be obtained for each such vertex. (Forms of it have already been used in the example of the field of a line current and the example of the parallel plate condenser.) At these vertices a finite change in z occurs due to an infinitely small change in t through the point corresponding with the vertex (there are poles of dz/dt at these points in the t-plane) and the related changes can be equated by integrating eqn (6.1) between the appropriate limits.

Consider a vertex of angle 0 at the point corresponding to $t = n$, giving a term $(t - n)^{-1}$ in eqn (6.1), and let

$$t - n = R e^{j\theta}.$$

Differenting this equation gives

$$dt = jR e^{j\theta} d\theta,$$

and, substituting for t and dt, eqn (6.1) becomes

$$dz = Sf(R e^{j\theta}, n, a, b, c, \ldots) d\theta. \tag{6.25}$$

Now if R is infinitely small, a change in θ from 0 to 2π causes t to pass through the value n and this change corresponds with one in z equal to the distance, D, between the parallel lines. Further, when $R \to 0$, $t \to n$ and eqn (6.25) takes a simple form for, in it, θ only occurs due to the terms dt and $(t - n)$ and the factors other than $(t - n)$ remain constant as θ changes from 0 to π. The form is

$$dz = \frac{S}{R e^{j\theta}} (n - a)^{(\alpha/\pi) - 1} (n - b)^{(\beta/\pi) - 1} \ldots jR e^{j\theta} d\theta,$$

and this, when simplified and integrated between $\theta = 0$ and $\theta = \pi$, gives

$$D = j\pi S(n - a)^{(\alpha/\pi) - 1} (n - b)^{(\beta/\pi) - 1} \ldots, \tag{6.26}$$

in which the factor corresponding to $(t - n)$ does not appear.

When the points $t = \pm \infty$ correspond with the meeting point at infinity of two parallel sides of the polygon containing the angle 0, eqn (6.25) can again be used, this time to relate the distance between the parallel lines directly with the scale constant. Then the term $(t - n)$ does not appear in eqn (6.24) which, with the substitution $t = R e^{j\theta}$, becomes

$$dz = S(R e^{j\theta} - a)^{(\alpha/\pi) - 1} (R e^{j\theta} - b)^{(\beta/\pi) - 1} \ldots jR e^{j\theta} d\theta.$$

[†]See footnote on next page.

As $R \to \infty$ the constants $a, b, \ldots$, become negligible, and the equation may be rewritten

$$dz = jSR\,e^{j\theta[\Sigma\alpha/\pi - (N-1) + 1]}\,d\theta$$

since there are $(N-1)$ terms in the equation, where N is the number of vertices. However, the sum of the interior angles of the polygon is $\pi(N-2)$ and, since the interior angle at vertex n is zero, this is equal to $\Sigma\alpha/\pi$. Thus,

$$dz = jS\,d\theta,$$

which, when integrated, gives

$$S = \frac{D}{j\pi}, \qquad (6.27)$$

where D is the distance between the parallel sides. The fact that when $R \to \infty$ eqn (6.25) reduces to a form independent of the constants $a, b, c, \ldots$, is to be expected since a circle of infinite radius corresponds, regardless of its centre, with the same path in the z-plane. Both eqns (6.26) and (6.27), which relate the constants of the transformation equation with the dimensions of the polygon, are of considerable importance and they are used frequently.

The method just discussed of writing eqn (6.25) in polar form and determining the value of the integral as $R \to 0$ or ∞ is equivalent to the evaluation of the integral by the method of residues.[†] The discussion so far has been restricted to an expression of the distance between parallel lines meeting at a vertex of angle of *zero*. Whilst this case is by far the most important one, it must be noted that parallel lines can also meet at an angle of $-\pi$ (this is explained in section 6.2.7) when eqn (6.1) contains a term $(t-n)^{-2}$. This gives a pole of order 2 at n, so that the above method of integration fails.

The distance between parallel lines is, however, always given by

$$D = j\pi \times \text{residue at pole}, \qquad (6.28)$$

and for a pole of order 2 the residue has to be found as the coefficient of $(t-n)^{-1}$ when the function $Sf(t, a, b, c, \ldots)$ of eqn (6.1) is expressed as a series, or as

$$\underset{t \to n}{\text{Lt}} \frac{d}{dt}[(t-n)^2 Sf(t, a, b, c, \ldots)]. \qquad (6.29)$$

6.2.6 The field of a current in a slot

Very many practical problems have been investigated using the boundary shape shown in Fig. 6.7(b): Jeans (p. 277), for example, described the calculation of the field at the corner of a Leyden jar; Carter [4] examined the field of a rectangular salient pole; and

---

[†]This is discussed fully in any book dealing with the theory of functions of a complex variable but the definitions of residue, pole and zero are given here. A pole of the function $f(z)$ (with finite principal part in the Laurent expansion) is a point at which $f(z)$ is infinite; if $f(z)$ contains a term $1/(z-a)^n$ the pole at $z = a$ is said to be of order n. The residue of $f(z)$ at the pole a is the coefficient of the term $1/(z-a)$ in the Laurent expansion of $f(z)$ about $z = a$. A zero of the function $f(z)$ is a point at which $f(z)$ is zero; if $f(z)$ contains a term $(z-b)^n$ it is said to have a zero of order n at $z = b$.

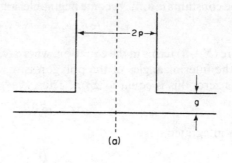

(a)

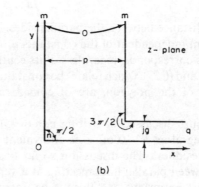

(b)

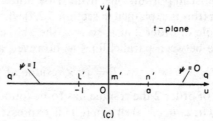

(c)

Fig. 6.7

Stein [5] investigated the flux distribution at the corner of a transformer core. In this section, the same boundary shape is used in the calculation of flux density and of permeance factors for the field of a current in a slot of an electrical machine. The same solutions apply, though with less satisfactory approximation, to the field of a salient pole [4]. (The solution, by conformal transformation, for the field of a salient pole with a rectangular tip is possible but difficult [6].)

Since the air gap in most electrical machines is small compared with the width of an open slot, it is sufficient in the analysis to consider a single slot, of width $2p$, separated by an air gap of length g from a plane surface, see Fig. 6.7(a). Also, so far as the field in the air gap is concerned, the slot can be treated as infinitely deep with the current at the 'bottom' of it. For a current $2I$ the potentials of the adjacent teeth are I and $-I$ with respect to the plane surface at zero potential, and the potential on the line of symmetry down the slot centre is also 0. Thus, one half of Fig. 6.7(a) can be represented, in the z-plane, as shown in Fig. 6.7(b) where the boundary mnq has a potential differing by I from that of the boundary mlq.

The polygon has four vertices at l, m, n, and q and, by choosing the limits of the real axis of the t-plane, Fig. 6.7(c), to correspond with q, it is defined by the angles $3\pi/2, 0$ and $\pi/2$ at l, m, and n respectively. Of the three finite points in the t-plane, l, m, and n, corresponding with these three vertices, two may be given convenient values but the third must be determined to give the required ratio of (g/p). Let the pairs of corresponding points be:

$$\text{at } l, \quad z = p + jg \quad \text{and} \quad t = -1;$$
$$\text{at } m, \quad z = j\infty \quad \text{and} \quad t = 0;$$
and
$$\text{at } n, \quad z = 0 \quad \text{and} \quad t = a.$$

Then the Schwarz–Christoffel equation giving the transformation between the planes is

$$\frac{dz}{dt} = S(t+1)^{1/2}t^{-1}(t-a)^{-1/2}, \tag{6.30}$$

and, making the substitution

$$u^2 = \frac{t-a}{t+1},$$

this can be simply integrated to yield

$$z = 2S\left[\frac{1}{\sqrt{a}}\tan^{-1}\frac{u}{\sqrt{a}} + \frac{1}{2}\log\left(\frac{1+u}{1-u}\right)\right] + k.$$

By choosing the origin in the z-plane to correspond with the point $t = a$ the constant k in the last equation is made zero, and the transformation equation reduces to

$$z = 2S\left[\frac{1}{\sqrt{a}}\tan^{-1}\frac{u}{\sqrt{a}} + \frac{1}{2}\log\left(\frac{1+u}{1-u}\right)\right]. \tag{6.31}$$

The constants S and a are best evaluated by the method of residues. At the vertex q the parallel sides are a distance jg apart and the point in the t-plane corresponding to this vertex is $t = \infty$. Therefore, eqn (6.27) gives

$$S = \frac{g}{\pi}. \tag{6.32}$$

At the vertex m the parallel lines are a distance p apart and the point in the t-plane corresponding to m is $t = 0$. Therefore eqn (6.26) gives

$$p = \frac{j\pi S}{\sqrt{-a}},$$

and, substituting for S from eqn (6.32), gives

$$a = \left(\frac{g}{p}\right)^2. \tag{6.33}$$

Then, putting the above values for S and a in eqn (6.31), the complete transformation equation becomes

$$z = \frac{2g}{\pi}\left[\frac{p}{g}\tan^{-1}\frac{pu}{g} + \frac{1}{2}\log\left(\frac{1+u}{1-u}\right)\right],$$ (6.34)

where

$$u^2 = \frac{t-(g/p)^2}{t+1}.$$ (6.35)

To obtain the required field in the z-plane it is necessary to have in the t-plane a potential I along the part of the real axis between $-\infty$ and 0, and a potential 0 along the part between 0 and ∞. This gives a field in the t-plane described by

$$w = \frac{I}{\pi}\log t,$$ (6.36)

and elimination of t and u between eqns (6.34)–(6.36) gives the solution for the field in the z-plane. The field map shown in Fig. 6.8 has been plotted from this solution, taking a value of 5 for the ratio p/g.

Flux density

The flux density at any point in the field is given by $|dw/dz|$. Equation (6.36) gives

$$\frac{dw}{dt} = \frac{I}{\pi}\frac{1}{t},$$

and this may be combined with eqns (6.30) and (6.32) to give

$$\frac{dw}{dz} = \frac{I}{g}\sqrt{\frac{t-a}{t+1}}.$$ (6.37)

It is not possible to evaluate dw/dz directly in terms of z. Instead, a curve of flux density variation with z is obtained by (a) substitution of values of t in eqn (6.37) to give flux density and (b) in eqns (6.34) and (6.35) to give the corresponding values of z. Taking

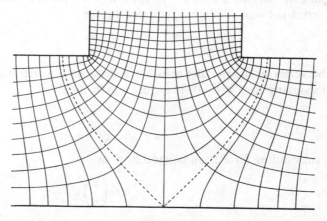

Fig. 6.8

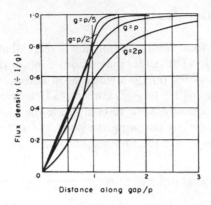

Fig. 6.9

a range of values of t between a and ∞ yields the curve for the distribution of flux density along the plane surface, shown in Fig. 6.9. This curve is of importance for the determination of the shape of the flux wave inducing losses in the surface of poles.

Leakage flux

The main and leakage fluxes may be considered to be separated by the fictitious flux line (shown dotted in Fig. 6.8) which leaves the tooth and just touches the plane surface at $z = 0$ in Fig. 6.7(a). In design calculations of their magnitudes it is convenient, as it is in many other problems, to make use of permeance coefficients. These are used to express a quantity of flux in a non-uniform part of the field in terms of an equivalent length of a uniform field; thus, in Fig. 6.7(b) the total main flux leaving the tooth can be calculated by assuming the uniform distribution of field to exist everywhere between q and l, and by representing the quantity of flux which fringes from the tooth side (Fig. 6.8) as a certain length of the uniform field.

The value of the 'fringe flux' is taken as the difference between the integral of the flux density along the plane surface from n to q and the flux which would pass between q and l, if the density everywhere there were uniform. Both of these quantities of flux are infinite, but their difference is finite and it can be found numerically by calculating the quantities as being bounded, not at q, but by a line at some convenient point sufficiently far down the gap for the density there to be uniform. In some problems, the difference in the flux densities can be expressed in such a form that the integration yields a simple analytical result. An example of this kind occurred in connection with the capacitor fringe field, section 6.2.3, and another is given in Reference [7]. This is not possible in the present example, but an approximate analytical result is obtained by assuming everywhere under the tooth, a uniform field which is bounded by a straight line—from the tooth corner, normal to the plane surface—and by calculating the fringe flux as that crossing the plane surface between $x = 0$ and $x = p$. The result in the form given by Carter, expresses the fringe flux as a constant, λ, times the air-gap length, where

$$\lambda = 0.72 \log\left[\frac{1}{4}\left(1 - \frac{p^2}{g^2}\right) + \frac{1}{90}\left(\frac{p}{g}\tan^{-1}\frac{g}{p}\right)\right],$$

and the logarithm is a common one and the angle is expressed in degrees.

6.2.7 Negative vertex angles

When two adjacent sides of a polygon diverge, they meet at infinity at a vertex having a negative angle. (The reader may readily confirm this by considering the sum of all the interior angles of the polygon which must add up to $\pi(N-2)$, N being the number of vertices.) Angles of this type give, in eqn (6.1), terms which have a power less than minus one and, consequently, tend to complicate the integration. Therefore, if possible, they should be avoided by choosing the point $t = \infty$ to correspond with the vertex having the negative angle (see the following section).

Opposite parallel plates

The polygonal boundary, shown in the z-plane, Fig. 6.10(a), consists of two thin, semi-infinite, parallel plates and it has two negative angles, one of which must be accounted for explicitly in forming the transformation equation. Apart from its being a good example for the discussion of negative angles and the evaluation of the residue at a second order pole [see eqn (6.29)], particular forms of this boundary can be used to derive solutions which are frequently transformed to obtain solutions for more complex problems.

The polygon has four vertices l, m, n, and h. Those at l and n have interior angles of

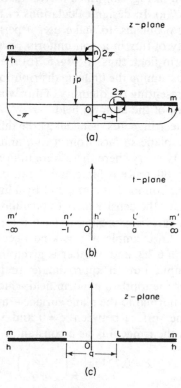

Fig. 6.10

2π, whilst those at m and h have angles of $-\pi$. The angles at m and h can be visualized by considering the movement of two sides at a vertex as the angle there is *decreased* through the value 0 until the lines become parallel again.

To transform the real axis of the t-plane, Fig. 6.10(b), into the polygon, let the point $t = \infty$ correspond with the vertex m, and let the remaining pairs of corresponding points be

$$\text{at } l, \quad z = q \quad \text{and} \quad t = a;$$
$$\text{at } h, \quad z = \infty \quad \text{and} \quad t = 0;$$
and
$$\text{at } n, \quad z = jp \quad \text{and} \quad t = -1.$$

Then the Schwarz–Christoffel equation is

$$\frac{dz}{dt} = S(t+1)t^{-2}(t-a), \tag{6.38}$$

and this integrates simply as

$$z = S\left(t + (1-a)\log t + \frac{a}{t}\right) + k. \tag{6.39}$$

The term t^{-2} in eqn (6.38) means that there is a second order pole of dz/dt at the point $t = 0$, and so eqn (6.28) must be used to express the distance, jp, between the parallel lines, with the constants in eqn (6.38). The residue of dz/dt at $t = 0$ is the coefficient of $1/t$ in the expansion of the right-hand side of eqn (6.38) and it is seen to be $S(1-a)$. [This result is obtained equally well using eqn (6.29)]. Hence, eqn (6.28) gives

$$jp = j\pi S(1-a). \tag{6.40}$$

Substitution for pairs of corresponding points in eqn (6.39) yields two additional equations which, together with eqn (6.40), are used to evaluate the constants S, a and k. Substitution for the vertex l gives

$$q = S\{(1+a) + (1-a)\log a\} + k, \tag{6.41}$$

and for the vertex n

$$jp = S\{-(1+a) + (1-a)j\pi\} + k. \tag{6.42}$$

Eliminating S and k between eqns (6.40)–(6.42) yields the equation

$$\pi\left(\frac{q}{p}\right) = \frac{2(1+a)}{(1-a)} + \log a, \tag{6.43}$$

which, however, must be solved graphically or by same suitable numerical method to obtain the value of a for a given value of (q/p). When a is known, eqns (6.41) and (6.42) are used to evaluate S and k.

Both plates on the real axis

When $p = 0$ the two plates are in line with each other and they can be represented on the real axis of the z-plane, symmetrically placed with respect to the origin, as shown in Fig. 6.10(c). Because of the symmetry, a has the value 1 and eqn (6.39) becomes

$$z = S\left(t + \frac{1}{t}\right) + k.$$

The values of the constants S and k are given directly by substitution for the pairs of corresponding points $z = -q/2$, $t = -1$, and $z = q/2$, $t = 1$ and are

$$S = q/4 \quad \text{and} \quad k = 0.$$

Hence, the transformation equation can be written

$$z = \frac{q}{4}\left(t + \frac{1}{t}\right). \tag{6.44}$$

The solution for the field due to a potential difference ψ_1 maintained between the two plates, is often used (see, for example, section 6.2.9) and is obtained by combining eqn (6.44) with the equation for the field in the t-plane,

$$w = \frac{\psi_1}{\pi} \log t.$$

It takes the most convenient form when $q = 2$ and, with this substitution, is

$$w = \frac{\psi_1}{\pi} \log(z + \sqrt{z^2 - 1})$$

$$= \frac{\psi_1}{\pi} \cosh^{-1} z. \tag{6.45}$$

An alternative derivation of eqn (6.45) is to use eqn (6.18) and to transform directly into the w-plane.

6.2.8 The forces between the armature and magnet of a contactor

A number of methods of determining the forces experienced by magnetized boundaries are developed in section 1.7 and two of them are discussed here in their application to a typical problem. Consider Fig. 6.11(a) which represents the armature and magnet of a contactor or the stator and rotor of a machine. A magnetic potential difference established between the two parts gives rise to two components of force on each: first, there is an attractive force and, secondly, there is a force tending to align the parts symmetrically with respect to one another. An exact treatment of the boundary shape requires the use of mathematics not discussed until the next section, but completely satisfactory results for many purposes can be achieved by consideration of the portion of the boundary shown in Fig. 6.11(b). This is possible because the field over most of the length of the narrow air gap, in both figures, is uniform and uninfluenced by effects at the ends of the elements, so that the boundary of Fig. 6.11(a) can be synthesized from two boundaries (each with an appropriate length of the parallel gap) of the type shown in Fig. 6.11(b).

Before the forces can be determined, the solution for the field must be obtained. Let the boundary of Fig. 6.11(b) be in the z-plane and let it be transformed into the real

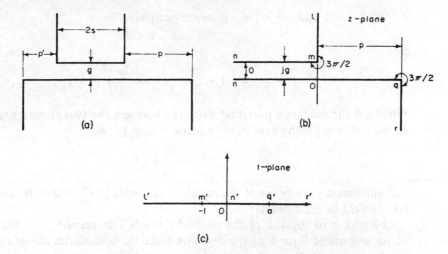

Fig. 6.11

axis of the t-plane, Fig. 6.11(c). Further, let the pairs of corresponding points in the two planes be:

$$z = j\infty, \qquad t = -\infty;$$
$$z = jg, \qquad t = -1;$$
$$z = -\infty, \qquad t = 0;$$
$$z = p, \qquad t = a;$$
$$z = p - j\infty, \quad t = \infty.$$

Then, since the vertex angles at the points jg, ∞, and p are, respectively, $3\pi/2, 0$, and $3\pi/2$, the Schwarz–Christoffel equation connecting the planes is

$$\frac{dz}{dt} = S \frac{\sqrt{(t+1)(t-a)}}{t}. \tag{6.46}$$

This equation, after multiplication of the numerator and the denominator of the right-hand side by $\sqrt{t-a}$, can be integrated to give

$$z = S\left[\frac{R(a+1)}{(R^2-1)} + (1-a)\tanh^{-1} R + j\sqrt{a}\log \frac{(R\sqrt{a}-j)}{(R\sqrt{a}+j)} \right] + k, \tag{6.47}$$

where

$$R = \sqrt{\frac{t+1}{t-a}}. \tag{6.48}$$

Application of eqn (6.26) to the point $t = 0$ yields

$$S = -\frac{jg}{\pi\sqrt{a}}; \tag{6.49}$$

and in eqn (6.47), using this value of S, substitution for the pairs of corresponding points

$z = jg, t = -1$ and $z = p, t = a$ gives, respectively,

$$k = 0$$

and

$$a = 1 + \frac{2p}{g} \pm \sqrt{\left(1 + \frac{2p}{g}\right)^2 - 1}. \tag{6.50}$$

Finally, if the magnetic potential difference between the two elements in the z-plane is ψ, the solution for the field in the t-plane is seen to be

$$w = \frac{\psi}{\pi} \log t, \tag{6.51}$$

and elimination of t between this equation and eqns (6.47) and (6.48) gives the solution for the field in the z-plane.

Consider now, by use of the method described in section 1.7.4, the determination of the 'alignment' force F on the elements. Since the boundaries are infinitely permeable, this force acts wholly on the vertical faces and is given by the integral, over the vertical surfaces, of the square of the flux density; that is, for Fig. 6.11(b),

$$F = \tfrac{1}{2}\mu_0 \int_{jg}^{j\infty} B_z^2 \, dz.$$

This equation may be written as

$$F = \frac{1}{2}\mu_0 \int_{jg}^{j\infty} \left(\left|\frac{dw}{dz}\right|\right)^2 dz$$

$$= \frac{1}{2}\mu_0 \int_{jg}^{j\infty} \left(\left|\frac{dw}{dt}\right|\right)^2 \left(\left|\frac{dt}{dz}\right|\right)^2 dz,$$

or, most conveniently, with a change of the variable of integration, as

$$F = \frac{1}{2}\mu_0 \int_{-1}^{-\infty} \left(\left|\frac{dw}{dt}\right|\right)^2 \left(\left|\frac{dt}{dz}\right|\right) dt. \tag{6.52}$$

Differentiation of eqn (6.51) gives

$$\frac{dw}{dt} = \frac{\psi}{\pi} \frac{1}{t},$$

and dt/dz is given by eqns (6.46) and (6.49) so that substituting for these derivatives in eqn (6.52) yields

$$F = \frac{\mu_0 \psi^2 \sqrt{a}}{2\pi g} \int_{-1}^{-\infty} \frac{dt}{t\sqrt{(t-a)(t+1)}}.$$

Integration of this equation leads to

$$F = \frac{\mu_0 \psi^2}{2\pi g} \left[\sin^{-1} \frac{t(1-a) - 2a}{t(1+a)} \right]_{-1}^{-\infty}$$

$$= \frac{\mu_0 \psi^2}{2\pi g} \left[\sin^{-1}\left(\frac{1-a}{1+a}\right) - \frac{\pi}{2} \right]. \tag{6.53}$$

This is the force on one side of the upper element in Fig. 6.11(a) so that the resultant force on this element is

$$\frac{\mu_0 \psi^2}{2\pi g} \left[\sin^{-1}\left(\frac{1-a}{1+a}\right) - \sin^{-1}\left(\frac{1-a'}{1+a'}\right) \right], \tag{6.54}$$

a' being given by eqn (6.50) in which p is replaced by p', the projection at the other side of the array.

As an alternative to the above method, the alignment force can be calculated by determining the rate of change, with respect to p (and p'), of the total flux entering the upper element—see section 1.7.3. The equations involved are such that this cannot be done analytically—it is not possible to form the equation $w = f(z)$—but it can be done numerically. In the computation, it is convenient to consider the actual boundary divided into two boundaries, of the type shown in Fig. 6.11(b), by the centre line of the upper element. The total flux Φ entering one half of the upper element is taken to be that crossing the boundary in the z-plane between $-s + jg$, where $2s$ is the width of the upper element, and some point jY, where Y is sufficiently large for the flux between jY and $j\infty$ to be negligible. The total flux is calculated for, and is plotted against, a range of values of p. Then, applying eqn (1.64), the difference between the slopes of the graph for the values p and p', multiplied by $\psi/2$, gives the resultant alignment force on the upper element; thus,

$$F = \frac{\psi}{2}\left(\frac{d\Phi}{dp} - \frac{d\Phi}{dp'}\right).$$

The forces of attraction can also be found using either of these methods. It is again convenient to consider the actual boundary divided into two portions by the centre line of the upper element. With the first method, the force on the upper element, for example, is given by

$$\frac{\mu_0}{2}\left\{ \int_{-s+jg}^{jg} B^2\,dz + \int_{-s+jg}^{jg} B'^2\,dz \right\},$$

where B is the flux density on the surface $z = jg$ for a projection p, and B' is the flux density on the same surface for a projection p'. Using the second method, the force is given by

$$\frac{\psi}{2}\left(\frac{d\Phi}{dg} + \frac{d\Phi'}{dg}\right),$$

where Φ and Φ' are the quantities of flux crossing the boundary, in the z-plane, between the points $-s + jg$ and jg, for projections p and p' respectively.

In general, of the two methods, the surface integral one is to be preferred. It is easier to apply and is the more accurate one when numerical techniques have to be used, for then, the rate of change of total flux method necessitates the determination of the slope of a curve.

6.2.9 A simple electrostatic lens

Many important problems involve the polygon (with five vertices) shown in Fig. 6.12(a): Dreyfus [8] has examined the electric field between the low and high voltage windings

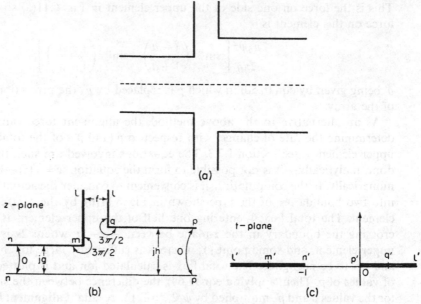

Fig. 6.12

of a transformer; Carter [9] and Kucera [10] have discussed the field between unequal, opposite slots; and Herzog [11] has analysed the field of an electrostatic lens. In a number of these problems the field is caused by a difference of potential maintained between the left- and right-hand sections of the polygon so that the line of symmetry is a flux line. It is shown now that the solution of such a field requires the transformation of the field between two semi-infinite plates (see section 6.2.7) and that only one choice of the t-plane constants is possible.

For a simple electrostatic lens with cylindrical symmetry the boundary of Fig. 6.12(a) represents a radial cross-section through the cathode and the anode. The anode usually consists of a thin plate (normal to the axis), but representing it as infinitely long modifies the inter-electrode field only slightly and makes the transformation a much easier one—all the constants may be evaluated by the method of residues. The field is symmetrical and to simplify the transformation it is convenient to consider one half of it as represented in the z-plane, Fig. 6.12(b); the anode with potential ψ_A is the section lmn, the cathode with potential 0 is the section pql, and np is the flux line along the axis of the lens.

To transform the z-plane field into a known field in the t-plane, the vertex l must be chosen to correspond with $t = \infty$. The t-plane field, Fig. 6.12(c), then has, on the real axis, two equipotential sections, $l'm'n'$, corresponding with lmn, and $p'q'l'$, corresponding with pql; that is, it is of the type discussed in section 6.2.7. Let the remaining pairs of corresponding points be:

$$m, \quad t = -a;$$
$$n, \quad t = -1;$$
$$p, \quad t = 0;$$
$$q, \quad t = b;$$

then the Schwarz–Christoffel equation gives

$$\frac{dz}{dt} = S \frac{\sqrt{(t+a)(t-b)}}{t(t+1)}.$$ (6.55)

The constants S, a, and b can all be evaluated simply by the method of residues; application of eqn (6.27) to the point $t = 0$ gives

$$S = -\frac{jf}{\pi};$$ (6.56)

and application of eqn (6.26) to the points 0 and -1 gives respectively

$$\frac{h}{f} = \sqrt{ab}$$ (6.57)

and

$$\frac{g}{f} = \sqrt{(b+1)(a-1)}.$$ (6.58)

Then, integrating eqn (6.55), and ignoring the position of the origin in the z-plane, the transformation equation becomes

$$z = -\frac{jf}{\pi}\left\{ \cosh^{-1}\left[\frac{2t+a-b}{a+b}\right] + \frac{h}{f}\cos^{-1}\left[\frac{(a-b)t-2ab}{(a+b)t}\right]\right.$$
$$\left. + \frac{g}{f}\cos^{-1}\left[\frac{2(a-1)(b+1)+(b-a+2)(t+1)}{(a+b)(t+1)}\right]\right\}.$$ (6.59)

The solution for the field in the t-plane is derived from eqn (6.45) by halving the scale in the t-plane (so that the plates are unit distance apart) and shifting the origin by 1. The result is

$$w = \frac{\psi_A}{\pi}\cosh^{-1}(2t+1),$$ (6.60)

and with eqn (6.59) this describes the field in the z-plane. The field strength in the z-plane, $|dw/dz|$, is evaluated from eqns (6.55)–(6.58) and from the expression for dw/dt, found by differentiating eqn (6.60). Its value along the axis of the lens leads to the determination of the forces acting on electrons emitted from the cathode.

6.3 TRANSFORMATION OF THE UPPER HALF PLANE INTO THE REGION EXTERIOR TO A POLYGON

6.3.1 The transformation

It is seen from the preceding section that the Schwarz–Christoffel eqn (6.1), can be used to obtain solutions for fields in regions 'exterior' to certain physical boundaries. However, these exterior regions must be assumed to have boundaries extending to infinity and, in fact, they correspond, mathematically, with the interiors of polygons closed at infinity. When it is necessary to analyse fields exterior to *finite*, polygonal boundaries a different transformation equation, now to be developed, is required.

This equation transforms the real axis of one plane, into the boundary of a polygon in another plane, in such a way that the upper half of the first plane becomes the region exterior to the polygonal boundary in the second plane. Consider a polygon in the z-plane, Fig. 6.13(a), which is to be transformed from the real axis of the t-plane, Fig. 6.13(b). The changes in direction at the vertices of the polygon are introduced by a function with the form of the right-hand side of eqn (6.1) and, therefore, it can be assumed that the desired transformation equation has the form

$$\frac{\mathrm{d}z}{\mathrm{d}t} = Sf(t)[(t-a)^{(\alpha/\pi)-1}(t-b)^{(\beta/\pi)-1}\cdots], \tag{6.61}$$

where the function $f(t)$ is to be chosen to satisfy the additional conditions of this transformation, and $\alpha, \beta, \gamma, \ldots$, are the *exterior* angles of the polygon (but the interior angles of the field).

To determine the form of $f(t)$ consider first the boundaries in the two planes. Because, in eqn (6.61), the part in square brackets completely defines the changes in direction of the z-plane boundary, $f(t)$ must cause no change in argument and, therefore, must be real, non-zero and finite, for all real values of t. Secondly, some point in the t-plane, not on the real axis, must correspond with $z = \infty$. Any non-real point in the t-plane may be chosen, but the simplest form of $f(t)$ is obtained by taking it to be $t = j$ and then, since there must be a pole (see footnote to p. 163) at the point, $f(t)$ must be a

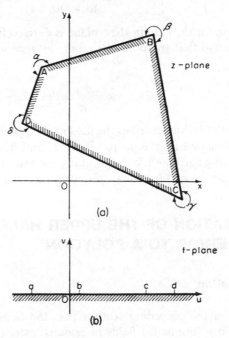

Fig. 6.13

function of $1/(t-j)$. But $f(t)$ must be real for all real values of t and this can be achieved only by introducing the product term $1/(t+j)^\dagger$ which makes

$$f(t) = g\left(\frac{1}{t^2+1}\right).$$

Again, because $t = \infty$ corresponds with a finite value of z, $f(t)$ must have a zero at $t = \infty$ and, since $\Sigma(\alpha/\pi - 1) = 2$ (from the sum of the exterior angles of the polygon), this zero must be of order three or more. Thus, it is necessary that

$$f(t) = h\left[\frac{1}{(t^2+1)^n}\right],$$

where n may take all integral values between 2 and ∞. In its most general form this may be expressed

$$f(t) = \frac{L}{(t^2+1)^2} + \frac{M}{(t^2+1)^3} + \frac{N}{(t^2+1)^4} + \cdots, \tag{6.62}$$

where $L, M, N, \ldots$ are all real constants. The values of these constants are derived from the requirement that an infinitely small circle surrounding $t = j$ corresponds with an infinite circle, *traced out once only*, in the z-plane. Now the small circle round the point $t = j$ is described by

$$t - j = R e^{j\theta},$$

and substituting this value in eqn (6.62), noting that as $R \to 0$, $t + j \to 2j$, gives

$$f(t) = -\frac{L}{4R^2 e^{2j\theta}} - \frac{M}{4R^3 e^{3j\theta}} - \frac{N}{4R^4 e^{4j\theta}} - \cdots.$$

Hence, as $dt = jR e^{j\theta} d\theta$, eqn (6.61) becomes

$$\frac{dz}{dt} = -\frac{jK}{4R e^{j\theta}}\left[L + \frac{M}{R e^{j\theta}} + \frac{N}{R^2 e^{2j\theta}} + \cdots\right],$$

where K is the value, as $t \to j$, of the terms in square brackets in eqn (6.61). In this last equation, as $R \to 0$, each term in the brackets gives an infinite circle in the z-plane corresponding with the circle in the t-plane but, as θ varies from 0 to 2π, all terms except the first cause the infinite circle to be traced out more than once. Thus, $M, N, \ldots$, must all be zero so that, finally,

$$f(t) = \frac{L}{(t^2+1)^2},$$

and the desired transformation equation is

$$\frac{dz}{dt} = S(t^2+1)^{-2}(t-a)^{(\alpha/\pi)-1}(t-b)^{(\beta/\pi)-1}\cdots. \tag{6.63}$$

$\dagger$The pole at $t = -j$ does not disturb the behaviour of the function in the upper half plane.

This equation differs from the Schwarz–Christoffel eqn (6.1), only by the additional factor $(t^2 + 1)^{-2}$. However, this term causes the integration, for given vertex angles, to be more difficult than that for eqn (6.1) with the same angles. Because of this, and because it cannot be used to derive a solution directly from the w-plane (see the example below), eqn (6.63) is never applied to exterior regions with infinite boundaries but only to those with finite boundaries. It was used by Bickley [12] to calculate the field distribution round a charged bus-bar with rectangular section.

Treatment of this boundary involves the use of non-simple functions, as indeed do all transformations using eqn (6.63) except those for a single, thin plate and various combinations of intersecting thin plates. These too are discussed briefly by Bickley [12] and the case of a single, charged plate is treated fully here.

6.3.2 The field of a charged, conducting plate

As an example of the use of eqn (6.63) consider the field of a conducting plate carrying a charge q. Let the plate lie along the real axis of the z-plane between the points c and $-c$, Fig. 6.14(a); and let the origin of the t-plane, Fig. 6.14(b), correspond with the point $z = c$ and the limits of the real axis with $z = -c$. The exterior angle of the polygon at $z = c$ is 2π and, therefore, eqn (6.63) gives for the transformation from the real axis of the t-plane into the z-plane boundary

$$\frac{dz}{dt} = S \frac{t}{(t^2 + 1)^2}.$$

When integrated this yields

$$z = -\frac{S}{2} \frac{1}{(t^2 + 1)} + k,$$

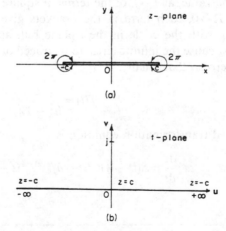

(a)

(b)

Fig. 6.14

from which the values of S and k are derived by substitution of the corresponding values of z and t. Hence, $k = -c$ and $S = -c$, and the equation of transformation is

$$z = c \frac{1 - t^2}{1 + t^2}. \tag{6.64}$$

The required field in the t-plane is different from any used previously in the book, but its form can be simply determined. First, the upper and lower surfaces of the plate correspond with the real axis in the t-plane and so a total charge q must be distributed over this real axis. And, further, since the plate is conducting, the real axis must be equipotential. Secondly, all the flux leaving the plate goes to infinity and so, as the point $t = j$ corresponds with infinity in the z-plane, there must be a charge $-q$ at $t = j$. Thus, the t-plane field is as shown in Fig. 6.15(a) and the reader will appreciate that it is one half of the field between equal unlike charges at the points $t = j$ and $t = -j$. The complex potential function describing it is, therefore (see section 2.2.1)

$$w = \frac{q}{2\pi} \log \frac{t - j}{t + j}.$$

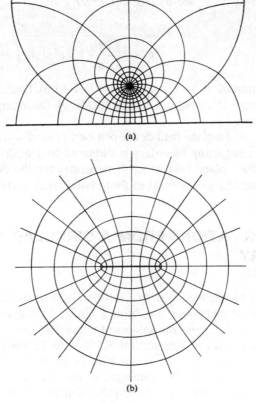

(a)

(b)

Fig. 6.15

Putting $jW = 2w\pi/q$, this equation may be rewritten

$$t = j\frac{1 + e^{jW}}{1 - e^{jW}}$$

$$= -\cot\left(\frac{W}{2}\right),$$

or, changing the origin of W by π,

$$t = \tan\left(\frac{W}{2}\right). \tag{6.65}$$

Thus, substituting this value for t in eqn (6.84), the solution of the field in the z-plane is

$$z = c\cos W,$$

or, writing W in terms of w and shifting the origin of the w-plane,

$$z = c\sin\left(\frac{-j2\pi w}{q}\right) = c\sin w'. \tag{6.66}$$

These are of the same form as eqns (5.78) and (5.79) developed as the solution for the field outside an equipotential elliptical boundary. The plate is of course the limiting ellipse of the family described by [see eqn (5.80)]:

$$\frac{x^2}{c^2\cosh^2\left(\dfrac{2\pi\psi}{q}\right)} + \frac{y^2}{c^2\sinh^2\left(\dfrac{2\pi\psi}{q}\right)} = 1,$$

and the map of its field is shown in Fig. 6.15(b). Equation (6.66) differs from eqn (5.79) only in that it includes the magnitude of the charge upon the plate (or elliptical boundaries).

The form of t-plane field developed here is used whenever the field exterior to a finite, charged, conducting boundary is obtained by transformation from the real axis. (The form of the t-plane field which is necessary for the determination of the fields of line currents or charges external to finite boundaries is discussed in section 7.2.2.)

6.4 TRANSFORMATIONS FROM A CIRCULAR TO A POLYGONAL BOUNDARY

In certain problems it is desirable to transform the boundary of a polygon, not from an infinite straight line, but from a circle. This may be because it is then easier to determine and express the solution for the field (see sections 6.4.2 and 7.2), or because the Poisson integral (see section 7.7.1) is to be used. There are four transformation equations which connect circular with polygonal boundaries—the interior or exterior of one boundary may be transformed into the interior or exterior of the other—and they may be derived by combining the bilinear transformation (see section 5.3) with eqn (6.1) (Schwarz–Christoffel) or eqn (6.63).

6.4.1 The transformation equations

Consider first the derivation of the equation which transforms the perimeter of the unit circle into a polygonal boundary, in such a way that the *interior* regions of the two boundaries correspond with one another. Let the polygon be in the z-plane, Fig. 6.16(a), and the circle in the t-plane, Fig. 6.16(c). The circle is first transformed into the real axis of a third complex plane, the p-plane, Fig. 8.16(b), and then the real axis of the p-plane is transformed into the polygonal boundary. The equation transforming the interior of the circle into the upper half of the p-plane is [see eqn (5.48)]

$$p = j\frac{1-t}{1+t},\tag{6.67}$$

which may be differentiated to give

$$\frac{dp}{dt} = -\frac{2j}{(1+t)^2}.\tag{6.68}$$

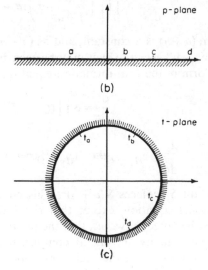

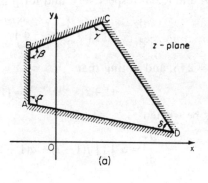

Fig. 6.16

Also, the equation transforming the upper half of the p-plane into the interior of the polygon is [see eqn (6.1)]

$$\frac{dz}{dp} = S(p-a)^{(\alpha/\pi)-1}(p-b)^{(\beta/\pi)-1}\ldots,$$

which may be written for brevity as

$$\frac{dz}{dp} = S\prod(p-a)^{(\alpha/\pi)-1},\tag{6.69}$$

where the symbol $\prod$ is taken to mean the product of terms of the form $(p-a)^{(\alpha/\pi)-1}$. Now the equation which transforms the circle into the polygon is given by the evaluation of

$$\frac{dz}{dt} = \frac{dz}{dp}\frac{dp}{dt},$$

in terms of the above expressions for dz/dp and dp/dt: substituting for these derivatives from eqn (6.69) and (6.68) respectively, and for p from eqn (6.67) gives

$$\frac{dz}{dt} = S\prod\left(j\frac{1-t}{1+t}-a\right)^{(\alpha/\pi)-1}\frac{2j}{(1+t)^2}.\tag{6.70}$$

By making $S'=2jS$, and noting that

$$(1+t)^{-\Sigma((\alpha/\pi)-1)} = (1+t)^2,$$

eqn (6.70) may be written

$$\frac{dz}{dt} = S'\prod[j(1-t)-a(1+t)]^{(\alpha/\pi)-1},$$

or

$$\frac{dz}{dt} = S'\prod\left[\left(\frac{j-a}{j+a}-t\right)(a+j)\right]^{(\alpha/\pi)-1}.$$

But, as the term $(a+j)$ is a constant, and as $(j-a)/(j+a)$ is the point in the t-plane, t_a, corresponding with $p=a$ (which may be seen by writing eqn (6.67) for t, in terms of p), the final form of the transformation equation is

$$\frac{dz}{dt} = S\prod(t_a-t)^{(\alpha/\pi)-1}$$

or

$$\frac{dz}{dt} = S(t_a-t)^{(\alpha/\pi)-1}(t_b-t)^{(\beta/\pi)-1}(t_c-t)^{(\gamma/\pi)-1}\ldots,\tag{6.71}$$

where the constant S replaces $S'(a+j)^2$. This equation is identical in form with the Schwarz–Christoffel equation, except that the order of subtraction in each term is reversed. It must be noted, however, that the points t_a, t_b, t_c, ..., corresponding with the vertices of the polygon, are, in general, complex with unit modulus.

The equation transforming the region *exterior* to the circle into that *exterior* to a polygon may be derived in a similar way to the above by using eqns (6.63) and (6.67), and by noting that eqn (6.67) transforms the exterior of the circle into the *lower* half of the *p*-plane. The equation is

$$\frac{dz}{dt} = St^{-2}(t - t_a)^{(\alpha/\pi)-1}(t - t_b)^{(\beta/\pi)-1}(t - t_c)^{(\gamma/\pi)-1}\ldots, \tag{6.72}$$

where the angles, $\alpha, \beta, \gamma, \ldots$, are the exterior angles of the polygon and, again, the points $t_a, t_b, \ldots$, are complex with unit modulus.

The remaining two equations connecting the unit circle with a polygon—the one from the interior of a polygon into the exterior of the circle, and the one from the exterior of a polygon into the interior of the circle—are very rarely used and are not given here. If required, they may be derived simply in a similar manner to that demonstrated above.

The use of the transformations from the unit circle does not make possible the solution of any problem which cannot be solved by transformation from the real axis; and, for both methods, similar difficulties are experienced in the integration of the differential equations—for a given polygon, the same class of function is involved in each. For the interior region of a polygon it is usually preferable to transform from the real axis because of the much simpler form of the field. For the exterior regions of a polygon, however, it is often preferable to use the transformation from the circle, as the necessary field solutions are easier to visualize. This point is demonstrated in the following example, and is discussed further in section 7.2. It is also demonstrated in the solutions for the field of two finite, intersecting, charged plates [13].

The general solution for the field inside the unit circle, due to any distribution of potential on its perimeter, is given by the Poisson integral which is discussed in section 7.7.1. With this integral it is possible to obtain the solution for the field within a polygon, due to any distribution of potential on the boundary, provided that the polygon can be transformed into the unit circle.

6.4.2 The field of a line current and a permeable plate of finite cross-section

Consider the solution for the field of a line current near a permeable plate of finite cross-section, Fig. 6.17, by transformation of the region exterior to the plate into that exterior to the unit circle. (The field exterior to the circle is of the form shown in Fig. 5.17.) Let the boundary of the plate be in the *z*-plane, with its ends at $z = l + jm$ and $z = -l - jm$, Fig. 6.18(a), and let the unit circle be in the *t*-plane, Fig. 6.18(b). The exterior angles of the polygon (at the ends of the plate) are both 2π, and the points, t_a and t_b, lying on the circle and corresponding with the vertices of the polygon, have coordinates which are, by symmetry, equal but of opposite sign. Thus, from eqn (6.72),

$$\frac{dz}{dt} = St^{-2}(t - t_a)(t - t_b),$$

and this, when integrated, gives for the equation connecting the planes

$$z = S\left(t + \frac{t_a^2}{t}\right) + k.$$

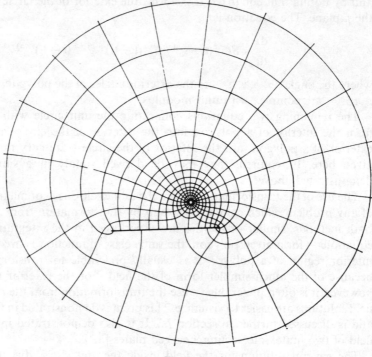

Fig. 6.17

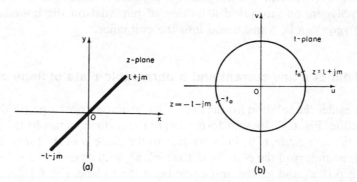

Fig. 6.18 (Reproduced by permission of the Institution of Electrical Engineers.)

Substituting, in this equation, the corresponding values of z and t for the vertices of the polygon, yields $k = 0$ and

$$S = \frac{l + jm}{2t_a},$$

where t_a may have any convenient value. Hence the transformation from the circle to the plate is

$$z = \frac{l + jm}{2t_a}\left(t + \frac{t_a^2}{t}\right). \tag{6.73}$$

[Compare this equation with eqn (5.67).]

To obtain the field generated by a line current in the z-plane it is clear that the required field in the t-plane is also generated by a line current at the point corresponding, through eqn (6.73), with the point of the current in the z-plane. The solution takes its simplest form when the current is positioned on the real axis of the z-plane at some point, c, chosen so that, in conjunction with the choice of $l + jm$, the z-plane field has the desired proportions; for then, the current in the t-plane also lies on the real axis at some point c'. From eqn (5.71) the solution for the field in the t-plane is

$$\phi + j\psi = \frac{i}{2\pi}\log\left(\frac{t^2 + 1}{t} - \frac{c^2 + 1}{c}\right), \tag{6.74}$$

and elimination of t between eqns (6.74) and (6.73) gives the solution in the z-plane.

The same solution could be obtained by using eqn (6.63) to transform the boundary of the plate from the real axis of the t-plane [14]. Then, though the transformation equation is simple [it is developed in section 6.3.2, eqn (6.64)], the reader will appreciate that the necessary form of the t-plane field is not evident without careful consideration. It is discussed in section 7.2.2.

The solution to this field is of value in the estimation of the influence which the plate has on the inductance of the current. The inductance can be found by comparing the difference of flux function between two points, one lying on the conductor surface and the other at a great distance from the conductor and the plate, for the two cases, with and without, the plate [14].

6.5 CLASSIFICATION OF INTEGRALS

In the light of the examples just considered it is helpful to examine briefly, in general terms, the relationship between a given type of boundary and the integral which is involved in deriving its transformation equation. The difficulty in performing the integration is significantly dependent upon the values of angle at the vertices and, in general, it increases with the number of vertices. Three broad classes of integral are encountered and they are discussed below, assuming that the boundary under consideration has, when possible, been reduced, by the use of symmetry, to include only the minimum number of vertices.

Integrals expressible in terms of simple functions

The first type of integral yields a solution that is expressible wholly in terms of simple functions and all the examples of this chapter are typical of it. The general form of this class of integral may be written

$$\int \frac{P(t)}{Q(t)}\sqrt{\frac{t - a}{t - b}}\,dt = \int \frac{P'(t)}{Q(t)}\frac{dt}{\sqrt{(t - a)(t - b)}} \tag{6.75}$$

where P, P', and Q are rational, algebraic functions of t. Inspection shows that a boundary having any number of angles of value $-\pi$, 0, and 2π, and up to two angles which are odd multiples of $\pi/2$, may be expressed in this, or simpler, form. The constants of the transformation equation for such a boundary can usually be determined simply though it has been seen (section 6.2.7) that, in certain cases, graphical or numerical methods may be required.

Integrals expressible in terms of non-simple functions

In practice, polygons with three or four vertex angles which are odd multiples of $\pi/2$ (and any number of value $-\pi$, 0, or 2π) frequently occur. They lead to integrals which are termed *elliptic*. Such integrals may be evaluated using elliptic functions and they are discussed in the next chapter.

Also, a number of polygons having two vertex angles which are not multiples of $\pi/2$ give rise to integrals which may be expressed in the form of Euler or, in particular cases, elliptic integrals. They are discussed in section 7.4.

Integrals requiring numerical evaluation

When the boundary is such as to give an integral not falling into any of the above classes, analytical methods fail, and numerical quadrature is used. See section 7.6.

REFERENCES

[1] H. A. Schwarz, Über einige Abbildungsaufgaben, *J. reine Angew. Math.*, **70**, 105 (1869).

[2] E. B. Christoffel, Sul Problema delle Temperature Stazionarie e la Rappresentazione di una Data Superficie, *Ann. Mat. Pura Appl.*, **1**, 95 (1867).

[3] W. Rogowski, Die elektrische Festigkeit am Ronde des Plattenkondensators, ein Beitrag zur Theorie der Funkenstrecken and Durchführungen, *Arch. Elektrotech.*, **12**, 1 (1923).

[4] F. W. Carter, Air-gap and interpolar induction, *J. Instn. Elect. Engrs.*, **29**, 925 (1900).

[5] G. M. Stein, Influence of the core form upon the iron losses of transformers, *Trans. Am. Instn. Elect. Engrs.*, **67** I, 95 (1948).

[6] I. A. Terry and E. G. Keller, Field pole leakage flux in salient pole dynamo electric machines, *J. Instn. Elect. Engrs.*, **83**, 845 (1938).

[7] F. W. Carter, Air-gap induction, *Elect. World, NY*, **38**, 884 (1901).

[8] L. Dreyfus, Die Anwendung des Mehrphasenfrequenzumformers zur Kompensierung von Drehstromasynchronmotoren, *Arch. Elektrotech.*, **13**, 507 (1924).

[9] F. W. Carter, The magnetic field of the dynamo-electric machine, *J. Instn. Elect. Engrs.*, **64**, 115 (1926).

[10] J. Kucera, Magnetische Zahnstreuungen bei elektrischen Maschinen, *Electrotech. Maschinenb.*, **58**, 329 (1940).

[11] R. Herzog, Berechnung des Streufeldes eines Kondensators, dessen Feld durch eine Blende begrenzt ist, *Arch. Elektrotech.*, **29**, 790 (1935).

[12] W. G. Bickley, Two-dimensional problems concerning a single closed boundary, *Phil. Trans.*, **228** A, 235 (1929).

[13] W. B. Morton, Electrification of two intersecting planes, *Phil. Mag.*, **1**, 337 (1926), and Irrotational flow past two intersecting planes, *Phil. Mag.*, **2**, 900 (1926).

[14] P. J. Lawrenson, A note on the analysis of the fields of line currents and charges, *Proc. Instn. Elect. Engrs.*, **109** C, 86 (1962).

Additional references

Cohn, S. B., Shielded coupled strip transmission line, *Inst. Radio Engrs. Trans.*, MTT-3, 29 (1955).

Cohn, S. B., Thickness correction for capacitive obstacles and strip conductors, *Inst. Radio Engrs. Trans.*, MTT-8, 638 (1960).

Dahlman, B. A., A double ground plane strip line for microwaves, *Inst. Radio Engrs.*, MTT-3, 52–7 (Oct. 1955).

Douglas, J. H. F., Reluctance of some irregular magnetic fields, *Trans. Am. Instn. Elect. Engrs.*, **34** I, 1067 (1915).

Getsinger, W. J., A coupled strip line configuration using printed-circuit construction that allows very close cupling, *Inst. Radio Engrs. Trans.*, MTT-9, (1961).

Kreyszig, E., Schlitzblendenkondensatoren. Ein Beitrag zur Praxis der Brechung von Potentialfeldern, *Z. angew. Phys.*, **7**, 13, 17 (1955).

7 GENERAL CONSIDERATIONS

7.1 INTRODUCTION

Each of the earlier chapters (5 and 6) is devoted, by the detailed presentation of simple examples, to introducing the reader to the basic techniques of transformation methods. In contrast, this chapter is concerned with more advanced topics, and a minimum of detail is given in the worked examples. The possible extensions and generalizations of the basic methods are examined, and these can be divided into two groups—those concerned with boundary shape and those concerned with the type of field source.

The boundary shapes discussed so far are either wholly curved or wholly polygonal with angles which are integral multiples of $\pi/2$. It is shown here how boundaries, which are partly curved and partly polygonal, or which are polygonal with angles not multiples of $\pi/2$, can be treated. A significant extension in the range of boundary shapes can also be achieved by the use of numerical integration of functions of a complex variable. The use of this technique for the determination, by iteration, of the constants in the Schwarz–Christoffel equation, and also for the subsequent calculation of the field, is described.

The discussion of field type gives particular emphasis to the fields of line currents near boundaries, to fields exterior to closed boundaries, and to the general solution of the Dirichlet problem (i.e. the problem in which potential is specified at all points on a closed boundary). The latter involves the use of the Schwarz complex potential function or the Poisson integral.

The solutions be conformal transformation of problems of practical importance involve elliptic functions. These are general functions of which trigonometric and hyperbolic functions are particular cases, and they can be used to evaluate integrals which are of an elliptic form. In this chapter elliptic functions are discussed only briefly to illustrate further the application of conformal transformation techniques. This is because the development of highly efficient numerical techniques has largely rendered obsolete the use of tables and the slowly convergent series involved with elliptic functions.

7.2 FIELD SOURCES

There is no previous general discussion of line sources (i.e. currents, charges, and poles) or their combination (e.g. in doublets) in the presence of boundaries, and this type of field is now considered in terms of the magnetic field of line currents. There are two reasons for discussing such fields. First, they have immediate applications in the solution of the fields of currents near complicated boundaries [1]. Secondly, a knowledge of them is necessary for the solution of certain fields exterior to closed boundaries.

First, consider a line current i remote from all boundaries, the field of which is expressed by

$$w = \frac{i}{2\pi} \log(t - \alpha). \tag{7.1}$$

By superposition it is possible to express the field of a combination of m currents, with magnitude i_n as

$$w = \frac{1}{2\pi} \sum_{n=0}^{m} i_n \log(t - \alpha_n). \tag{7.2}$$

Consideration of these equations raises the matter of the existence of currents at infinity, a point best explained in terms of the hydrodynamical analogy of the above field. The equation of the laminar flow of liquid between a combination of sources and sinks is the sum of logarithmic terms, as in eqn (7.2) and, it is easily seen that the algebraic sum of the sources at finite points is equal, in magnitude, to the sink at infinity. The corresponding relationship is also true for currents, which must have a return path at infinity except, of course, when the algebraic sum of the currents at finite points is zero. This return path does not affect the field equation therefore since it is at infinity; but, when transformed, it may occur at a finite point and then it must be taken into account. This feature has already appeared in the discussion of two currents inside a permeable tube, section 5.1.3. It is discussed further in section 6.3.2, where the field of a charged plate is obtained; the charge at infinity, in the z-plane, is transformed to the point $t = +j$, in the t-plane.

In general, solutions can be simply obtained for the fields of currents (or charges) near boundaries which are impermeable, infinitely permeable or partly impermeable and partly infinitely permeable. This is conveniently done by transformation from the infinite straight line, and the only limitation is that it must be possible to find the appropriate transformation equation. Consideration is now given to the field solutions for the above types of boundary, and it will be appreciated from the last paragraph that a distinction must be made between infinite and finite boundaries.

7.2.1 Infinite boundaries

For convenience, the infinite straight-line boundary is taken to lie along the real axis and the line current at a point on the imaginary axis.

Line current near an infinitely permeable plane

As shown in section 2.2.1, the field of a line current, at a distance a from an infinitely permeable plane extending to infinity, may be expressed as

$$w = \frac{1}{2\pi} \log(t^2 + a^2). \tag{7.3}$$

Line current near an impermeable plane

If the plane is impermeable to flux, the field equation becomes, see section 2.2.1

$$w = \frac{1}{2\pi} \log \frac{t - ja}{t + ja}.$$ (7.4)

The field of a current near an impermeable plate, for example, is found by combining eqn (7.4) and eqn (6.4) and a map of this field is shown in Fig. 7.1.

Line current near partly permeable and partly impermeable boundary

The field of a current near a boundary consisting of two parts, one permeable and the other impermeable (Fig. 7.2) may be obtained from the field of four line currents—two positive and two negative. Figure 7.3 shows, in the p-plane, the position of these currents the field of which is given by

$$w = \frac{1}{2\pi} \log \frac{(p - a)^2 + b^2}{(p + a)^2 + b^2}.$$ (7.5)

It is seen that there are two lines of symmetry—one a flux line and the other an equipotential line, and any quadrant in the p-plane can be transformed to give the upper half of the t-plane by the equation

$$p = 2\sqrt{t}.$$ (7.6)

Thus the field expression in the t-plane is

$$w = \frac{1}{2\pi} \log \frac{(2\sqrt{t} - a)^2 + b^2}{(2\sqrt{t} + a)^2 + b^2}.$$ (7.7)

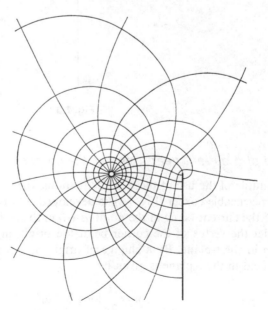

Fig. 7.1 (Reproduced by permission of the Institution of Electrical Engineers.)

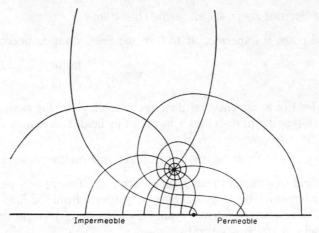

Fig. 7.2

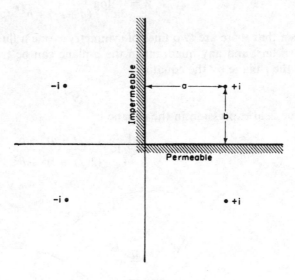

Fig. 7.3

The field of a current near a permeable corner

As an example of the use of these field equations, the field of a line current close to an infinitely permeable corner (Fig. 7.4), is developed. Let the t- and z-planes be as shown in Fig. 7.5 the current being at the point $(a + j)$ in the t-plane, and $(c - jb)$ in the z-plane. Further, let the vertex of the corner be at the origin in the z-plane, corresponding to the origin in the t-plane. By a change of origin of $+a$ in eqn (7.3), it is seen that the required field in the t-plane is given by

$$w = \frac{i}{2\pi} \log\left[(t - a)^2 + 1\right]. \tag{7.8}$$

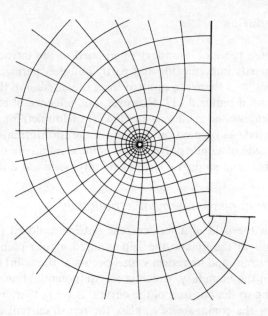

Fig. 7.4 (Reproduced by permission of the Institution of Electrical Engineers.)

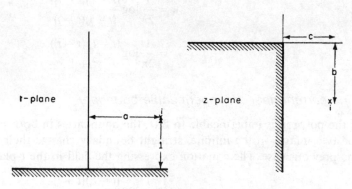

Fig. 7.5

The Schwarz–Christoffel equation relating the two planes is

$$\frac{dz}{dt} = S\sqrt{t},$$ (7.9)

which, when integrated, gives

$$z = \tfrac{2}{3} S t^{3/2}.$$ (7.10)

The constant of integration is zero since the origins in the two planes correspond, and the remaining constants, S and a, are found in terms c and b by substitution in eqn (7.10) for corresponding points. Then, eliminating t from eqns (7.8) and (7.10) gives the required field solution.

7.2.2 Finite boundaries

Of the three types of boundary mentioned in the introduction to this section the one which is partly impermeable and partly infinitely permeable is of little interest for finite boundaries. It is therefore omitted from the discussion though the field solution can be easily found if required [1]. However, the influence of closed boundaries on a uniform applied field is often of interest and the use of doublets for the treatment of such influences is considered. As pointed out earlier, in the transformation from the infinite half-plane to the outside of a finite polygon, the point at infinity in one plane becomes a finite point in the other, and so field source at infinity occurs at a finite point when transformed.

Line current near a permeable boundary

To obtain the field of a line current outside a closed polygonal boundary of infinite permeability in the z-plane, the field required, in the t-plane, is that due to the corresponding line current and its return which occurs at the point $t = +j$, and the images of both of these in the boundary, which is an equipotential line. If the point in the t-plane corresponding to the position of the current is $t = t_i$, there is an image of like sign at $t = \bar{t}_i$, where $\bar{t}_i$ is the conjugate of t_i; also, the return current at $t = +j$ has an image of like sign at $t = -j$. Hence the field equation is

$$w = \frac{1}{2\pi} \log \frac{(t - t_i)(t - \bar{t}_i)}{(t + j)(t - j)}$$

$$= \frac{1}{2\pi} \log \frac{(t - t_i)(t - \bar{t}_i)}{t^2 + 1}. \tag{7.11}$$

Line current near an impermeable boundary

If the polygon is impermeable to flux, the boundaries in both planes are flux lines and the two images in the infinite, straight boundary change their sign as compared with the previous case. The equation expressing the field in the t-plane is, then,

$$w = \frac{1}{2\pi} \log \frac{(t - t_i)(t + j)}{(t - \bar{t}_i)(t - j)}. \tag{7.12}$$

Charged boundaries

The field of a charged plate is obtained earlier, see section 6.3.2. As explained there, the expression for the t-plane field used to obtain the field of finite, charged boundaries is

$$w = \frac{1}{2\pi} \log \frac{t - j}{t + j}. \tag{7.13}$$

Uniform applied field: use of doublet

To obtain the field of a polygonal boundary in an applied uniform field, from the infinite, straight-line boundary, requires the use of doublets, see section 2.3.2: the applied uniform field in the z-plane is achieved by placing a doublet at infinity. This doublet is transformed

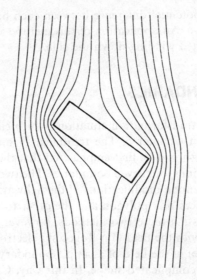

Fig. 7.6   (Reproduced by permission of Professor W. G. Bickley and the London Mathematical Society.)

to the point $t = +j$ in the t-plane where it also has an image at the point $t = -j$. Hence, if the boundary is equipotential the field in the t-plane is given by

$$w = \frac{1}{t+j} + \frac{1}{t-j}$$

$$= \frac{2t}{t^2 + 1}.$$

$$(7.14)$$

Figure 7.6 shows a map of the equipotential lines for a rectangular conductor in an applied uniform electric field. If the boundary is a flux line the field takes the form

$$w = \frac{1}{t+j} - \frac{1}{t-j}$$

$$= \frac{-2j}{t^2 + 1}.$$

$$(7.15)$$

7.2.3   Distributed sources

It is possible to represent a source, which is distributed over an area or along a line, by a finite number of line sources and, in this way, to obtain a good approximation for certain practical problems. However, because of the difficulty of locating points away from the boundaries, see section 7.5 the method is of most value for problems in which the currents are distributed along the boundary. For example, boundaries which are not equipotential may be represented as consisting of a number of small equipotential segments, the potential difference between each being obtained by a line current at the point of division. An application of this method would be to the field of a salient pole with a distributed winding down the side.

7.3 CURVED BOUNDARIES

Several simple transformation equations which can be used to give curved boundaries are described in Chapter 5. The treatment of more difficult transformations is given here, though it should be first emphasized that there is no general method of obtaining analytical solutions for curved boundaries. The two principal methods which can be used are both based upon modifications of the Schwarz–Christoffel equation: one is used to round the vertices of a polygon and the other to curve some or all of its sides.

Before discussing these two methods, however, it is useful to recall the device used in connection with the field of a Rogowski electrode, see section 6.2.3: an equipotential line (or flux line) in the field of a simple boundary shape is used to represent an actual boundary of a complicated shape. In this way, Carter [2] examined the field near the salient pole-tip by using the simple boundary shown in Fig. 7.7 representing the pole shape by an equipotential line shown dotted. In a similar way he also represented, in the same paper, the field of a semi-closed slot. There is no routine method of approach to such problems; instead, a combination of intuition and trial and error is required, choosing first a simplified boundary and then an appropriate field line.

Curved boundaries are, in general, of limited practical application in electric and magnetic field problems, but exceptions to this occur in the study of breakdown of insulation under high voltages and of special forms of high frequency transmission lines. For the interested reader, a large number of more specialized transformations for boundaries connecting entirely of curved segments are given in the books by Köber and Koppenfels and Stallman.

7.3.1 Rounded corners

It is possible to obtain a transformation equation which makes the vertices of a polygon rounded instead of sharp, by modifying the appropriate product terms in the Schwarz–Christoffel equation. The shape of the curve(s) cannot be chosen precisely to

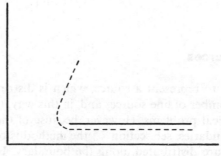

Fig. 7.7

correspond with any given curve(s) but, by trial and error, a good approximation to the required curve can usually be obtained. The method consists in rounding one or more corners by replacing the terms of the form $(t - a)^{\alpha/\pi - 1}$ (which give rise to sharp corners) by terms of the form

$$[(t - a')^{\alpha/\pi - 1} + \lambda(t - a'')^{\alpha/\pi - 1}].$$

The points a' and a'' in the t-plane correspond to the limits of the curve, and λ is a factor which determines the shape of the curve—for $t > a'$ or $t < a''$, $\mathrm{d}z/\mathrm{d}t$ is constant, but for $a' > t > a''$, $\mathrm{d}z/\mathrm{d}t$ varies continuously with t. The total change in direction over the curved section is the same as that produced by the equivalent, normal vertex factor. By consideration of the new term, it is clear that, providing $\alpha \geqslant \pi$, it does not introduce poles, zeros or sudden changes in $\mathrm{d}z/\mathrm{d}t$ and so it does not give rise to any additional vertices or discontinuities in the boundary. For problems having $\alpha < \pi$, it is necessary, in order to avoid the introduction of poles, to transform the negative imaginary half plane, for which the corresponding vertex angle is $(2\pi - \alpha)$.

As a demonstration of the method, consider the field between an infinite charged plane and a conducting block with a rounded corner, used by Drinker [3] in the analysis of the performance of magnetic reading heads. Let this configuration be in the z-plane with the dimensions shown in Fig. 7.8; it is seen that the corner is rounded between the points $z = a + l$ and $z = a - jl$. The real axis of the t-plane is to be transformed into this configuration, the precise shape of the rounded corner being fixed by the choice of the constant λ. In this case, the value is chosen to give a minimum value of the maximum field strength on the curved boundary. (The shape of corner so chosen makes breakdown between the two surfaces occur at a maximum voltage.)

Let the corresponding points in the two planes be:

$$z = 0, \qquad t = -1$$
$$z = -j\infty, \quad t = 0;$$
$$z = a - jl, \quad t = b;$$
and
$$z = a + l, \quad t = c.$$

If the corner were sharp instead of rounded, the Schwarz–Christoffel equation would

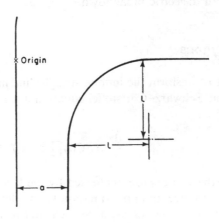

Fig. 7.8

be of the form

$$\frac{dz}{dt} = \frac{S}{t}(t - \alpha)^{1/2}. \tag{7.16}$$

The rounding of the corner is thus achieved by modifying the term $(t - \alpha)^{1/2}$ to give the equation

$$\frac{dz}{dt} = \frac{S}{t}(\sqrt{t - b} + \lambda\sqrt{t - c}), \tag{7.17}$$

integration of which leads to

$$z = 2S\left[\sqrt{t - b} - \sqrt{b}\tan^{-1}\left(\frac{t - b}{b}\right)^{1/2} + \lambda\left(\sqrt{t - c} - \sqrt{c}\tan^{-1}\left(\frac{t - c}{c}\right)^{1/2}\right)\right] + c_2. \tag{7.18}$$

Substituting for corresponding points in the two planes and using the method of residues leaves one relationship connecting the constants to be found from a consideration of the field gradient.

To obtain a potential difference between the two sections of the boundary in the z-plane requires a field in the t-plane of the form

$$w = \frac{1}{\pi}\log t. \tag{7.19}$$

The flux density distribution in the z-plane is given by

$$\left|\frac{dw}{dt}\frac{dt}{dz}\right|$$

and λ is chosen to give the least value of the maximum density on the curved boundary. This maximum is found by differentiating dw/dz with respect to λ and equating the result to zero.

A transformation involving rounded corners has been employed by Cockroft [4] in connection with dielectric breakdown.

7.3.2 Curvilinear polygons

It is possible to transform the infinite straight line into a polygon with curved sides by including in the Schwarz–Christoffel equation a 'curve factor', $C(t)$, giving an equation of form

$$\frac{dz}{dt} = SC(t)\prod(t - t_n)^{(\alpha/\pi) - 1}. \tag{7.20}$$

The angles at the vertices are unaffected by the factor $C(t)$ which has to be chosen, by trial and error, to give an approximation to the desired shapes of the curved sections. The form of the curve factor must be such that it introduces no singularities or zeros into the equation, for these would result in additional vertices or in discontinuities. It is not possible to give any standard procedure for finding the appropriate curve factor, but several types of factor have been considered by Page [5] and Leathem [6].

A simple boundary of this type is considered in section 5.2.1, where a semicircular boss on a straight line is discussed. It is shown that the equation

$$z = t + \sqrt{t^2 - 1} \tag{7.21}$$

transforms the real axis of the t-plane into the required boundary shape, the part corresponding to $|t| \leqslant 1$ being circular and the part corresponding to $|t| \geqslant 1$ being straight. The curve factor is of special type, since it makes only one side of the polygon curved. Its form can be seen by differentiating eqn (7.21) and writing the resulting expression as

$$\frac{\mathrm{d}z}{\mathrm{d}t} = (t + \sqrt{t^2 - 1})(t + 1)^{-1/2}(t - 1)^{-1/2}; \tag{7.22}$$

the first term is the curve factor $C(t)$, the remaining two terms giving the defining vertex angles of $\pi/2$ at $t = +1$ and $t = -1$. A more general form of the curve factor,

$$C(t) = t + \lambda\sqrt{t^2 - 1}, \tag{7.23}$$

gives the semi-elliptical boss [compare eqn (5.65)].

In conclusion, it should be noted that the search for a suitable factor to give a satisfactory approximation to any particular curvilinear polygon is often lengthy and sometimes unrewarding.

7.4 ANGLES NOT MULTIPLES OF $\pi/2$

The treatment of polygonal boundaries given in the earlier chapters is restricted to cases in which all the angles are multiples of $\pi/2$ (including the zero multiple). Whilst these include the great majority of practically useful boundary shapes, there are a number of useful and interesting boundaries which have vertex angles with other values. They have not received attention earlier because, apart from the trivial case of polygons with only one vertex, the transformation equation cannot be expressed in terms of simple functions.

All polygons defined by two vertices involve integrals of the Euler type (see Copson) and some of these can be reduced to elliptic form. When the polygon is defined by more than two vertices, analytical methods fail, but it can be handled by using numerical integration (see section 7.5).

Two-vertex problems

Consider any polygon defined by two vertex angles α and β, the length of the side joining them being l. For a reason which appears later, let this side be on the real axis in the z-plane. Then the polygon, which may be closed at infinity or at a finite point, can be transformed from the real axis of the t-plane by the Schwarz–Christoffel equation

$$\frac{\mathrm{d}z}{\mathrm{d}t} = St^{(\alpha/\pi) - 1}(t - 1)^{(\beta/\pi) - 1}, \tag{7.24}$$

where the points $t = 0$ and $t = 1$ correspond to the vertices in the z-plane. When integrated this is of the form

$$z = S \int t^{(\alpha/\pi)-1}(t-1)^{(\beta/\pi)-1} \, dt, \tag{7.25}$$

the constant of integration being zero when the origins in the two planes are made to correspond. This integral is Eulerian and of the first kind, and the definite form of it, between real limits of t, can be expressed in terms of the gamma function, which is defined by

$$\Gamma(u) = \int_0^\infty e^{-t}t^{(u-1)} \, dt. \tag{7.26}$$

(For the evaluation of the gamma function the reader should consult Copson, p. 209.) To determine the value of the scale constant S, the definite integral is most simply evaluated between the limits 0 and 1, and it is for this reason that the points in the t-plane corresponding to the vertices are so chosen. Hence,

$$l = S \int_0^1 t^{(\alpha/\pi)-1}(t-1)^{(\beta/\pi)-1} \, dt. \tag{7.27}$$

The solution of this standard definite integral (which is the beta function) (see Copson, p. 213) gives

$$S = \frac{l}{(-1)^{(\beta/\pi)-1}} \frac{\Gamma[(\alpha+\beta)/\pi]}{\Gamma(\alpha/\pi)\Gamma(\beta/\pi)}, \tag{7.28}$$

so that the scale constant of the transformation can be found simply in terms of the finite dimension of the polygon. However, to find the field in the z-plane, other than on the boundary, it is necessary to employ the method of numerical integration of a complex variable.

It can be shown that the Eulerian integral, eqn (7.25) becomes elliptic for particular values of the vertex angles. All of the known cases occur when both the angles are multiples of either $\pi/3$, $\pi/4$, or $\pi/6$ and many of them are discussed in Reference [7]. When all the vertices occur in the finite region, the boundary is, of course, triangular, and the four cases which are elliptic have angles of:

$$\frac{\pi}{2}, \frac{\pi}{4}, \text{ and } \frac{\pi}{4}; \qquad \frac{\pi}{6}, \frac{\pi}{3}, \text{ and } \frac{\pi}{2};$$

$$\frac{\pi}{3}, \frac{\pi}{3}, \text{ and } \frac{\pi}{3}; \qquad \frac{\pi}{6}, \frac{\pi}{6}, \text{ and } \frac{2\pi}{3}.$$

An important application of the analysis of two-vertex problems occurs in the study of the breakdown of liquids in an electric field between two sharp corners or between one sharp corner and a plane (see Ollendorf; Buchholz, p. 122; and Reference [8]).

7.5 THE USE OF ELLIPTIC FUNCTIONS

Elliptic functions are so called because they were first encountered in the calculation of the length of the arc of an ellipse. Consider this calculation for an ellipse, Fig. 7.9,

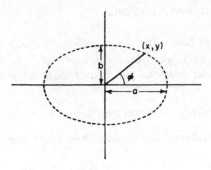

Fig. 7.9

described by the parametric equations

$$x = a \sin \phi \quad \text{and} \quad y = b \cos \phi.$$

Now the length L of any arc is given by $\int \sqrt{dx^2 + dy^2}$, where the limits of the integral define the ends of the arc. Thus for an elliptical arc defined by the angles ϕ_1 and ϕ_2,

$$L = \int_{\phi_1}^{\phi_2} \sqrt{a^2 \cos^2 \phi + b^2 \sin^2 \phi} \, d\phi$$

$$= \int_{\phi_1}^{\phi_2} [a^2 - \sin^2 \phi(a^2 - b^2)]^{1/2} \, d\phi.$$

This can be written as

$$L = a \int_{\phi_1}^{\phi_2} (1 - k^2 \sin^2 \phi)^{1/2} \, d\phi, \tag{7.29}$$

where

$$k = \frac{\sqrt{a^2 - b^2}}{a},$$

is the eccentricity of the ellipse and is less than unity. This integral cannot be evaluated in terms of elementary functions, except by a series, and it is known as an elliptic integral of the second kind. Since $k < 1$ and $\sin \phi \leqslant 1$, the integrand can be expanded by the binomial theorem: thus

$$(1 - k^2 \sin^2 \phi)^{1/2} = 1 - \frac{k^2}{2} \sin^2 \phi - \frac{k^4}{8} \sin^4 \phi - \cdots.$$

This series is uniformly convergent, and consequently it can be integrated term by term, using a recursion formula for $\int \sin^n x \, dx$.

   In similar ways other types of elliptic integral, and the elliptic functions to which they are related, can be evaluated. Tables of the values of elliptic functions, prepared from such series, are readily available and are frequently used in calculation. However, when a digital computer is employed, it is usually preferable to use the series directly rather than to feed the information from the tables into the machine store. A list of available tables is given in Appendix 3.

7.5.1   Elliptic integrals and functions

There are three basic *elliptic integrals* with which all others can be connected and they were defined by Legendre as being of the first, second, and third kinds. The *elliptic functions* are derived from the first of these integrals as explained below.

The elliptic integral of the first kind

The simplest and most useful of the three basic integrals is that of the first kind, defined by

$$z = \int_0^\phi \frac{d\phi}{\sqrt{(1 - \sin^2\theta \sin^2\phi)}}, \tag{7.30}$$

where $|\sin\theta| < 1$, θ is called the *modular angle*, and ϕ the *amplitude*. Putting $t = \sin\phi$ and $k = \sin\theta$ in eqn (7.30) gives

$$z = \int_0^t \frac{dt}{\sqrt{(1 - t^2)(1 - k^2 t^2)}}. \tag{7.31}$$

This form of the elliptic integral of the first kind is due to Jacobi, and k is called the *modulus*. For certain values of k (or $\sin\theta$) z reduces to an elementary function of t: if $k = 0$, $z = \sin^{-1} t$ and if $k = 1$, $z = \tanh^{-1} t$.

For a general upper limit, the integral is said to be *incomplete*, and it is given the symbol $F(t, k)$ or $F(\phi, \theta)$. When the upper limit is given by $t = 1$ ($\phi = \pi/2$) the integral is *complete* and it is given the symbol $K(k)$ or $K(\theta)$.

Derived from the modulus k is the *complementary modulus k'*, defined by

$$k' = \sqrt{1 - k^2}. \tag{7.32}$$

The complete integral to modulus k' is denoted by the symbol K', and it can be shown, using eqn (7.31) and substituting for k' in terms of k, that

$$K' = \int_1^{1/k} \frac{dt}{j\sqrt{(1 - t^2)(1 - k^2 t^2)}}, \tag{7.33}$$

and thus that

$$K + jK' = \int_0^{1/k} \frac{dt}{\sqrt{(1 - t^2)(1 - k^2 t^2)}}. \tag{7.34}$$

This equation can be useful when determining the constants of a transformation, and is employed later in section 7.5.2.

The principal Jacobian elliptic functions

It was shown by Jacobi that certain elliptic functions, very useful in analytical work, could be derived simply from the first elliptic integral. Consider again eqn (7.31): z is a fnction of t and k; but equally t is a function of z and k and, in the notation of Jacobi,

$$t = \mathrm{sn}(z, k). \tag{7.35}$$

This equation defines a new function sn[†] which is one of the principal Jacobian elliptic functions. When the value of the modulus is obvious, this equation is usually abbreviated to

$$t = \operatorname{sn} z.$$

There are two other principal functions, cn z and dn z,[‡] defined such that

$$\operatorname{sn}^2 z + \operatorname{cn}^2 z = 1 \tag{7.36}$$

and

$$\operatorname{dn}^2 z = 1 - k^2 \operatorname{sn}^2 z. \tag{7.37}$$

sn z, cn z, and dn z are the three principal Jacobian elliptic functions.

The elliptic integral of the second kind

Legendre defined the elliptic integral of the second kind, $E(\phi, \theta)$, by

$$E(\phi, \theta) = \int_0^\phi \sqrt{1 - \sin^2 \theta \sin^2 \phi}\, d\phi. \tag{7.38}$$

Putting

$$t = \sin \phi \quad \text{and} \quad k = \sin \theta$$

gives the Jacobian form

$$E(t, k) = \int_0^t \frac{\sqrt{1 - k^2 t^2}}{\sqrt{1 - t^2}}\, dt. \tag{7.39}$$

When the upper limit of the integral is $\phi = \pi/2$ (or $t = 1$), the integral is said to be complete, and it is given the symbol $E(\theta)$ [or $E(k)$].

The elliptic integral of the second kind can be expressed in terms of Jacobian elliptic functions by the substitution

$$\sin \phi = \operatorname{sn} \alpha.$$

It can be shown (see Whittaker and Watson, p. 492) that

$$\frac{d}{d\alpha} (\operatorname{sn} \alpha) = \operatorname{cn} \alpha \operatorname{dn} \alpha,$$

and, hence, by differentiating eqn (7.40), that

$$\cos \phi\, d\phi = \operatorname{cn} \alpha \operatorname{dn} \alpha\, d\alpha.$$

This gives

$$d\phi = \frac{\operatorname{cn} \alpha \operatorname{dn} \alpha}{\cos \phi}\, d\alpha, \tag{7.41}$$

[†] sn is an abbreviation for sine amplitude or sin am, and is pronounced 'ess-en'.
[‡] cn and dn are pronounced 'see-en' and 'dee-en'.

and substituting in eqn (7.38) from eqn (7.41) yields

$$E(\phi, \theta) = \int_0^\phi \frac{\text{cn}\,\alpha\,\text{dn}\,\alpha}{\cos \phi} \sqrt{1 - \sin^2 \theta \sin^2 \phi}\,\text{d}\alpha. \qquad (7.42)$$

Also, from the definitions of dn α and cn α and using eqn (7.40) it is evident that

$$\text{dn}\,\alpha = \sqrt{1 - \sin^2 \theta \sin^2 \phi} \quad \text{and} \quad \text{cn}\,\alpha = \cos \phi.$$

Finally, therefore, substituting in eqn (7.14) gives (7.42)

$$E(\phi, \theta) = \int_0^\phi \text{dn}^2\,\alpha\,\text{d}\alpha, \qquad (7.43)$$

which expresses the second elliptic integral in terms of Jacobian elliptic functions.

7.5.2 Two finite charged plates

The application of elliptic functions to the analysis of a physical problem is first illustrated by considering the field of two charged conducting plates of equal length and lying on a straight line (Fig. 7.10). This field is often transformed into more complicated ones: see, for example, References [9] and [10].

Figure 7.11(a) shows the two conducting plates placed along the real axis of the t-plane. Figure 7.11(b) shows, in the w-plane, a rectangular boundary two opposite sides of which correspond to the conducting plates, the other two sides being flux lines. The uniform field,

$$w = \psi + j\phi,$$

inside the rectangle is to be transformed into the field in the upper half of the t-plane. Let the plates in the t-plane lie between $t = +1$ and $+1/k$ and between $t = -1$ and $-1/k$, so that the proportions of the boundaries in that plane are fixed by the choice of

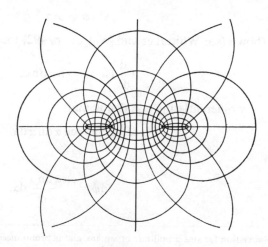

Fig. 7.10

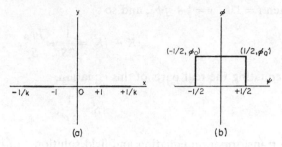

Fig. 7.11

the value of k. (This choice of the t-plane constants leads, as demonstrated below, to definite integrals which can be directly expressed in terms of complete elliptic integrals, k being the modulus of the integrals.) Further, let the potential between the plates be unity, and let the corresponding points in the w- and t-planes be as follows:

$$t = +1, \ \phi = 0 \quad \text{and} \quad \psi = \tfrac{1}{2};$$

$$t = +\frac{1}{k}, \ \phi = \phi_0 \quad \text{and} \quad \psi = \tfrac{1}{2};$$

$$t = -1, \ \phi = 0 \quad \text{and} \quad \psi = -\tfrac{1}{2};$$

and

$$t = -\frac{1}{k}, \ \phi = \phi_0 \quad \text{and} \quad \psi = -\tfrac{1}{2}.$$

The Schwarz–Christoffel equation connecting the planes is

$$\frac{dw}{dt} = \frac{S}{\sqrt{1 - t^2}\sqrt{1 - k^2 t^2}}, \tag{7.44}$$

which can be written

$$w = \int_0^t \frac{S \, dt}{\sqrt{1 - t^2}\sqrt{1 - k^2 t^2}}. \tag{7.45}$$

This is an elliptic integral of the first kind and it can thus be expressed in terms of Jacobian functions as

$$t = \text{sn}\left(\frac{w}{S}\right), \tag{7.46}$$

a form which is neat and simple to use for determining a relationship between the w- and t-plane constants. Writing eqn (7.34) in terms of Jacobian functions

$$\text{sn}(K + jK') = \frac{1}{k}. \tag{7.47}$$

When $t = 1/k$, eqns (7.46) and (7.47) are equivalent, so that

$$\frac{w}{S} = K + jK'. \tag{7.48}$$

But, when $t = 1/k$, $w = \frac{1}{2} + j\phi_0$, and so

$$K + jK' = \frac{1}{2S} + \frac{j\phi_0}{S}. \tag{7.49}$$

Hence, equating the real parts of this equation,

$$S = \frac{1}{2K}, \tag{7.50}$$

and the transformation equation and field solution can be written

$$t = \text{sn}(2Kw, k). \tag{7.51}$$

This solution can be used, for example, to find the capacitance between plates.
Equating the imaginary parts of eqn (7.48)

$$\frac{\phi_0}{S} = K',$$

and substituting for S from eqn (7.50) gives

$$\phi_0 = \frac{K'}{2K}. \tag{7.52}$$

Since the potential difference between the plates is unity, the flux ϕ_0 passing between these plates is equal to the capacitance between them. The variation of capacitance with the ratio of plate separation to plate length is shown in Fig. 7.12.

A map of the flux and equipotential lines can be determined by calculating the real and imaginary parts of sn $2Kw$ and substituting values of ϕ and ψ. This is facilitated by writing sn $2Kw$ in the form sn $(\alpha + j\beta)$ and using the expansion (see Whittaker and Watson)

$$\text{sn}\,(\alpha + j\beta) = \frac{\text{sn}\,\alpha \cdot \text{dn}\,\beta}{1 - \text{sin}^2\,\beta\,\text{dn}^2\,\alpha} + j\,\frac{\text{cn}\,\alpha\,\text{dn}\,\alpha\,\text{sn}\,\beta\,\text{cn}\,\beta}{1 - \text{sin}^2\,\beta\,\text{dn}^2\,\alpha}. \tag{7.53}$$

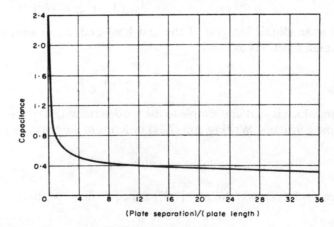

Fig. 7.12

The above and similar transformations, involving boundaries consisting of finite plates, have been used in the calculation of the impedance of waveguides and transmission lines [3–6].

7.5.3 Elliptic integrals of the third kind

In practice, integrals are sometimes encountered which involve three variables (as opposed to two, t and k, in the integrals of the first and second kinds). One of these integrals was defined by Legendre as the elliptic integral of the third kind and given the symbol

$$\prod(t, k_1, k),$$

where

$$\prod(t, k_1, k) = \int_0^t \frac{dt}{(1 - k_1^2 t^2) \sqrt{(1 - t^2)(1 - k^2 t^2)}}. \tag{7.54}$$

Putting $t = \operatorname{sn} u$ and $k_1 = k \operatorname{sn} \alpha$ gives the same integral in terms of Jacobian functions as

$$\prod(u, \alpha) = \int_0^u \frac{du}{1 - k^2 \operatorname{sn}^2 \alpha \operatorname{sn}^2 u}, \tag{7.55}$$

to modulus k. Jacobi, however, defined the elliptic integral of the third kind in a different way as

$$= k^2 \operatorname{sn} \alpha \operatorname{cn} \alpha \operatorname{dn} \alpha \int_0^u \frac{\operatorname{sn}^2 u \, du}{1 - k^2 \operatorname{sn}^2 \alpha \operatorname{sn}^2 u}, \tag{7.56}$$

which is here given the symbol $\prod_J(u, \alpha)$. This form is the more convenient for evaluation, but integrals are more easily recognized in Legendrian form. The two forms are related by

$$\prod = u + \frac{\operatorname{sn} \alpha}{\operatorname{cn} \alpha \operatorname{dn} \alpha} \prod_J. \tag{7.57}$$

In general the evaluation of integrals of the third kind is very difficult, and the only feasible way of doing this involves the use of various auxiliary functions. The development of numerical quadrature techniques has largely rendered this type of analysis uneconomical and unnecessary.

However, there are some solutions which are of historic interest and which have been frequently applied to practical problems in the past. The field outside a charged rectangular conductor was first analysed by Bickley [11]. Transformation of the boundary of the conductor can be achieved either from the infinite straight line or from the unit circle.

Another example of the use of elliptic integrals of the third kind occurs in the analysis of the influence of slots or grooves, on the field in the air gap of electrical machines. The transformation required for the solution of this problem has the same form as that required for the solution of the field of a succession of infinitely deep slots in a machine air gap, and both cases have been discussed by Coe and Taylor [12]. The example involving a succession of slots has been described in detail by Gibbs (p. 190).

7.5.4   The occurrence of elliptic function

Any integral of the form

$$\int \frac{R_1(t)dt}{R_2(t)\sqrt{(t-a)(t-b)(t-c)}}$$

or

$$\int \frac{R_2(t)dt}{R_1(t)\sqrt{(t-a)(t-b)(t-c)(t-d)}},$$

in which $R_1(t)$ and $R_2(t)$ are rational functions of t, is elliptic. It can be expressed in terms of the basic elliptic integrals, though the required manipulation and subsequent numerical evaluation may be extremely difficult and laborious. The first of the above two forms of integral arises in the transformation of a boundary having three right angles, and the second four right angles. An integral containing more than four rooted factors in the denominator (occurring when the boundary has more than four right angles) is said to be hyperelliptic, and it cannot be expressed in terms of the basic elliptic integrals. Elliptic integrals also occur in other transformations, e.g. in the treatment of some boundaries having rounded corners [13] and of others having angles not multiples of $\pi/2$ see reference [7]. It is not possible to discuss here the general problem of manipulation, and instead the reader is referred to the excellent treatises by Edwards and by Tannery and Molk.

When attempting to assess the degree of difficulty associated with the analysis of a particular problem, one cannot in general apply any simple test. However, it is possible to distinguish broadly the values of the three methods of handling an elliptic integral which have been described, its expression in terms of Jacobian elliptic functions, its reduction to Legendre's three integrals, and the expansion of the integrand as a series. Of these, the first is the best for analytical manipulation. The second is of very limited value both for manipulation and evaluation, and the third is unsuitable for manipulation but is sometimes good for numerical evaluation.

The considerations involved in the numerical evaluation of an analytical solution are rather more straightforward. When the evaluation is carried out by hand, the only generally practical method involves the use of the tables; the chief problem is that there are then large intervals over certain ranges and as a result interpolation is very difficult. This process has, in effect, been superseded. The alternative to the above method involves the computation of the function from a series expansion, see Appendix 2. This is preferable to feeding tables into the computer store. Freeman [14] has applied the series developed by King, which are most suitable for evaluation of elliptic integrals and functions and which converge more rapidly than the other better-known series. However, because of the development of numerical techniques of quadrature, any problem that involves, in its solution, the use of tables or the summation of series is more efficiently treated by numerical methods.

7.6   NUMERICAL METHODS

It has already been pointed out that analytical methods are often inadequate for polygonal boundaries with many vertices, particularly when these are not integral multiples of $\pi/2$. For these boundaries a numerical method can be used to make possible a significant extension in the range of tractable problems.

Numerical methods of determining the constants in a conformal transformation and the use of numerical procedures to integrate the Schwarz–Christoffel equation were first developed in the study of electrical machine problems [15, 16, 17]. The transformation used for one of these problems and the equations expressing the values of the constants are discussed in section 7.5.3. Later, the application of combined analytical and numerical techniques to the calculation of the characteristic impedance of waveguides was proposed [18] and the treatment of arbitrary singly and doubly connected regions reported [19, 20]. Finally, direct-search techniques of minimization [21] have been applied to the determination of constants and integration routines with automatically varying step-length employed [22, 23].

There are two aspects to the solution of a problem—the numerical integration of the Schwarz–Christoffel equation for any complex value of t, and the determination of the t-plane constants for any set of values of the desired dimensions.

The Schwarz–Christoffel equation may be expressed in integral form as

$$Z_{nm} = S \int_{t}^{t_m} f(t)\,\mathrm{d}t, \tag{7.58}$$

and the values of the transformation constants can be obtained by integrating eqn (1) to give

$$|Z_{i,i+1}| = \left| K \int_{t_i}^{t_{i+1}} f(t)\,\mathrm{d}t \right|. \tag{7.59}$$

The explicit relationships, obtained by relating the distance between parallel lines to the values of the residues at the appropriate poles, can either be solved directly or by an iteration process to give some of the constants. The remaining implicit relationships,

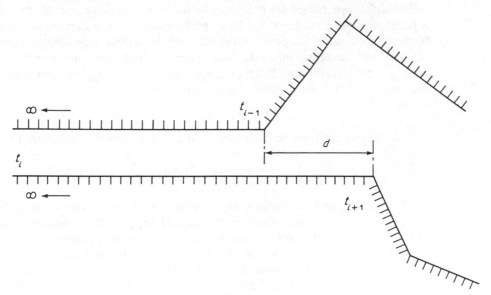

Fig. 7.13

where a formal numerical integration is needed to relate the chosen dimensions to the t-plane constants, are rearranged to form a set of equations, elements of which tend to zero at the required solution. A typical element, F_j, is given by

$$F_j = |Z_{i,i+1}| - \left| K \int_{t_i}^{t_{i+1}} f(t)\,dt \right|. \tag{7.60}$$

An error function composed of the elements of F_j is defined as

$$E_r = \sum_j F_j^2.$$

Initial values of the unknown constants t_1, t_2, etc. are chosen and are varied by a direct search program until E_r has been made sufficiently small. The method of Powell [21] can be used but has largely been superseded by Peckham's method [24].

7.6.1 Classes of integral that arise in numerical evaluation

If a boundary involves parallel lines meeting at infinity, the integral has a pole of order 1 at the relevant vertex. It is convenient to separate the discussion into two parts according to this criterion.

(a) Integrals not containing poles of order 1

This class of integral is of the form

$$Z = \left| K \int_{t_i}^{t_i+1} (t - t_i)^\alpha (t_{i+1} - t)^\beta H(t)\,dt \right|, \tag{7.61}$$

where $\alpha, \beta > -1$.

The difficulties caused by poles can be overcome by commencing the integration at a point removed from the poles by a small amount, δ. The numerical integrations can be performed using Simpson's formulae with a suitable step length. However, more reliable ˙and accurate methods have been evolved and one approach is to use Gauss–Jacobi quadrature. In this method the interval $|t_i, t_{i+1}|$ is first transformed into the interval $|-1, 1)$ by the linear transformation

$$t = \tfrac{1}{2}[t_i + t_i + 1 + X(t_{i+1} - t_i)]. \tag{7.62}$$

The integral can then be evaluated by the Gauss–Jacobi quadrature formula

$$\int_{-1}^{1} (1 + x)^\alpha (1 - x)^\beta H(x)\,dx = \sum_{K=1}^{M} A_K H(x_K). \tag{7.63}$$

The nodes x_K and the coefficients A_K which are related to the Jacobi polynomial are found numerically as described by Stroud and Secrest. The consistency of results for different values of M is used to check the accuracy of the solution.

A third approach known as the Clenshaw–Curtis method [25, 26], consists of expanding the integral in a finite Chebyshev series and integrating the terms in the series one by one. Smith and Imhof have derived a quadrature formula for the method and Smith and O'Hara, after tests on different functions, give an expression for a reliable error estimate. The Chebyshev series is chosen in preference to other series, because it is rapidly convergent. This is important both in the reduction of computational time and in the estimation of an accurate upper bound for the error. An approach related to the Clenshaw–Curtis method has been implemented by Branders and Piessens [27] and this is available as a computer package in which the degree of accuracy required can be specified.

(b) Integrals containing poles of order 1

When obtaining solutions for boundaries which have some parallel sides, as shown in Fig. 7.13, it is necessary to find an expression from which to evaluate dimensions, such as 'd'. It can be shown that this involves the Cauchy principal value denoted by

$$d = P \int_{t_{i-1}}^{t_{i+1}} \frac{|K|(t - t_{i-1})^{\alpha}(t_{i+1} - t)^{\beta} G(t)\, dt}{t - t_i}. \tag{7.64}$$

There are two convenient methods for the accurate evaluation of this integral. The first involves the use of the Chebyshev series and the second the removal of the singularity.

Evaluation using the Chebyshev series

The interval $|t_{i-1} t_{i+1}|$ can as before the transformed into the interval $|-1, 1)$ by a linear transformation, the geometry shown in Fig. 7.14 will be discussed. It can be shown that the field in induction machines can be evaluated by treating the air gap field in sections which are subsequently sewn together, that is connected through common boundary conditions. The slotted region shown in Fig. 7.14 is transformed from the

Evaluation by substracting out the singularity

This technique of evaluating Cauchy principal values is based on the method of removing the singularity.

Consider the function

$$d = P \int_{-1}^{+1} S\, \frac{(1 + x)^{\alpha}(1 - x)^{\beta} G(x)}{x - A}\, dx. \tag{7.65}$$

This can be expanded as a Taylor series about the point $x = A$ and the components evaluated by methods already discussed.

The field in the region of an electrical machine air gap with unequal slot openings

As an example of the numerical evaluation of integrals arising in the use of conformal transformation, the geometry shown in Fig. 7.14 will be discussed. It can be shown that the field in induction machines can be evaluated by treating the air gap field in sections which are subsequently sewn together, that is connected through common boundary conditions. The slotted region shown in Fig. 7.14 is transformed from the

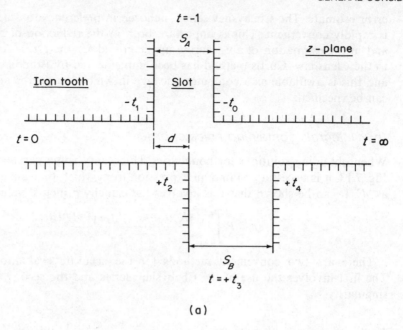

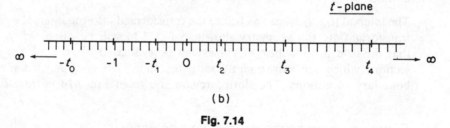

Fig. 7.14

real-axis of the t-plane by the Schwarz–Christoffel equation. The appropriate boundary conditions are imposed in the t-plane and then transformed to the Z-plane.

The transformation equation is given by

$$\frac{dz}{dt} = \frac{K\sqrt{[(t+t_0)(t+t_1)(t-t_2)(t-t_4)]}}{t(t+1)(t-t_3)}, \tag{7.66}$$

where K, t_0, t_1, t_2, t_3 and t_4 are constants which have to be found before the equation can be usefully employed. To find the six constants, six equations have to be derived. Four of the six can be obtained by evaluating the residuals at the poles of eqn (7.66) and these are:

$$K = \frac{1}{\pi}$$

$$t_3^2 = t_0 \cdot t_1 \cdot t_2 \cdot t_4 \tag{7.67}$$

$$S_A^2(1+t_3)^2 = (1-t_0)(t_1-1)(t_2+1)(t_4+1) \tag{7.68}$$

$$S_B^2(1+t_3)^2 t_3^2 = (t_3+t_0)(t_3+t_1)(t_3-t_2)(t_4-t_3) \tag{7.69}$$

in which S_A is the width of one slot and S_B is the width of the other as shown in Fig. 7.14(a). The other two equations are derived as follows: the condition that the two surfaces on either side of a slot are collinear is given by

$$\left| P \int_{-t_1}^{-t_0} \frac{S\sqrt{[(t+t_0)(t+t_1)(t-t_2)(t-t_4)]}}{t(t+1)(t-t_3)} \cdot dt \right| = 0. \tag{7.70}$$

The slot displacement, d, is given by

$$d = P \int_{-1_1}^{t_2} \frac{S\sqrt{[(t+t_0)(t+t_1)(t-t_2)(t-t_4)]}}{t(t+1)(t-t_3)} dt. \tag{7.71}$$

To solve the six non-linear equations direct search methods are employed.

It is possible to reduce the number of constants that have to be found in the search process to two. Suppose arbitrary values of the constants t_3 and t_4 are chosen, the constants t_0, t_1 and t_2 can be obtained as the roots of a cubic equation formulated by suitable rearrangement of eqns (7.66)–(7.69). Equations (7.70) and (7.71) are rearranged to give the sum of squares function E_r. The constants t_3 and t_4 are varied in the direct search programme until E_r is made sufficiently small. Solving the equations in this manner means that sets of constants can be found for different values of S_A and S_B, and the slot displacement d. Equations (7.70) and (7.71) have been solved using both methods of integration discussed in section 7.6.1(b), integrals containing poles of order 1.

For the purpose of checking the accuracy of numerical evalution, it is useful to consider the one finite value of slot displacement for which the constants can be found analytically. This occurs when the slot centers are aligned, so that

$$d = (S_B - S_A)/2.$$

It can be shown that the constants are given by the following equations

$$t_1 = \frac{x + \sqrt{(x^2 - 4)}}{2} \tag{7.72}$$

$$t_0 = \frac{1}{t_1} \tag{7.73}$$

$$t_4 = \frac{y + \sqrt{(y^2 - 4)}}{2} \tag{7.74}$$

$$t_2 = \frac{1}{t_4} \tag{7.75}$$

and $t_3 = 1$, where

$$x = \frac{S_A^2 - S_B^2 + \sqrt{[(S_A^2 - S_B^2 + 4) + 16S_B^2]}}{2}$$

and

$$y = x + S_B^2 - S_A^2.$$

Taking three typical sets of values of S_A and S_B the constants were computed from eqns (7.72)–(7.75). Equations (7.70) and (7.71) were then evaluated using methods

Table 7.1

$S_A = 1.0$, $S_B = 1.25$

$t_0 = 0.23218932 \times 10^{+1}$		$t_1 = 0.43068303 \times 10^0$
$t_2 = 0.33563320 \times 10^0$	$t_3 = 1$	$t_4 = 0.29794430 \times 10^{+1}$
Method 1		
$n = 14$	$d = 0.12500000$	$E_{nT} = 0.20 \times 10^{-9}$
$n = 25$	$H = 0.00000000$	$E_{nT} = 0.20 \times 10^{-9}$
Method 2		
	$d = 0.12500000$	
	$H = 0.18189894 \times 10^{-10}$	

$S_A = 2.0$, $S_B = 2.5$

$t_0 = 0.36781859 \times 10^{+1}$		$t_1 = 0.27179926 \times 10^0$
$t_2 = 0.16569204 \times 10^0$	$t_3 = 1$	$t_4 = 0.60352931 \times 10^2$
Method 1		
$n = 12$	$d = -0.25000000$	$E_{nT} = 0.81 \times 10^{-9}$
$n = 33$	$H = 0.00000000$	$E_{nT} = 0.12 \times 10^{-7}$
Method 2		
	$d = 0.25000000$	
	$H = 0.29103830 \times 10^{-6}$	

$S_A = 5.0$, $S_B = 7.5$

$t_0 = 0.44126434 \times 10^{+1}$		$t_1 = 0.22662127 \times 10^0$
$t_2 = 0.27885148 \times 10^{-1}$	$t_3 = 1$	$t_4 = 0.35861385 \times 10^{+2}$
Method 1		
$n = 10$	$d = -1.25000000$	$E_{nT} = 0.94 \times 10^{-8}$
$n = 44$	$H = 0.00000001$	$E_{nT} = 0.34 \times 10^{-7}$
Method 2		
	$d = 0.25000000$	
	$H = 0.00000000$	

---

1 and 2. The results of the calculations are presented in Table 7.1. In the table, n is the number of Chebyshev coefficients used in method 1, and H is the result obtained from evaluating eqn (7.70). For method 1, the error term E_{nT}, was taken as the sum of the absolute values of the last three terms in the Chebyshev series multiplied by I_0, that is

$$E_{nT} = (|b_n| + b_{n-1}| + |b_{n-2}|) \cdot I_0 \qquad (7.76)$$

In method 2, the integral was evaluated using the Gauss four and six point formulas to a specified accuracy of 10^{-6}.

The integrals have been computed to a higher accuracy than would normally be required in a direct-search process; however, the results illustrate that high accuracy can be obtained using either method.

The authors have had considerable experience with both methods and have found that the second is more efficient both in terms of computer storage and execution time.

For example, when finding the constants for the slot combinations $S_A = 1.0$, $S_B = 1.25$, the slot centres being displaced at intervals of 0.1 units from 0 to 2 units. Extensive use of this numerical analysis procedure has been made for the analysis of the magnetic field in the air gap of induction machines [28, 29, 30, 31].

7.7 NON-EQUIPOTENTIAL BOUNDARIES

Consideration previously in the book has been mainly restricted to equipotential or flux line boundaries, though a variety of boundaries which are combinations of both has been considered. This section discusses the three basic classes of problem as defined by the field conditions on the (closed) boundary: the first class has a potential distribution specified at all points on the boundary, the second has the normal component of field gradient specified, and the third is a combination of the two previous classes. The first class of problem is treated in section 7.6.1 and the two previous classes. The first class of problem is treated in section 7.6.1 and the second and third classes in section 7.6.2.

7.7.1 Boundary value problems of the first kind

A problem in which values of potential are specified at all points along the boundary is said to be of the *first kind* (or of the *Dirichlet* type). It is possible to obtain a solution to any such problem for the interior of the unit circle or the infinite half-plane, and then to use this solution, by means of transformation methods, for a wide range of other boundary shapes.

Field inside the unit circle

The solution inside the unit circle is obtained by means of the *complex potential function of Schwarz*. If $\psi'(\theta')$ describes the specified potential distribution along the periphery of the circle, θ' being the angle subtended at the centre, and if p is the complex variable of position, then the solution for the field inside the unit circle is expressed by

$$\psi + j\phi = \frac{1}{2\pi} \int_0^{2\pi} \frac{e^{j\theta'} + p}{e^{j\theta'} - p} \psi'(\theta') \, d\theta'. \qquad (7.77)$$

The real part of this equation can also be put in a standard form by putting $p = r e^{j\theta}$ to give

$$\psi = \frac{1}{2\pi} \int_0^{2\pi} \frac{1 - r^2}{1 - 2r\cos(\theta - \theta') + r^2} \psi'(\theta') \, d\theta'. \qquad (7.78)$$

This is the *Poisson integral* and it can be used for evaluating potential distributions for the case of a single, circular boundary.

In using either of these equations it is required only to express the potential distribution along the circular boundary in terms of θ' and to solve the resulting definite integral for every point p at which the field is to be determined. This integral often has no analytical solution, but numerical methods can be used (see section 7.6.1). If the complex potential function has to be evaluated at many points, the numerical method involves much computation, though it is unlimited in its scope.

A useful problem, capable of analytical solution, arises when the boundary consists of equipotential sections. If there are m of these, each with potential ψ_n, and if the limits of the nth one subtend angles θ'_n and θ'_{n-1} at the centre, the complex potential function is, from eqn (7.77)

$$\psi + j\phi = \frac{1}{2\pi} \sum_{n=1}^{m} \psi_n \int_{\theta'_{n-1}}^{\theta_n} \frac{e^{j\theta'} + p}{e^{j\theta'} - p} \, d\theta'. \tag{7.79}$$

This gives, when integrated,

$$\psi + j\phi = \frac{-1}{2\pi} \sum_{n=1}^{m} (\theta'_n - \theta'_{n-1})\psi_n - \frac{j}{\pi} \sum_{n=1}^{m} \psi_n \log \frac{e^{j\theta'_n} - p}{e^{j\theta_{n-1}} - p}, \tag{7.80}$$

which may be simply used to give immediately a solution to any problem of this type. For example, the solution for the unit circle with two equal equipotential sections, of potential $+\psi_A$ and $-\psi_A$, is given, by substituting $\theta'_0 = 0$, $\theta'_1 = \pi$ and $\theta'_2 = 2\pi$ as

$$\psi + j\phi = -\frac{2j}{\pi} \psi_A \log \frac{p+1}{p-1} + 2\psi_A. \tag{7.81}$$

Field in the upper half-plane

The above solution obtained by Schwarz can be modified to give the solution of the field in the upper half-plane due to a potential distribution along the real axis. Thus, by applying the bilinear transformation

$$t = j\left(\frac{1-p}{1+p}\right), \tag{7.82}$$

the complex potential function becomes

$$\psi + j\phi = \frac{j}{\pi} \int_{-\infty}^{+\infty} \frac{1 + v't}{(1 + v'^2)(t - v')} \psi(v') \, dv', \tag{7.83}$$

in which $\psi(v')$ is the potential distribution along the real axis of t. The real part of eqn (7.83) leads to the equivalent of the Poisson integral for the infinite half-plane, namely,

$$\psi = \frac{1}{\pi} \int_{-\infty}^{\infty} \frac{v}{(u - v')^2 + v^2} \psi(v') \, dv'. \tag{7.84}$$

Again, if there are m equipotential sections along the real axis, the integrals can be performed analytically; and if the nth section is at potential ψ_n and lies between the points v'_n and v'_{n+1}, the potential function becomes

$$\psi + j\phi = \frac{j}{\pi} \sum_{n=1}^{m} (\psi_n - \psi_{n+1}) \log \frac{\sqrt{1 + v'^2_n}}{t - v'_n} + \psi_m. \tag{7.85}$$

This equation could be obtained by considering the field of a combination of line sources, at the points of potential division, along the real axis. For two sections, at potentials $+\psi_A$ and $-\psi_A$, divided at the origin, eqn (7.85) reduces to the familiar form

$$\psi + j\phi = -\frac{2j}{\pi} \psi_A \log t. \tag{7.86}$$

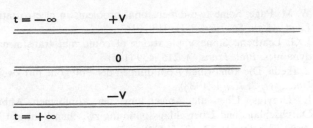

$t = -\infty$ +V

0

—V

$t = +\infty$

Fig. 7.15

Equation (7.85) has been used (see Weber) to give the electric field of a transfomer winding, represented as three sheets (Fig. 7.15). This boundary can be transformed simply from the real axis of the t-plane. If the relative electric potentials on the windings are $+ V$, 0, and $- V$, the field in the t-plane is given by eqn (7.85) as

$$w = \psi + j\phi = \frac{j}{\pi}\left(V \log \frac{\sqrt{2}}{t+1} + V \log \frac{\sqrt{1+q^2}}{1-q} \right) - V, \qquad (7.87)$$

the sections being separated by the points $t = -1$ and $t = +q$.

7.7.2 Boundary value problems of the second and mixed kinds

If the boundary conditions specify the normal components of gradient, the problem is said to be of the *second* kind (or of the *Newmann* type). If the boundary conditions describe potentials on some parts of the boundary and potential gradient over the rest, the problem is said to be of *mixed* kind. In general, transformation methods are unsuited to both these classes of problem, which for simple boundary shapes are best treated by the methods of Chapter 3. However, in the special case where the boundary is formed partly by flux lines (that is, lines with a constant normal gradient) and partly by equipotential lines, solutions are often available and appear frequently elsewhere in the book. For example, in section 7.2 the solution for the fields of currents in the infinite half-plane is given for second and mixed kinds of boundary condition on the real axis; and in section (7.2.4) the solution is given for two equal finite plates (mixed conditions on the real axis). Similar solutions have also been obtained for the unit circle, see Weber and Reference [32], but these are not of great practical interest.

REFERENCES

[1] P. J. Lawrenson, A note on the analysis of the fields of line currents and charges, *Proc. Instn. Elect. Engrs.*, **109C**, 86 (1962).

[2] F. W. Carter, The magnetic field of the dynamo-electric machine, *J. Instn. Elect. Engrs.*, **359**, 1115 (1926).

[3] S. Drinker, Short wavelength response of magnetic reproducing heads with rounded gap edges, *Phillips Res. Rep.*, **16**, 307 (1961).

[4] J. D. Cockroft, The effect of curved boundaries on the distribution of electrical stress round conductors, *J. Instn. Elect. Engrs.*, **66**, 385 (1928).

[5] W. M. Page, Some two-dimensional problems in electrostatics and hydrodynamics, *Proc. Lond. Math. Soc.*, *ser. 2*, II, 313 (1912).

[6] J. G. Leathem, Some applications of conformal transformation to problems in hydrodynamics, *Phil. Trans. A*, **215**, 439 (115).

[7] Y. Ikeda, Die konformen Abbildungen der Polygone mit zwei Ecken, *J. Fac. Sci. Hokkaido Univ.*, *ser. 2*, **2**(2), 1 (1938).

[8] L. Dreyfus, Uber die Anwendung der konformen Abbildung zur Berechnung der Durchschlags und Uberschlagsspannung zwischen kantigen Konstruktionsteilen unter 01, *Arch. Elektrotech.*, **13**, 123 (1924).

[9] F. W. Carter, The magnetic field of the dynamo-electric machine, *J. Instn. Elect. Engrs.*, **359**, 1115–1139 (1926).

[10] J. B. Izatt, Characteristic impedance of two special forms of transmission line, *Proc. Instn. Elect. Engrs.*, **111**, 1551 (1964).

[11] W. G. Bickley, Two-dimensional potential problems for the space outside a rectangle, *Proc. Lond. Math. Soc.*, *ser. 2*, **37**(2), 82 (1932).

[12] R. T. Coe and H. W. Taylor, some problems in electrical machine design involving elliptic functions, *Phil. Mag.*, **6**, 100 (1928).

[13] J. D. Cockroft, The effect of curved boundaries on the distribution of electrical stress round conductors, *J. Instn. Elect. Engrs.*, **66**, 385 (1928).

[14] E. M. Freeman, The calculation of harmonics due to slotting in the flux-density waveform of a dynamo-electric machine, *Proc. Instn. Elect. Engrs.*, **109C**, 581 (1962).

[15] K. J. Binns, The magnetic field and centring force of displaced ventilating ducts in machine cores, *Proc. Instn. Elect. Engrs.*, **108C**, 64 (1961).

[16] K. J. Binns, Pole-entry flux pulsations, *Proc. Instn. Elect. Engrs.*, **109C** (1962).

[17] K. J. Binns, Calculation of some basic flux quantities in induction and other doubly-slotted electrical machines, *Proc. Instn. Elect. Engrs.*, **111**, 1847 (1964).

[18] P. A. Laura, Characteristic impedance of a rectangular transmission line, *Proc. Instn. Elect. Engrs.*, **113**, 1595 (1956).

[19] P. A. Laura, Conformal mapping of a class of doubly connected regions, *NASA Tech. Rep.*, Grant No. G125-61 (1965).

[20] M. K. Richardson, A numerical method for the conformal mapping of finite doubly connected regions with application to the torsion problem of hollow bars, PhD thesis, Univ. of Alabama (1965).

[21] M. J. D. Powell, An efficient method for finding the minimum of a function of several variables without calculating derivatives, *Comput. J.*, **7**, 115 (1964).

[22] P. J. Lawrenson and S. K. Gupta, Conformal transformation employing direct-search techniques of minimisation, *Proc. Instn. Elect. Engrs.*, **115**, 427 (1968).

[23] K. J. Binns, Numerical methods of conformal transformation, *Proc. Instn. Elect. Engrs.*, **118**, 909 (1971).

[24] G. Peckham, A new method of minimising a sum of squares without calculating gradients. *Computer J.*, **13**, 413–420 (1970).

[25] K. J. Binns, G. R. Rees and P. Kahan, The evaluation of improper integrals encountered in the use of conformal tansformation. Inter. *Jour. New Methods in Engineering*, **14**, 567–580 (1979).

[26] J. P. Imhof, On the method of numerical integration of Clenshaw and Curtiss, *Num. Math.*, **5**, 138 (1963).

[27] M. Branders and R. Piessens, An extension of Clenshaw Curtis Quadrature, *J. Comp. and Appl. Math*, **1**, 55 (1975).

[28] K. J. Binns and G. R. Rees, Main flux pulsations and tangential tooth ripple forces in induction motors, *Proc. IEE*, Vol. 122, No. 3, March, 1975.

[29] K. J. Binns and G. R. Rees, Radial tooth ripple forces in induction motors due to the main flux, *Proc. IEE*, **125**, (11) (November, 1978).

[30] K. J. Binns and P. A. Kahan, Effect of load on the flux pulsations and radial force pulsations of induction motor teeth, *Proc. IEE*, **127B** (4) (July, 1980).

[31] K. J. Binns and P. A. Kahan, Effect of load on the tangential force pulsations and harmonic torques of squirrel-cage induction motors, *Proc. IEE*, **128B** (4) (July, 1981).

[32] J. Hodgkinson, A note on a two-dimensional problem in electrostatics, *Q. J. Math.*, **9**, 5 (1938).

Additional references

Anderson, G. M., The calculation of the capacitance of coaxial cylinders of rectangular cross-sections, *Trans. Am. Instn. Elect. Engrs.*, **69** II, 728.

Bates, R. H. T., The characteristic impedance of the shielded slab line, *Trans. Instn. Rad. Engrs.*, **MTT-4**, 28 (1956).

Bergmann, S., *Math. Z.*, **19**, 8 (1923).

Chen, T. S., Determination of the capacitance, inductance, and characteristic impedance of rectangular lines, *Trans. Instn. Rad. Engrs.*, **MTT-8**, 510 (1960).

Chisholm, R. M., Characteristic impedance of through and slab lines, *Trans. Instn. Rad. Engrs.*, **MTT-4**, 166 (1956).

Cohn, S. B., Analysis of the metal-strip delay structure for microwave lenses, *J. App. Phys.*, **20**, 251 (1949).

Cohn, S. B., Charateristic impedance of broadside coupled strip transmission line, *Trans. Instn. Rad. Engrs.*, **MTT-8**, 633 (1960).

Cohn, S. B., Thickness corrections for capacitance obstacles and strip conductors, *Trans. Instn. Rad. Engrs.*, **MTT-8**, 638 (1960).

Davy, N., The field between equal semi-infinite rectangular electrodes or magnetic pole-pieces, *Phil. Mag.*, ser. 7, **35**, 819 (1944).

Foster, K., The characteristic impedance and phase velocity of high line, *Br. Instn. Rad. Engrs.*, **18**, 718 (1958).

Getzinger, W. J., A coupled strip line configuration using printed-circuit construction that allows very close coupling, *Trans. Instn. Rad. Engrs.*, **MTT-9**, 535 (1961).

Getzinger, W. J., Coupled rectangular bars between parallel plates, *Trans. Instn. Rad. Engrs.*, **MTT-10**, 65, 1 (1962).

Greenhill, A. G., Solution by means of elliptic functions of some problems in the conduction of electricity and heat in plane figures, *Q. J. Pure Appl. Math.*, **17**, 289 (1881).

Herbert, C. M., *Phys. Rev.*, **II**, 17, 157 (1921).

Ikeda, Y., Die konformen Abbildungen der Polygone mit zwei Ecken, *J. Fac. Sci. Hokkaido Univ.*, ser. 2, **2**(2), 1 (1938).

Joyce, W. B., Separation of variables solution from the Schwarz–Christoffel transformation. *Q. J. Appl. Math.*

Langton, N. H. and Davy, N., Two-dimensional field in a semi-infinite slot terminated by a semi-circular cylinder, *Br. J. Appl. Phys.*, **4**, 134 (1953).

Langton, N. H. and Davy, N., Two-dimensional field above and below an infinite corrugated sheet, *Br. J. Appl. Phys.*, **3**, 156 (1952).

Langton, N. H. and Dary, N., Two-dimensional field above and below an infinite corrugated sheet, *Br. J. App. Phys.*, **3**, 156 (1952).

Levy, R., New coaxial to strip transformers using rectangular lines, *Trans. Instn. Rad. Engrs.*, MTT-9, 273 (1961).

Love, A. E. H., Some electrostatic distributions in two dimensions, *Proc. Lond. Math. Soc., ser. 2*, **22**, 337 (1923).

Morsztyn, K., The application of conformal transformations and elliptic functions to the analysis of a synchronous non-salient-pole machine, *Monash Univ. Eng. Report*, MEE 66-1 (1966).

Morton, N. B., Two-dimensional fields specified by elliptic functions, *Phil. Mag.*, **2**, 827 (1926).

Palmer, H. B., Capacitance of a parallel-plate capacitor, *Elect. Engrs., Lond.*, **56**, 363 (1937).

Peterson, H., Electrostatic problems, *Z. Physik.*, **38**, 727 (1926).

Powell, M. J. D., A method of minimising the sum of squares of non-linear function without calculating derivatives, *Comput. J.*, **7**, 303 (1965).

Richmond, H. W., On the electrostatic field of a plane of circular grating formed of thick rounded bars, *Proc. Lond. Math. Soc., ser. 2*, **22**, 389 (1924).

Rose, M. E., Magnetic field corrections in the cyclotron, *Phys. Rev.*, **53**, 715 (1938).

Terry, A. and Keller, E. G., Field pole leakage flux in salient-pole dynamo-electric machines, *J. Instn. Engrs.*, **83**, 845 (1938).

COMPUTATIONAL MODELLING— BASIC METHODS

8.1 INTRODUCTION AND ENGINEERING OBJECTIVES

The last ten years, not withstanding the continuing importance of the methods discussed in the earlier chapters of this book, has seen the widespread use of numerical techniques in electromagnetic device design. Today, efficient numerical solutions can be obtained for a wide range of problems that are beyond the scope of analytical methods. In particular the limitations imposed by analytic methods, their restriction to homogeneous, linear and steady-state problems, can be overcome by the use of numerical methods. For example, it has been possible for several years now to compute the magnetic forces acting on the components of an electrical machine, taking into account the three-dimensional geometry including the slots, conductors, etc. as well as the saturation effects of the material. By way of illustration see Fig. 8.1 which shows a three-dimensional view of a permanent magnet machine analysed [1] using a general purpose electromagnetic computer program. To model the geometry alone required many thousands of numbers quite apart from the data needed to represent the non-linear properties of the magnetic materials. Thus it is now a matter of routine for engineers to solve many of the complex problems arising during the design phase of sophisticated modern electromagnetic devices by using largely automatic and general purpose procedures; examples include electrical machines of all types as well as the large magnets used in medical diagnostic equipment, particle accelerators, electron beam lenses, fusion magnets and industrial processes—the range of applications is extensive. Such capabilities are due largely to the enormous advances that have been made in the power of the digital computer; and although the basis of numerical analysis (and the construction of problem solving procedures or algorithms as they are usually called) goes back a long way, there is also a steady stream of new developments allowing for more and more difficult problems to be solved. With the present level of activity it is possible to predict that in the next five years the efficiency of computer methods will reach the stage when genuine computer aided design (CAD) procedures are practical for three-dimensional systems, i.e., design 'lay-out' software will be integrated with electromagnetic field analysis programs.

Numerical analysis is the technique for solving mathematical problems by numerical approximation. By this means solutions are arrived at with the aid of rational numbers

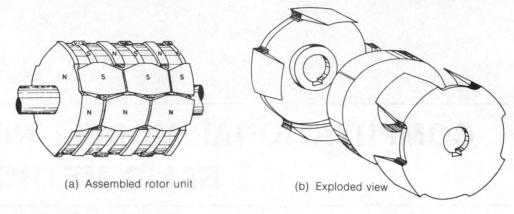

(a) Assembled rotor unit   (b) Exploded view

Fig. 8.1 3D model of a permanent magnet generator.

(numbers of the form p/q where p and q are integers) and, inevitably, since digital computers are used to perform the computations, only a finite set of these numbers will be available. It follows that the use of the digital computer will introduce errors because of the limited precision of the numbers stored in a particular machine, i.e. a real number is rounded to match the number of binary digits or 'word length' available. In addition to this 'rounding error' the size of the computer memory (i.e. the number of 'words') will limit the complexity of the problem that can be solved which in turn means that it will be impossible to achieve a perfect representation of the geometry; thus the actual problem will usually be replaced by a 'computer model', built by the engineer to represent the critical features under investigation. For example, in Fig. 8.2(a) there is shown a model of a 'bending magnet' used for deflecting beams of particles from an accelerator used in high energy physics experiments. This model is already an abstraction from the real magnet since only the most magnetically sensitive components are shown, and in order to investigate the fields produced under the poles and at the ends at least two simpler models can be analysed before dealing with the full three-dimensional geometry, see Figs 8.2(b) and 8.2(c).

As will become clear later the use of numerical technique will introduce further 'modelling errors' called discretization or truncation errors arising when the mathematical description, a continuous partial differential equation, is replaced by an approximate numerical description characterized by discrete points. The discretization and solution of continuum problems, in this case electrical engineering devices, have been approached differently by mathematicians and engineers, see the text by Zienkiewicz and Taylor (p. 2) [2]. The former have concentrated on general techniques, directly applicable to the field equations, such as finite differences approximations [3], see also the text by Southwell [4], and weighted residual procedures (Crandall [5] and Finlayson [6]), whereas the latter have often used a more intuitive approach exploiting analogues between real discrete elements and finite regions of the continuum. Indeed, electrical engineers have often used circuit analogues to model their problem both experimentally [7] and numerically [8]; and before the widespread use of the digital computer many other analogues were used experimentally such as 'resistive paper' and 'electrolytic tanks' (Webber [9], ch. 5). The intuitive, direct analogy, approach by engineers, particularly in

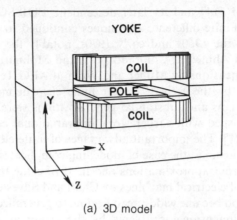

(a) 3D model

(b) Two 2D models

Fig. 8.2 Idealization of the geometry for a particle beam bending magnet.

the area of structural mechanics led to the development of the finite element method [10]; and by 1960 this technique was widely used in other disciplines. It is no accident that this thrust came at this time since in the early 1960s occurred the rapid development of the digital computer as a universal tool for engineers. In the meantime electrical engineers had, in the main, followed and applied the developments in finite differences, now a highly developed discipline of mathematics (Smith [11]), and were able to write elegant computer codes particularly for simple static two-dimensional configurations with linear media—for example, Hornsby [12] at CERN developed a successful code used extensively in the design of particle physics magnets.

An important milestone in the solution of electromagnetics field problems came in 1963 with the seminal work of Winslow [13] at the Lawrence Livermore Laboratory, California; he developed a discretization scheme based on an irregular grid of plane triangles, not only by using a generalized finite difference scheme but also by introducing a variational principle which he showed led to the same result. This latter approach can be seen as being equivalent to the finite element method and is accordingly one of the earliest examples of this technique used for electromagnetics. The resulting computer code TRIM [14] and its later descendent POISSON [15] have been used all over the world. Finite difference techniques continued to be applied by electrical engineers throughout the 1960s and early 1970s, notably the work of Trutt at Delaware [16], Erdelyi and Ahmed at Colorado [17], and Molinari and Viviani [18] in Genoa, and in three dimensions by Muller and Wolff at AEG Telefunken, Germany [19].

However by the early 1970s the finite element method was under scrutiny by the mathematicians and substantial generalizations were made [20] and many cross links were established with earlier work on variational calculus and generalized weighted residuals [21]. The important advantages of finite elements over finite differences were being exploited, i.e., the ease of modelling complicated boundaries and the extendibility to higher order approximations and in 1970 came the first application of the method to rotational electrical machines by Chari and Silvester [22]. From this time on the use of the method became widespread leading to generalized applications for time dependent and three-dimensional problems by the group at Rutherford Laboratory with the production of the codes TOSCA [23, 24] and CARMEN [25].

A parallel development to the above has been with integral methods; these integral formulations, unlike differential formulations which solve the defining partial differential equations (e.g. Poisson's equation), use the corresponding integral equation forms, e.g. equations based on Green's theorem—see section 1.8.2 and Appendix 4 (A4.17). The moment method is an example of an integral formulation, see the text by Harrington [26] for a basic treatment; yet another class of integral procedures are the so called boundary element methods [27, 28] based on applications of Green's integral theorems. Whilst these methods are often difficult to apply they can produce accurate economic solutions and have been used extensively in certain static and high frequency problems, an example of a general purpose program, first developed in 1971, is the magnetization integral equation code GFUN which was specifically designed for three-dimensional static problems [29].

In the rest of this chapter the basic ideas of numerical modelling for electromagnetic continuum problems are explained. The next section will deal with a basic mathematical model with particular reference to the two-dimensional linear case, then the following two sections are devoted to the finite difference and finite element methods in their simplest form. The chapter ends with a section on the unifying principle known as 'the weighted residual method' which will form the basis of much of the developments in later chapters.

8.2 MATHEMATICAL MODELS

8.2.1 Differential equations

The appropriate model for an electromagnetic field problem is derived from the basic defining equations. The fundamental behaviour of the electromagnetic field was reviewed in Chapter 1, and it was noted there (section 1.7) that the electrostatic field is uniquely defined if (a) the electrostatic potential satisfies a partial differential equation of the Poisson type (see eqn (1.36)) and (b) if the problem is bounded by a surface over which boundary conditions are prescribed. It was also shown that the magnetostatic and electric current flow cases could be similarly formulated; the magnetics case requiring some care

Table 8.1 Mathematical model equations for static fields.

Problem type	Potential	Defining equation
Electrostatics	Scalar	$\nabla \cdot \varepsilon \nabla V = -\rho$
Current flow	Scalar	$\nabla \cdot \sigma \nabla V = 0$
Magnetostatics	Reduced scalar	$\nabla \cdot \mu \nabla \phi = -\nabla \cdot \mu \mathbf{H}_s$
	Total scalar	$\nabla \cdot \mu \nabla \psi = 0$
	Vector	$\nabla \cdot \dfrac{1}{\mu} \nabla A_z = -J_z$

when scalar potentials are used because of the multi-value nature of the potential due to current sources—see section 1.4. For convenience, Table 8.1 lists the various mathematical models of the Poisson type (also see section 1.7); all these models are examples of differential formulations because only differential operators appear in the equations.

That is, for all the common static problems in two dimensions, a suitable mathematical model is the Poisson equation:

$$\nabla \cdot \kappa \nabla u + Q = 0 \qquad (8.1)$$

where u is a potential defined in the interior of a domain Ω spanned by a surface Γ. Over the surface Γ either:

$$u = u_o \quad \text{Dirichlet boundary condition} \qquad (8.2)$$

or,

$$\frac{\partial u}{\partial n} = q \quad \text{Neumann boundary condition,} \qquad (8.3)$$

it is shown in Appendix 6 that if the Newmann condition is imposed everywhere then to ensure uniqueness it is necessary to specify u at least at one point. The possibility of a mixed boundary condition, i.e. a linear combination of eqns (8.2) and (8.3), is sometimes used, for example, over Γ,

$$a_o u + b_o \frac{\partial u}{\partial n} = c_o. \qquad (8.4)$$

In general the constitutive parameter κ is a function of position and field, however for the time being only linear problems will be considered so κ is restricted to be piecewise constant, i.e. the possibility of regions each with a different but constant value of κ are allowed. The potential u must satisfy, in addition to the boundary conditions, the appropriate interface conditions between contiguous interior regions with differing constitutive parameters, to make these ideas clear consider the T-bar magnetostatics problem shown in Fig. 8.3.

The geometric model illustrated in Fig. 8.3 is already an approximation of what, in reality, could be a detail from a three-dimensional electrical machine highlighting the slot geometry. Figure 8.3(b) shows the mathematical model complete with boundary conditions allowing for symmetry and approximating the external effects by assigning

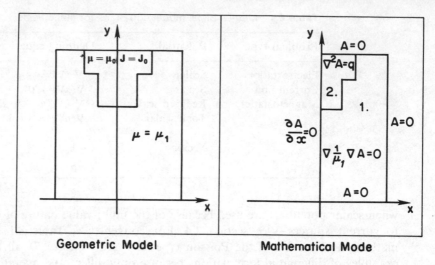

Fig. 8.3 Example of geometric and mathematical models.

flux boundaries. In this case the interface conditions between region 1 with permeability μ_1 and region 2 with permeability $\mu = \mu_0$ are given by:

$$\mathbf{A}_1 - \mathbf{A}_2 = 0 \tag{8.5}$$

$$(\mathbf{B}_1 - \mathbf{B}_2) \cdot \mathbf{n} = 0 \tag{8.6}$$

$$(\mathbf{H}_1 - \mathbf{H}_2) \times \mathbf{n} = 0 \tag{8.7}$$

when $\mathbf{n}$ is the normal vector at the interface.

8.2.2 Integral equations

The integral analogue of the field equations can also be used as a mathematical model for problems in electromagnetics. For example the equivalent integral equation to the Poisson differential equation for constant constitutive parameters, see section 1.6, can be written for the magnetostatic case as follows:

$$\phi(r) = \frac{1}{4\pi} \int_\Gamma \phi \frac{\partial}{\partial n} \left(\frac{1}{R} \right) d\Gamma - \frac{1}{4\pi} \int_\Gamma \frac{\partial \phi}{\partial n} \left(\frac{1}{R} \right) d\Gamma, \tag{8.8}$$

where ϕ is the magnetostatic reduced scalar potential. In contrast with the differential approach eqn (8.8) does not require additional boundary conditions; the integral equation is complete as it stands, since the boundary values appear explicitly in the integrands. Thus if both ϕ and $\partial\phi/\partial n$ are known over the boundaries then the field problem is essentially solved everywhere because interior values can be calculated easily by numerically integrating eqn (8.8). However, in a well posed problem, only one or the other is known at a boundary point and the integral equation must be solved to determine the unknown boundary value, either ϕ or $\partial\phi/\partial n$. The major advantage of integral equations can be immediately inferred from eqn (8.8), where it can be seen that the domain of the integration is confined to surfaces. Even the case where non-linear materials are present only these regions would appear in the equation; thus in general just the 'active parts' of the mathematical model have to be discretized. Unlike the situation with differential equations the free space regions do not appear in the integral formulations, and this in turn means a far simpler discretization to a degree that has the effect of reducing the dimensionality of the problem by one! There are drawbacks, however, as will be discussed in Chapters 9, 12, and 13.

8.2.3 Finite difference methods

The finite difference method (FD) has been extensively used for solving electromagnetic field problems of all kinds with enormous success; the method itself is of considerable mathematical importance and continues to play a central role in numerical analysis. As has already been said in the introduction to this chapter, finite difference techniques formed the basis of the first general purpose computer codes in electromagnetics [12, 13, 16] and in other branches of computational modelling, like fluid mechanics, the method is still highly relevant. The object of any numerical method is to obtain a numerical solution of the defining equation or mathematical model; a defining equation like eqn (8.1) is a continuous partial differential equation valid everywhere over the domain of the problem. However, from the discrete nature of numbers, it follows that a numerical solution can only be obtained for a finite number of points. The implies the necessity of constructing an additional model, the so called numerical model or discretized model which consists of a number of strategically chosen points distributed in such a way as to adequately represent the problem under consideration.

The finite difference method consists essentially of overlaying the problem with a mesh of lines 'parallel' to the coordinate system used, and an approximate solution to the defining equation is then found at the mesh points defined by intersections of the lines—see Fig. 8.4. The approximation consists of replacing each derivative of the equation by a finite difference expression relating the value of the unknown variable at a point with its value at neighbouring points. For example, consider the Poisson eqn (8.1)

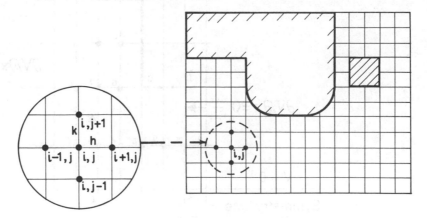

Fig. 8.4 Example of a finite difference mesh.

for the electrostatics case in two dimensions i.e.

$$\frac{\partial^2 V}{\partial x^2} + \frac{\partial^2 V}{\partial y^2} = -\rho. \tag{8.9}$$

With reference to the grid in Fig. 8.4 consider the mesh point labelled i, j and its immediate neighbours, then by use of Taylor's theorem for two variables (see Smith [11]) the value of V at a point can be expressed in terms of its neighbouring value and separation distance h as follows:

$$V(x + h) = V(x) + hV'(x) + \tfrac{1}{2}h^2 V''(x) + \tfrac{1}{6}h^3 V'''(x) + \cdots \tag{8.10}$$

and

$$V(x - h) = V(x) - hV'(x) + \tfrac{1}{2}h^2 V''(x) - \tfrac{1}{6}h^3 V'''(x) + \cdots \tag{8.11}$$

hence

$$V(x + h) + V(x - h) = V_{i+1,j} + V_{i-1,j}$$
$$= 2V_{i,j} + h^2 V''(x) + O(h^4)$$

and

$$V(x + h) - V(x - h) = V_{i+1,j} - V_{i-1,j}$$
$$= 2hV'_{i,j} + O(h^3)$$

therefore,

$$V''_{i,j} = (V_{i+1,j} - 2V_{i,j} + V_{i-1,j})/h^2 \tag{8.12}$$

and

$$V'_{i,j} = (V_{i+1,j} - V_{i-1,j})/2h. \tag{8.13}$$

Equations (8.12) and (8.13) are approximations to the derivatives and have been obtained by neglecting powers of h higher than two; approximations to the derivatives in the y direction are obtained in terms of k in a similar way.

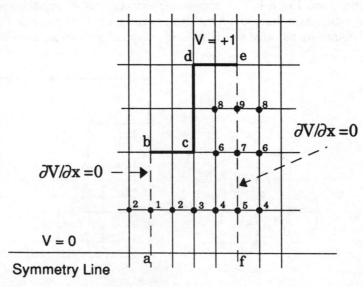

Fig. 8.5 Simple electrostatic problem to illustrate finite difference procedure.

These approximations have a leading error of order h^2 or k^2 and their use will introduce a numerical modelling error, already referred to dependent upon the mesh cell size; this error is sometimes called the 'truncation error' since it arises by truncating the infinite Taylor's series or sometimes the 'discretization error' because it depends upon the discretization density. If these finite difference expressions are now substituted into the defining equation, eqn (8.9), the following local approximation is obtained.

$$\frac{1}{h^2}(V_{i+1,j} - 2V_{i,j} + V_{i-1,j}) + \frac{1}{k^2}(V_{i,j+1} - 2V_{i,j} + V_{i,j-1}) + \rho = 0. \tag{8.14}$$

if eqn (8.14) is applied to each mesh point in turn, including the boundary points for which the prescribed values of V are inserted, a linear set of algebraic equations for the unknown values of V results; the following example should make the procedure clear. Consider the electrostatic problem shown in Fig. 8.5.

If the finite difference formula eqn (8.14) with $h = 1$, $k = 2$ and $\rho = 0$ is applied to each mesh point in turn, see Fig. 8.5, the following set of linear equations is obtained:

$$
\begin{aligned}
2V_2 - 2V_1 + \tfrac{1}{4}(1 - 2V_1) &= 0 \\
V_3 - 2V_2 + V_1 + \tfrac{1}{4}(1 - 2V_2) &= 0 \\
V_4 - 2V_3 + V_2 + \tfrac{1}{4}(1 - 2V_3) &= 0 \\
V_5 - 2V_4 + V_3 + \tfrac{1}{4}(V_6 - 2V_4) &= 0 \\
2V_4 - 2V_5 + \tfrac{1}{4}(V_7 - 2V_5) &= 0 \\
V_7 - 2V_6 + 1 + \tfrac{1}{4}(V_8 - 2V_6 + V_4) &= 0 \\
2V_6 - 2V_7 + \tfrac{1}{4}(V_9 - 2V_7 + V_5) &= 0 \\
V_9 - 2V_8 + 1 + \tfrac{1}{4}(1 - 2V_8 + V_6) &= 0 \\
2V_8 - 2V_9 + \tfrac{1}{4}(1 - 2V_9 + V_7) &= 0.
\end{aligned}
\tag{8.15}
$$

In the above equations the symmetry has been taken into account by including points outside the region of interest, see Fig. 8.5 at points labelled 4, 6, 8 and 2, these points are symmetric with respect to the normal derivative boundaries ef and ab respectively. The set of mesh-point equations after grouping like terms can be expressed in matrix form as follows:

$$
\begin{pmatrix}
-10 & 8 & 0 & 0 & 0 & 0 & 0 & 0 & 0 \\
4 & -10 & 4 & 0 & 0 & 0 & 0 & 0 & 0 \\
0 & 4 & -10 & 4 & 0 & 0 & 0 & 0 & 0 \\
0 & 0 & 4 & -10 & 4 & 1 & 0 & 0 & 0 \\
0 & 0 & 0 & 8 & -10 & 0 & 1 & 0 & 0 \\
0 & 0 & 0 & 1 & 0 & -10 & 4 & 1 & 0 \\
0 & 0 & 0 & 0 & 1 & 8 & -10 & 0 & 1 \\
0 & 0 & 0 & 0 & 0 & 1 & 0 & -10 & 4 \\
0 & 0 & 0 & 0 & 0 & 0 & 1 & 8 & -10
\end{pmatrix}
\begin{pmatrix}
V_1 \\ V_2 \\ V_3 \\ V_4 \\ V_5 \\ V_6 \\ V_7 \\ V_8 \\ V_9
\end{pmatrix}
=
\begin{pmatrix}
-1 \\ -1 \\ -1 \\ 0 \\ 0 \\ -4 \\ 0 \\ -5 \\ -1
\end{pmatrix}.
\tag{8.16}
$$

eqn (8.16) is the standard form for a set of linear algebraic equations, i.e.

$$Ax = B. \tag{8.17}$$

It can be seen that the numerical model introduced by the finite difference method has the effect of transforming the continuous partial differential equation, eqn (8.9) to the discrete set of algebraic equations, eqn (8.16), regardless of the complexity of the physical problem. This will always be the case. Note that, for a problem to be properly posed the system matrix $\mathbf{A}$ corresponding to the differential operator, in this case ∇^2, must be a square matrix, i.e., the number of unknowns must equal the number of equations and also for solutions to exist the inverse of the matrix $\mathbf{A}$ must exist. For the special case of eqn (8.16) the solution is readily obtained by simple methods, e.g. $V = 0.491$ compared with $V = 0.484$ for a convergent solution obtained by increasing the number of mesh points. A feature of the matrix in eqn (8.16) to notice is the 'sparsity' of entries other than zero; this characteristic is quite general with differential formulations and is exploited in techniques for solving systems of equations like eqn (8.16) to achieve economy in computation time. For cases involving large numbers of unknowns the choice of solution procedures becomes critical, it depends upon the equation type to be solved, the detailed structure of the matrix and size of the problem. There is an enormous literature on this subject and for the moment the reader is referred to the text by Smith (1969) [11]. However, a detailed discussion of two methods, a direct method and an iterative method, will be given in later chapters in connection with the finite element method—see sections 9.4, and 10.2.2. Also a general review of these techniques is given in Chapter 13.

A major difficulty with the finite difference method is that the discretization schemes (mesh, grids, etc.) have a fixed topology (order and arrangement) although the actual geometric distributions of mesh lines can be irregular. For example a regular grid ruled

Fig. 8.6 Quadrupole magnet meshed with triangular elements.

upon a rubber sheet which is subsequently deformed to match the boundaries of a problem will provide a fixed topology mesh for finite difference formulae, and at the same time introduce irregularity to allow a proper matching of boundary shapes—see, for example, Winslow's TRIM algorithm [13]. This can only be achieved if the complexity of the model is not too severe since 'pathological' meshes can easily arise violating the fixed topology. Another way of overcoming the difficulty of irregular boundary shapes is to use special formulae for mesh points near boundaries, this is necessary in any case for applying symmetry boundary conditions and for introduction of the interface conditions between differing media. A further limitation arises when it is required to introduce higher order terms for the basic Taylor's series to improve the accuracy of approximation. All of these difficulties are neatly surmounted by the finite element approach which is introduced in the next section.

8.3 THE FINITE ELEMENT METHOD

8.3.1 Elements

In this section the development of the finite element method is introduced by considering how some of the limitations of the finite difference could be overcome. For example, if instead of the regular topological grid the domain of the problem is discretized into elements then the boundary segments will conform to the discretization in a natural manner; see, for example, Fig. 8.6 in which triangular elements are shown matching the boundary of a magnetic pole profile of a quadrupole magnet. The second limitation in connection with the finite difference method is that the extension to a higher order approximation by including higher order terms in the Taylor's series expansion, is not very easy to apply; on the other hand the finite element approach offers a very natural extension. Thus consider a single element, and assume that the solution variable ϕ obeys a simple relation like:

$$\phi = \alpha_1 + \alpha_2 x + \alpha_3 y + \alpha_4 xy + \alpha_5 x^2 + \cdots \tag{8.18}$$

where α_i are constants. Furthermore, since the solution should be continuous when moving across an element boundary, it is desirable to express eqn (8.18) in terms of the potential values at the element vertices or nodes. Thus for the triangle elements shown in Fig. 8.6 eqn (8.18) can be truncated to:

$$\phi = \alpha_1 + \alpha_2 x + \alpha_3 y$$

and ϕ expressed as

$$\phi = N_1 \phi_1 + N_2 \phi_2 + N_3 \phi_3. \tag{8.19}$$

The N_1, N_2 etc. are functions of the spatial coordinates which can easily be evaluated for simple shapes providing the number of nodes in eqn (8.19) is the same as the number of terms in eqn (8.18). Thus a triangle with three nodes would yield a linear variation of solution locally and a triangle with six nodes, three at the vertices and three at mid sides, would yield a second order variation. By this means the finite element concept offers a simple method of extending the order of the local solution variable. The N_i are called shape functions (see Zienkiewicz and Taylor, p. 25[2]) and can either be determined directly by substituting in eqn (8.18) for each node in turn, then solving the small system of equations for α_i etc. or sometimes more easily by inspection, for example it should be noted that according to eqn (8.19) N_i must be unity at the ith node and zero at all other nodes. A detailed analysis for actual elements is dealt with in the next chapter.

8.3.2 Variational method

In order to establish the global behaviour of the problem the local element description, characterized by eqn (8.19), must be integrated over the entire domain in some way in order to take account of the boundary and interface conditions. One approach is to apply a variational principle corresponding to the defining equation (see Finlayson [6]), for instance a necessary condition for the potential energy in an electrostatic system to be a minimum is that the electrostatic potential must satisfy the Laplace equation. For the special case of the Laplace equation in electrostatics this energy can be expressed as:

$$W = \int_\Omega \frac{1}{2} \mathbf{E} \cdot \mathbf{D} \, d\Omega = \frac{1}{2} \int_\Omega \epsilon E^2 \, d\Omega$$

$$= \frac{1}{2} \int_\Omega \epsilon \left(\left[\frac{\partial \phi}{\partial x} \right]^2 + \left[\frac{\partial \phi}{\partial y} \right]^2 \right) d\Omega. \tag{8.20}$$

Substitution of the element shape functions from eqn (8.19) into eqn (8.20) after summing over all the n elements leads to:

$$W = \sum^n \frac{1}{2} \epsilon \int_{elem} \left[\left(\sum_i^3 \frac{\partial N_i}{\partial x} \phi_i \right)^2 + \left(\sum_i^3 \frac{\partial N_i}{\partial y} \phi_i \right)^2 \right] d\Omega. \tag{8.21}$$

In order to minimize the total energy expression, eqn (8.21) must be differentiated with respect to a typical value of ϕ_j and equated to zero in the usual way. Thus,

$$\frac{\partial W}{\partial \phi_j} = \sum^n \epsilon \int_{elem} \left[\left(\sum_i^3 \frac{\partial N_i}{\partial x} \phi_i \right) \frac{\partial N_j}{\partial x} + \left(\sum_i^3 \frac{\partial N_i}{\partial y} \phi_i \right) \frac{\partial N_i}{\partial y} \right] d\Omega = 0. \tag{8.22}$$

Taking care to merge contributions for elements with shared nodes, eqn (8.22) is seen to be of the form,

$$K\phi = 0 \tag{8.23}$$

with

$$K_{ij} = K_{ji} = \int_{elem} \epsilon \left(\frac{\partial N_i}{\partial x} \frac{\partial N_j}{\partial x} + \frac{\partial N_i}{\partial y} \frac{\partial N_j}{\partial y} \right) d\Omega. \tag{8.24}$$

The above formulation as it stands will only apply to the Dirichlet problem where assigned potentials are specified at the boundaries, however the energy functional can easily be extended to include the Neumann problem case and indeed also to the Poisson problem with sources. The system set of equations, eqn (8.23), will become non-homogeneous and well posed when the boundary values are assigned. The variational method will not be pursued further here since an equivalent finite element formulation based on the Galerkin procedure which has a wider applicability and is easier to apply is preferred and will be discussed in the next section.

8.4 METHOD OF WEIGHTED RESIDUALS

A very direct approach that leads to the numerical solution of a partial differential equation of the Poisson type is to use a local element approximation like eqn (8.19) containing unknown parameters and substitute into the defining equation, see for example Finlayson [6]. Suppose that locally,

$$u = \sum N_i a_i, \tag{8.25}$$

where N_i are suitable chosen shape functions (in the context of weighted residuals these are often called trial functions) and that a_i are a set of parameters (field values) to be determined. Substitute eqn (8.25) into eqn (8.1) to obtain,

$$R = (\nabla \cdot \kappa \nabla \sum N_i a_i + Q) = 0, \tag{8.26}$$

where, in general, R will only vanish if a_i is the exact solution. Equation (8.26) is the residue equation and is a measure of the error introduced by using the approximate solution eqn (8.25). Direct application of eqn (8.26) will produce a rudimentary solution, provided the N_i satisfy the boundary conditions on eqn (8.1). This method is known as point matching. The solution is matched to a number of points within the domain of the problem equal to the number of parameters. These points are sometimes called collocation points and the method the collocation method. A far better (the term stronger is sometimes used) approximation is obtained if the residue error is forced to zero in an average sense over the domain of the problem rather than a series of unrelated local operations. To this end eqn (8.26) is written,

$$\int_\Omega \omega_i R d\Omega = 0, \tag{8.27}$$

where ω_i is a specially chosen function called a weighting function. Equation (8.27) is known as the weighted residual approximation to the solution. It should be clear that if eqn (8.27) is satisfied for any ω_i then eqn (8.1) also must be satisfied at all points in the domain, otherwise if $\nabla \cdot \kappa \nabla u \neq -Q$ at one or more points then it would be possible to chose a function ω_i which would violate eqn (8.27). If the number of weighting functions is chosen equal to the number of trial parameters a set of linear equations is obtained, i.e.

$$\sum \left(\int_\Omega \omega_i \nabla \cdot \kappa \nabla N_j d\Omega \right) a_j = - \int_\Omega \omega_i Q d\Omega \tag{8.28}$$

which is of the form.

$$K_{ij} a_j = g_i. \tag{8.29}$$

This approach enables nearly all of the commonly used methods for determining numerical solutions to be unified; particular techniques are available for eqn (8.28) by particular choices of weighting function. To illustrate the more commonly used weighting functions consider the problem of determining the field inside a highly permeable rectangular conductor which can be solved by analytic means, see Appendix 2 and Fig. 8.7. This example satisfies the linear Poisson equation, so that

$$\nabla^2 A_z = -\mu\mu_0 J_z = -Q, \tag{8.30}$$

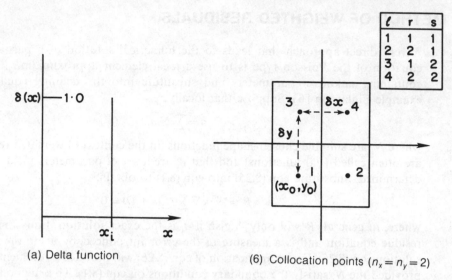

(a) Delta function   (6) Collocation points ($n_x = n_y = 2$)

Fig. 8.7 Rectangular conductor example.

where A_z is the magnetic vector potential. The boundary condition for this problem is that the normal component of flux density at the conductor surface is zero. First the simplest methods involving the use of collocation points already referred to above will be dealt with from the standpoint of weighting functions. Secondly, the so callled Galerkin method will be introduced in which the weighting functions are identified with the trial functions of eqn (8.25).

8.4.1 Collocation methods

Point collocation

In order to construct the weighted residual expression, eqn (8.28), it is first necessary to select a suitable trial or basis function N, see eqn (8.25). There are many possible choices, and one technique is to choose a set of global functions that satisfy the problem boundary condition

$$B_n = \begin{cases} \dfrac{\partial A}{\partial y} = 0 & \text{if } x = \pm a \\[2mm] -\dfrac{\partial A}{\partial x} = 0 & \text{if } y = \pm b. \end{cases} \tag{8.31}$$

This problem can be solved analytically by means of the double Fourier series, see section 4.3.3 and Appendix 2,

$$A = \sum_r \sum_s \sin\left[\frac{r\pi(x + a)}{2a}\right] \sin\left[\frac{s\pi(y + b)}{2b}\right] a_{rs}, \tag{8.32}$$

where r, s have integer value > 0, and by comparing eqn (8.32) with eqn (8.25) suitable shape functions can be identified as:

$$N_{rs} = \sin rX \sin sY \tag{8.33}$$

with

$$X = \frac{\pi(x + a)}{2a} \quad \text{and} \quad Y = \frac{\pi(y + b)}{2b}. \tag{8.34}$$

By substituting eqn (8.32) into eqn (8.30) the relation

$$\frac{-\pi^2}{4} \sum \sum \frac{r^2 b^2 + s^2 a^2}{a^2 b^2} N_{rs} a_{rs} = -\mu\mu_0 J \tag{8.35}$$

is obtained.

As stated already the point-matching method eqn (8.26) can be used directly. However, in the context of weighted residuals the appropriate weighting function to insert into the general expression, eqn (8.27) is the delta function—see Fig. 8.7

$$\omega_i = \delta(x - x_i, y - y_i), \tag{8.36}$$

where ω_j is unity at the collocation point (x_j, y_j) and zero everywhere else.

There now follows two examples with a single collocation point and secondly, an array of collocation points.

Example 1: For a single collocation point at the origin from eqn (8.33), $N_{11} = 1$, and from eqn (8.35),

$$A = \frac{4\mu\mu_0}{\pi^2} \frac{Ja^2 b^2}{a^2 + b^2} \sin X \sin Y. \tag{8.37}$$

Example 2: A rectangular array of points (x_j, y_j) on a regular grid, see Fig. 8.7. The rectangle is divided into n_x regions in the x direction and n_y regions in y direction with the collocation points at the region centres.

From eqn (8.35) a system of linear equations is obtained of the form:

$$G_{ij} a_j = q_i \tag{8.38}$$

in which G_{ij} is the coefficient matrix given by

$$G_{ij} = -(r_i^2 b^2 + s_i^2 a^2) \sin r_i X_j \sin s_i Y_j. \tag{8.39}$$

The right-hand side vector is q, given by

$$q_i = -\frac{4\mu\mu_0 Ja^2 b^2}{\pi^2}. \tag{8.40}$$

For the rectangular grid case

$$r_i = (i - 1) \bmod n_x + 1$$
$$s_i = (i - 1) \operatorname{div} n_x + 1, \tag{8.41}$$

where here mod and div should be interpreted as the modulus and division operators on integers—as used in PASCAL for example; thus on integer division p/q, p div q gives the rounded down integer and p mod q the remainder.

For the example shown in Fig. 8.7, setting $a = b$, and writing $\alpha = \pi/4$ the coefficient matrix equation (8.39) and right-hand side eqn (8.40) lead to:

$$\begin{pmatrix} 2\sin\alpha & 5\sin 2\alpha\sin\alpha & 5\sin\alpha\sin 2\alpha & 8\sin 2\alpha \\ 2\sin 3\alpha\sin\alpha & 5\sin 6\alpha\sin\alpha & 5\sin 3\alpha\sin 2\alpha & 8\sin 6\alpha\sin 2\alpha \\ 2\sin\alpha\sin 3\alpha & 5\sin 2\alpha\sin 3\alpha & 5\sin\alpha\sin 6\alpha & 8\sin 2\alpha\sin 6\alpha \\ 2\sin 3\alpha & 5\sin 6\alpha\sin 3\alpha & 5\sin 3\alpha\sin 6\alpha & 8\sin 6\alpha \end{pmatrix}$$

$$\begin{pmatrix} a_1 \\ a_2 \\ a_3 \\ a_4 \end{pmatrix} = \begin{pmatrix} -1 \\ -1 \\ -1 \\ -1 \end{pmatrix} \frac{4\mu\mu_0 J a^2}{\pi^2}. \tag{8.42}$$

By using the weighted residual method, eqn (8.27) with four trial functions evaluated at the four collocation points, the original problem involving a partial differential equation, eqn (8.30) has been transformed to a simple algebraic system of four equations. The accuracy of this solution and for other values of $a:b$ are explored in Fig. 8.8.

Sub-domain collocation

A better solution may be obtained if, instead of using a point delta function for the weighting in eqn (8.27), a function is used which essentially makes the integral of the residue zero over a sub-region of the problem space, then the method is called sub-domain collocation. Thus for the sub-domain Ω eqns (8.27) and (8.35) lead to:

$$\left[\int_{\Omega_j} \sum_r \sum_s (r^2 b^2 + s^2 a^2) \sin rX \cos sY \, d\Omega \right] a_{rs} + \int_{\Omega_j} \frac{4\mu\mu_0 J}{\pi^2 a^2 b^2} d\Omega = 0 \tag{8.43}$$

and for the rectangular array of points case,

$$G_{ij} = (r_i^2 b^2 + s_i^2 a^2) \int_{\Omega_j} \sin r_i X_j \sin s_i Y_j d\Omega \tag{8.44}$$

and

$$q_i = - \int_{\Omega_j} \frac{4\mu\mu_0 J_i}{\pi^2 a^2 b^2} d\Omega, \tag{8.44}$$

problems similar to this are fully worked out in the text by Zienkiewicz and Taylor, p. 215 [2].

8.4.2 Galerkin method (global)

If the trial functions are chosen to be the weighting functions then the weighted residual procedure becomes the Galerkin method [21]. The Galerkin method is one of the most widely used and successful techniques for the numerical solution of 'field' problems and has played an important part in the development of finite-element methods. As with the previously discussed weighted-residual approaches this method will be illustrated by means of the rectangular conductor problem.

Example 1: By inspection a very simple trial function for the Poisson equation, with just one unknown parameter A_i, is

$$A = (x^2 - a^2)(y^2 - b^2)A_1 = N_1 A_1 \tag{8.46}$$

which clearly satisfies the boundary conditions eqn (8.31). Therefore by applying Galerkin weighting $\omega_j = N_j$, eqn (8.27) becomes,

$$\left(\int N_1 \nabla^2 N_1 \, dx dy \right) A_1 = - \mu\mu_0 J \int N_1 dx \, dy \tag{8.47}$$

and substituting for N_1, using eqn (8.46) leads to

$$\left(2 \int (x^2 - a^2)(y^2 - b^2)[(x^2 - a^2)(y^2 - b^2)] \, dx \, dy \right) A_1 = - \mu\mu_0 J \int (x^2 - a^2)(y^2 - b^2) \, dx dy. \tag{8.48}$$

After performing the integrations and simplification the value of A_1 is found and then by substituting A_1 into eqn (8.46) the value of A is found to be

$$A = \frac{5}{8} \mu\mu_0 J \frac{(x^2 - a^2)(y^2 - b^2)}{a^2 + b^2}. \tag{8.49}$$

Example 2: As a further example consider a two parameter trial function of the form

$$A = (x^2 - a^2)(y^2 - b^2)A_1 + x^2(x^2 - a^2)(y^2 - b^2)A_2, \tag{8.50}$$

that is $A = N_1 A_1 + N_2 A_2$.

Applying Galerkin weighting to eqn (8.50) the following pair of algebraic equations result:

$$k_{11} A_1 + k_{12} A_2 = f_1 \tag{8.51}$$
$$k_{21} A_1 + k_{22} A_2 = f_2 \tag{8.52}$$

with

$$k_{ij} = \int N_i \nabla^2 N_j dx \, dy \tag{8.53}$$

$$f_i = - \mu\mu_0 J \int N_i dx \, dy. \tag{8.54}$$

By substituting the appropriate shape functions N_i from eqn (8.50) into eqns (8.53) and (8.54); and then integrating, the following 2×2 matrix system is obtained:

$$\begin{bmatrix} (a^2 + b^2) & a^2 \left(\dfrac{a^2}{7} + \dfrac{b^2}{5} \right) \\ \left(\dfrac{a^2}{7} + \dfrac{b^2}{5} \right) & a^2 \left(\dfrac{a^2}{7} \left(\dfrac{a^2}{3} + \dfrac{11b^2}{5} \right) \right) \end{bmatrix} \begin{pmatrix} A_1 \\ A_2 \end{pmatrix} = \begin{pmatrix} 5 \\ 1 \end{pmatrix} \frac{\mu\mu_0 J}{8}. \tag{8.55}$$

Results for the special cases of $b = a$ and $b = 2a$ are shown in Fig. 8.8 compared with the other weighted residual methods previously discussed.

Example 3: If instead of algebraic trial functions the Fourier functions and eqn (8.33)

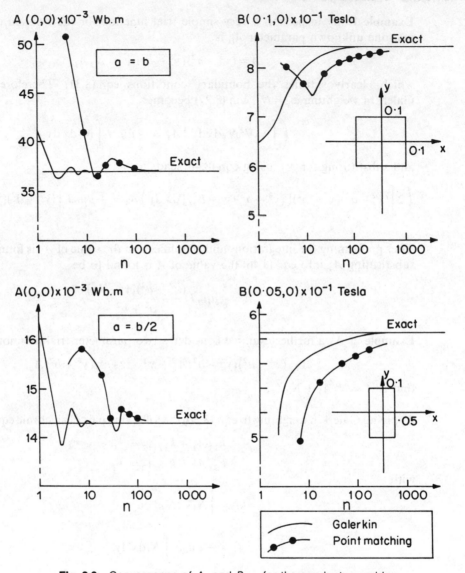

Fig. 8.8 Convergence of A_0 and B_{max} for the conductor problem.

are used the Galerkin method tends to the exact solution as the number of parameters increases to infinity.

For example: the Galerkin weighted residual expression corresponding to eqn (8.32)

$$\left[\frac{\pi^2}{4a^2b^2} \int_{-b}^{b} \int_{-a}^{a} (r^2b^2 + s^2a^2) \sin^2 rX \sin^2 sY \, dx dy \right] a_{rs}$$

$$= -\mu\mu_0 J \int_{-b}^{b} \int_{-a}^{a} \sin rX \sin sY \, dx dy. \tag{8.56}$$

Since the trigonometric functions are orthogonal, for $r \neq s$ they have zero value, only the diagonal terms in the double series are non-zero. Thus the values of a_{rs} are the solution of an uncoupled set of equations; thus after integration:

$$a_{rs} = -\frac{64\mu\mu_0 J}{\pi^4} \frac{a^2 b^2}{rs(b^2 r^2 + a^2 s^2)} \tag{8.57}$$

and the approximate solution is

$$A = \sum_r \sum_s a_{rs} \sin rX \sin sY. \tag{8.58}$$

Equations (8.57) and (8.58) are precisely the same as would be obtained by applying the Fourier series method and as the number of terms is increased the sum will tend to the exact solution. Thus for $b = 2a$ at point $(0, 0)$:

$$A = -\frac{44.032\mu\mu_0 J a^2}{\pi^2}. \tag{8.59}$$

Consideration of the local and subdomain Galerkin procedures will be left until the next chapter where they will be used as the basis of a detailed presentation of the finite element method.

8.4.3 Comparison of weighting functions and generalization

So far the method of weighted residuals has been applied to the very simple problem of the ferromagnetic conductor in order to illustrate the basic method of weighted residuals and to gain some information on the effectiveness of different weights. Comparative results for the computed value of the vector potential at the centre of the bar and the maximum values of the fields at edges are plotted in Fig. 8.8. Two points of interest emerge; first, that all cases converge as the number of parameters is increased and secondly, the Galerkin method converges faster than point collocation. These conclusions prove to be quite general, and the reader will find support for this in the literature, see for example, Zienkiewicz, p. 215 [2]. In the ferromagnetic conductor example, global basis functions were used and these must satisfy both the defining differential equation and the boundary conditions. This imposes a severe restriction on their range applicability and it is only for quite simple geometries that global functions can be applied routinely.

The weighted residual method itself is generic and can be applied not only to differential equations but to integral equations as well; in fact it is usual in the literature (Harrington [26], Chapter 1) to use a quite general nomenclature in developing the theory. Most of the equations that arise in electromagnetic field theory are of the form

$$L(u) = g, \tag{8.60}$$

where $L(u)$ stands for a linear differential operator or linear operator such as

$$L(u) = \nabla^2$$

$$\text{or} = \int d\Omega. \tag{8.61}$$

An operator is linear if

$$L(\alpha u_1 + \beta u_2) = \alpha L(u_1) + \beta L(u_2) \tag{8.62}$$

where α and β are scalar quantities. Thus the weighted residual expression, eqn (8.27) can be written as

$$\int_\Omega \omega [L(u) - g] \, d\Omega = 0. \tag{8.64}$$

Many other choices of weighting functions have been used, for example, the so called 'method of moments' or moment method [26] in which

$$\omega_i = x^i, i = 0, 1, 2, 3 \dots \tag{8.64}$$

The moment method has frequently been used in computing numerical solutions of integral equations [28, 29].

The variational or Rayleigh Ritz procedure briefly covered in section 8.3.2 can be shown to be equivalent to the Galerkin method, see Zienkiewicz and Taylor (p. 235) [2]. An alternative minimization method is known as 'least squares' in which the weighting function is chosen as

$$\omega_i = [L(u) - q] \tag{8.65}$$

such that

$$R = \int_\Omega [(u) - q]^2 \, d\Omega = 0 \tag{8.66}$$

that is

$$R = \int_\Omega [\sum L(N_j \phi_j) - g_i]^2 \, d\Omega = 0. \tag{8.67}$$

The approximate value of ϕ are computed such that R_i is a minimum, i.e. (least) by setting

$$\frac{\partial R_j}{\partial \phi_i} = 0 \tag{8.68}$$

Table 8.2 Table of weighting functions.

Weight ω		Type of method
Point δ_j function	Point collocation	Unity at point, zero elsewhere
Area Δ_j function	Sub-domain collocation	Unity over sub-domain, zero elsewhere
$1, x, x^2, \dots$	Method of moments	
$(Lu - g)$	Least Squares	Minimize with respect
$u = \sum N_j a_j$	Method	to parameters a_j
N(global)	Galerkin global	
N(sub-domain or or element)	Galerkin local	Finite element method or finite difference if grid is regular

For convenience the common choices of weighting functions are listed in Table 8.2. For a full account the reader should consult one of the standard texts, e.g. Zienkiewicz Chapter 9 [2].

8.5 CONCLUDING REMARKS

In this chapter some of the basic concepts for developing useful procedures for the numerical solution of field problems have been presented. The aim has been to explain without rigour some of the concepts which will be used in the later chapters. The modelling stages involved in formulating a problem to make it amenable to computer solutions are essentially three; first, the appropriate mathematical model must be identified, the field equations and boundary conditions must be determined. Secondly, a geometric (idealized) model must be built by concentrating upon the significant aspects of the problem in order to minimize the amount of data. Thirdly, a discretization scheme (numerical method) must be selected (exploiting any special features) in order to create a numerical method that the computer can process as efficiently as possible. It has been stressed that the use of numerical techniques and digital computers will by necessity introduce 'modelling' errors as well as the usual truncation and precision errors arising from the numerical analysis techniques and computer word length used, respectively. Engineering judgement is needed to abstract suitable limiting models from a complex geometric object for computing purposes.

The classic technique of finite differences has been briefly introduced, and though this method is widely used in fluid dynamics and semiconductor modelling, it has been replaced by finite element procedures in many other branches of engineering. A major drawback with finite differences is that the discretization schemes have fixed topology which causes considerable difficulties when complex boundaries have to be modelled (see Fig. 8.1). Also, the extension of the method to higher order approximations involve considerable complications. Both of these difficulties are readily overcome in the Galerkin weighted residual approach which forms the basis of the finite element method. On the other hand finite difference methods allow error estimates to be made readily, still a research area with the finite element approach—see Chapter 13, and they are particularly useful in developing *ad-hoc* computer solutions for problems defined by complicated equations.

In line with modern practice a generic approach has been adopted by concentrating on the weighted residual method; this technique can be viewed as a unifying principle with the many standard methods arising as special cases. Furthermore, it can be applied directly to the problem defining equilibrium equations (unlike the variational method which requires, *ab-initio*, knowledge of an appropriate energy functional). This information for many applications is not readily available, e.g. semi-conductor analysis. However, for most cases where the energy functional is known the Galerkin weighted residual technique leads to exactly the same numerical model, i.e. identical set of algebraic equations.

It has been shown that a Galerkin weighted residual method is more effective than the collocation methods. The literature of such comparisons for many different subject areas, is extensive—see the text by Zienkiewicz and Taylor [2] already cited, and for a very detailed account of both weighted residual and variational methods the book by

Finlayson [6] is recommended for further information. In formulating field problems emphasis has been given to unifying principles based on weighted residuals; and in the next chapter these ideas will be used as the basis of the finite element method. Furthermore, both differential and integral formulations for the electromagnetic field has been introduced and it has been stated that the formulations of the weighted residual technique in principle can be applied to both forms.

REFERENCES

[1] K. J. Binns and T. S. Low, Performance and application of multi-stacked imbricated permanent magnet generators, *Proc. IEE Part B*, **130**, 407 (1983).

[2] O. C. Zienkiewicz and R. Taylor, *The Finite Element Method, 4th Edition, Volumes 1 and 2* Maidenhead: McGraw-Hill (1991).

[3] L. F. Richardson, The approximate solution by finite differences of physical problems, *Trans. R. Soc. (London) A*, **210**, 307 (1910).

[4] R. V. Southwell, *Relaxation Methods in Theoretical Physics*. Oxford: Clarendon Press (1946).

[5] S. H. Crandall, *Engineering Analysis*. New York: McGraw-Hill (1956).

[6] B. A. Finlayson, *The Method of Weighted Residuals and Variational Principles*. New York: Academic Press (1972).

[7] K. Oberretl, The determination of complex magnetic fields, eddy currents and forces using rl-network analogues, *Achiv fur Elektrotechnik*, **48**, 297 (1965).

[8] C. J. Carpenter, A network approach to the numerical solution of eddy current problems, *IEEE Trans. on Magnetics*, **MAG-16**, 1517–1522 (1975).

[9] E. Weber, *Electromagnetic Fields, Vol 1*. New York: John Wiley (1950).

[10] M. J. Turner *et al.*, Stiffness and deflection analysis of complex structures, *J. Aero. Sci*, **23**, 805 (1956).

[11] G. D. Smith, *Numerical Solution of Partial Differential Equations*. Oxford: Oxford University Press (1965).

[12] J. S. Hornsby, A computer program for the solution of elliptic partial differential equations, *Tech. Rep. 63-7*, CERN (1967).

[13] A. A. Winslow, Numerical solution of the quasi-linear Poisson equation in a non-uniform triangular mesh, *J. Comput. Phys.*, **1**, 149 (1971).

[14] J. Colonias, Calculation of 2-dimensional fields by digital display techniques, *Tech. Rep. 17340*, UCRL (1967).

[15] K. Halbach and R. Holsinger, Poisson user manual, *Tech. Rep.*, Lawrence Livermore Laboratory, Berkeley (1972).

[16] F. C. Trutt, *Analysis of homopolar inductor alternators*. PhD thesis, University of Delaware (1982).

[17] E. A. Erdelyi and S. V. Ahmed, Non-linear theory of synchronous machines on load, *IEEE Trans. on Power Apparatus and Systems*, **85**, 792 (1966).

[18] G. Molinari and A. Viviani, Grid and metric optimisation procedures in finite difference and finite element method, *Proc. of IEEE PES Winter Meeting, New York*, **A78**, 289–1, 1–9 (1978).

[19] W. Muller and W. Wolff, General numerical solution of the magnetostatics equations, *Tech. Rep. 49(3)*, AEG Telefunken (1976).

[20] J. T. Oden, A general theory of finite elements, *Int. J. Num. Meth. Eng.*, **1**(1), (1969).

[21] B. G. Galerkin, Series solution of some problems of elastic equilibrium of rods and plates, *Vestn. Inyh. Tech.*, **19**, 897 (1915).

[22] M. V. K. Chari and P. P. Silvester, Finite element analysis of magnetically saturated dc machines, *IEEE Trans. PAS*, **90**, 2362 (1971).

[23] J. Simkin and C. W. Trowbridge, On the use of the total scalar potential in the numerical solution of field problems in electromagnets, *IJNME*, **14**, 423 (1979).

[24] J. Simkin and C. W. Trowbridge, Three dimensional non-linear electromagnetic field computations using scalar potentials, *Proc. IEEE*, **127** (6) 368–374 (1980).

[25] C. R. I. Emson and J. Simkin, An optimal method for 3D eddy currents, *IEEE Trans. on Magnetics*, **MAG-19,** 2450 (November 1983).

[26] R. F. Harrington, *Field Computation by Moment Methods*. Macmillan (1968).

[27] J. Simkin and C. W. Trowbridge, Magnetostatic fields computed using an integral equation derived from green's theorem, in *Compumag Conference on the Computation of Magnetic Fields* (Chilton, Didcot, Oxon.), p. 5, Rutherford Appleton Laboratory (1976).

[28] C. W. Trowbridge, *Applications of Integral Equation Methods for the Numerical Solution of Magnetostatic and Eddy Current Problems*, in *Finite Elements in Electrical and Magnetic Field Problems*, Ch. 10, Chichester: Wiley (1979).

[29] M. J. Newman, L. R. Turner and C. W. Trowbridge, G-FUN: An Interactive Program as an aid to Magnet Design, in *Proc. Int. Conf. on Magnet Tech. (MT4)*, (Y. Winterbottom, ed.), pp. 617–626, Brookhaven National Laboratory (1972).

9

TWO-DIMENSIONAL STATIC LINEAR PROBLEMS

In Chapter 8 some of the basic concepts of numerical methods were discussed with a particular emphasis on using a set of trial or shape functions with unknown parameters to represent the solution to a problem (see eqn (8.25)), and to show how to obtain the values of these parameters by means of the weighted residual method (see section 8.4). This approach was applied to a simple problem using global trial functions specially chosen to satisfy the boundary conditions. Also, in the previous chapter, the method of finite differences was briefly introduced to obtain an approximation to the solution of a partial differential equation by using a Taylor's series (see section 8.2.3) and this was followed by an outline of the derivation of the finite element method using variational principles (see section 8.3.2). Both these procedures in fact incorporate local rather than global trial approximations and allow for considerable geometric flexibility. In this chapter, the use of local trial functions which lead naturally to the concept of finite elements is developed further, beginning with relatively simple situations which are useful in their own right but also lead to the treatment of more substantial problems, specifically two-dimensional static solutions.

9.1 DISCRETIZING POISSON'S EQUATION BY FINITE ELEMENTS

In this section a general method for the solution of the large group of electromagnetic problems listed in Table 8.1 will be derived. All of these problems are characterized by the mathematical model of eqn (8.1) with boundary conditions given by eqns (8.2) and (8.3). The concept of subdividing the problem space into finite elements over which the solution is assumed to follow a simple local approximating trial function (shape functions) will be explored and it will be shown how the Galerkin method allows these local elements to be integrated into a numerical or computational model for the whole problem.

9.1.1 Galerkin method and local basis functions

Consider Fig. 9.1 in which a problem domain has been discretized into a number of triangular finite elements. Focusing attention on one particular element, say element 6, it can be seen that, for the numbering system used, the three nodes defining this element are the nodes 5, 6 and 9. To define the local behaviour of the field, suppose that, within a particular element, the solution of eqn (8.1) u is given by

$$u \sim U = \alpha_1 + \alpha_2 x + \alpha_3 y. \tag{9.1}$$

Thus it is postulated that within the element the potential can be approximated by a linear law, and since triangles have three nodes it is quite natural to select an equation which is linear in x and y with three other parameters. In this case the representation is said to be complete because eqn (9.1) contains all the terms necessary for a linear variation in two dimensions, namely three, which is equal to the number of nodes specifying the triangle. If rectangles had been selected an appropriate choice satisfying conditions of completeness would be

$$u \sim U = \alpha_1 + \alpha_2 x + \alpha_3 y + \alpha_4 xy. \tag{9.2}$$

From eqn 9.1 three equations can be constructed, one for each node, and using a local

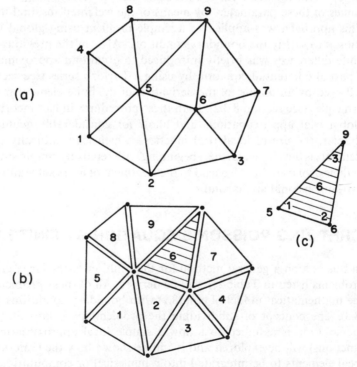

Fig. 9.1 Problem space discretized into finite elements.

numbering convention (see Fig. 9.1(c));

$$U_1 = \alpha_1 + \alpha_2 x_1 + \alpha_3 y_1$$
$$U_2 = \alpha_1 + \alpha_2 x_2 + \alpha_3 y_2 \quad (9.3)$$
$$U_3 = \alpha_1 + \alpha_2 x_3 + \alpha_3 y_3$$

or in matrix notation,

$$\begin{pmatrix} U_1 \\ U_2 \\ U_3 \end{pmatrix} = \begin{pmatrix} 1 & x_1 & y_1 \\ 1 & x_2 & y_2 \\ 1 & x_3 & y_3 \end{pmatrix} \begin{pmatrix} \alpha_1 \\ \alpha_2 \\ \alpha_3 \end{pmatrix} \quad (9.4)$$

i.e.

$$\mathbf{U} = \mathbf{C}\alpha. \quad (9.5)$$

Solving eqn (9.5) for the vector α by Cramer's rule leads to:

$$\alpha_1 = \frac{1}{2A} \begin{vmatrix} U_1 & x_1 & y_1 \\ U_2 & x_2 & y_2 \\ U_3 & x_3 & y_3 \end{vmatrix}; \quad \alpha_2 = \frac{1}{2A} \begin{vmatrix} 1 & U_1 & y_1 \\ 1 & U_2 & y_2 \\ 1 & U_3 & y_3 \end{vmatrix}$$

$$(9.6)$$

$$\alpha_3 = \frac{1}{2A} \begin{vmatrix} 1 & x_1 & U_1 \\ 1 & x_2 & U_2 \\ 1 & x_3 & U_3 \end{vmatrix}; \quad 2A = \begin{vmatrix} 1 & x_1 & y_1 \\ 1 & x_2 & y_2 \\ 1 & x_3 & y_3 \end{vmatrix}$$

where A is the area of the triangle. Furthermore, by defining

$$a_1 = x_2 y_3 - x_3 y_2$$
$$b_1 = y_2 - y_3 \quad (9.7)$$
$$c_1 = x_3 - x_2,$$

and $a_2, b_2, \ldots,$ etc. similarly defined by cyclic permutation of suffices, the solution of eqn (9.4) may be conveniently written as

$$\alpha_1 = (a_1 U_1 + a_2 U_2 + a_3 U_3)/2A$$
$$\alpha_2 = (b_1 U_1 + b_2 U_2 + b_3 U_3)/2A \quad (9.8)$$
$$\alpha_3 = (c_1 U_1 + c_2 U_2 + c_3 U_3)/2A$$

with

$$2A = a_i + b_i x_i + c_i y_i. \quad (9.9)$$

Hence, for a general point, using eqns (9.1) and (9.8),

$$U = \frac{1}{2A} [(a_1 + b_1 x + c_1 y)U_1 + (a_2 + b_2 x + c_2 y)U_2 + (a_3 + b_3 x + c_3 y)U_3], \quad (9.10)$$

or more concisely,

$$U = \sum N_i U_i \quad (9.11)$$

with U_i the solution value at node i and N_i the local shape functions, given by,

$$N_i = \frac{a_i + b_i x + c_i y}{2A}. \quad (9.12)$$

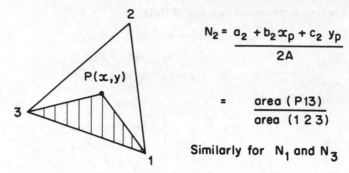

Fig. 9.2 Area coordinates for a triangle [note that these expressions satisfy eqn (9.13)].

Also from eqn (9.9) it can be seen that in particular $N_1 = 1$ at node 1 and from eqn (9.7) $N_1 = 0$ at nodes 2 and 3; and in general

$$N_i(x_k, y_k) = \delta_{ik}. \tag{9.13}$$

Equation (9.12) is in fact, the transformation from cartesian to a natural system of coordinates usually called area coordinates (Zienkiewicz and Taylor, p. 128) [12] where for example $N_2 = \text{area} \, (P13)/A$, see Fig. 9.2. This allows a simple coordinate transformation for a point inside the element, expressed in terms of its area coordinate, to the global cartesian system, i.e.

$$
\begin{aligned}
x &= N_1 x_1 + N_2 x_2 + N_3 x_3 \\
y &= N_1 y_1 + N_2 y_2 + N_3 y_3 \\
1 &= N_1 + N_2 + N_3.
\end{aligned}
\tag{9.14}
$$

The last equation of (9.14) indicates that only two of the three area coordinates or basis functions are independent.

9.1.2 Element continuity and element coefficients

The next stage is formulating the finite element procedure is to substitute the expressions for the triangular basis functions defined by eqn (9.11) into the weighted residual formula for a discretization point j, eqn (8.27), which for the Poisson equation is

$$\int_\Omega w_j [\nabla \cdot \kappa \nabla U + Q] \, d\Omega = 0. \tag{9.15}$$

However, the local basis function for U, given by eqn (9.11) applies only to an element, and so

$$\int_\Omega w_j [\nabla \cdot \kappa \nabla U + Q] \, d\Omega \sim \sum_{m=1}^{M} \int_{\text{elem}} w_j [\nabla \cdot \kappa \nabla U + Q] \, d\Omega = 0, \tag{9.16}$$

where the integral over the problem domain has been replaced by the sum of the M

elements, i.e.

$$\int_\Omega d\Omega = \sum_{m=1}^{M} \int_{elem} d\Omega. \tag{9.17}$$

It should be clear that eqn (9.16) will only be valid if all the integrands on the r.h.s. are finite. This is ensured by the definitions of the shape functions for points inside elements, (see eqn (9.12)), but not guaranteed valid in crossing inter-element boundaries because of the discontinuities arising when U is differentiated. Consider the case illustrated in Fig. 9.3 at an element boundary, where (a) U is forced to be continuous because of the basis function definition, (b) its first derivative is discontinuous but finite, whereas (c) the second derivative is not only discontinuous but infinite. Thus there will be problems in eqn (9.16) if the second derivative is present hence the integrand should not contain derivatives higher than the first. This level of continuity is known as C_0 in the literature (see for example Zienkiewicz and Taylor [2] p. 212) and the choice of shape functions defined by eqns (9.11) and (9.12) are said to have C_0 continuity. In general if the higest derivative in the weighted residual equation is $(s-1)$, where s is an integer, then shape

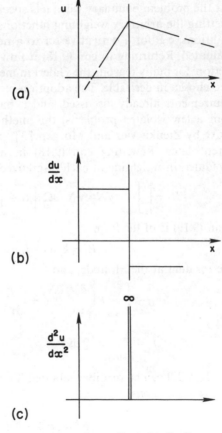

Fig. 9.3 Slope discontinuity C_0.

functions of C_s continuity are required to avoid singular (infinite) integrands. In the case of the Poisson equation under consideration the highest derivative present in the second, so either shape and weighting functions with first derivative continuity must be chosen or alternatively, the second derivative must be removed. In order to retain the simplicity of shape functions with C_0 continuity the second derivative can be removed by integrating eqn (9.15) by parts using Green's theorem (see Appendix 4). That is

$$\int_\Omega w_i[\nabla \cdot \kappa \nabla U + Q]\,d\Omega = \int_\Omega \nabla w_i \cdot \kappa \nabla U\,d\Omega + \int_\Omega w_i Q\,d\Omega - \int_\Gamma w_i \kappa \frac{\partial U}{\partial n}\,d\Gamma = 0. \qquad (9.18)$$

In eqn (9.18) the continuity requirements on U have been reduced at the expense of increasing the continuity requirements on the weighting functions ω_i—in the literature this is often called the weaker form of the original expression, eqn (9.15) (Zienkiewicz, and Taylor [2] p. 212).

For the class of problem under discussion, i.e. Poisson equation where U is identified as a scalar potential, it should be noted that the surface integral in eqn (9.18) includes the physical continuity conditions pertaining at the interfaces between materials characterized by the constitutive parameter κ, see section 8.2.1, and also when this equation is applied element by element as required by eqn (9.16) this surface term will vanish identically because of the physical interface condition requiring continuity of normal flux. At the problem boundary when u is specified this surface term can be made to vanish by setting the arbitrary weighting functions there to zero. For other types of boundary conditions, e.g. for a derivative set to a non-zero value, this integral would have to be evaluated. Returning to eqn (9.18) it now remains to select an appropriate weighting function; for many reasons the Galerkin method in which the shape functions themseleves are chosen in desirable. The adoption of this procedure will not violate the continuity requirements already discussed, and it has been established by experience, that apart from a few isolated problems, this method gives the best results, see for example the text by Zienkiewicz and Morgan [3] (p. 105), and other advantages will become apparent later. Rewriting eqn (9.18) in an element-by-element form and specializing to Galerkin weighting at each discretized point i leads to

$$R_i = \left[\int_{elem} \nabla N_i \kappa \nabla N_j\,d\Omega\right] u_j + \int_{elem} N_i Q\,d\Omega. \qquad (9.19)$$

The element eqn (9.19) is of the form

$$R_i = k_{ij} u_j + f_i, \qquad (9.20)$$

where R_i is the residual at the ith node, and

$$k_{ij} = \int_{elem} \kappa\left(\frac{\partial N_i}{\partial x}\frac{\partial N_j}{\partial x} + \frac{\partial N_i}{\partial y}\frac{\partial N_j}{\partial y}\right)d\Omega \qquad (9.21)$$

$$f_i = \int_{elem} N_i Q\,d\Omega. \qquad (9.22)$$

The integrals (9.21, 9.22) can be readily evaluated for the basis functions defined by eqn (9.12) by using the expression

$$\int_{elem} N_1^a N_2^b N_3^c\,d\Omega = 2A\frac{a!b!c!}{(a + b + c + 2)!} \qquad (9.23)$$

i.e.

$$k_{ij} = \int_{\text{elem}} \frac{\kappa(b_i b_j + c_i c_j)}{4A^2} \, d\Omega = \frac{\kappa(b_i b_j + c_i c_j)}{4A}, \tag{9.24}$$

and

$$f_i = \int_{\text{elem}} \frac{(a_i + b_i x + c_i y)}{2A} Q \, d\Omega = \frac{AQ}{3}. \tag{9.25}$$

Thus the local system is:

$$\begin{bmatrix} R_1 \\ R_2 \\ R_3 \end{bmatrix} = \frac{\kappa}{4A} \begin{bmatrix} b_1^2 + c_1^2 & b_1 b_2 + c_1 c_2 & b_1 b_3 + c_1 c_3 \\ & b_2^2 + c_2^2 & b_2 b_3 + c_2 c_3 \\ \text{symmetric} & & b_3^2 + c_3^2 \end{bmatrix} \begin{bmatrix} u_1 \\ u_2 \\ u_3 \end{bmatrix} + \frac{AQ}{3} \begin{bmatrix} 1 \\ 1 \\ 1 \end{bmatrix}. \tag{9.26}$$

The system is symmetric which is a consequence of the choice of Galerkin weighting and the symmetric form of the Poisson equation, and this important advantage of the Galerkin method is carried over into the system matrix discussed next.

9.2 ASSEMBLING THE MATRIX

So far the expressions for the local element coefficients for the first order triangular elements appropriate to the Poisson problem have been obtained, and in the next step the local elements coefficient have to be merged in order to form the system matrix. For example, in Fig. 9.1 element 6 has global nodes 5, 6 and 9 but global node 5 is also shared by elements 1, 2, 5, 8 and 9. In order to clarify all of the above procedures the rectangular conductor problem, already considered extensively in section 8.4, will be used. Equation (8.30) will be solved for the simple rectangular conductor $2a \times 2b$ as shown in Fig. 9.4. Referring to Fig. 9.4(a) the first example is for two elements only, with a total of four nodes and it is easy to write down the system matrix by inspection using the notation

$$k_{ij}^{(n)} = \frac{\kappa}{4A} (b_i^{(n)} b_j^{(n)} + c_i^{(n)} c_j^{(n)}) \tag{9.27}$$

for the nth element coefficient, and

$$f_i^{(n)} = \frac{AQ}{3} \tag{9.28}$$

for the nth r.h.s. Thus, for element 1 (nodes 1, 2 and 4), using eqn (9.26)

$$\begin{bmatrix} R_1 \\ R_2 \\ R_4 \end{bmatrix} = \begin{bmatrix} k_{11}^{(1)} & k_{12}^{(1)} & k_{14}^{(1)} \\ & k_{22}^{(1)} & k_{24}^{(1)} \\ & & k_{44}^{(1)} \end{bmatrix} \begin{bmatrix} u_1 \\ u_2 \\ u_4 \end{bmatrix} + \begin{bmatrix} f_1^{(1)} \\ f_2^{(1)} \\ f_4^{(1)} \end{bmatrix} \tag{9.29}$$

and for element 2 (nodes 1, 4 and 3)

$$\begin{bmatrix} R_1 \\ R_4 \\ R_3 \end{bmatrix} = \begin{bmatrix} k_{11}^{(2)} & k_{14}^{(2)} & k_{13}^{(2)} \\ & k_{44}^{(2)} & k_{43}^{(2)} \\ & & k_{33}^{(2)} \end{bmatrix} \begin{bmatrix} u_1 \\ u_4 \\ u_3 \end{bmatrix} + \begin{bmatrix} f_1^{(2)} \\ f_4^{(2)} \\ f_3^{(2)} \end{bmatrix}. \tag{9.30}$$

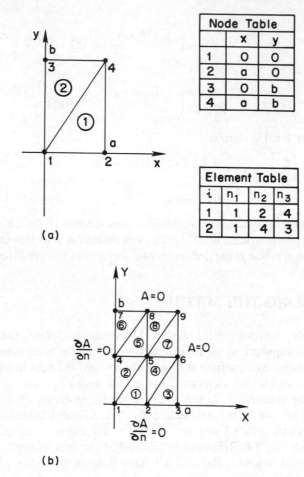

Fig. 9.4 Rectangular conductor subdivided into elements. (a) Discretized into two elements, (b) discretized into eight elements.

Now since the total residual must sum to zero, see eqn (9.15), merging of all the element contributions by nodes leads to:

$$
\begin{bmatrix}
k_{11}^{(1)} + k_{11}^{(2)} & k_{12}^{(1)} & k_{13}^{(2)} & k_{14}^{(1)} + k_{14}^{(2)} \\
k_{12}^{(1)} & k_{22}^{(1)} & 0 & k_{24}^{(1)} \\
k_{13}^{(2)} & 0 & k_{33}^{(2)} & k_{43}^{(2)} \\
k_{14}^{(1)} + k_{14}^{(2)} & k_{24}^{(1)} & k_{43}^{(2)} & k_{44}^{(1)} + k_{44}^{(2)}
\end{bmatrix}
\begin{bmatrix}
u_1 \\ u_2 \\ u_3 \\ u_4
\end{bmatrix}
= -
\begin{bmatrix}
f_1^{(1)} + f_1^{(2)} \\
f_2^{(1)} \\
f_3^{(2)} \\
f_4^{(1)} + f_4^{(2)}
\end{bmatrix}.
\tag{9.31}
$$

Equation (9.31) is a system of linear equations; and this assembly and merging process is straightforward and is easy to automate. As a second example, consider the discretization shown in Fig. 9.4(b). The system matrix, by inspection, is shown in Table 9.1 in which the coefficients are indicated by element numbers, e.g. $K = (r + s + t)$.

These two examples confirm that symmetry is preserved after merging elements and this enables very efficient methods to be used for solving the system. Also in this case

Table 9.1 System matrix and r.h.s. for the Fig. 9.4(b) example.

(a) System matrix

$$
\begin{bmatrix}
(1+2) & (1) & 0 & (2) & (1+2) & 0 & 0 & 0 & 0 \\
 & (1+3+4) & (3) & 0 & (1+4) & (3+4) & 0 & 0 & 0 \\
 & & (3) & 0 & 0 & (3) & 0 & 0 & 0 \\
 & & & (2+5+6) & (2+5) & 0 & (6) & (5+6) & 0 \\
 & & & & (1+2+4+5+7+8) & (4+7) & 0 & (5+8) & (7+8) \\
 & & & & & (3+4+7) & 0 & 0 & (7) \\
 & & & & & & (6) & (6) & 0 \\
\text{Symm} & & & & & & & (5+6+8) & (8) \\
 & & & & & & & & (7+8)
\end{bmatrix}
$$

(b) System r.h.s.

$$
-
\begin{bmatrix}
(1+2) \\
(1+4+3) \\
(3) \\
(2+5+6) \\
(1+2+4+5+7+8) \\
(3+4+7) \\
(7) \\
(5+6+8) \\
(7+8)
\end{bmatrix}
$$

the sparse nature of the system matrix appears. This property becomes increasingly apparent as the system increases in size. Sparsity arises because of the lack of coupling between nodes, e.g. nodes 1, 3 are not coupled in the example given and this coefficient is zero. The lack of coupling is a consequence of the differential nature of the operator; thus sparsity goes with a differential operator that this is not the case of integral equations will be seen later where every node is coupled to every other and gives rise to a fully populated matrix. It will be shown later that sparsity effects the choice of solution procedure for solving the linear system of equations, see section 13.4.

9.3 BOUNDARY CONDITIONS

After assembling the system matrix the essential boundary conditions for the problem have to be introduced. In general these are prescribed values of potential or field, valid on the boundary surfaces, and maybe, continuous functions. In the discretized model the boundary conditions are applied to the boundary nodes as point values. For example in the simple example shown in Fig. 9.4 a value of $A = 0$ is assigned to nodes, 3, 6, 7, 8, 9 and $\partial A/\partial n = 0$ to nodes 1, 2 and 4 to allow for symmetry. The first condition $A = 0$ is termed the essential boundary condition and is known by mathematicians as a Dirichlet boundary. The second, the 'zero normal-derivative' condition, is termed natural since it is the default state, arising by virtue of the surface integral term in eqn (9.18); this term vanishes for all surfaces, and by implication at all nodes on these surfaces, whenever the normal derivative is zero. This latter condition is also a special case of a general class of derivative condition known as a Neumann boundary. The Dirichlet conditions can be imposed for the discretized problem by simply deleting the nodes in the system matrix corresponding to each boundary node and evaluating the right-hand sides for the remaining nodes by inserting the appropriate value for each of the assigned nodes. Thus, for the problem of the rectangular conductor with the discretization shown in Fig. 9.4, delete nodes 3, 6, 7, 8 and 9 and for the case $b = 2a$ eqn (9.31) becomes:

$$
\begin{bmatrix}
5 & -4 & -1 & 0 \\
 & 10 & 0 & -2 \\
\text{Symm} & & 10 & -8 \\
 & & & 20
\end{bmatrix}
\begin{bmatrix}
A_1 \\
A_2 \\
A_3 \\
A_4
\end{bmatrix}
= -\frac{Qa^2}{3}
\begin{bmatrix}
2 \\
3 \\
3 \\
6
\end{bmatrix}
\tag{9.32}
$$

This case is particularly simple because the values of A on the boundary are zero, and are known as homogeneous conditions. The reflecting conditions along the lines $y = 0$, and $x = 0$, i.e. $\partial A/\partial n = 0$, were taken into account, because of the natural boundary conditions pertaining at all nodes when the surface integral in eqn (9.18) vanishes. Furthermore, note that for all internal points the interface physical conditions between neighbouring elements ($\kappa(\partial U)/\partial n$ continuous) are satisfied. This is because the opposite directions of the local normals ensure proper cancellation of the element surface integrals. The above scheme is not very suitable in general because of the matrix and right-hand side reorganization required. For machine computation it is customary to retain the original matrix size by a systematic adjustment of coefficients. This simplifies the programming data structures with only a small overhead of adding relatively few unknowns to the problem. Consider for example the 3×3 system of linear equations

$$
\begin{bmatrix}
k_{11} & k_{12} & k_{13} \\
k_{21} & k_{22} & k_{23} \\
k_{31} & k_{32} & k_{33}
\end{bmatrix}
\begin{bmatrix}
u_1 \\
u_2 \\
u_3
\end{bmatrix}
=
\begin{bmatrix}
q_1 \\
q_2 \\
q_3
\end{bmatrix}
\tag{9.33}
$$

and suppose that u_3 is known so that

$$
u_3 = U_3.
\tag{9.34}
$$

Now, instead of deleting the third row and adjusting the values of q to obtain the 2×2 system

$$
\begin{bmatrix}
k_{11} & k_{12} \\
k_{21} & k_{22}
\end{bmatrix}
\begin{bmatrix}
u_1 \\
u_2
\end{bmatrix}
=
\begin{bmatrix}
q_1 - k_{13}U_3 \\
q_2 - k_{23}U_3
\end{bmatrix}
\tag{9.35}
$$

with its attendant reorganization, the 3×3 structure can be preserved as follows

$$
\begin{bmatrix}
k_{11} & k_{12} & 0 \\
k_{21} & k_{22} & 0 \\
0 & 0 & 1
\end{bmatrix}
\begin{bmatrix}
u_1 \\
u_2 \\
u_3
\end{bmatrix}
=
\begin{bmatrix}
q_1 - k_{13}U_3 \\
q_2 - k_{23}U_3 \\
U_3
\end{bmatrix}.
\tag{9.36}
$$

Thus the solution of (9.36) will yield the values of u_1 and u_2 and preserve the value of u_3 to U_3 precisely. This scheme can be generalized and readily implemented with the result that no node-column reorganization of the original matrix is necessary. For the case where the ith value of U is prescribed the following procedure may be adopted:

1. Subtract from the jth member of the r.h.s. the product of k_{ji} and U_i.
2. Zero the ith row and column of K.
3. Assign $k_{ii} = 1$.
4. Set jth member of r.h.s. equal to U_i.

Thus for the example shown in Fig. 9.4(b) (see Table 9.1) with homogeneous boundary conditions $A = 0$ the following system of equations is obtained:

$$
\begin{bmatrix}
5 & -4 & 0 & -1 & 0 & 0 & 0 & 0 & 0 \\
 & 10 & 0 & 0 & -2 & 0 & 0 & 0 & 0 \\
 & & 1 & 0 & 0 & 1 & 0 & 0 & 0 \\
 & & & 10 & -8 & 0 & 0 & 0 & 0 \\
 & & & & 20 & 0 & 0 & 0 & 0 \\
 & & & & & 1 & 0 & 0 & 0 \\
 & & & & & & 1 & 0 & 0 \\
 & \text{Symm} & & & & & & 1 & 0 \\
 & & & & & & & & 1
\end{bmatrix}
\begin{bmatrix}
A_1 \\ A_2 \\ A_3 \\ A_4 \\ A_5 \\ A_6 \\ A_7 \\ A_8 \\ A_9
\end{bmatrix}
= -\frac{Qa^3}{3}
\begin{bmatrix}
2 \\ 3 \\ 0 \\ 3 \\ 6 \\ 0 \\ 0 \\ 0 \\ 0
\end{bmatrix}.
\tag{9.37}
$$

In this case, because of the inhomogeneity of the boundary conditions, the result is trivial with the r.h.s. needing only minimal adjustment.

9.4 LINEAR ALGEBRA AND SOLVING THE SYSTEM EQUATIONS

The above illustrates how the Poisson equation $(\nabla \cdot \kappa \nabla u = -Q)$, defined interior to a domain Ω spanned by a boundary Γ on which either u or $\partial u/\partial n$ is prescribed, has been transformed to a discretized set of algebraic equations

$$Ku = q \tag{9.38}$$

using the Galerkin weighted residual method of finite elements. The remaining problem is to solve the linear set of eqn (9.38) to obtain the nodal values of u. Many different approaches have been used to solve systems algebraic equations and there is an extensive literature on the subject, see for example the text by Strang [4]. In this section the classical direct method of Gaussian elimination will be applied to symmetric and sparse systems; however, later in the book a method using an iterative procedure which has significant advantages for large systems will be presented (see sections 10.2.2 and 13.4).

In the method of Gaussian elimination the unknowns are eliminated successively by dividing each equation in turn by the diagonal or leading coefficient and then making a 'forward substitution', e.g. in the system eqn (9.32):

$$
(a) \qquad
\begin{pmatrix}
5u_1 & -4u_2 & -u_3 & \\
-4u_1 & 10u_2 & & -2u_4 \\
-u_1 & & 10u_3 & -8u_4 \\
 & -2u_2 & -8u_3 & 20u_4
\end{pmatrix}
= -
\begin{pmatrix}
2 \\ 3 \\ 3 \\ 6
\end{pmatrix}
\tag{9.39}
$$

becomes the system,

$$(b) \quad \begin{pmatrix} u_1 & -\frac{4}{5}u_2 & -\frac{1}{5}u_3 & \\ -4u_1 & 10u_2 & & -2u_4 \\ -u_1 & & 10u_3 & -8u_4 \\ & -2u_2 & -8u_3 & 20u_4 \end{pmatrix} = -\begin{pmatrix} \frac{2}{5} \\ 3 \\ 3 \\ 6 \end{pmatrix} \tag{9.40}$$

Forward substitute u_1 given by b.1 into b.2, b.3 and b.4, i.e.

$$(c) \quad \begin{pmatrix} u_1 & -\frac{4}{5}u_2 & -\frac{1}{5}u_3 & \\ & \frac{34}{5}u_2 & -\frac{4}{5}u_3 & -2u_4 \\ & -\frac{4}{5}u_2 & \frac{49}{5}u_3 & -8u_4 \\ & -2u_2 & -8u_3 & 20u_4 \end{pmatrix} = -\begin{pmatrix} \frac{2}{5} \\ \frac{23}{5} \\ \frac{17}{5} \\ 6 \end{pmatrix} \tag{9.41}$$

and repeat for c.2, etc.,

$$(d) \quad \begin{pmatrix} u_1 & -\frac{4}{5}u_2 & -\frac{1}{5}u_3 & \\ & u_2 & -\frac{4}{34}u_3 & \frac{10}{34}u_4 \\ & & 165u_3 & -140u_4 \\ & & -140u_3 & 330u_4 \end{pmatrix} = -\begin{pmatrix} \frac{2}{5} \\ \frac{23}{34} \\ 67 \\ 125 \end{pmatrix} \tag{9.42}$$

and for d.3 etc.

$$(e) \quad \begin{pmatrix} u_1 & -\frac{4}{5}u_2 & -\frac{1}{5}u_3 & \\ & u_2 & -\frac{4}{34}u_3 & \frac{10}{34}u_4 \\ & & u_3 & -\frac{140}{165}u_4 \\ & & & 6970u_4 \end{pmatrix} = -\begin{pmatrix} \frac{2}{5} \\ \frac{23}{34} \\ \frac{67}{165} \\ 6001 \end{pmatrix} \tag{9.43}$$

Thus u is explicitly determined and the other values are determined by a simple back substitution thus:

$$\begin{matrix} u_4 \\ u_3 \\ u_2 \\ u_1 \end{matrix} = -\begin{pmatrix} 0.8610 \\ 1.1366 \\ 1.0634 \\ 1.4780 \end{pmatrix}, \tag{9.44}$$

and $A_1 = (Qa_2/3)u_1$, etc., hence at the origin $A_1 = -0.0154$ for $a = 0.05m$, see eqn (8.59).

This method can be applied providing, at each step, the diagonal coefficients are non-zero. This is the case for all matrices that arise in the Galerkin discretization of the Poisson problem and for a large range of other equations encountered in electromagnetics. Even if a zero diagonal coefficient is encountered it may be possible to interchange some of the remaining rows to avoid dividing by zero. If this is not possible it means that the determinant is zero and that there is no solution to the equations.

The general form of the Gaussian elimination algorithm is as follows. For the lth step in the forward substitute procedure, the elimination of u is achieved by expressing:

$$k_{ij}^l = k_{ij}^{l-1}/k_{ii}^{l-1}$$
$$q_i^l = q_i^{l-1}/k_{ii}^{l-1}$$

(9.45)

and

$$k_{lj}^l = k_{ij}^{l-1} - k_{il}^{l-1} k_{lj}^l$$
$$q_i^l = q_i^{l-1} - k_{il}^{l-1} q_l^l.$$

(9.46)

After applying the above $n - 1$ times the original set of equations is reduced to a single equation,

$$k_{nn}^{n-1} u_n = q_n^{n-1}$$

(9.47)

and the remainder of the unknowns are then computed by the back substitution,

$$u_l = q_l^l - \sum_{j=l+1}^{n} k_{lj}^l u_j.$$

(9.48)

Symmetry can easily be exploited by only storing the upper or lower triangle of coefficients and use the relation $k_{ij} = k_{ji}$. Sparsity can be exploited by adopting a compressed storage scheme for the coefficients illustrated in Fig. 9.5. It is very easy then to write a computer program to solve a banded symmetric system embodying the compressed storage scheme of Fig. 9.5 and the Gaussian elimination algorithm eqns (9.45) to (9.48). All the necessary sub-systems to construct a complete system for solving the

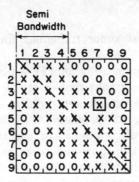

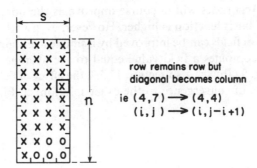

Fig. 9.5 Storage scheme for banded symmetric matrices.

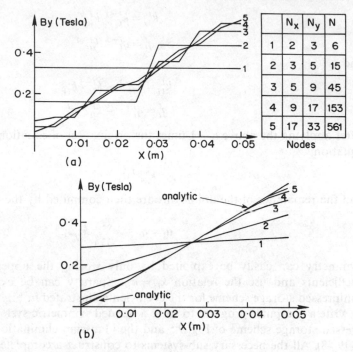

Fig. 9.6 Results for rectangular bar problem. (a) Distribution of B_y for five discretization levels, (b) distribution of B_y after averaging.

linear Poisson problem by first order triangular finite elements have now been constructed.

In Fig. 9.6 results from a computer program [5], which uses each of the sub-systems discussed so far, are shown. The graphs show a set of solutions at different levels of discretization for the rectangular conductor problem—starting with six nodes, see Fig. 9.4. The effect on smoothness of the gradients as the discretization is increased is very marked. The reason for the discontinuities in field are simply a consequence of the choice of basis functions eqn (9.1); the potentials were selected as linear in x, y therefore the fields, which are the potential gradients, will be constant. Thus the fields across the domain will be 'piecewise' constant and will exhibit the 'staircase' effect shown in Fig. 9.6(a). Accuracies will of course improve as the number of elements increases or if the order of basis function is higher. However, even with first order basis functions, the smoothness of fields can be improved by a simple averaging procedure which for a particular node computes a field value equal to the average from all elements sharing the node. In Fig. 9.6(b), see also section 13.5.2, the effect of field averaging is shown together with the rate of convergence to the exact solution, cf. Fig. 8.8.

9.5 SUMMARY

The basic steps have now been described to compute numerical solutions for the Poisson equation using triangular finite elements and the Galerkin weighted residual procedure

Modelling levels in finite elements		
Level	Model	Illustration
1	**Physical** Ferromagnetic conductor permeability μ current density **J**	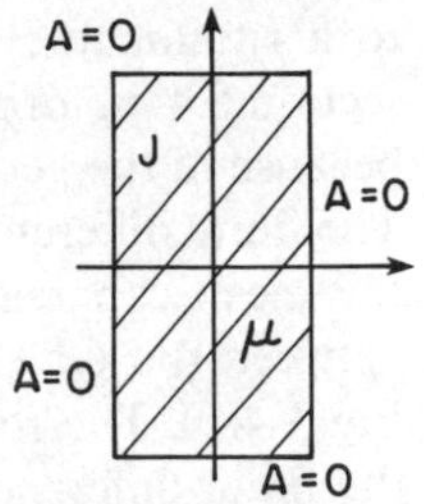
2	**Mathematical** $\nabla^2 A = -\mu\mu J = Q$ Symmetry exploited and boundary conditions assigned	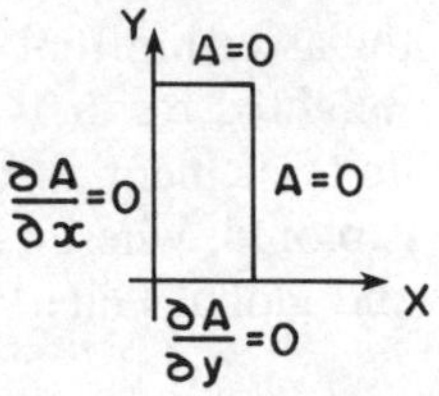
3	**Finite-Element** $(k_{ij}A_i = f_j)$ Linear base $A = N_1 A_1 + N_2 A_2 + N_3 A_3$ Galerkin $$k_{ij} = \int_{elem} \kappa \left(\frac{\partial N_i}{\partial x}\frac{\partial N_j}{\partial x} + \frac{\partial N_i}{\partial y}\frac{\partial N_j}{\partial y} \right) d\Omega$$ $$f_i = \int_{elem} N_i Q d\Omega$$	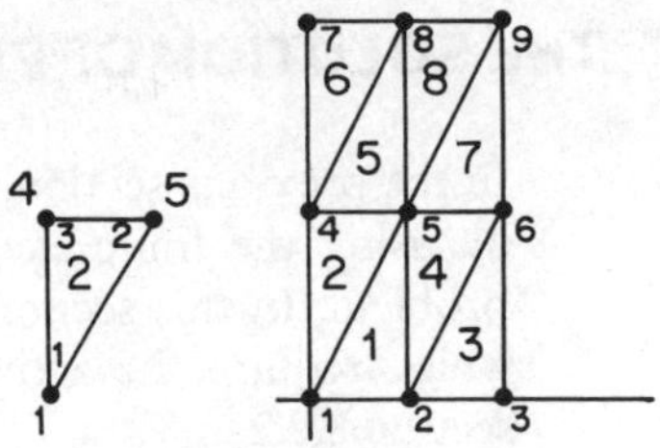
4	**Algebraic** Element matrices merged to obtain banded symmetric system $KA = F$ with boundary conditions included. Sparsity and banded nature utilized by storing in compact form: $(b \times n)$	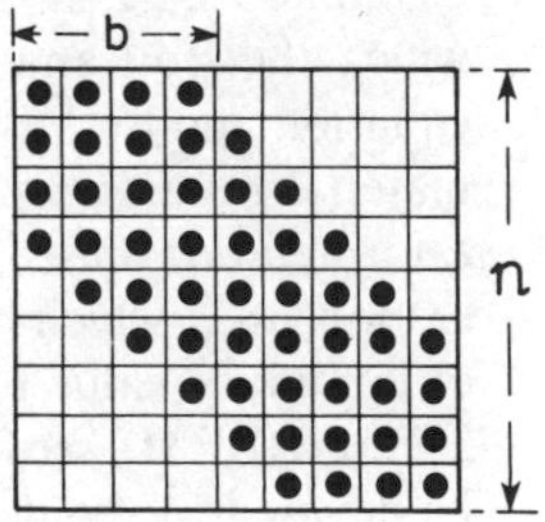
3	**Computer** System solved by Gaussian elimination $A = K^{-1}F$	

Fig. 9.7 Five stages in computing a numerical solution.

and these stages are summarized in Fig. 9.7. The finite element method provides a systematic technique for replacing a continuous partial differential with a set of discrete algebraic equations which can then be solved by standard procedures. The system matrix has a banded and symmetric structure which permits the use of compressed numbering schemes to minimize computer storage requirements. The method has been applied here to a rudimentary problem for illustrative purposes. This same example was used in section 8.4 in order to illustrate the global weighted residual method. In that case, because of the geometric simplicity of the rectangular shape, the convergence was faster. The finite difference method could also be used extremely simply for this problem since the mesh can readily be made to match the geometry. Indeed there is no difference between the Galerkin finite element and Taylor series finite difference for the mesh in Fig. 9.4(b). The reader can easily verify that the system matrix coefficients obtained by the finite difference approximation, for the rectangular grid illustrated in Fig. 9.4(b) is the same as eqn (9.37) obtained by Galerkin finite elements. This goes towards justifying the assertion that the finite difference method is a subset of the more general Galerkin weighted residual approach (see Zienkiewicz and Taylor [2], p. 256). The power of the finite element method will be made manifest though as more complex problems are explored, where complicated boundaries and higher order bases functions are needed and global Galerkin techniques would not be possible.

9.6   THE SOLUTION OF POISSON EQUATION BY BOUNDARY ELEMENTS

In the previous section the Poisson partial differential equation was solved numerically by using the finite element method which involved a complete discretization of the problem. In this section the use of the alternative integral formulations are considered which require the active material regions of the problem to be discretized only, see section 8.2.2.

The use of integral methods can, in some cases, result in considerable economies, for example, Fig. 9.8(a) shows a finite element model, (b) an integral equation model in which the iron and outer boundaries are modelled by elements, and (c) an integral equation model for iron problems having constant permeability in which only the material boundaries are discretized. Two approaches will be considered; the first uses a formulation based on a direct application of Green's second theorem equation, see section 1.8.2, which leads to a systematic solution technique, applicable to a wide class of boundary value problems, and is generally known by the name **Boundary Element Method** (BEM), see for example the text by Brebbia [6]. The first three subsections will be devoted to the development of this procedure. The second approach, which avoids the inclusion of a normal derivative term in the formulnation, is a special case and will be introduced by physical considerations. This latter method will be referred to by the name **Boundary Integral Method** (BIM) (Trowbridge [7], p. 191 and will be considered in section 9.8; however, its range of application is limited to pure ferromagnetic or pure dielectric problems. In section 12.3 a variation of the BIM, using the magnetization vector **M**, is applied to three-dimensional static problems. The chapter concludes with a brief introduction to the advantages of integral methods when used in conjunction with parallel computers.

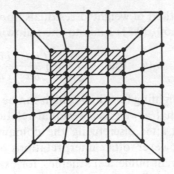

(a) Differential Model with far
field boundary at a finite
distance from the magnet

(b) Integral Model, variable μ
and far field boundary at
infinity

(c) Integral Model, constant μ and
far field boundary at infinity

Fig. 9.8 Comparison of differential and integral discretization schemes.

9.7 WEIGHTED RESIDUAL AND BOUNDARY ELEMENTS

A general formula, useful for numerical purposes, involving the integrals of potentials
and normal flux densities on problem boundaries can be derived by means of the
weighted residual method [6]. The problem to be solved is the Poisson equation defined
by the partial differential eqn (8.1), thus in a domain Ω bounded by a surface Γ a
potential u satisfies;

$$\nabla \cdot \kappa \nabla u + Q = 0 \qquad (9.49)$$

with the following boundary conditions,

$$u = \bar{u} \qquad \in \Gamma_1, \qquad (9.50)$$

$$\frac{\partial u}{\partial n} = \frac{\partial \bar{u}}{\partial n} \qquad \in \Gamma_2 \qquad (9.51)$$

where

$$\Gamma = \Gamma_1 + \Gamma_2. \qquad (9.52)$$

Equation (9.50) defines the essential conditions and eqn (9.51) defines the natural conditions (see section 9.3). The method of weighted residuals applied to eqn (9.49) and to the boundary conditions eqns (9.50), (9.51) leads to the following relation:

$$\int_\Omega w_i(\kappa\nabla^2 u - Q)d\Omega + \int_{\Gamma_1} \bar{w}_i(u - \bar{u})d\Gamma + \int_{\Gamma_2} \bar{\bar{w}}_i\left(\frac{\partial u}{\partial n} - \frac{\partial \bar{u}}{\partial n}\right)d\Gamma = 0, \qquad (9.53)$$

where w_i, $\bar{w}_i$ and $\bar{\bar{w}}_i$, are arbitrary weighting functions which can be chosen expeditiously. Now in section 9.1 it was shown that a weighted residual relation of the above type leads to the finite element method after discretizing the domains of integration and choosing simple basis functions. However, other interpretations of weighting functions are possible which will lead to different numerical schemes; for example if the first term of eqn (9.53) is transformed by means of Green's second theorem, see Appendix 4 (A4.17), i.e. it may be written as:

$$\kappa\int_\Omega w_i\nabla^2 u\,d\Omega = \kappa\int_\Omega u\nabla^2 w_i\,d\Omega + \kappa\int_\Gamma w_i\frac{\partial u}{\partial n}d\Gamma - \kappa\int_\Gamma u\frac{\partial w_i}{\partial n}d\Gamma \qquad (9.54)$$

and also if in eqn (9.53) the two surface weights are set by writing

$$\left.\begin{aligned} \bar{w}_i &= \kappa\frac{\partial w_i}{\partial n} \\[2mm] \bar{\bar{w}}_i &= -\kappa w_i \end{aligned}\right\} \qquad (9.55)$$

and

eqn (9.53) finally becomes

$$\int_\Omega u\nabla^2 w_i\,d\Omega - \int_\Omega w_i q\,d\Omega = -\int_{\Gamma_2} w_i\frac{\partial \bar{u}}{\partial n}d\Gamma - \int_{\Gamma_1} w_i\frac{\partial u}{\partial n}d\Gamma + \int_{\Gamma_2}\frac{\partial w_i}{\partial n}u\,d\Gamma + \int_{\Gamma_1}\frac{\partial w_i}{\partial n}\bar{u}\Gamma,$$

$$(9.56)$$

where use has been made of eqn (9.52) and κ has been absorbed into q, i.e. $q = Q/\kappa$. The effect of this transformation is to integrate the unknown function u at the expense of differentiating the weight w, and more importantly it is now possible to select w such that the volume integral term of eqn (9.56) can be integrated. This can be achieved by specifying w_i to be the 'fundamental solution' corresponding to the Laplacian operator $\nabla^2 w$, i.e. choose w_i to satisfy

$$\nabla^2 w_i + \delta_i = 0, \qquad (9.57)$$

where δ_i is the Dirac delta function. For example w_i is the potential corresponding to a point charge concentrated at the origin, and

$$\kappa\int_\Omega u\nabla^2 w_i\,d\Omega = -\kappa u_i, \qquad (9.58)$$

where u_i represents the value of the unknown function at the singular point. Therefore, if w_i in eqn (9.56) is selected to satisfy eqn (9.58) then the value of the unknown function u_i at an interior point of the domain is given by

$$u_i = -\int_\Omega w_i q\,d\Omega + \int_{\Gamma_1} w_i\frac{\partial u}{\partial n}d\Gamma - \int_{\Gamma_1}\frac{\partial w_i}{\partial n}\bar{u}d\Gamma + \int_{\Gamma_2} w_i\frac{\partial \bar{u}}{\partial n}d\Gamma - \int_{\Gamma_2}\frac{\partial w_i}{\partial n}u\,d\Gamma \qquad (9.59)$$

in which the fundamental solution w_i is a function of both the source and field point positions. The solutions of eqn (9.57) are well known [6], thus for an isotropic three-dimensional medium the fundamental solution w satisfies

$$\frac{\partial^2 w}{\partial r^2} + \frac{2}{r}\frac{\partial w}{\partial r} = \delta \tag{9.60}$$

i.e.

$$w = \frac{1}{4\pi r}, \tag{9.61}$$

and for a two-dimensional medium w satisfies

$$\frac{\partial^2 w}{\partial x^2} + \frac{\partial^2 w}{\partial y^2} = -\delta \tag{9.62}$$

i.e.

$$w = \frac{1}{2\pi}\ln\frac{1}{r}. \tag{9.63}$$

These two results can be verified by substitution for $r \neq 0$ and by integrating over a small sphere or cylinder and proceeding to a limit for $r = 0$ (see for example, Jaswon and Symm [8]). In order to use eqn (9.59) as a solution technique it is first necessary to consider the solution for points on the boundaries. It will be shown that it is possible to develop a method to solve either for u or $\partial u/\partial n$ for a number of discrete points on the boundary itself. As has been frequently emphasized, the problem as originally posed (eqns (9.49) to (9.51)), demands that either u or $\partial u/\partial n$ is prescribed on the boundary (not both quantities) thus eqn (9.59) can be used to solve for $u \in \Gamma_2$ and $\partial u/(\partial n) \in \Gamma_1$ respectively. Subsequent to finding all the boundary unknowns the value of u at an interior point can then be determined by a direct application of eqn (9.59). As can be seen from eqn (9.63) the fundamental solution w itself becomes singular when the field point is moved on to the boundary and so the appropriate limits have to be found. The simplest case occurs whenever the boundary is smooth.

In Fig. 9.9, the surface in the neighbourhood of the point in question is distorted locally to include a small cylinder (sphere in 3D), i.e.

$$\int_{\Gamma_2} \frac{\partial w}{\partial n} u \, d\Gamma = \int_{\Gamma_{2-\rho}} \frac{\partial w}{\partial n} u \, d\Gamma + \int_{\Gamma_\rho} \frac{\partial w}{\partial n} u \, d\Gamma \tag{9.64}$$

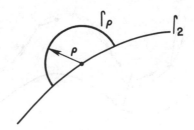

Fig. 9.9 Limiting the singular point on a smooth boundary.

and in the limit, using eqn (9.63), of $\rho \Rightarrow 0$

$$\int_{\Gamma_\rho} \frac{\partial w}{\partial n} u \, d\Gamma \Rightarrow -\int_{\Gamma_\rho} \frac{u \, d\Gamma}{2\pi\rho} = -\frac{u}{2} \qquad (9.65)$$

and

$$\int_{\Gamma_2} w \frac{\partial u}{\partial n} \, d\Gamma \Rightarrow \frac{1}{2\pi} \int_{\Gamma_2} \frac{\partial u}{\partial n} \, d\Gamma \ln \frac{1}{\rho} = 0 \qquad (9.66)$$

with similar results for a point on Γ_1. Hence the limit adds an additional $u/2$ term to eqn (9.59), and the general solution to Poisson's equation for a domain spanned by a surface $\Gamma = \Gamma_1 + \Gamma_2$ where $u = \bar{u} \in \Gamma_1$ and $\partial u/\partial \bar{u} = \partial \bar{n}/(\partial n) \in \Gamma_2$ is given by

$$\kappa c u_i - \int_\Omega Q_i \, d\Omega + \int_{\Gamma_2} u \frac{\partial w_i}{\partial n} \, d\Gamma = \int_{\Gamma_1} \frac{\partial u}{\partial n} w_i \, d\Gamma + \int_{\Gamma_2} w_i \frac{\partial \bar{u}}{\partial n} \, d\Gamma + \int_{\Gamma_1} \frac{\partial w_i}{\partial n} \bar{u} \, d\Gamma, \qquad (9.67)$$

where $c = 1$ for an interior point and $c = 1/2$ for a point on a smooth boundary.

9.7.1 Boundary element coefficients and system matrix

The boundary element integral equation could in principle be solved by a full Galerkin procedure as was carried out for the case of the differential operator formulation described in section 9.1. However, this would involve performing integrations of integral operators for the matrix coefficients and would prove to be very expensive to compute. It is customary in the classic formulations of the BEM, to use a sub-domain point collocation method and this approach will be adopted here. Initially, consideration will be given to constant basis functions in which the unknown functions will be located at the centre of each element and constant along the length. See, for example, Fig. 9.10. Thus the boundary is discretized into n elements of which n_1 are on Γ_1 and n_2 are on Γ_2. The values of u and $\partial u/\partial n$ are assumed to be constant on each element and equal to the value at the mid node of the element. Thus the discretized form of eqn (9.67) is

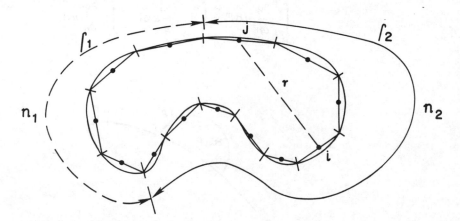

Fig. 9.10 Boundary elements matching at centroids.

given by

$$\tfrac{1}{2}u + B_i + \sum_j^n u_j \int_\Gamma \frac{\partial w}{\partial n}\, d\Gamma = \sum_j^n \frac{\partial u_j}{\partial n} \int_\Gamma w\, d\Gamma, \tag{9.68}$$

where B_i is a source term involving an integration over the volume for each node. For the Laplace equation B_i vanishes and for many problems this term can be removed if a particular integral is known. See the rectangular bar problem, section 9.4 and Appendix 2, also the example at end of this section. The integrals in eqn (9.68) can be carried out analytically for elements with constant basis functions. However, for higher order elements this becomes difficult and so a numerical technique will be used for all nodes except when $i = j$ which is straightforward analytically. Equation (9.68) can be written out for each of the nodes in turn thus obtaining a set of n equations, i.e.

$$B_i + \sum_{j=1}^n H_{ij} u_j = \sum_{j=1}^n G_{ij} \frac{\partial u_j}{\partial n} \tag{9.69}$$

in which

$$H_{ij} = \int_{\Gamma_j} \frac{\partial w}{\partial n}\, d\Gamma + \tfrac{1}{2}\delta_{ij} \tag{9.70}$$

and

$$G_{ij} = \int_{\Gamma_j} w\, d\Gamma. \tag{9.71}$$

It should be observed that n_1 values of u and n_2 values of $\partial u/\partial n$ are known, hence there are precisely n unknowns in system eqn (9.69); and after reordering, in such a way that all the unknowns are on the left-hand side, yields the system of equations:

$$\mathbf{AX} = \mathbf{Y}. \tag{9.72}$$

See Table 9.2 for an example of the system matrix. Note the system is fully populated and non-symmetric in contrast to the finite element discretizations of section 9.3. The Gaussian elimination algorithm (see section 9.4) can now be used to solve the system of equations for the unknown values of potential and normal gradient in the boundaries. After the system eqn (9.72) has been solved the values of u and $\partial u/\partial n$ are now known over the whole boundary, and the interior values can be calculated by eqn (9.67), thus

$$u_i = \sum_{i=j}^n \frac{\partial u_j}{\partial n} G_{ij} - \sum_{i=j}^n u_j H_{ij} \tag{9.73}$$

and the interior fluxes $\partial u/\partial x$, $\partial u/\partial y$ by differentiating eqns (9.67) to (9.71). The integrals in eqns (9.70) and (9.71) can be evaluated using a 4 point Gauss quadrature rule (see Appendix 5) for all elements except at the self point ($i = j$). For this particular point the integrand in the first term of eqn (9.70) is zero because r and n are orthogonal, i.e. $\partial/\partial n(\ln 1/r) = 0$, thus leaving only the singular term:

$$H_{ii} = \tfrac{1}{2} \tag{9.74}$$

and

$$G_{ii} = \int_{\Gamma_i} w\, d\Gamma = \frac{1}{2\pi} \int_{\Gamma_i} \ln 1/r\, dr \tag{9.75}$$

Table 9.2   System matrix $Ax = Y$ for $n = 5$, $n_2 = 3$, $n_1 = 2$.

$$
\begin{bmatrix}
H_{11} & H_{12} & -G_{13} & -G_{14} & -G_{15} \\
H_{21} & H_{22} & -G_{23} & -G_{24} & -G_{25} \\
H_{31} & H_{32} & -G_{33} & -G_{34} & -G_{35} \\
H_{41} & H_{42} & -G_{43} & -G_{44} & -G_{45} \\
H_{51} & H_{52} & -G_{53} & -G_{54} & -G_{55}
\end{bmatrix}
\begin{bmatrix}
u_1 \\
u_2 \\
\dfrac{\partial u_3}{\partial n} \\
\dfrac{\partial u_4}{\partial n} \\
\dfrac{\partial u_5}{\partial n}
\end{bmatrix}
$$

$$
=
\begin{bmatrix}
G_{11}\dfrac{\partial u_1}{\partial n} + G_{12}\dfrac{\partial u_2}{\partial n} - H_{13}u_3 - H_{14}u_4 - H_{15}u_5 - B_1 \\
G_{21}\dfrac{\partial u_1}{\partial n} + G_{22}\dfrac{\partial u_2}{\partial n} - H_{23}u_3 - H_{24}u_4 - H_{25}u_5 - B_2 \\
G_{31}\dfrac{\partial u_1}{\partial n} + G_{32}\dfrac{\partial u_2}{\partial n} - H_{33}u_3 - H_{34}u_4 - H_{35}u_5 - B_3 \\
G_{41}\dfrac{\partial u_1}{\partial n} + G_{42}\dfrac{\partial u_2}{\partial n} - H_{43}u_3 - H_{44}u_4 - H_{45}u_5 - B_4 \\
G_{51}\dfrac{\partial u_1}{\partial n} + G_{52}\dfrac{\partial u_2}{\partial n} - H_{53}u_3 - H_{54}u_4 - H_{55}u_5 - B_5
\end{bmatrix}
$$

As with finite elements it is useful to change to a local coordinate system in which

$$r = l\xi \tag{9.76}$$

where $2l$ is the length of the element. The collocation point is at the origin, see Fig. 9.11, and it follows that:

$$G_{ii} = \frac{1}{\pi}\int_0^b \ln 1/r \, dr = \frac{l}{\pi}\left[\ln 1/l + \int_0^1 \ln 1/\xi \, d\xi\right] = \frac{l}{\pi}[\ln 1/l + 1] \tag{9.77}$$

and for $i \neq j$ using Gauss quadrature

$$H_{ij} = \int \frac{\partial w}{\partial n} d\Gamma = \frac{l}{2\pi}\int \frac{\partial}{\partial r}\left[\ln(1/r)\right]\frac{\partial r}{\partial n} d\xi$$

$$= -\frac{l}{2\pi}\int_{-1}^1 \frac{d}{r^2} d\xi = -\frac{l}{\pi}\sum_{k=1}^m W_k \frac{d}{r^2(\xi_k)}, \tag{9.78}$$

where W_k and ξ_k are the weight, and Gauss points respectively in an m-point Gauss

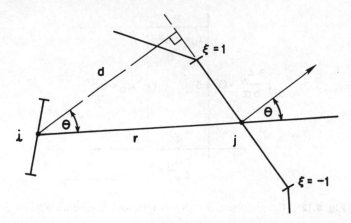

Fig. 9.11 Geometry for element coefficients.

quadrature formula (see Appendix 5). Similarly

$$G_{ij} = \int w\,d\Gamma = \frac{l}{2\pi}\int_{-1}^{1}\ln(1/r)\,d\xi = \frac{l}{2\pi}\sum_{k=1}^{m}W_k\ln\left(\frac{1}{\xi_k}\right). \qquad (9.79)$$

The boundary element method with constant elements will now be applied to the rectangular bar case. However, instead of solving eqn (8.30) directly it will be easier to subtract a particular integral and solve the homogeneous Laplace equation; this is possible in this case because the right-hand side term is a constant. The removal of the source term, will simplify the boundary element equations since the integral for the first term in eqn (9.69) will not be required. A particular integral satisfying the boundary conditions and symmetry for this problem is

$$A = A_0 - 1/2\mu\mu_0 J x \qquad (9.80)$$

and A_0 satisfies the Laplace equation

$$\nabla^2 A_0 = 0 \qquad (9.81)$$

with

$$\left.\begin{aligned}
A_0 &= qa^2/2 & :x = a\\[2mm]
A_0 &= qx^2/2 & :y = b\\[2mm]
\frac{\partial A_0}{\partial x} &= 0 & :x = 0\\[2mm]
\frac{\partial A_0}{\partial y} &= 0 & :y = 0
\end{aligned}\right\}, \qquad (9.82)$$

where $q = \mu\mu_0 J$.

A simple discretization scheme is shown in Fig. 9.12. The system matrix has the

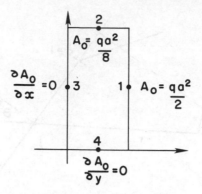

Fig. 9.12 Rectangular bar problem with four boundary elements.

following structure

$$
\begin{bmatrix}
-G_{11} & -G_{12} & H_{13} & H_{14} \\
-G_{21} & -G_{22} & H_{23} & H_{24} \\
-G_{31} & -G_{32} & H_{33} & H_{34} \\
-G_{41} & -G_{42} & H_{43} & H_{44}
\end{bmatrix}
\begin{bmatrix}
\dfrac{\partial A_0}{\partial x_1} \\[4pt]
\dfrac{\partial A_0}{\partial y_2} \\[4pt]
A_{03} \\[4pt]
A_{04}
\end{bmatrix}
= -\frac{qa^2}{2}
\begin{bmatrix}
H_{11} + \dfrac{H_{21}}{4} \\[4pt]
H_{21} + \dfrac{H_{22}}{4} \\[4pt]
H_{31} + \dfrac{H_{32}}{4} \\[4pt]
H_{41} + \dfrac{H_{42}}{4}
\end{bmatrix}.
\tag{9.83}
$$

The coefficients can now be evaluated by numerical quadrature using the expressions in eqns (9.78) and (9.79) with the weights and Gauss points given in Appendix 5.

In Fig. 9.13 results for 4 to 80 boundary elements are given for values of the vector potential A at the origin and for the vertical component of field at $x = a$. The boundary element solutions displayed in Fig. 9.13 apparently converge very rapidly to the exact values; for comparative purposes the corresponding, albeit slower, convergence values obtained for the finite element method are also shown. In making such comparisons however, due account must be taken of the amount of computation actually needed. In the finite element (differential case) the integrands in the coefficient integrals are trivial and only involve simple algebraic expressions, and hence the computations involved in assembling the matrix are relatively cheap. Also the system matrix is sparse and symmetric again ensuring a relatively cheap calculation. On the other hand the boundary element method coefficient integrands contain more complicated functions and the system matrix is asymmetric and full—both features leading to expensive computation. These computational considerations become far more acute as the size of the problem increases with the finite element method becoming very competitive, but at the level of the rudimentary problem illustrated the boundary element offers a clear advantage. The comparison becomes more realistic if instead of plotting the finite element predictions against the total number of discretized point, including the interior points, the results

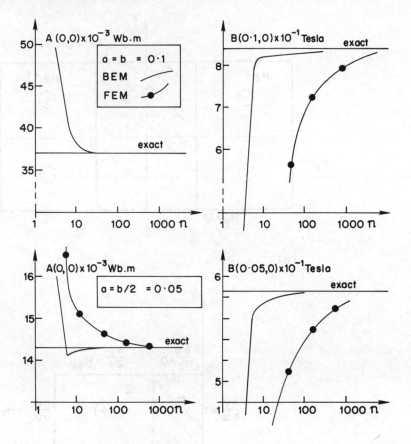

Fig. 9.13 Convergence as a function of the number of boundary elements.

are plotted instead against the number of boundary points as with the case of the boundary element method. The strengths and weaknesses of the two procedures will be considered in more detail later in the chapter.

9.7.2 Multiple regions and open boundary problems

The example given in the last section contained only one region and for single region Laplacian ($Q = 0$) problems the constitutive constant κ cancels out (see eqn (9.59)).

A more interesting case is that with regions of different permeabilities or permitivities. To fix ideas consider the simple magnetostatic example illustrated in Fig. 9.14. Figure 9.14(a) shows a symmetric dipole conductor system for which the field H_s can be calculated directly by quadrature of the Biot–Savart law—see eqn (1.24) etc. and a slab of iron. The external boundaries are to be flux boundaries, i.e. there is no magnetic

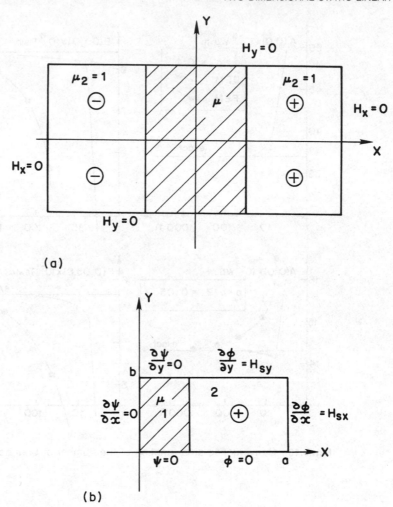

Fig. 9.14 Dealing with multi-regions. (a) A simple magnetostatic problem, (b) symmetry taken into account.

field component normal to the surface, so this problem can be reduced to one having two regions, see Fig. 9.14(b). Permeabilities μ_1 and μ_2 are ascribed to regions 1 and 2 respectively and since in region 1 there are no sources the total scalar potential ψ may be used, see Table 8.1, but in region 2 it is better to use the reduced potential ϕ, see section 1.4, in order to avoid the problem of multivalued functions. Without too much loss in generality region 2 may be taken as free space with $\mu_2 = \mu_0$. Hence the following equations and boundary conditions apply:

Region 1

$$\nabla \cdot \mu \nabla \psi = 0 : \in \Omega_1$$

$$\psi = 0 : y = 0 \quad \text{Essential boundary conditions}$$

$$\frac{\partial \psi}{\partial y} = 0 \; : \; y = b \quad \text{Natural boundary conditions}$$

$$\frac{\partial \psi}{\partial x} = 0 \; : \; x = 0$$

Region 2

$$\nabla^2 \phi = 0 \; : \; \in \Omega_2$$

$$\phi = 0 \; : \; y = 0 \quad \text{Essential boundary conditions}$$

$$\frac{\partial \phi}{\partial y} = H_{sy}(x) \; : \; y = b \quad \text{Natural boundary conditions}$$

$$\frac{\partial \phi}{\partial x} = H_{sx}(y) \; : \; x = a$$

The potentials and normal derivatives along the interface still have to be taken into account, otherwise a direct application of the boundary element method would lead to an undetermined system. However sufficient extra equations can be obtained directly from the physical interface conditions (see section 1.2). Thus normal **B** has to be continuous which implies

$$-\mu \frac{\partial \psi}{\partial x} = -\frac{\partial \phi}{\partial x} + H_{sx}, \tag{9.84}$$

and also tangential **H** has to be continuous which implies

$$-\frac{\partial \psi}{\partial y} = -\frac{\partial \phi}{\partial y} + H_{sy}, \tag{9.85}$$

also by integration we have

$$\psi = \phi - \int_0^y H_{sy}(a, y) \, \mathrm{d}y. \tag{9.86}$$

Thus the two potentials are related by a 'potential jump' given by eqn (9.86) across the region interface.

In order to form the boundary element equation the two regions are first decoupled as shown in Fig. 9.15 with boundary elements in each region. If eqn (9.69) is now applied to region 1 for each node in turn then:

$$\left.
\begin{aligned}
H_{11}\psi_1 + H_{12}\psi_2 + H_{14}\psi_4 &= G_{13}\frac{\partial \psi_3}{\partial n} + G_{14}\frac{\partial \psi_4}{\partial n} \\[2mm]
H_{21}\psi_1 + H_{22}\psi_2 + H_{24}\psi_4 &= G_{23}\frac{\partial \psi_3}{\partial n} + G_{24}\frac{\partial \psi_4}{\partial n} \\[2mm]
H_{31}\psi_1 + H_{32}\psi_2 + H_{34}\psi_4 &= G_{33}\frac{\partial \psi_3}{\partial n} + G_{34}\frac{\partial \psi_4}{\partial n} \\[2mm]
H_{41}\psi_1 + H_{42}\psi_2 + H_{44}\psi_4 &= G_{43}\frac{\partial \psi_3}{\partial n} + G_{44}\frac{\partial \psi_4}{\partial n}
\end{aligned}
\right\} \tag{9.87}$$

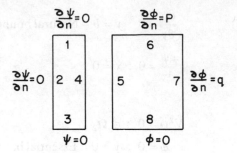

Fig. 9.15 Decoupled region.

thus for region 1 there are four equations but five unknowns. Similarly for region 2 there are four equations and five unknowns, as follows:

$$
\left.
\begin{aligned}
H_{55}\phi_5 + H_{56}\phi_6 + H_{57}\phi_7 &= G_{55}\frac{\partial\phi_5}{\partial n} + G_{56}p + G_{57}q + G_{58}\frac{\partial\psi_8}{\partial n}\\[2mm]
H_{65}\phi_5 + H_{66}\phi_6 + H_{67}\phi_7 &= G_{65}\frac{\partial\phi_5}{\partial n} + G_{66}p + G_{67}q + G_{68}\frac{\partial\psi_8}{\partial n}\\[2mm]
H_{75}\phi_5 + H_{76}\phi_6 + H_{77}\phi_7 &= G_{75}\frac{\partial\phi_5}{\partial n} + G_{76}p + G_{77}q + G_{78}\frac{\partial\psi_8}{\partial n}\\[2mm]
H_{85}\phi_5 + H_{86}\phi_6 + H_{87}\phi_7 &= G_{85}\frac{\partial\phi_5}{\partial n} + G_{86}p + G_{87}q + G_{88}\frac{\partial\psi_8}{\partial n}
\end{aligned}
\right\} \tag{9.88}
$$

Now when the two systems are coupled two more equations will be needed to ensure a well posed problem. The two missing equations are the physical interface conditions defined for this problem by eqns (9.85) and (9.86). The unknowns at nodes 4 and 5 in Fig. 9.15 are connected by:

$$
\psi_4 = \phi_5 - \int_0^{y_4} H_{sy}(a, y)\,\mathrm{d}y = \phi_5 + c \tag{9.89}
$$

say, and

$$
\mu\frac{\partial\psi_4}{\partial n} = \frac{\partial\phi_5}{\partial n} - H_{sx}(a, y_4) = \frac{\partial\phi_5}{\partial n} + b \tag{9.90}
$$

say. Hence the solution of the problem can be obtained by solving the following 10×10

system:

$$
\begin{bmatrix}
H_{11} & H_{12} & H_{14} & -G_{13} & -G_{14} & 0 & 0 & 0 & 0 & 0 \\
H_{21} & H_{22} & H_{24} & -G_{23} & -G_{24} & 0 & 0 & 0 & 0 & 0 \\
H_{31} & H_{32} & H_{34} & -G_{33} & -G_{34} & 0 & 0 & 0 & 0 & .0 \\
H_{41} & H_{42} & H_{44} & -G_{43} & -G_{44} & 0 & 0 & 0 & 0 & 0 \\
0 & 0 & 1 & 0 & 0 & 0 & -1 & 0 & 0 & 0 \\
0 & 0 & 0 & 0 & \mu & -1 & 0 & 0 & 0 & 0 \\
0 & 0 & 0 & 0 & 0 & H_{55} & H_{56} & H_{57} & -G_{55} & -G_{58} \\
0 & 0 & 0 & 0 & 0 & H_{65} & H_{66} & H_{67} & -G_{65} & -G_{68} \\
0 & 0 & 0 & 0 & 0 & H_{75} & H_{76} & H_{77} & -G_{75} & -G_{75} \\
0 & 0 & 0 & 0 & 0 & H_{85} & H_{86} & H_{87} & -G_{85} & -G_{88}
\end{bmatrix}
\begin{bmatrix}
\psi_1 \\
\psi_2 \\
\psi_3 \\
\dfrac{\partial \psi_3}{\partial n} \\
\dfrac{\partial \psi_4}{\partial n} \\
\dfrac{\partial \psi_5}{\partial n} \\
\phi_5 \\
\phi_5 \\
\phi_5 \\
\dfrac{\partial \phi_8}{\partial n}
\end{bmatrix}
=
\begin{bmatrix}
0 \\
0 \\
0 \\
0 \\
c \\
b \\
G_{56}p + G_{57}q \\
G_{66}p + G_{67}q \\
G_{76}p + G_{77}q \\
G_{86}p + G_{87}q
\end{bmatrix}
$$

It can be seen therefore that a two region problem is characterized by the following matrix structure:

$$
\begin{bmatrix}
R_1 & \vdots & 0 \\
\cdots & \text{Interface} & \cdots \\
0 & \vdots & R_2
\end{bmatrix}
\begin{bmatrix}
U_1 \\
\\
U_2
\end{bmatrix}
=
\begin{bmatrix}
q_1 \\
\\
q_2
\end{bmatrix}
\tag{9.91}
$$

and that a multiregion problem can then be treated systematically. In the example worked out above, defined by Fig. 9.14(a), the external boundaries were assumed to be near the problem, this is very often not the case in practice where free space is a magnetic 'material' ($\mu = \mu_0$) and the true boundaries are essentially at infinity. This class of problem is termed an open boundary problem and can be treated elegantly by integral methods.

Consider Fig. 9.16 in which we show a symmetric magnet surrounded by free space and a 'far field boundary' which may be at infinity. In applying the BEM to this problem symmetry is utilized to reduce the number of unknowns. For example, in Fig. 9.16 the whole magnet is illustrated but, because the potentials in the second, third and fourth quadrants are identical to those in the first quadrant, it is only necessary to compute the potentials in the first quadrant explicitly. Furthermore the far field boundary shown in the figure can be expanded to infinity because there are no boundary connections between it and the magnet; the far field boundary has no effect on the problem whatsoever. This is obvious for real problems where the potential and its normal derivatives at the far field boundary are zero. There appears to be a difficulty, at first sight, for the two-dimensional limit because of the behaviour of the logarithmic potential at large

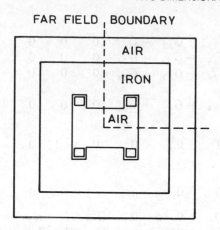

Fig. 9.16. *Infinite boundary.* Boundary element method using equivalent elements and symmetry—the far field is shown, but it can be at infinity.

distances, see for example eqn (9.63). However, the line integral of the field for a complete surface must be zero therefore the contributions for all elements of a surface will cancel to produce zero potential at infinity.

9.8 ALTERNATIVE BOUNDARY INTEGRAL METHODS

The presence of the normal derivative term in eqn (9.68) can cause problems at discontinuous corners since at these exceptional points the direction of the surface normal is undefined. This becomes a serious difficulty in generalizing the standard boundary element method to higher order shape functions as it then becomes necessary to introduce element end-points as nodes instead of single collocation points at the centre. However, for cases where there are no prescribed potentials, such as the symmetric magnetic problem shown in Fig. 9.16 where all external boundaries are at infinity, the following procedure avoids the normal derivative term entirely [9]. The total field vector $\mathbf{H}$ can be expressed as the sum

$$\mathbf{H} = \mathbf{H}_s + \mathbf{H}_m \tag{9.92}$$

in which H_s is the field for the prescribed sources, and H_m is the field produced by the induced dipoles in the ferromagnetic region. Now since curl($\mathbf{H}$) = $\mathbf{J}$ by definition eqn (9.92) implies that curl ($\mathbf{H}_m$) = O and $\mathbf{H}$ may be written as

$$\mathbf{H}_m = -\nabla\phi \tag{9.93}$$

where ϕ is the reduced scalar potential. The reduced potential may also be expressed in terms of $\mathbf{M}$ the dipole magnetization per unit volume, see eqn (1.39), i.e.

$$\phi = \frac{1}{4\pi}\int_\Omega \mathbf{M}\cdot\nabla\left(\frac{1}{R}\right)d\Omega \tag{9.94}$$

with

$$\mathbf{M} = (\mu - 1)\mathbf{H} = \chi\mathbf{H} = -\chi\nabla\psi, \tag{9.95}$$

where χ is the susceptibility. Furthermore, the total scalar potential ψ is given by a relation similar to eqn (9.92), i.e.

$$\psi = \phi + \psi_s \tag{9.96}$$

with ψ_s the scalar potential of the prescribed sources given by

$$\psi_s = \int_0^t h_{st}dt \tag{9.97}$$

which is the line integral of the tangential component of the source field along a suitably chosen contour which must not intersect a current loop to ensure uniqueness; accordingly ψ_s vanishes for a complete contour in order to satisfy Ampère's law. The reduced potential is now eliminated from eqn (9.96) by eqns (9.94), (9.95) to obtain

$$\psi(\mathbf{r}) = -\frac{1}{4\pi}\int_\Omega \nabla\psi(\mathbf{r}')\chi(\mathbf{r}')\cdot\nabla\left(\frac{1}{R}\right)d\Omega + \psi_s. \tag{9.98}$$

Equation (9.98) can be transformed by Green's theorem, see Appendix 4 (A4.16) to eliminate the $\nabla\psi$ term, i.e.

$$\psi(\mathbf{r}) = \frac{1}{4\pi}\int_\Omega \psi(\mathbf{r}')\nabla\chi(\mathbf{r}')\cdot\nabla\left(\frac{1}{R}\right)d\Omega - \frac{1}{4\pi}\int_\Gamma \psi(\mathbf{r}')\nabla\chi(\mathbf{r}')\left(\frac{1}{R}\right)d\Gamma + \psi_s. \tag{9.99}$$

The volume term in eqn (9.99) for constant μ reduces further since:

$$\int_{\Omega_0} \nabla^2\left(\frac{1}{R}\right)d\Omega = -1 \tag{9.100}$$

for points r inside a sufficiently small domain Ω_0 enclosing the point r' and

$$\int_{\Omega_0} \nabla^2\left(\frac{1}{R}\right)d\Omega = 0 \tag{9.101}$$

for points outside. Thus for constant permeability eqn (9.99) becomes

$$\mu\psi(\mathbf{r}) = -\frac{1}{4\pi}\int_\Gamma (\mu_r - 1)\psi(\mathbf{r}')\frac{\partial(1/R)}{\partial n}d\Gamma + \psi_s(\mathbf{r}) \tag{9.102}$$

for points on the surface Γ of the domain. In the two-dimensional limit the $1/R$ term must be replaced by $-\log R$ the appropriate singular solution of the two-dimensional form of Laplace equation—see section 9.7.

In order to solve eqn (9.102) numerically the boundary can be discretized into line elements similar to the procedure adopted for the boundary element method—see section 9.7, but in this case, these are no derivative terms in the equation and the usual finite element practice of specifying the unknown potential at element nodes can be employed in order to represent a higher order variation. Figure 9.17 shows a boundary segment approximated by straight line elements, consider a particular element having a linear variation of potential along its length; the potential at any intermediate point may be written as follows:

$$\psi = N_1\psi_1 + N_2\psi_2 \tag{9.103}$$

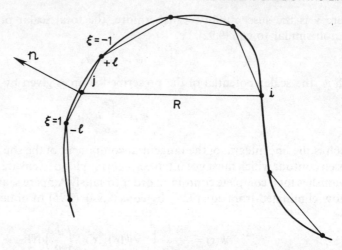

Fig. 9.17 Geometry and coordinates for first order line elements.

where N_1 and N_2 are linear functions of position. In terms of the normalized line coordinate

$$\xi = s/l \tag{9.104}$$

$$N_1 = (1 + \xi)/2 \quad N_2 = (1 - \xi)/2. \tag{9.105}$$

To allow for higher order variation internal nodes may be included, thus for a quadratic variation in potential a midside node is introduced and

$$\psi = N_1 \psi_1 + N_2 \psi_2 + N_3 \psi_3 \tag{9.106}$$

with

$$\left.\begin{array}{l} N_1 = \xi(1 + \xi)/2 \\ N_2 = \xi(1 - \xi)/2 \\ N_3 = (1 + \xi)(1 - \xi) \end{array}\right\}. \tag{9.107}$$

Thus the discretized form of eqn (9.99) using a point matching (see section 8.4.1) technique is

$$\mu \psi_i = (K_{ij} \psi_j + \psi_s) \tag{9.108}$$

with

$$K_{ij} = \sum \frac{(1 - \mu_r)}{2\pi} \int_\Gamma N_j \frac{\partial(-\log R)}{\partial n} \, d\Gamma, \tag{9.109}$$

where the sum is over all elements containing j.

This leads to a set of equations with one equation for each node. The matrix K is dense and unsymmetric; however its components can be evaluated analytically using only elementary functions [9].

This technique provides a very efficient method of solving two-dimensional linear

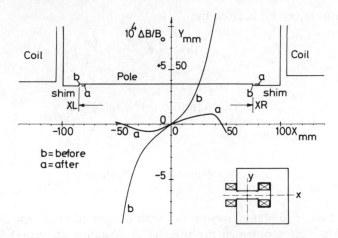

Fig. 9.18 Results for the scalar integral equation applied to a C-magnet.

magnetostatic problems and has been used extensively in the design of linear electromagnetic devices requiring accurate field shapes produced by pole face shims [10]; see for example Fig. 9.18 in which some results are shown for the optimization of shims for a particle accelerator magnet requiring field accuracies one part in 10^4. Because of the simplicity of the matrix coefficients, Eqn (9.109), the extension to a full Galerkin weighting is feasible and should produce even better results [11]. Alternatively this boundary formulation can be derived mathematically by extending the weighted residual procedure of section 12.7, see Reference [12].

Another variant of the integral approach which uses the magnetization as the unknowns will be discussed in Chapter 12 on three-dimensional applications.

9.9 AXISYMMETRIC FORM OF POISSON'S EQUATION

In this section the rotationally or axisymmetric case for the solution of the Poisson and Laplace equations will be considered by extending the finite element method developed for the plane symmetry case in section 9.1. The axisymmetric forms of eqn (8.1) for a cylindrical coordinate system are required and these forms depend upon whether the independent variable u is a scalar or vector potential so the two cases will be dealt with in turn. Thus the axisymmetric total scaler potential ψ satisfies eqn (1.28) where the field components are given by:

$$H_r = -\frac{\partial \psi}{\partial r} \tag{9.110}$$

$$H_z = -\frac{\partial \psi}{\partial z}. \tag{9.111}$$

The scalar Poisson problem (other examples include the electrostatic potential V, see eqn (1.36)) which in cylindrical polar coordinates (see Stratton [13], p. 51) using the unit vector $(\mathbf{e}_1, \mathbf{e}_2, \mathbf{e}_3)$ is given by:

$$\text{div}\left[\mu\left(\mathbf{e}_1\frac{\partial\psi}{\partial r} + \mathbf{e}_3\frac{\partial\psi}{\partial z}\right)\right] = q \tag{9.112}$$

i.e.

$$\frac{1}{r}\frac{\partial(\mu r(\partial\psi)/\partial r)}{\partial r} + \frac{\partial(\mu(\partial\psi)/\partial z)}{\partial z} = q \tag{9.113}$$

or

$$\frac{\partial(\mu r(\partial\psi)/\partial r)}{\partial r} + \frac{\partial(\mu r(\partial\psi)/\partial z)}{\partial z} = rq, \tag{9.114}$$

i.e. similar to plane xy symmetry with μ replaced by μr and source q replaced by rq.

The vector potential problem (magnetostatics, see eqn (1.34)), which in cylindrical coordinates (ibid.) the field components are given by:

$$B_r = \frac{\partial A_\phi}{\partial z}, \tag{9.115}$$

$$B_z = -\frac{1}{r}\frac{\partial(rA_\phi)}{\partial r}, \tag{9.116}$$

and eqn (1.34) becomes

$$\text{curl}\frac{1}{\mu}\left(-\frac{\partial A_\phi}{\partial z}\mathbf{e}_1 + \frac{1}{r}\frac{\partial rA_\phi}{\partial r}\mathbf{e}_3\right) = \mu_0 J_\phi \tag{9.117}$$

i.e.

$$\frac{\partial(1/\mu r)(\partial rA/\partial z)}{\partial z} + \frac{\partial(1/\mu r)(\partial rA/\partial r)}{\partial r} = -\mu_0 J_\phi, \tag{9.118}$$

which is similar to the plane xy symmetry case with (A) replaced by (rA). Note that rA is the flux flowing through a surface spanned by a circular filament of radius r. Also note $1/\mu$ is replaced by $1/\mu r$.

9.10 SCALAR POTENTIAL SOLUTION

The scalar axisymmetric form of the Poisson problem can be solved by the finite element method starting from the general weighted residual expression eqn (8.27). From eqns (9.114) and (8.27):

$$\int_\Omega \omega_j\left(\frac{\partial(\mu r(\partial\psi/\partial r))}{\partial r} + \frac{\partial(\mu r(\partial\psi/\partial z))}{\partial z}\right)r\,dr\,dz = \int_\Omega \omega_j qr^2\,dr\,dz. \tag{9.119}$$

Integrating the LHS by parts to eliminate the second order derivative to weaken the continuity requirements to C_0, see section 9.1, i.e.

$$\int_\Omega \nabla\omega_j\mu r\nabla\psi\,dr\,dz = \int_\Omega \omega_j qr^2\,dr\,dz + \int_\Gamma \omega_j\mu r\frac{\partial\psi}{\partial n}\,d\Gamma. \tag{9.120}$$

The surface integral in eqn (9.120) can be made to vanish by setting the weight to zero for cases where ψ is specified on the external boundaries and also vanishes between elements because of the physical interface conditions—see section 9.1.2. If in eqn (9.120) Galerkin weighting is chosen ($\omega_j = N_j$), and the boundary conditions either, $\psi = \psi_0$ or, $\partial\psi/\partial n = 0$, are imposed and, furthermore if the shape functions

$$\psi = \sum N_i \psi_i \tag{9.121}$$

are introduced then eqn (9.120) for each element becomes for the ith discretized node

$$R_i = \left[\int_{\text{elem}} \mu \nabla N_i \cdot \nabla N_j r^2 \, dr \, dz \right] \psi - \int_{\text{elem}} N_i q r^2 \, dr \, dz, \tag{9.122}$$

eqn (9.122) is of the form:

$$R_i = k_{ij} \psi_j + f_i \tag{9.123}$$

with

$$k_{ij} = \int_{\text{elem}} \mu \left(\frac{\partial N_i}{\partial r} \frac{\partial N_j}{\partial r} + \frac{\partial N_i}{\partial z} \frac{\partial N_j}{\partial z} \right) r^2 \, dr \, dz, \tag{9.124}$$

and

$$f_i = \int_{\text{elem}} N_i q r^2 \, dr \, dz. \tag{9.125}$$

For the triangular area shape functions eqn (9.12), eqns (9.124) and (9.125) become

$$k_{ij} = \int_{\text{elem}} \frac{\mu(b_i b_j + c_i c_j) r^2 \, dr \, dz}{4A^2}, \tag{9.126}$$

and,

$$f_i = -\int_{\text{elem}} \frac{(a_i + b_i r + c_i z) q r^2 \, dr \, dz}{2A}, \tag{9.127}$$

While it is obviously possible to carry out the integrals in the above equations analytically they are more complicated than for the planar case and it is usual to use a numerical quadrature scheme instead. For planar triangels a n-point Gaussian quadrature formula; see Appendix 5, may be used, and eqns (9.126) and (9.127) become

$$k_{ij} = \frac{\mu(b_i b_j + c_i c_j)}{4A^2} \sum_k W_k r^2(\xi_k) \tag{9.128}$$

and

$$f_i = -\frac{q}{2A} \sum_k W_k [a_i + b_i r(\xi_k) + c_i z(\xi_k) r^2(\xi_k)], \tag{9.129}$$

where ξ_k and W_k are the Gauss points and weights respectively.

Therefore to obtain an approximate solution for the axisymmetric scalar equation the domain can be discretized into triangles and element coefficients evaluated by eqns (9.128) and (9.129). The system matrix is then constructed and solved as in section 9.2 to 9.4.

9.11 VECTOR POTENTIAL SOLUTION

The solution to the vector equation, eqn (9.118), can be approached in a similar way but in this case a vector quantity is involved, and so it is necessary to use the vector form of Green's theorem (see Appendix 4) in eqn (9.118) to eliminate the second derivatives, i.e. the weighted residual form of eqn (9.118) becomes

$$-\int_{\Omega} \frac{1}{\mu}\left[\frac{\partial \omega_j}{\partial z}\frac{\partial A_\phi}{\partial z} + \frac{1}{r^2}\frac{\partial(r\omega)}{\partial r}\frac{\partial(rA_\phi)}{\partial r}\right] r\,dr\,dz = \int_{\Omega} \omega_j J_\phi r\,dr\,dz \qquad (9.130)$$

The subsequent finite element discretization follows the same procedure as used for the scalar form with the element coefficients given by:

$$k_{ij} = \int_{\text{elem}} \frac{1}{4A^2\mu}\left(c_i c_j + \frac{[(a_j + 2b_j r + c_j z)(a_i + 2b_i r + c_i z)]}{r^2}\right) r\,dr\,dz \qquad (9.131)$$

and

$$f_i = -J\int_{\text{elem}} (a_i + b_i r + c_i z) r\,dr\,dz, \qquad (9.132)$$

where

$$R_i = k_{ij}A_j + f_i. \qquad (9.133)$$

Alternatively a simpler form can be derived by setting

$$\Phi = rA, \qquad (9.134)$$

when Φ is the magnetic flux, i.e.

$$k_{ij} = \int_{\text{elem}} \frac{(b_i b_j + c_i c_j)\,dr\,dz}{4A^2\mu r}, \qquad (9.135)$$

$$f_i = \int_{\text{elem}} \frac{J(a_i + b_i r + c_i z)\,dr\,dz}{2A}. \qquad (9.136)$$

All of these coefficients can be evaluated by Gaussian quadrature and assembled into a system matrix exactly as described in section 9.1.

Illustrative examples using the above procedures are shown in the next two figures; in Fig. 9.19 potentials for high voltage bush were obtained using the axisymmetric scalar potential and Fig. 9.20 shows the field map of flux in the objective lens of an electron microscope using the axisymmetric vector potential. The finite element discretization is shown in both cases also, and since good accuracy was a requirement elements with a quadratic shape function were used, see section 11.5. The coefficients for the vector forms have singular terms at $r = 0$, see eqns (9.131) and (9.135). This problem can be avoided by using a numerical quadrature scheme for which the Gauss points are interior to the element. However, when the fields are recovered there is a problem in evaluating gradients for points on the axis. This arises because the axial field component is given by eqn (9.116), and even though the gradient is well defined, B_z at $r = 0$ is not. It is usual to choose r to be at the centroid of elements and for first order elements this is a good compromise for elements off axis. For axial elements the effect will be exacerbated, and it may be necessary to use very small elements in order to achieve even a tolerable accuracy. In the next section a recently developed procedure is referred to which enables the fields to be recovered accurately for all points.

9.12 ANALYTIC TRANSFORMATIONS TO IMPROVE THE ACCURACY NEAR THE AXIS

The problem of recovering fields for points near the axis has been alluded to in the last section. One way to avoid the singularity in the stiffness matrix is to use $\mathbf{A}/r$ as the solution potential [14], however more generally Lowther and Silvester [15] considered modified potentials of the form $r^{-p}A_\phi$ and they advocate the use of this method with a factor $p > 1/2$, because then the singularities in the system matrix integrals are eliminated.

Nevertheless recent work by Melissen and Simkin [16] have reported results for some simple examples where very poor accuracy for the field on axis can be obtained even when the above methods are used. This is because the form of the potential is not well adapted to the low order polynomial interpolations used as basis functions. They introduce a novel formulation which consists of transforming the potential variable as well as applying a coordinate transformation.

9.13 CONCLUDING REMARKS

Formulations based on partial differential equations and on integral equations have now been considered. This chapter is concluded by summarizing the strengths and weaknesses of these two main methods, see Fig. 9.6. For the differential formulation the following two points apply:

- The application of the Galerkin method generates an algebraic system of equations which is sparse and symmetric.
- The entire problem space, including the far-field region, has to be discretized or approximated by boundary conditions.

The alternative integral formulation leads to an algebraic system matrix of a very different character, i.e.

- The system matrix is fully populated, since every point is coupled to every other point.
- The material regions only of the problem space have to be discretized. This follows because the integrals are only defined for the active regions, which further implies that boundary conditions are intrinsically satisfied. Thus the far-field region has been automatically taken into account. Another important consequence is that for linear problems only the surfaces of active regions have to be discretized.

At first sight it may appear that the integral approach will be superior to the differential

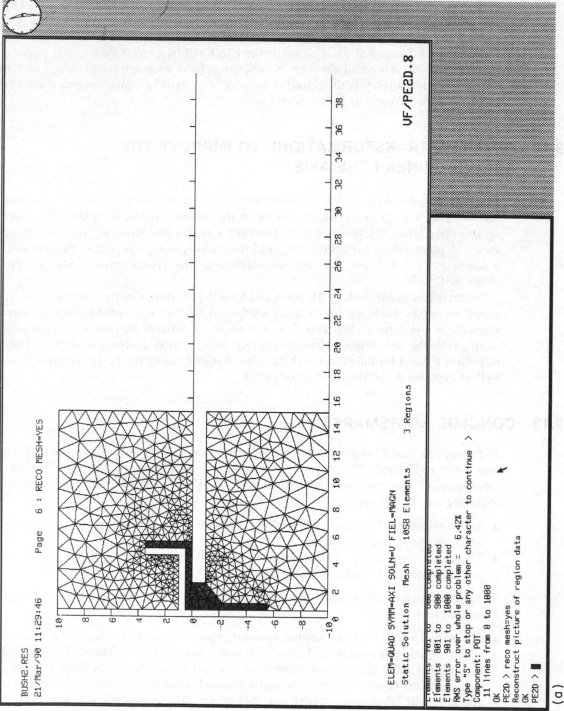

(a)

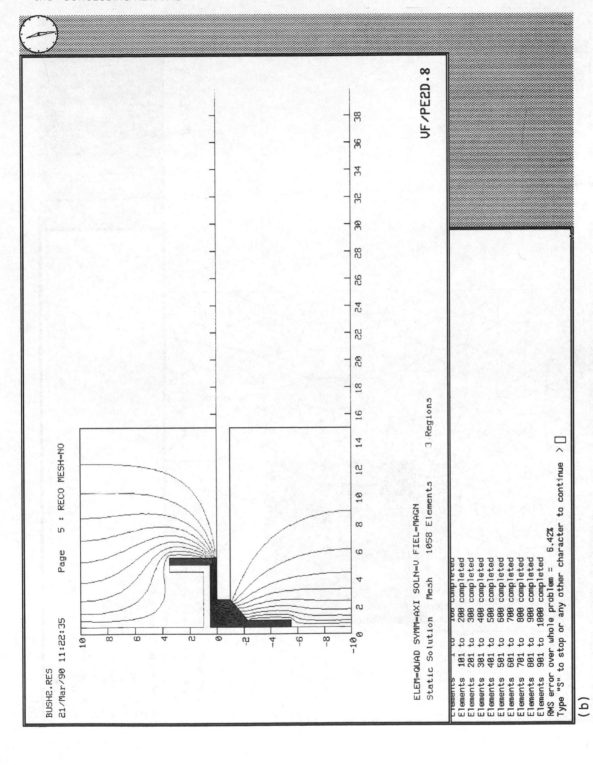

Fig. 9.19 High voltage 'Bush' insulator [1].

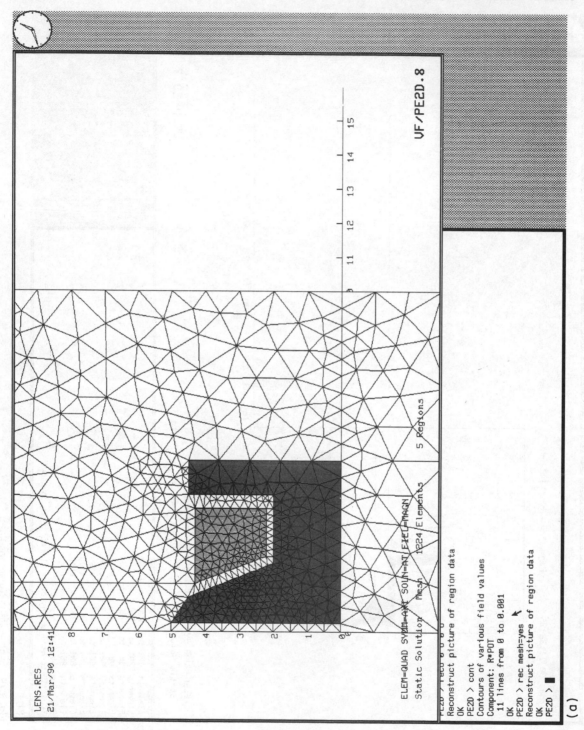

(a)

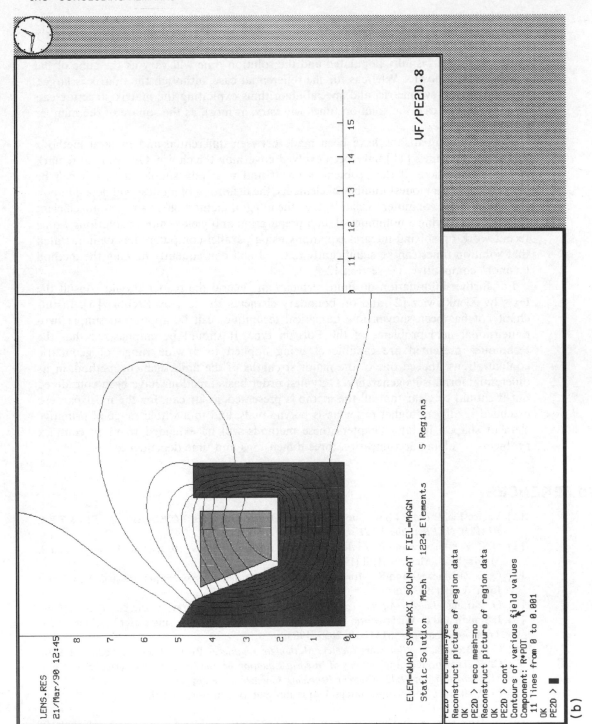

Fig. 9.20 Electron microscope objective lens [1].

approach since the mesh generation problem is far simpler—the empty space does not have to be filled with elements and the far-field boundary conditions are correctly imposed by the formulation. However there are significant difficulties because the matrix, although smaller, is fully populated and the solution time will vary as the cube of the number of unknowns. Whereas for the differential case, although the matrix is larger, it is sparse and symmetric, and special algorithms exploiting the matrix structure can be used. In this case the solution times will vary, at most, as the square of the number of unknowns.

Extensive comparisons have been made between differential and integral methods and some guidelines [11] have been evolved governing the choice. One general remark is appropriate here. If the problem is linear and relatively simple in that it can be discretized into a modest number of elements, the definition of modest will depend upon the power of the computer available, then the integral method offers an accurate efficient technique involving a minimum of data preparation and yields smooth solutions. Some recent work on solving integral equations using parallel computers has demonstrated that solution times can be significantly reduced and consequently making the method far more competitive, see section 12.4.

For further information on finite elements in general the reader should consult the texts by Zienkiewicz [2] and on boundary elements the text by Brebbia [6]. In this chapter it has been shown how numerical techniques can be applied to simple two-dimensional field problems of the Poisson type. It should be emphasized that the techniques presented are capable of being applied to a wide range of geometric configurations. Indeed one of the major strengths of the finite element method, in its differential form, is its generality. Only first order basis functions have been considered but it should be clear that all the methods presented so far can, for the most part, be extended readily to higher order basis polynomials, and to a whole range of primitive element shapes. In later chapters these methods will be extended to more complex problems involving non-linearity, three dimensions and time dependence.

REFERENCES

[1] Vector Fields Ltd, 24 Bankside, Kidlington, Oxford OX5 1JE, *TOSCA, GFUN, ELEKTRA, BIM2D, OPERA, and PE2D User Manuals*, 1990.

[2] O. C. Zienkiewicz and R. Taylor, *The Finite Element Method, 4th Edition, Volumes 1 and 2*. Maidenhead: McGraw-Hill (1991).

[3] O. C. Zienkiewicz and K. Morgan, *Finite Elements and Approximation Method*. New York: John Wiley (1983).

[4] G. Strang, *Linear Algebra and its Applications*. New York: Academic Press (1976).

[5] J. Simkin and C. W. Trowbridge, Electromagnetics CAD using a single user machine, *IEEE Trans on Magnetics*, **MAG-19**, 2655 (1983).

[6] C. A. Brebbia, *The Boundary Element Method for Engineers*. Printec Press, 2nd edition (1980).

[7] C. W. Trowbridge, *Applications of Integral Equation Methods for the Numerical Solution of Magnetostatic and Eddy Current Problems*. Chichester: Wiley, (1979).

[8] M. A. Jaswon and Symm, Integral equation methods in potential theory, *Proc. Roy. Soc. A, 275*, p. 23 (1963).

[9] A. Armstrong *et al.*, The solution of 3D magnetostatic problems using scalar potentials, in *Compumag Grenoble* (J. C. Sabbonadiere, ed.), (Grenoble, France), Laboratoire d'Electrotechnique, ENSEGP (1978).

[10] A. G. A. M. Armstrong *et al.*, Automated optimisation of magnet design using a boundary integral method, *IEEE Trans. on Magnetics*, **MAG-18**(2) (1982).

[11] J. Simkin, A comparison of integral and differential solutions for field problems, *IEEE Trans. on Magnetics*, **MAG-18** (1982).

[12] C. F. Bryant, M. H. Roberts, and C. W. Trowbridge, Implementing a boundary integral method on a transputer system, in *Compumag Conference on the Computation of Electromagnetic Fields, Tokyo, September 1989 Proceedings, IEEE Trans. on Magnetics*, **MAG-26** (2), March, 819–812 (1990).

[13] J. A. Stratton, *Electromagnetic Theory*. New York: McGraw-Hill, (1941).

[14] D. Lowther *et al.*, Newton–Raphson finite element programs for axisymmetric vector fields, *IEEE Trans. on Magnetics*, **MAG-19**, 2523–2526 (1983).

[15] D. Lowther and P. P. Silvester, *Computer Aided Design in Magnetics*. Berlin: Springer-Verlag (1986).

[16] H. Melissen and J. Simkin, A new coordinate transformation for the finite element solution of axisymmetric problems in magnetostatics, in *Compumag Conference on the Computation of Electromagnetic Fields, Tokyo, September 1989 Proceedings*, **MAG-26**(2), March, 391–394 (1990).

NON-LINEAR EFFECTS IN TWO-DIMENSIONAL FIELDS

10.1 INTRODUCTION

In the last chapter the finite element method was used to develop general algorithms for solving the two-dimensional and axisymmetric forms of the Laplace and Poisson equation applicable to linear problems. The finite element method is particularly suitable for dealing with non-linear devices like electromagnets with ferromagnetic materials which exhibit anisotropy and saturation phenomena and the results of Chapter 9 are extended here to include these non-linear effects. The constitutive relationships between the magnetic flux (**B**) and magnetic field (**H**) are, in general, non-linear, furthermore, some materials are hysteretic, that is the **B–H** relationship is not single valued and the **BH** 'state' at a particular point in space and time depends upon its previous magnetic history. However in this chapter only the simplest non-linear situations will be considered; in particular problems with non-hysteretic materials, which are single valued over the range of **B–H** values required and can be represented, typically, by the anhysteretic part of the curve shown in Fig. 10.1. The simplest case of isotropic (non-linear) media is considered first but subsequently situations involving permanent magnets and also anisotropy are introduced.

10.2 SATURATION EFFECTS

10.2.1 Introduction

In order to extend the finite element procedures to include non-linear material properties (saturation in the magnetic case) a mathematical model describing the magnetic properties of the material is needed. This model will depend upon the actual form of the Poisson equation required and the several variants are given in Table 8.1. For the magnetic scalar potential form of the equation the appropriate constitutive parameter is the permeability μ (see eqn (1.6)) which can be expressed in terms of a relative

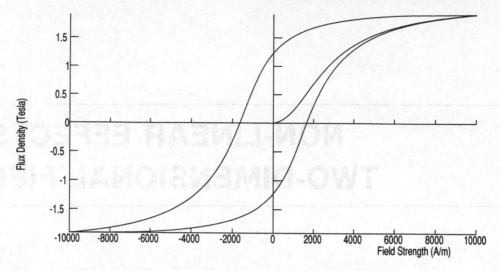

Fig. 10.1 Typical **B–H** curve for magnetic materials.

permeability μ_r (a dimensionless number) by:

$$\mu = \mu_0 \mu_r, \tag{10.1}$$

where μ_0 is the permeability of free space ($= 4\pi \times 10^{-7}$). The permeability is a function of **B** or **H** in a non-linear material. For the vector potential formulation the reciprocal of permeability (usually known as the reluctivity, γ) is appropriate, and is given by:

$$\gamma = \frac{1}{\mu} = f(|\mathbf{B}|) = f\left[\sqrt{\left(\frac{\partial A}{\partial x}\right)^2 + \left(\frac{\partial A}{\partial y}\right)^2}\right]. \tag{10.2}$$

These two models have the general form in the Poisson eqn (8.1) of

$$\kappa = f(u), \tag{10.3}$$

where the function f, for isotropic non-hysteretic materials, is assumed to be a continuous single valued function of the potential u. It is now necessary to include the above non-linear relationships in the Galerkin method discussed in Chapter 9. It was shown there that this method, applied to linear materials, led to the discretized form of eqn (8.1), eqn (9.38), i.e.

$$\mathbf{Ku} = \mathbf{Q}. \tag{10.4}$$

The coefficients of the banded sparse matrix **K** were assembled from the individual element coefficients, see eqns (9.21) and (9.22), sections 9.1 to 9.2. In eqn (9.21) the parameter κ (permeability) was assumed to be constant over the element. It is also often the practice to retain this assumption in non-linear problems as well, but now the value of κ must be allowed to vary from element to element. It should be mentioned though that, in many cases, substantial improvements in accuracy can be achieved by forcing κ to follow a linear or higher order form within each element. In either case the coefficients for a particular element depend upon the actual value of κ and hence by eqn (10.3), they are dependent upon the solution vector u. Furthermore it then follows that the system coefficients K_{ij} are functions of u also and that u itself will satisfy a system of non-linear algebraic equations, i.e.

$$\mathbf{K}(\mathbf{u})\mathbf{u} = \mathbf{Q}. \tag{10.5}$$

10.2.2 Simple iterative methods

Equation (10.5) may be solved by a simple iterative technique that proceeds by successively solving a linear system of equations starting with an estimated value for κ with a new value of κ being obtained at the end of each iteration from the constitutive relation, eqn (10.3). Suppose $u^{(0)}$ is the solution to eqn (10.5) obtained by the Gaussian elimination method described in section 9.4, using an assumed value of $\kappa^{(0)}$, then an improved value of κ is given by eqn (10.3) i.e.

$$\kappa^{(1)} = f(u^{(0)}), \tag{10.6}$$

and this value in turn can be used to obtain an improved value of $u^{(1)}$.

Then after (m) iterations the $(m + 1)$th solution from eqn (10.5) is given by

$$\mathbf{u}^{(m+1)} = \mathbf{\kappa}^{-1}(\mathbf{u}^{(m)})\mathbf{Q}. \tag{10.7}$$

The simple iteration process for a single unknown is illustrated in Fig. 10.2(a); the method is said to converge in a practical sense if after m iterations

$$\mathbf{u}^{(m+1)} - \mathbf{u}^{(m)} = \mathbf{e} \tag{10.8}$$

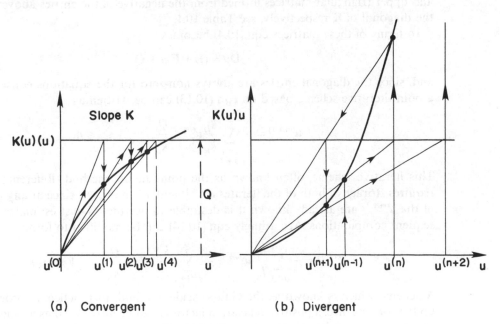

(a) Convergent   (b) Divergent

Fig. 10.2 Simple iterative method.

where **e** is an array of error values such that

$$|\mathbf{e}| < \alpha \tag{10.9}$$

with α set to a small positive number.

The error array **e** is also known as the 'error norm' and is sometimes taken to be the largest error but more usually the RMS value defined by

$$|\mathbf{e}| = \sqrt{(\mathbf{e}\mathbf{e}^T)} \tag{10.10}$$

is taken.

Figure 10.2(b) illustrates a non-convergent situation which can occur whenever

$$F'(\mathbf{u}^{(n)}) = \left| \frac{\mathrm{d}}{\mathrm{d}u}(\mathbf{K}(\mathbf{u})u - Q) \right| \geqslant 1, \tag{10.11}$$

see the text by Wait (p. 75) [1]. The simple iterative method requires a full solution of the latest updated system to be carried out at each iteration. It can be seen however that it is unnecessary to compute these solutions to very high accuracy during the earlier iterations and, in any case the improvement transferred from one iteration to the next arises only from the updated values of the constitutive parameter and not from the intermediate solutions explicitly. Thus instead of a direct method like Gaussian elimination a completely iterative technique combining both the linear and non-linear procedures may be more effective, e.g. the Gauss–Seidel algorithm (Varga [2], p. 58) in which the system matrix K is decomposed into:

$$\mathbf{K} = \mathbf{D} - \mathbf{E} - \mathbf{F}, \tag{10.12}$$

where **D** is the matrix formed from the diagonal entries only, and **E** and **F** are lower and upper triangular matrices formed from the negative of the entries above and below the diagonal of **K** respectively, see Table 10.1.

In terms of these matrices eqn (10.4) becomes

$$\mathbf{D}u = (\mathbf{E} + \mathbf{F})u + \mathbf{Q} \tag{10.13}$$

and, since the diagonal entries are always non-zero for the equations considered here, a point iterative scheme based on eqn (10.13) can be written as:

$$u_i^{(m+1)} = - \sum_{\substack{j=1 \\ j \neq i}} \frac{k_{ij}}{k_{ii}} u_j^{(m)} + \frac{Q_i}{k_{ii}}, \qquad 1 \leqslant i \leqslant n. \tag{10.14}$$

This iterative scheme, often known as the point Jacobi method (Reference [2], p. 57) requires storage of both of the iterates $u^{(m+1)}$ and $u^{(m)}$ however, since at any point some of the $u^{(m+1)}$ are already known it is desirable to use these latest estimates in all subsequent computations. Accordingly eqn (10.14) can be modified to form:

$$u_i^{(m+1)} = - \sum_{j=1}^{i-1} \frac{k_{ij}}{k_{ii}} u_j^{(m+1)} - \sum_{j=i+1}^{n} \frac{k_{ij}}{k_{ii}} u_j^{(m)} + \frac{Q_i}{k_{ii}}, \qquad 1 \leqslant i \leqslant n. \tag{10.15}$$

A scheme which is known as the Gauss–Seidel method. In practice, in order to speed up the convergence, a positive relaxation factor ω can be introduced as a weighted mean of two successive iterations, a value of ω in which $1 > \omega \geqslant 0$ corresponds to under-

Table 10.1 Triangular decomposition.

(a) General decomposition into triangular matrices

$$\mathbf{K} = \begin{pmatrix} k_{11} & k_{12} & k_{13} & k_{14} \\ k_{21} & k_{22} & k_{23} & k_{24} \\ k_{31} & k_{32} & k_{33} & k_{34} \\ k_{41} & k_{42} & k_{43} & k_{44} \end{pmatrix} = \mathbf{D} - \mathbf{E} - \mathbf{F}$$

$$= \begin{pmatrix} k_{11} & 0 & 0 & 0 \\ 0 & k_{22} & 0 & 0 \\ 0 & 0 & k_{33} & 0 \\ 0 & 0 & 0 & k_{44} \end{pmatrix} - \begin{pmatrix} 0 & 0 & 0 & 0 \\ -k_{21} & 0 & 0 & 0 \\ -k_{31} & -k_{32} & 0 & 0 \\ -k_{41} & -k_{42} & -k_{43} & 0 \end{pmatrix} - \begin{pmatrix} 0 & -k_{12} & -k_{13} & -k_{14} \\ 0 & 0 & -k_{23} & -k_{24} \\ 0 & 0 & 0 & -k_{34} \\ 0 & 0 & 0 & 0 \end{pmatrix}$$

Note—If $\mathbf{K}$ is symmetric then $\mathbf{E} = \mathbf{F}^T$.

(b) LU decomposition

$$\mathbf{E} = \begin{pmatrix} 0 & 0 & 0 & 0 \\ -k_{21} & 0 & 0 & 0 \\ -k_{31} & -k_{32} & 0 & 0 \\ -k_{41} & -k_{42} & -k_{43} & 0 \end{pmatrix} = \mathbf{DL}$$

$$= \begin{pmatrix} k_{11} & 0 & 0 & 0 \\ 0 & k_{22} & 0 & 0 \\ 0 & 0 & k_{33} & 0 \\ 0 & 0 & 0 & k_{44} \end{pmatrix} \cdot \begin{pmatrix} 0 & 0 & 0 & 0 \\ -k_{21}/k_{22} & 0 & 0 & 0 \\ -k_{31}/k_{33} & -k_{32}/k_{33} & 0 & 0 \\ -k_{41}/k_{44} & -k_{42}/k_{44} & -k_{43}/k_{44} & 0 \end{pmatrix}$$

$$\mathbf{F} = \begin{pmatrix} 0 & -k_{21} & -k_{13} & -k_{14} \\ 0 & 0 & -k_{23} & -k_{24} \\ 0 & 0 & 0 & -k_{34} \\ 0 & 0 & 0 & 0 \end{pmatrix} = \mathbf{DU}$$

$$= \begin{pmatrix} k_{11} & 0 & 0 & 0 \\ 0 & k_{22} & 0 & 0 \\ 0 & 0 & k_{33} & 0 \\ 0 & 0 & 0 & k_{44} \end{pmatrix} \begin{pmatrix} 0 & -k_{12}/k_{11} & -k_{13}/k_{11} & -k_{14}/k_{11} \\ 0 & 0 & -k_{23}/k_{22} & -k_{24}/k_{22} \\ 0 & 0 & 0 & -k_{34}/k_{33} \\ 0 & 0 & 0 & 0 \end{pmatrix}$$

Note—If $\mathbf{K}$ is symmetric then $\mathbf{L} = \mathbf{D}^{-1}\mathbf{U}^T\mathbf{D}$.

relaxation and a value of $\omega > 1$ corresponds to over-relaxation. Introducing such a factor into eqn (10.15) leads to the following iterative scheme:

$$u_i^{(m+1)} = u_i^{(m)} + \omega \left[-\sum_{j=1}^{i-1} \frac{k_{ij}}{k_{ii}} u_j^{(m+1)} - \sum_{j=i+1}^{n} \frac{k_{ij}}{k_{ii}} u_j^{(m)} + \frac{Q_i}{k_{ii}} - u_i^{(m)} \right]. \qquad (10.16)$$

This modified Gauss–Seidel scheme is often known as the Liebmann method (Reference [2] p. 58), or successive overlaxation method SOR, and of course reduces to a straight Gauss–Seidel method if ω is set to unity. In matrix notation if

$$\mathbf{L} = \mathbf{D}^{-1}\mathbf{E}$$
$$\mathbf{U} = \mathbf{D}^{-1}\mathbf{F}, \qquad (10.17)$$

see Table 10.1, eqn (10.16) can be written as

$$\mathbf{u}^{(m+1)} = (\mathbf{I} - \omega\mathbf{L})^{-1}[(1-\omega)\mathbf{I} + \omega\mathbf{U}]\mathbf{u}^{(m)} + \omega(\mathbf{I} - \omega\mathbf{L})^{-1}\mathbf{D}^{-1}\mathbf{Q}. \qquad (10.18)$$

The selection of suitable values of ω is context dependent and has been extensively covered in the literature (loc cit, pp. 109 and 283), and for a classic study relevant to the Poisson equation the work of Carre is recommended [3]. For more recent work the paper by Browne and Lawrenson [4], which is appropriate to solving time-harmonic problems (see next chapter) associated with electrical machines, is also recommended. It is possible to combine the linear iteration scheme of eqn (10.18) with the non-linear scheme of eqn (10.7). Note that only a single array of unknown vectors is needed since new values of $\mathbf{u}^{(m+1)}$ can be exchanged for old values of $\mathbf{u}^{(m)}$ as they are computed; and also updates of the non-linear constitutive parameter κ can also be done every Gauss–Seidel step if desired using a mix of old and new values, i.e. instead of eqn (10.3) the following can be used which will prevent excessive oscillations:

$$\kappa^{(m+1)} = f[\omega_r\mathbf{u}^{(m)} + (1-\omega_r)\mathbf{u}^{(m-1)}], \qquad (10.19)$$

where ω_r is the under-relaxation factor selected to have a value $0 < \omega_r \leqslant 1$. A flow chart of the combined linear and non-linear algorithm is shown in Fig. 10.3, in which the number of Gauss–Seidel steps per non-linear iteration can be varied as well as the two relaxation factors ω and ω_r depending upon the context.

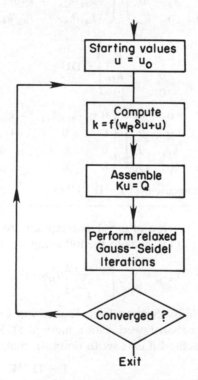

Fig. 10.3 Flow chart of iterative method with relaxed Gauss–Seidel non-linear iterations.

10.2.3 The Newton–Raphson methods

The simple iterative methods described in the last sub-section are not always stable and can take a long time to converge. A more common approach is to use one of the so-called Newton–Raphson methods (Wait [1], p. 90) which will converge rapidly provided sufficient good starting values are used. Thus if $u^{(0)}$ represents an array of initial values to the solution, an improved solution of eqn (10.5) $u^{(1)}$ can be expressed as

$$\mathbf{u}^{(1)} = \mathbf{u}^{(0)} + \delta\mathbf{u}^{(0)}, \tag{10.20}$$

where $\delta\mathbf{u}^{(0)}$ is an array of error values. Also be re-expressing eqn (10.5) as

$$(\mathbf{K}(\mathbf{u})\mathbf{u} - \mathbf{Q}) = \mathbf{F}(\mathbf{u}) = 0, \tag{10.21}$$

where Q is independent of u, the function F defined by eqn (10.21) can be expanded by Taylor's multidimensional series (loc cit, p. 90) as

$$F(\mathbf{u}^{(0)} + \delta\mathbf{u}^{(0)}) = F(\mathbf{u}^{(0)}) + \frac{\partial F_1}{\partial u_1}\delta u_1^{(0)} + \frac{\partial F_2}{\partial u_2}\delta u_2^{(0)} + \cdots \tag{10.22}$$

by neglecting second order terms and higher.

Thus for a 2×2 system, eqn (10.21) is

$$\left.\begin{array}{l} F_1 = K_{11}u_1 + K_{12}u_2 - Q_1 = 0 \\ F_2 = K_{21}u_1 + K_{22}u_2 - Q_2 = 0 \end{array}\right\} \tag{10.23}$$

and by applying eqn (10.22) to each equation in turn and setting $F(u^{(1)}) = 0$ for an improved solution leads to the system:

$$\left.\begin{array}{l} 0 = F_1(u_1^{(0)}, u_2^{(0)}) + \dfrac{\partial F_1}{\partial u_1}\delta u_1^{(0)} + \dfrac{\partial F_1}{\partial u_2}\delta u_2^{(0)} \\[2ex] 0 = F_2(u_1^{(0)}, u_2^{(0)}) + \dfrac{\partial F_2}{\partial u_1}\delta u_1^{(0)} + \dfrac{\partial F_2}{\partial u_2}\delta u_2^{(0)} \end{array}\right\} \tag{10.24}$$

which can be solved for δu_1 and δu_2.

Equation (10.24) can be generalized to an $n \times n$ system readily using matrix notation, i.e.

$$J\delta u^{(0)} = -F(u^{(0)}), \tag{10.25}$$

where $\mathbf{J}$, called the Jacobian matrix, is given by

$$\mathbf{J} = \begin{bmatrix} \dfrac{\partial F_1}{\partial u_1} & \dfrac{\partial F_1}{\partial u_2} & \dfrac{\partial F_1}{\partial u_3} & \cdots \\[2ex] \dfrac{\partial F_2}{\partial u_1} & \dfrac{\partial F_2}{\partial u_2} & \cdots & \cdots \\[2ex] \cdots & \cdots & \cdots & \cdots \end{bmatrix}. \tag{10.26}$$

Therefore by using eqns (10.26) and (10.20) the improved solutions $u^{(1)}$ may be written as

$$\mathbf{u}^{(1)} = \mathbf{u}^{(0)} - \mathbf{J}^{(-1)}\mathbf{F}(\mathbf{u}^{(0)}). \tag{10.27}$$

Further improvement is obtained by iterating eqn (10.27), thus

$$\mathbf{u}^{(m+1)} = \mathbf{u}^{(m)} - \mathbf{J}^{-1}[\mathbf{K}^{(m)}[\mathbf{u}^{(m)}]\mathbf{u}^{(m)} - \mathbf{Q}], \tag{10.28}$$

where m is the iteration number. Equation (10.28) is the simple Newton–Raphson iterative scheme and is illustrated in Fig. 10.4. This scheme may not converge if the initial guess is not sufficiently close to the solution, see Fig. 10.4(b). For this reason it is usual in practice to obtain the initial starting values by performing a few simple iterative steps before changing to the Newton–Raphson method. The coefficients for the Jacobian matrix are obtained from the system matrix as follows: from eqns (10.26) and (10.21)

$$J_{ij} = \frac{\partial F_i}{\partial u_j} = K_{ij}(u_j) + \frac{\partial K_{ij}}{\partial u_j} u_j. \tag{10.29}$$

The first term is simply the system matrix coefficient already dealt with in section 9.1. The second term involves further analysis. Consider the Galerkin weighted residual expression

$$K_{ij} = \int_{\Omega} \nabla \mathbf{N}^T \kappa \nabla \mathbf{N}_j \, d\Omega \tag{10.30}$$

which is valid for the ith node and implies that elements have been merged at this stage without the element type specified. Now from eqn (10.3) κ is a smooth function of u and for the electromagnetic case considered, u is actually a smooth function of $|\nabla u|$—see eqns (10.1) and (10.2). Therefore

$$\frac{\partial K_{ij}}{\partial u_j} u_j = \int_{\Omega} \nabla \mathbf{N}^T \frac{\partial \kappa}{\partial u_j} (\nabla \mathbf{N}_1 u_1 + \nabla \mathbf{N}_2 u_2 + \cdots) \, d\Omega. \tag{10.31}$$

Now if

$$p^2 = \nabla u^T \nabla u = (p_x^2 + p_y^2) \tag{10.32}$$

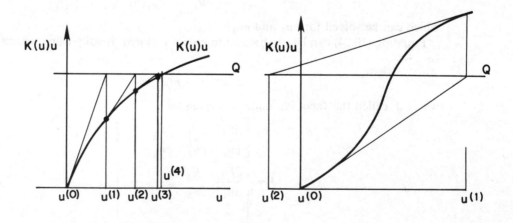

(a) Convergent   (b) Possible Divergence

Fig. 10.4   Newton–Raphson method.

and,

$$p = \nabla u = \nabla N_1 u_1 + \nabla N_2 u_2 + \cdots \tag{10.33}$$

e.g. if u is equated to vector potential then,

$$p = |\mathbf{B}| = \sqrt{\left(\frac{\partial A_z}{\partial x}\right)^2 + \left(\frac{\partial A_z}{\partial y}\right)^2}, \tag{10.34}$$

or if u is a scalar potential

$$p = |\mathbf{H}| = \sqrt{\left(\frac{\partial \phi}{\partial x}\right)^2 + \left(\frac{\partial \phi}{\partial y}\right)^2}. \tag{10.35}$$

Differentiating p^2 with respect to u_j and using eqn (10.32) leads to

$$\mathbf{p} \cdot \frac{\partial p}{\partial u_j} = p_x \frac{\partial p_x}{\partial u_j} + p_y \frac{\partial p_y}{\partial u_j} = \mathbf{p}^T \frac{\partial \mathbf{p}}{\partial u_j} = \mathbf{p}^T \nabla N_j. \tag{10.36}$$

Returning to eqn (10.31), eqns (10.36) and (10.33) yields

$$\frac{\partial \kappa}{\partial u_j} = \frac{\partial \kappa}{\partial \mathbf{p}} \frac{\partial \mathbf{p}}{\partial u_j}, \tag{10.37}$$

and

$$\frac{\partial \kappa}{\partial \mathbf{p}} \frac{\partial \mathbf{p}}{\partial u_j} = \frac{1}{p} \frac{\partial \kappa}{\partial p} \mathbf{p}^T \nabla N_j = 2 \frac{\partial \kappa}{\partial (\mathbf{p})^2} \nabla N^T \mathbf{u} \nabla N \tag{10.38}$$

by eqns (10.36). Hence, finally from eqns (10.31) and (10.29)

$$J_{ij} = K_{ij} + \int_\Omega 2 \frac{\partial \kappa}{\partial (\mathbf{p})^2} \nabla N^T (\nabla N \cdot \mathbf{u})(\nabla N \cdot \mathbf{u}) \nabla N_j \, d\Omega. \tag{10.39}$$

The second term is symmetric and can be formed directly from the system matrix $\mathbf{K}$ itself. For the plane triangles considered in section 9.1.2 this procedure is particularly straight forward since the gradients are constant within an element, so using the expressions for element coefficients, eqn (9.24), eqn (10.39) becomes for a particular element:

$$J_{ij}^e = K_{ij}^e + \frac{A^3}{8} \frac{\partial \kappa}{\partial (\mathbf{p})^2} \sum_k^3 \sum_l^3 (b_i b_k + c_i c_k)(b_j b_l + c_j c_l) u_k u_l \tag{10.40}$$

and after merging elements to reform the Jacobian J, it can be readily seen from eqn (10.39), that if the system matrix $K_{ij} = \kappa K_{ij}^*$ and by writing

$$\sum K_{mk}^* u_k = R_m$$

the merged Jacobian can be written as

$$J = \kappa K_{ij}^* + \frac{2}{A} \frac{\partial \kappa}{\partial (\mathbf{p})^2} R_i R_j. \tag{10.41}$$

The Jacobian in this case is symmetric; this is a consequence of the dependence of κ on $|\nabla u|$, other dependencies may involve asymmetric Jacobians. The necessary sub-systems are now available for a Newton–Raphson algorithm, the procedure of which is illustrated by the flow chart shown in Fig. 10.5. It can be seen that a completely new system of equations has to be solved at each iteration, so frequently an approximation to the Jacobian is made by writing

$$J_{ij}^{(n)} = J_{ij}^{(0)} \tag{10.42}$$

and modifying eqn (10.27) to read

$$u^{(m+1)} = u^{(m)} - [J^{(0)}]^{-1} [K^{(m)}(u^{(m)})u^{(m)} - Q] \tag{10.43}$$

where only one full solution is needed. This approach is known as the modified Newton–Raphson method (Wait, p. 121). This approach, though very economical at each step, will need more iterations to converge to the solution. However in many cases an overall economy is possible; another variation on this theme would be to update

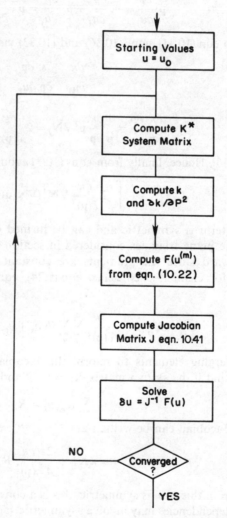

Fig. 10.5 Flow chart for Newton–Raphson method.

the Jacobian only at intervals after several iterations. Further improvements can be made by the introduction of an over-relaxation factor ω as was done in the simple iterative scheme of section 10.2.2; here the correction term is multiplied by a relaxation factor ω, set often near to 2, to improve the convergence rate; it should be stressed however, that such choices usually have to be made on the basis of experiment.

To illustrate the methods described above the simple finite element computer code used in Chapter 9 has been extended to include both simple and Newton iterations and the two methods are compared for the rectangular bar problem in Fig. 10.6. For the sake of simplicity the non-linear characteristics of the material are assumed to follow an analytic form known as the Fröhlich equation. The Fröhlich expression is:

$$B = \frac{H}{a + bH} + cH, \tag{10.44}$$

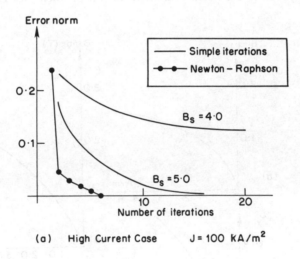

(a)   High Current Case   J = 100 kA/m²

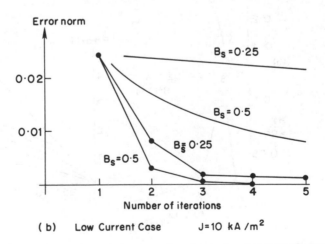

(b)   Low Current Case   J=10 kA /m²

Fig. 10.6 Comparison of convergence using simple iterative and Newton methods.

where the constants in eqn (10.44) can be determined by fitting against measurements giving quite reasonable results in many cases; however for the purposes of comparison of Newton and simple iteratives the exact nature of the B–H relation is unimportant. The constant 'a' in eqn (10.44) is essentially equal to the initial reluctivity; likewise 'b' is the reciprocal of the saturation flux density B_s and so for large values of H eqn (10.44) becomes

$$B = B_s + cH \qquad (10.45)$$

thus the constant 'c' is related to the additional field provided by the source current. For the results shown in Fig. 10.6 'a' is taken to be $1/(1000\,\mu_0)$ and $c = \mu_0$ with various values of B_s. Since the B–H relation is analytic the derivative $\partial\gamma/\partial B^2$ is smooth, see Fig. 10.7(a), (b) and (c). The value of the maximum field is plotted in Fig. 10.7 as a function of B_s and source current density; the saturation effects are clear and should be compared to the linear behaviour for this problem, see Fig. 9.6. In Fig. 10.6 the error

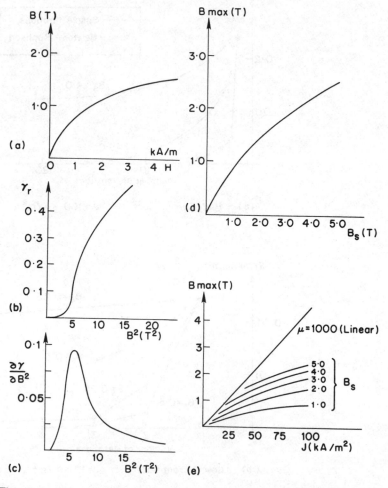

Fig. 10.7 Saturation effects in the bar problem using the Fröhlich relation.

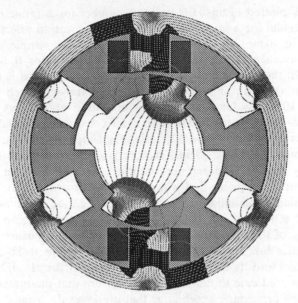

(a) Flux map

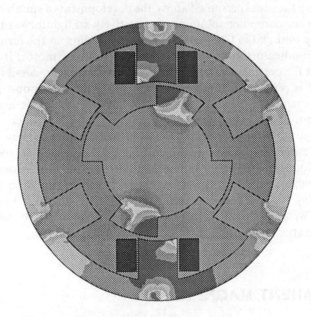

(b) Magnetic saturation levels.

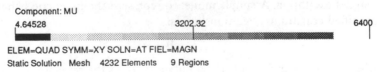

Component: MU

4.64528   3202.32   6400

ELEM=QUAD SYMM=XY SOLN=AT FIEL=MAGN

Static Solution   Mesh   4232 Elements   9 Regions

Fig. 10.8   Flux map of a switched reluctance motor.

norm is plotted against the iterative number for a range of values for B_s; Fig. 10.6(a) gives results for a large current in which saturation affects are very pronounced (see also Fig. 10.6(e). It will be noticed that the simple iterative method becomes non-convergent for values of $B_s < 4$ tesla and even for $B_s = 5$ tesla it takes a very long time. On the other hand the Newton–Raphson method is very stable for this problem and converges for all values of B_s, always following much the same convergence path. Figure 10.6(b) is for a low current condition and shows a similar benefit with the Newton–Raphson method which is again convergent and stable whilst the simple method diverges for values of $B_s < 0.25$ tesla. The simple method will become more competitive if the modified Gauss–Seidel scheme is used, see section 10.2.2, but the problems of stability remain. These results indicate that the Newton–Raphson method is more economic and reliable and whilst the simple method may be necessary in some cases to calculate good initial values it should not be used on its own. In the Newton–Raphson method it is essential that the constitutive curves are smooth to allow an accurate estimate of the derivatives to be made; in order to ensure smoothness it is necessary, in practice for 'real' magnetic materials, to fit the B–H data to smooth curves and evaluate κ and its derivatives as functions of B^2 or H^2. If polynomial or cubic splines fitting is used care should be taken to ensure that the material data is not distorted by the fitting procedure, a technique that preserves the 'shape' of the original curve should be used [5].

Using the ideas presented above the development of suitable algorithms for monotonic non-linear computer solutions is a relatively straightforward extension of the methods already outlined in Chapter 9. In Fig. 10.8 results, in the form of a flux map, are shown for a non-linear solution for a switched reluctance motor. It is not appropriate here to discuss the pertinent design features of this machine in detail [6], but the example shows that it is possible to model practical problems with non-trivial geometries and non-linear materials. The unshaded areas are non-magnetic and in practice the air gap between the rotor and stator would be very much smaller than shown. Figure 10.8(b) shows the distribution of permeability resulting from the non-linear magnetic properties of the materials used. Leaving until later (Chapter 13) detailed considerations of mesh generation and error estimation, it can be said that provided the elements have a reasonable shape, i.e. aspects ratios of the order unity in regions where the field is changing rapidly, good quantitative results are achieved. For the case illustrated 4000 nodes were used and the problem needed approximately ten Newton–Raphson steps to converge.

10.3 PERMANENT MAGNETS

The above procedures can be relatively easily extended to allow treatment of permanent magnet excitation. A simple model to represent the magnetic behaviour is given by the modified constitutive relationship, i.e.

$$\mathbf{B} = \mu^1 \mathbf{H} + \mathbf{M}_0, \tag{10.46}$$

where μ^1 is the apparent permeability and M_0 is the remanence. The coercive force term in eqn (10.46) leads to an additional source term for the Poisson equation in both vector and sclar potential solutions, e.g. the vector potential eqn (8.30) becomes:

$$\nabla \times \frac{1}{\mu} \nabla \times \mathbf{A} + \nabla \times \frac{\mathbf{M}_0}{\mu_1} = \mathbf{J} \tag{10.47}$$

and after application of the Galerkin procedure, see section 9.1, there is an extra term

$$\int_\Omega \mathbf{N}_i \nabla \times \frac{\mathbf{M}_0}{\mu^1} d\Omega$$

and this can be integrated by parts, in an exactly similar way as the main terms, in order to restore natural inter-element continuity conditions. For the two-dimensional

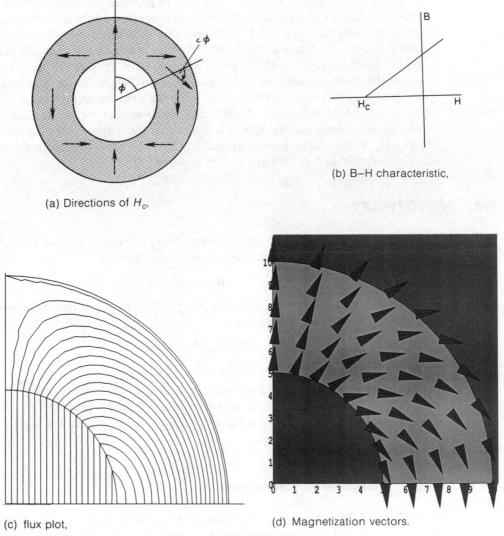

(a) Directions of H_c,

(b) B–H characteristic,

(c) flux plot,

(d) Magnetization vectors.

Fig. 10.9 Permanent magnet cylinder.

case this can be written as:

$$\int_\Omega N_i \left(\frac{\partial M_{0x}}{\partial y} - \frac{\partial M_{0y}}{\partial x} \right) d\Omega = \int_\Omega \left(\frac{\partial N_i}{\partial y} M_{0x} - \frac{\partial N_i}{\partial x} M_{0y} \right) d\Omega - \int_\Gamma N_i M_{0t} d\Gamma. \quad (10.48)$$

The second term on the RHS of eqn (10.48) ensures inter-element continuity naturally by virtue of the continuity of the tangential component of **H**. The additional term appears as a source term to be added to eqn (9.25); thus by using eqn (9.12)

$$f_i = \frac{c_i M_{0x} - b_i M_{0y}}{2\mu^1} \quad (10.49)$$

if M_0 is assumed to be a constant within an element. As an example consider the permanently magnetized cylinder shown in Fig. 10.9; the direction of the magnetization is shown by the arrows in Fig. 10.9(a) and (d). In practice this configuration would be achieved by placing a current dipole inside the cylinder to produce uniform magnetization distribution with direction 2ϕ [7]. A simplified linearized BH characteristic for a magnetic material is shown in Fig. 10.9(b). For this material, the apparent permeability is of the order of 1.001 in the negative quadrant for $H_c < H < 0$ and $0 < B_c < M_0$. However, the finite element method does permit predictions in the non-linear regions if details of the exact curve are known by using the iterative techniques of section 10.2. The only additional data required will be the direction of H_c or M_0. In each permanent magnet finite element, see Fig. 10.9(a), it is seen that the magnets are magnetized so that M_0 and H_c vary in the magnet sinusoidally in 2ϕ. The resultant flux in the bore of the cylinder is shown in Fig. 10.9 and is very uniform [7].

10.4 ANISOTROPY

For anisotropic materials the magnetic permeability μ is not in general a scalar quantity and a general form of the constitutive relation is as follows:

$$\mathbf{B}_i = \mu_{ik}\mathbf{H}_k, \quad (10.50)$$

where μ_{ik} is a tensor of rank 2 (Smythe [8], p. 351); i.e. in the two-dimensional static case eqn (10.50) can be expanded to

$$B_x = \mu_{xx}H_x + \mu_{xy}H_y$$
$$B_y = \mu_{yx}H_x + \mu_{yy}H_y. \quad (10.51)$$

The actual values of the tensor components depend upon the exact nature of the material; thus for the polymer-bonded rare-earth magnets the model proposed by Binns *et al.*, which takes into account fields in the non-preferred direction as well as the preferred one [9], has been very effective.

10.4.1 Permanent magnets with anisotropy

Virtually all of the important developments in permanent magnets which have occurred in recent years [10, 11] have involved anisotropic magnets, whether of the rare earth type (samarium–cobalt, neodymium–iron–boron etc.) or the relatively cheap ferrites (those with stronger fields being anisotropic). Figure 10.10 shows one of the most useful configurations [12] which normally makes use of rare earth or ferrite magnets. A four pole geometry is shown making use of four non-radial magnets. It can be seen that the adjacent magnets are perpendicular and the magnets have a simple rectangular section.

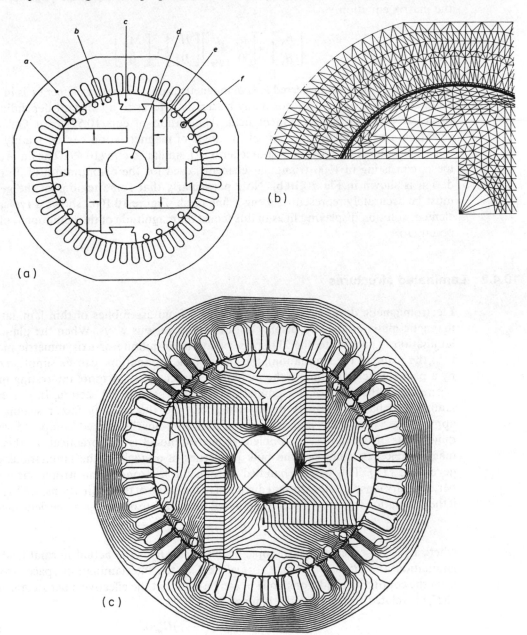

Fig. 10.10 Computer analysis of permanent magnet machine. (a) Geometric model, (b) Finite element model, (c) flux plot. (Reproduced from *Proc. IEE Part B*, **131** (November 1984), p255.)

The finite-element method can readily model both the geometry and the physical properties of the permanent magnets. The 2D computer codes used for this analysis allowed elements containing anisotropic materials to be orientated in any direction relative to the global axes for the main geometry [12] representation for anisotropic permanent magnet materials is proposed and discussed in reference [9] expressed by the matrix equation.

$$
\begin{bmatrix} B_x \\ B_y \end{bmatrix} = \begin{bmatrix} \mu_x & c \\ 0 & \mu_y \end{bmatrix} \begin{bmatrix} H_x \\ H_y \end{bmatrix} + \begin{bmatrix} M_x \\ 0 \end{bmatrix}.
\tag{10.52}
$$

If the x direction is the preferred axis of magnetization, the matrix element c is in general a function of H_x and H_y but for many materials reduces to a constant depending upon the local direction and remanent field. Application of eqn (10.52) to a Galerkin or variation method taking account of the angle of magnetization can be readily carried out and results in an additional source term [9], similar to eqn (10.49). The finite element mesh, consisting of 1500 triangular elements, used for the computation of an optimal design is shown in Fig. 10.10(b). Note particularly that some regions of the geometry must be accurately represented using a fine mesh. Figure 10.10(c) shows a typical finite element solution displaying lines of flux (constant magnitude of the component of vector potential).

10.4.2   Laminated structures

Electromagnetic devices are frequently formed from assemblies of thin laminations of magnetic material and can display anisotropy in various ways. When the plane of the lamination is parallel to the 2D cross-section of an infinite or axisymmetric model (**B** is in the plane of the lamination), the effect of the laminations can be simply modelled by a packing factor scaling of the materials characteristics. A more interesting problem is one where the plane of the laminations is normal to the 2D section. In this case the laminations have a strongly directional effect and the packing factor scaling is not appropriate. Where the laminations are thin and closely packed compared with the dimensions of the device, a complete geometric model is not practical. In this case a macroscopic description of the bulk properties of a stack of the laminations can be developed [13]. This can be handled introducing a strongly anisotropic macroscopic permeability tensor. In the normal direction to the plane of the laminations, see Fig. 10.11, if the lamination thickness is small compared with the rate of change of the field, then

$$
B_n^e = B_n^a = B_n^i,
\tag{10.53}
$$

where B_n^e is the effective macroscopic normal field, B_n^i is the actual normal field in the lamination and B_n^a is the actual normal field in the interlamination space. However, from the continuity conditions on **H**, it can be seen that the effective macroscopic normal $\mathbf{H}(H_n)$ is related to the **H** in the lamination by

$$
H_n^e = (pH_n^i + (1 - p)H_n^i \mu),
\tag{10.54}
$$

where p is the linear packing factor of the magnetic material and μ is the permeability of the material.

The equations relating the fields parallel to the plane of the laminations can be derived in a similar way and are

$$H_t^e = H_t^a = H_t^i \tag{10.55}$$

$$B_t^e = (pB_t^i + (1-p)B_t^i/\mu). \tag{10.56}$$

The only complication introduced in the calculation by these representations is that given the effective $\mathbf{B}^e$ or $\mathbf{H}^e$; in order to find the actual permeability in a non-linear material, eqn (10.56) has to be solved for B_t and similarly eqn (10.54) for H_n, since the value of $\mathbf{B}^i$ must be consistent with the non-linear material relationship eqn (10.2), i.e.

$$\frac{1}{\mu} = f(B_n^i). \tag{10.57}$$

and since the equations are non-linear an iterative method must be employed. In fact a simple Newton–Raphson iteration is all that is required with the correction to $\mathbf{B}^i$ given by $\delta\mathbf{B}^i$ expressed as:

$$\delta B_n^i = -\left(\frac{\partial F}{\partial B_n}\right)^{-1} F(B_n^i) \tag{10.58}$$

hence from eqns (10.56), (10.57),

$$F = [pB_n^i + (1-p)f(B_n^i)] - B_n^e = 0 \tag{10.59}$$

with similar expressions for H_n. The permeability tensor defined by eqn (10.51) has a diagonal form in the local (n, t) coordinate system, i.e.

$$\begin{bmatrix} B_n \\ B_t \end{bmatrix} = \begin{bmatrix} \mu_{nn} & 0 \\ 0 & \mu_{pp} \end{bmatrix} \begin{bmatrix} H_n \\ H_t \end{bmatrix}. \tag{10.60}$$

Thus from the solution of (10.59) the anisotropic permeability tensor is easily evaluated

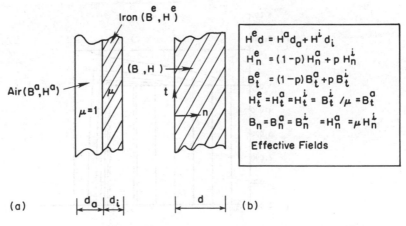

Fig. 10.11 Laminations and packing. (a) Laminated structure, $d_i/d =$ packing factor, (b) Macroscopic model.

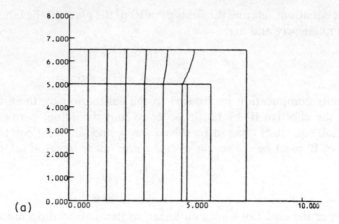

(a)

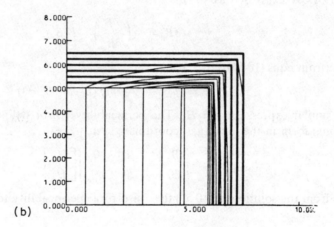

(b)

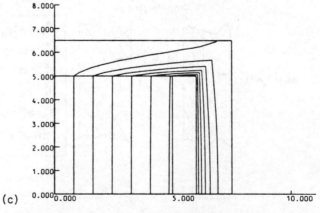

(c)

Fig. 10.12 The use of a macroscopic model for laminated structures. (a) Flux distribution with solid iron, (b) flux distribution with laminated iron, (c) flux distribution using macroscopic model.

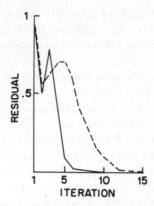

Fig. 10.13 A comparison of convergence rates. Solid line is modified Newton and dashed line is standard Newton.

from the expressions for the macroscopic fields, for example from eqns (10.51, 10.60)

$$\mu_{nn} = \frac{B_n^e}{H_n^e} = \frac{B_n^i}{(pH_n^i + (1-p)H_n^i\mu)}. \tag{10.61}$$

The macroscopic permeability is highly directional, and non-linear, and to solve the global system, eqn (10.28), after forming the Jacobian and residual eqns (10.29), (10.30) using the values of the permeability tensor given by eqn (10.60) etc., requires a specially modified form of the Newton–Raphson method first presented in section 10.2.3. The iterative scheme of eqn (10.43) is modified to include a relaxation parameter ω as follows:

$$\mathbf{A}^{(n+1)} = \omega \mathbf{J}^{-1}[\mathbf{K}^{(n)}\mathbf{A}^{(n)} - \mathbf{Q}]. \tag{10.62}$$

The relaxation factor ω, a positive parameter between zero and unity, is selected so that the residual

$$\mathbf{R} = \mathbf{K}^{(n+1)}\mathbf{A}^{(n+1)} - \mathbf{Q} \tag{10.63}$$

is minimized; i.e. a binary search technique can be used to find the value of ω corresponding to the smallest error norm of R using the latest values of $\mathbf{A}^{(n+1)}$ calculated for a range of ω using eqn (10.62). This process can be repeated at every Newton step or less often as required.

Figure 10.12 shows results for a simple test example in which it can be seen that the macroscopic model using the above relationships is very similar to the full lamination model. Figure 10.13 also shows the rate of convergence for this problem and compares the performance of the modified Newton scheme with the standard method.

10.5 SUMMARY

In this chapter some of the basic techniques for handling non-linearities in the context of two-dimensional fields have been presented. Only simple situations have been dealt with, namely, saturation phenomena, simple permanent magnets and laminated sheets. It has been shown how non-linear effects can be fairly readily incorporated as an extension of the linear solution procedures described in Chapter 9 with very little additional effort. The superiority of the Newton–Raphson iterative technique and its modifications over the simple iteration method has been strongly emphasized, and for the problems considered so far has been shown to lead to a symmetric Jacobian matrix. On the other hand, the essential information on the simple iterative approach has also been given and is easy to apply, and has proved very popular with various workers when used in conjunction with the Gauss–Seidel [14].

The search for satisfactory macroscopic magnetic models continues and is an important research topic [9]. The problem of hysteresis has not been considered here because an overall effective technique has not, so far, been established though recent work on hysteresis modelling by Kadar and Torre [15], Jiles and Thoelke [16], and Mayergoyz [17] is relevant and interesting. The full treatment of hysteresis will make major demands on computing power since in a truly hysteretic situation each point in the material has a minor loop to be taken into account. In this context the innovative work by Simkin [18] in which a phenomenological description of vector hysteresis, proposed by Mayergoyz [17], was applied to problems arising in the non-destructive testing area using the finite element method. This procedure allows the prediction of hysteretic effects in a practical situation.

REFERENCES

[1] R. Wait, *The Numerical solutions of Algebraic Equations*. Chichester: Wiley (1979).

[2] R. S. Varga, *Matrix Iterative Analysis*. New Jersey: Prentice-Hall (1962).

[3] B. A. Carre, The determination of the optimum accelerating factor for successive over-relaxation, *Computer Journal*, **4**(1), (1961).

[4] B. T. Browne and P. J. Lawrenson, Numerical solution of an elliptic boundary (complex variable) value problem, *J. Inst. Maths Applics.*, **17**, 311 (1976).

[5] A. C. Ahlin, On smooth interpolation by continuously connected piecewise polynomials, *Rendiconte del Circolo Matematico de Palermo*, **4**(11) (1971).

[6] P. J. Lawrenson *et al.*, Analysis of switched reluctance stepping motor, Private Communication (1979).

[7] H. Zijlstra, Permanent magnet systems for magnetic resonance imaging, *Proc 9th Magnet Technology Conference, Zurich, SIN*, p. 258 (1985).

[8] W. R. Smythe, *Static and Dynamic Electricity*. New York: McGraw-Hill, 3rd edition, (1968).

[9] K. J. Binns *et al.*, Behaviour of a polymer bonded rare earth magnet, *Proc. IEE Part B*, **130**, (January 1983).

[10] K. J. Binns and M. S. Jabber, High field self starting permanent magnet synchronous motor, *Proc. IEE Part B*, **128**(3), 157(1981).

[11] K. J. Binns, *Permanent Magnet Machines-Handbook of Electrical Machines*, Chapter 9, New York: McGraw-Hill (1987).

[12] K. J. Binns and T. M. Wong, Analysis and performance of a high field permanent magnet synchronous machine, *Proc. IEE Part B*, **131** (November 1984).

[13] C. Biddlecombe and J. Simkin, Enhancements to the pe2d package, *IEEE Trans. on Magnetics*, **MAG-19**, 2635 (1983).

[14] P. J. Lawrenson *et al.*, Analysis of switched reluctance stepping motor, Private Communication (1979).

[15] G. Kadar and E. D. Torre, Hysteresis modelling, in *Intermag Tokyo*, IEEE Trans on Magnetics (1987).

[16] D. Jiles and J. B. Thoelke, Theory of ferromagnetic hysteresis: Determination of model parameters from experimental hysteresis loops, in *Intermag Washington*, *IEEE Trans. on Magnetics* (1989).

[17] I. M. Mayergoyz, Mathematical models of hysteresis, *IEEE Trans. on Magnetics*, **MAG-22**, 603 (1986).

[18] J. Simkin, Modelling ferromagnetic materials for electromagnetic non destructive testing simulation, in *Third National Seminar on Non-Destructive Evaluation of Ferromagnetic Materials* (Houston, Texas), pp. 109–113, Western Atlas International, Inc. (1988).

TWO-DIMENSIONAL TIME-DEPENDENT FIELDS

11.1 INTRODUCTION

The numerical methods described in this chapter are applicable to the important group of time-dependent field problems usually referred to as eddy current effects. The development will follow closely the techniques already presented for solving static fields by the use of the finite element method and are a natural and logical extension of those techniques. By the end of this chapter sufficient information will have been given to enable the reader to understand the elements of a computer based time-dependent field solving system in two dimensions, or indeed, for him to be able to customize his own system if he so desires. The treatment begins by examining the basic equations, given in Chapter 1, that govern the behaviour of electromagnetic fields in eddy current problems. This is followed by a detailed presentation of the modifications needed to the finite element method in order to solve first, fields varying in a steady state mode at a fixed frequency and second, fields with a general variation in time. These two cases are known as time harmonic and transient respectively. In the later part of the chapter some account will be given of the treatment of eddy current problems involving the 'skin-effect' phenomena and the special techniques needed for modelling the 'external' circuit components which, in practice, are used to provide the source fields. Finally, an introduction to the modelling of problems involving moving conductors will be given.

11.2 EQUATIONS FOR LOW FREQUENCY FIELDS

The usual starting point for solving time-dependent low frequency electromagnetic field problems is with the quasi-static subset of Maxwell's equations in their normal differential form, see eqns (1.1) to (1.4). It was stated in Chapter 1 that for low frequency problems (e.g. where ω, the angular frequency, is so small that the wavelength is large compared with the dimensions of the problem that the displacement current ($\partial \mathbf{D}/\partial t$) term in Maxwell's equations can be neglected). At the surface of a conducting region the normal component of current density J_n will be assumed to be zero since any tendency for current to flow is opposed by the electric charges accumulating at the surface. In two dimensions there are two limiting cases to deal with, (a) one component of field with two components of vector potential, and (b) two components of field and one component of vector potential.

Case (a) follows from eqn (1.33) and uses the magnetic field intensity **H** directly, i.e.

$$\nabla^2 H_z = \mu\sigma\frac{\partial H_z}{\partial t} \tag{11.1}$$

and for case (b) the vector potential **A** is particularly useful; since there is only one component A_z, and the divergence condition ($\mathrm{div}\,A = \partial A_z/\partial z = 0$) is automatically satisfied, see section 1.2.1. Furthermore, the scalar potential in eqn (1.16) is at most a constant which can be adjusted to define the connectivity between 'ends' of conductors at infinity, see section 11.6. According eqns (11.6) and (11.7) reduce to a single partial differential equation (PDE) in A_z,

$$\nabla \cdot \frac{1}{\mu}\nabla A_z = \sigma\frac{\partial A_z}{\partial t} + J_s. \tag{11.2}$$

Both cases are PDEs of the diffusion type and can be solved by numerical methods. Problems of type case (a) are often solved by analytical methods or, failing that, by using the one-dimensional limits of the methods for case (b). The latter case is of greater practical interest and will now be considered in detail.

11.3 NUMERICAL SOLUTION OF STEADY STATE AC PROBLEMS

11.3.1 Time harmonic solutions

There are very many situations in electromagnetics requiring field solutions for time-harmonic or steady-state conditions, e.g. alternating current (AC) machines. The basic equation for the steady state AC case for linear problems is obtained from eqn (11.2) by introducing the concept of phasors, i.e. the vector potential **A** becomes a complex variable which may be written as

$$\mathbf{A}^C = \mathbf{A}^R + i\mathbf{A}^I \tag{11.3}$$

with similar expressions for all other field quantities (since the linear case is in question). The physical time varying values of A can then be obtained by using expressions of the type:

$$A = A^C \exp i\omega t, \tag{11.4}$$

and recovering the real part, note that from now on in this chapter A is a single component vector, e.g. A_z. The phasor angular frequency is ω and the superscripts C, R and I refer to the complex, real and imaginary parts respectively. If eqn (11.4) is now substituted into eqn (11.2) the following equation for A^C results:

$$\nabla \cdot \frac{1}{\mu}\nabla A^C + i\omega\sigma A^C = J_s^C. \tag{11.5}$$

The excitation J_S^C will be either purely imaginary or real and sets the reference for the phase angle of the solution, e.g:

$$J_S^C = J \cos \omega t \tag{11.6}$$

and the solution at any point will be given by

$$A = |A^C| \cos(\omega t + \phi), \tag{11.7}$$

where the modulus $|A^C|$ and phase shift ϕ are recoverable from the complex solution A^C for eqn (11.5), i.e.

$$|A^C| = \sqrt{(A^R)^2 + (A^I)^2} \tag{11.8}$$

$$\phi = \tan^{-1}\left[\frac{A^I}{A^R}\right]. \tag{11.9}$$

Accordingly solutions of eqn (11.5) will be sought within a finite domain of space Ω subject to boundary conditions on the surface Γ. The most commonly occurring situation in practice requires either the normal or tangential flux density to be zero respectively. In terms of the vector potential this requires on Γ the homogeneous boundary conditions:

$$\frac{\partial A}{\partial n} = 0 \tag{11.10}$$

or

$$A = 0. \tag{11.11}$$

The numerical procedure for the solution of problems governed by eqn (11.5) using the Galerkin finite element is exactly the same as for the time invariant case covered in sections 9.1 and 9.2. Thus, the weighted residual integral, eqn (8.26), is applied to eqn (11.5), which after integrating by parts to ensure C_0 continuity (see section 9.1.2) leads to:

$$-\int_\Omega \nabla W \frac{1}{\mu} \nabla A^C \, d\Omega + \int_\Gamma W \frac{\partial A^C}{\partial n} \, d\Gamma - \int_\Omega W(i\sigma\omega A^C - J_S^C) \, d\Omega = 0. \tag{11.12}$$

By introducing the basis functions

$$A^C = \sum N_i A_I^C \tag{11.13}$$

and setting $W_i = N_i$ for Galerkin weighting, the discretized form of eqn (11.12) becomes

$$\left[\int_{elem} \frac{1}{\mu}\left(\frac{\partial N_i}{\partial x}\frac{\partial N_j}{\partial x} + \frac{\partial N_i}{\partial y}\frac{\partial N_j}{\partial y}\right) dx \, dy - i \int_{elem} \sigma\omega N_i N_j \, dx \, dy\right] A_i^C = \int_{elem} N_i J_s^C \, dx \, dy. \tag{11.14}$$

This is of the form

$$k_{ij}^C = Q \tag{11.15}$$

which is similar to eqn (9.21) apart from the $N_i N_j$ term in the element matrix, however this new term can also be integrated analytically for triangular elements.

Using the linear triangular elements of section 9.1.2, yields

$$k_{ij}^C = p_{ij} + iq_{ij} = f_i + ig_i \tag{11.16}$$

in which, from eqn (9.24),

$$p_{ij} = \frac{1}{4S\mu}(b_ib_j + c_ic_j) \tag{11.17}$$

and from eqn (9.23),

$$q_{ij} = -\sigma\mu \int_{elem} N_iN_j \, dx \, dy$$

$$= \frac{S\sigma\omega}{12} \begin{bmatrix} 2 & 1 & 1 \\ 1 & 2 & 1 \\ 1 & 1 & 2 \end{bmatrix}, \tag{11.18}$$

where S is the area of the triangular element. Also from eqn (9.25)

$$\left.\begin{array}{l} f_i = \tfrac{1}{3}SJ_R \\ g_i = \tfrac{1}{3}SJ_I \end{array}\right\}, \tag{11.19}$$

where J_R and J_I are the real and imaginary parts of J_S respectively. For a particular element the local matrices are as follows:

$$\begin{bmatrix} (3(b_1^2 + c_1^2) - 2i\alpha S^2) & (3(b_1b_2 + c_1c_2) - i\alpha S^2) & (3(b_1b_3 + c_1c_3) - i\alpha S^2) \\ \text{(symmetric)} & (3(b_2^2 + c_2^2) - 2i\alpha S^2) & (3(b_2b_3 + c_2c_3) - i\alpha S^2) \\ & & (3(b_3^2 + c_3^2) - 2i\alpha S^2) \end{bmatrix} \begin{bmatrix} A_1^c \\ A_2^c \\ A_3^c \end{bmatrix}$$

$$= 4\mu S^2 \begin{bmatrix} J_R + iJ_I \\ J_R + iJ_I \\ J_R + iJ_I \end{bmatrix} \tag{11.20}$$

with

$$\alpha = \omega\sigma\mu. \tag{11.21}$$

The element matrices of eqn (11.20) are symmetric as they are for the case considered in Chapter 9. The assembly of the elemental matrices to the system matrix is also carried out in exactly the same manner as previously. For example, consider the two-element rectangular bar problem of Fig. 9.4(a) in section 9.2 with $b = 2a$. For the steady state time harmonic condition the system matrix is obtained by merging the two-element matrices as follows: for element 1

$$\begin{bmatrix} (15 - 2i\alpha) & -(12 + i\alpha) & (-i\alpha) \\ \text{(symmetric)} & (15 - 2i\alpha) & (3 - i\alpha) \\ & & (3 - 2i\alpha) \end{bmatrix} \begin{bmatrix} A_1^c \\ A_2^c \\ A_4^c \end{bmatrix} = 4\mu \begin{bmatrix} J_1^c \\ J_2^c \\ J_4^c \end{bmatrix} \tag{11.22}$$

and for element 2

$$\begin{bmatrix} (3 - 2i\alpha) & (-i\alpha) & -(3 + i\alpha) \\ \text{(symmetric)} & (12 - 2i\alpha) & (12 + i\alpha) \\ & & (15 - 2i\alpha) \end{bmatrix} \begin{bmatrix} A_1^c \\ A_4^c \\ A_3^c \end{bmatrix} = 4\mu \begin{bmatrix} J_1^c \\ J_4^c \\ J_3^c \end{bmatrix}. \tag{11.23}$$

Hence, by merging eqns (11.22), (11.23) and using eqn (9.31) the system matrix becomes

$$\begin{bmatrix} (18 - 4i\alpha) & -(12 - i\alpha) & -(3 - i\alpha) & (-2i\alpha) \\ -(12 + i\alpha) & (15 - 2i\alpha) & & (3 - i\alpha) \\ -(3 + i\alpha) & & (15 - 2i\alpha) & -(12 + i\alpha) \\ (-2i\alpha) & (3 - i\alpha) & -(12 - i\alpha) & (15 - 4i\alpha) \end{bmatrix} \begin{bmatrix} A_1^c \\ A_2^c \\ A_3^c \\ A_4^c \end{bmatrix} = 4\mu \begin{bmatrix} 2 \\ 1 \\ 1 \\ 2 \end{bmatrix}, \quad (11.24)$$

where $J^R = J$, and $J^I = 0$.

It should be noted that the system matrix in eqn (11.24) is symmetric and that the coefficients are complex, thus a computer program will need to utilize variables declared as complex. In other respects the steady state code will be similar to the codes needed for the statics case. If instead of using complex numbers the equivalent real matrix to eqn (11.24) is used; i.e. from eqns (11.15) and (11.16) where

$$K_{ij}(A_j^R + iA_j^I) = f_i + ig_i$$

and by equating real and imaginary parts, the elements of the real matrix are:

$$(p_{ij}A_j^R - q_{ij}A_j^I) = f_i,$$

and

$$(q_{ij}A_j^R + p_{ij}A_j^I) = q_i,$$

with a matrix double the size of eqn (11.24) having a structure of the following form:

$$\begin{bmatrix} p_{11} & -q_{11} & p_{12} & -q_{12} & \cdots \\ q_{11} & p_{11} & q_{12} & p_{12} & \cdots \\ p_{21} & -q_{21} & p_{22} & -q_{22} & \cdots \\ \cdot & \cdot & \cdot & \cdot & \cdots \\ \cdot & \cdot & \cdot & \cdot & \cdots \end{bmatrix} \begin{bmatrix} A_1^R \\ A_1^I \\ A_2^R \\ \cdot \\ \cdot \end{bmatrix} = \begin{bmatrix} f_1 \\ g_1 \\ f_2 \\ \cdot \\ \cdot \end{bmatrix}. \quad (11.25)$$

The matrix in eqn (11.25) is now not symmetric and it may be seen that there will be a gain in using complex arithmetic but of course there are computer overheads in performing complex arithmetic so the degree of gain will be set by the computer hardware used. The increase in computational time needed to solve a steady state eddy current problem over that for the equivalent magnetostatics case will be due to the increase in time required to perform the complex arithmetic. For example, using the Gaussian elimination method described in Chapter 9, this increase will be at least a factor of four because of the number of implied real operations in each complex multiplication. The very simple matrix structures generated when finite difference methods (see Chapter 8) are used can be solved very efficiently by iterative methods. One of the most effective techniques for this is the procedure of successive overrelaxation (SOR, see Chapter 10) which can be modified to include complex variables [1].

An illustrative example is shown in Fig. 11.1 which is a C-core magnet energized by a 50 cycles AC current which induces eddy currents in an horizontal aluminium plate; the forces generated by the interaction of eddy field and currents causes the plate to be levitated. Figure 11.1(a) shows the finite element mesh used for the computations, and Fig. 11.1(b) shows the magnetic flux contours [2, 3], and the 'in phase' (phase angle = 0) eddy current distribution in the aluminium plate drawn using 'grey scale' shading. In practice the mesh size in the air-gap would need to be very much smaller to obtain high accuracy.

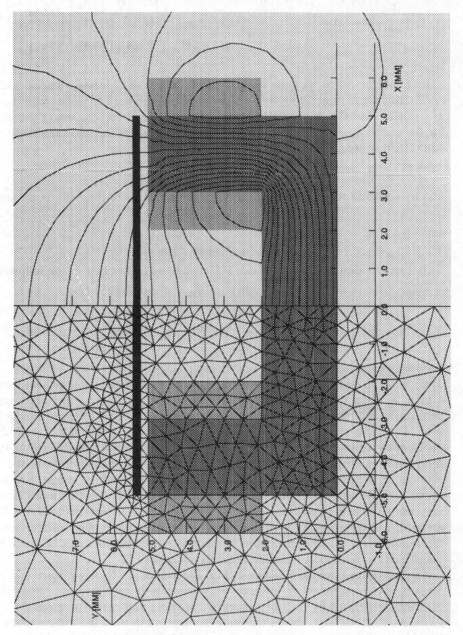

Fig. 11.1   Example of 2D steady state eddy current problem. (a) Mesh, (b) Magnetic flux and eddy current distribution

11.3.2 Skin effect problems and the surface impedance boundary condition

For many practical eddy current problems the magnetic flux penetration into a conductor is confined mainly to surface layers, indeed for a semi-infinite slab of conductor with an externally applied uniform alternating field, B_o, parallel to the slab the amplitude of flux decays exponentially, see Hammond [4], p. 208, given by:

$$B = B_o \exp\left[\frac{-x(1 + \imath)}{\delta} \right] \tag{11.26}$$

with analogous expressions for current density. The skin depth (Hammond [4], p. 206), δ, is given by

$$\delta = \sqrt{\frac{2}{\omega \sigma \mu_o \mu_r}},$$

where at a depth of δ the magnitude of B is $1/\varepsilon$ of the surface value. For problems when the skin depth is small (cf. for iron at 50 Hz, with $\mu_r = 1000$, and $\sigma = 10^7$ s/m, $\delta = 0.5$ mm) all of the activity in the problem is confined to a surface layer and would appear to render the discretization into elements in the bulk of the conducting material unnecessary, and therefore wasteful. This idea has been exploited by the use of a special boundary condition, applicable in these cases, known as the surface impedance condition [5]. This condition can be derived by considering the one-dimensional conducting slab model as above and so will be an approximation only. The condition in terms of the vector potential A for two-dimensional problems is as follows:

$$\frac{\partial A}{\partial n} = -\frac{(1 + \imath)}{\delta \mu_r} A = -\gamma A. \tag{11.27}$$

Equation (11.27) can now be used instead of boundary condition, eqns (11.10), (11.11) to approximate the effect of regions where the skin depth is small compared with the size of the conductor. The implementation of eqn (11.27) by the Galerkin method can be readily carried out by evaluating the second term in eqn (11.12) for all element edges that form the conductor surface. Thus from eqn (11.27) this term becomes

$$\int_\Gamma W_i \frac{\partial A^c}{\partial n} d\Gamma = -\gamma \left(\int_\Gamma N_i N_j d\Gamma \right) A_j. \tag{11.28}$$

The integral in eqn (11.28) can be easily evaluated analytically for linear elements, e.g. if the surface impedance condition is applied to edge Γ_{12} of a triangular element then the element matrix of eqn (11.20) is modified as follows:

$$\begin{bmatrix} k_{11}^c + \gamma l/3 & k_{12}^c + \gamma l/6 & k_{13}^c \\ & k_{22}^c + \gamma l/3 & k_{23}^c \\ \text{(symmetric)} & & k_{33}^c \end{bmatrix} \begin{bmatrix} A_1^c \\ A_2^c \\ A_3^c \end{bmatrix} = \begin{bmatrix} Q_1^c \\ Q_2^c \\ Q_3^c \end{bmatrix}. \tag{11.29}$$

The surface impedance approach has been used extensively for many aspects of power machine design and for more information the reader should consult the literature [6]. The advantage of the method is that computation of fields and eddy currents within conducting media is avoided. It has been shown that the technique yields results of comparable accuracy with other methods when the surfaces have no corners. Recently the method has been extended to include treatment of corners and also to three-dimensional geometries [7] with encouraging results.

11.4 TIME-DEPENDENT FORMULATIONS

11.4.1 Spatial and time discretization

The general time-dependent case requires numerical solutions of eqn (11.2) itself and, for completeness must allow for excitations of arbitrary form. It does, of course, involve considerably more computing than the above harmonic case since the explicit introduction of time adds an extra dimension. The Galerkin procedure can be applied to eqn (11.2) directly, i.e. for a particular element

$$\left[\int_{\text{elem}} \frac{1}{\mu}\left(\frac{\partial N_i}{\partial x}\frac{\partial N_j}{\partial x} + \frac{\partial N_i}{\partial y}\frac{\partial N_j}{\partial y}\right)dx\,dy\right]A_i - \left[\int_{\text{elem}} \sigma N_i N_j\,dx\,dy\right]\frac{\partial A_i}{\partial t}$$

$$- \int_{\text{elem}} N_i J_s\,dx\,dy = 0. \tag{11.30}$$

If the elements are assembled the following set of system equations are formed, i.e.

$$RA + S\frac{\partial A}{\partial t} + \mathbf{B} = 0, \tag{11.31}$$

where $\mathbf{R}$ and $\mathbf{S}$ are matrices. Equation (11.31) is the space discretized form of eqn (11.2), and it remains to deal with the additional matter of the time discretization. Because of the presence in eqn (11.31) of the two matrices $\mathbf{R}$ and $\mathbf{S}$ it is not possible to solve directly for $\partial A/\partial t$ which would then allow a simple time integration scheme to be used. One approach is to separate the conducting and non-conducting parts of the problem, i.e. partition the problem such that A_1 is the solution vector in conducting regions and A_2 the solution vector in the non-conducting regions by reordering the matrices. Thus eqn (11.31) becomes

$$R_{11}A_1 + R_{12}A_2 + S\frac{\partial A_1}{\partial t} + B_1 \tag{11.32}$$

and

$$R_{21}A_1 + R_{22}A_2 + B_2 = 0. \tag{11.33}$$

Equation (11.33) can first be solved for A_2 using starting values for A_1 at $t = t_0$; then eqn (11.32) is solved for $\partial A_1/\partial t$ using the new values of A_1. A time integration can then be carried out to determine the values of A_1 at $t = t_0 + \delta t$, etc., the process being repeated for each time step. However, this procedure is quite expensive in that it involves two matrix solutions and several matrix-vector multiplications to obtain $\partial A/\partial t$ before any estimate of A at a subsequent time can be made.

11.4.2 Time marching schemes

An alternative scheme is to apply the Galerkin procedure to the time domain as well as to the space domain. A first order shape function is chosen for A as a function of

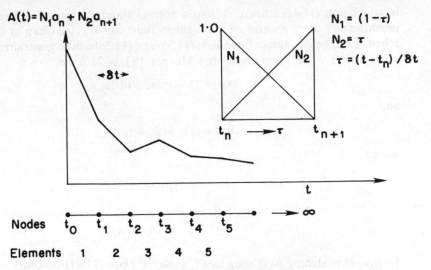

Fig. 11.2 Finite elements in time.

time, see Fig. 11.2, i.e. for

$$t_n \leqslant t \leqslant t_{n+1} \tag{11.34}$$

let

$$A(t) = (1 - \tau)a_n + \tau a_{n+1} \tag{11.35}$$

and

$$B(t) = (1 - \tau)b_n + \tau b_{n+1} \tag{11.36}$$

where

$$\tau = \frac{t - t_n}{(t_{n+1} - t_n)} = \frac{(t - t_n)}{\delta t}. \tag{11.37}$$

Note

$$\left.\begin{aligned} \tau = 0 \quad \text{at} \quad t = t_n \\ \tau = 1 \quad \text{at} \quad t = t_{n+1} \end{aligned}\right\} \tag{11.38}$$

and

thus satisfying the basic definition of shape functions. Furthermore from eqn (11.35)

$$\frac{\partial A}{\partial t} = -\frac{1}{\delta t}(a_n - a_{n+1}). \tag{11.39}$$

If the Galerkin procedure is now applied to eqn (11.31) with τ as the weighting function then

$$\int_0^1 \tau \left[\left[\frac{\mathbf{S}}{\delta t}(\mathbf{a}_n - \mathbf{a}_{n+1}) + \mathbf{R}[\mathbf{a}_{n+1}\tau + \mathbf{a}_n(1 - \tau)] + \mathbf{b}_{n+1}\tau + \mathbf{b}_n(1 - \tau) \right] \delta t \, d\tau = 0. \tag{11.40}$$

By integrating eqn (11.40) the following recurrence relation between a_{n+1} and a_n is obtained:

$$\left[\frac{R}{3} - \frac{S}{\delta t} \right] a_n + \left[\frac{2R}{3} + \frac{S}{\delta t} \right] a_{n+1} + \frac{b_n}{3} + \frac{b_{n+1}}{3} = 0 \tag{11.41}$$

in which eqn (11.41) corresponds to a normal standard central difference formula. By introducing a more general set of weighting functions W_n (see Chapter 8) other difference schemes can be generated. Equations (11.35) and (11.36) can be generalized by introducing a parameter θ (see Zienkiewicz and Morgan [8], p. 286), i.e.

$$\mathbf{A}(t) = (1 - \theta\tau)\mathbf{a}_n + (\theta\tau)\mathbf{a}_{n+1} \tag{11.42}$$

and

$$\mathbf{B}(t) = (1 - \theta\tau)\mathbf{b}_n + (\theta\tau)\mathbf{b}_{n+1} \tag{11.43}$$

where

$$\theta = \frac{\displaystyle\int_0^1 W\tau\,d\tau}{\displaystyle\int_0^1 W\,d\tau}.$$

For point matching weighting (see Chapter 11) eqn (11.41) becomes

$$\left[\mathbf{R}(1-\theta) - \frac{\mathbf{S}}{\delta t}\right]\mathbf{a}_n + \left[\mathbf{R}\theta + \frac{\mathbf{S}}{\delta t}\right]\mathbf{a}_{n+1} + \mathbf{b}_n(1-\theta) + \mathbf{b}_{n+1}\theta = 0. \tag{11.44}$$

The above equation reduces to the central difference form eqn (11.41) when $\theta = 2/3$, and other values of θ correspond to other well known methods for time stepping, including the Crank–Nicholson method, with $\theta = 1/2$, which is found to be more efficient with most problems. In order to obtain the value of the vector $\mathbf{a}_{n+1}$ it is necessary to solve the system defined by the $(\mathbf{R}\theta + \mathbf{S}/\delta t)$ matrix and exactly the same procedures at each time step can be used as already developed for the statics case. In eqn (11.44) the $\mathbf{b}$ vectors are known functions of time and the step length can be selected to maintain the relative error between bounds. Indeed, the larger the step length the larger the error in the approximation for that particular time interval. Thus by decreasing the step length the accuracy should improve and so by a suitable strategy this behaviour can be used to control the error to within an upper and lower bound set by the user. One method of achieving this is to compute the difference in the computed values for both a 'full' step and for two 'half' steps and then proceed further by either doubling or halving the basic step length to constrain this difference between prescribed bounds [3]. For other strategies, and for a fuller discussion on stability, see Zienkiewicz and Morgan [8], p. 289. The computer procedures used previously for the steady state and statics case can be extended by the introduction of a time marching loop—see Fig. 11.3 for a flow chart.

As an example of a general (transient) problem consider solid conducting iron cylinder placed in a uniform magnetic field which is suddenly switched on at $t = 0$. Figure 11.4(a) shows the finite element mesh used in this case together with a magnetic flux plot. Note the use of thin elements near cylinder surface to take account of skin effect. Analytic solutions are available [9], and this example then allows a precise verification of the method and computer program and is, therefore, an example of a 'bench-mark' problem. In Fig. 11.4(b) and (c) graphs of results from a computer code [3] are compared with the analytic field values at various times and positions. The evaluation of the transient field as it progresses towards the time invariant value (steady state) is clearly shown.

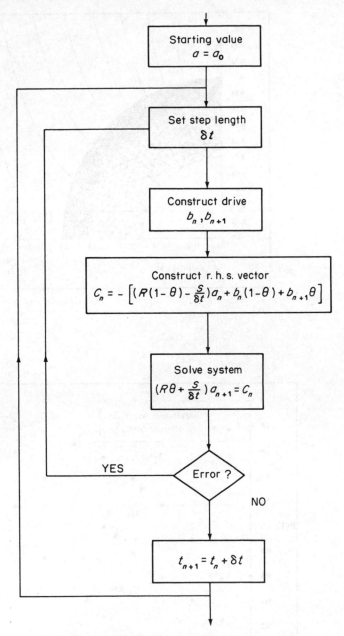

Fig. 11.3 Flow chart for transient field problems.

11.4.3 Non-linearity

The inclusion additionally of non-linear effects such as saturable iron components in eddy current fields poses many difficult problems. For example, the use of a complex phasor description for AC fields is strictly valid only for linear media although adequate solutions have been obtained by simply ignoring this fact by attempting to satisfy the

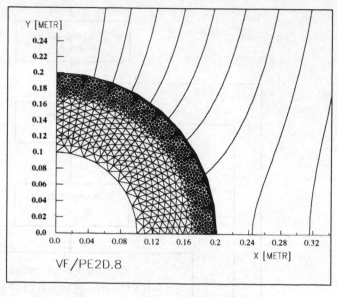

VF/PE2D.8

(a)

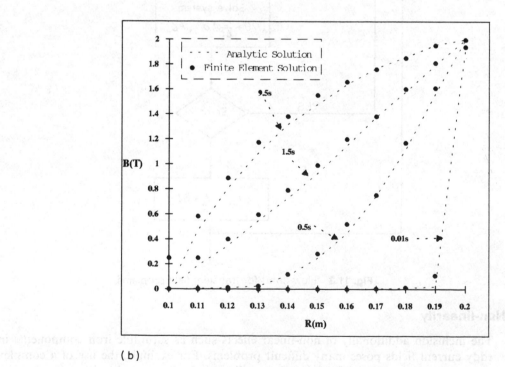

(b)

Fig. 11.4 *continued*

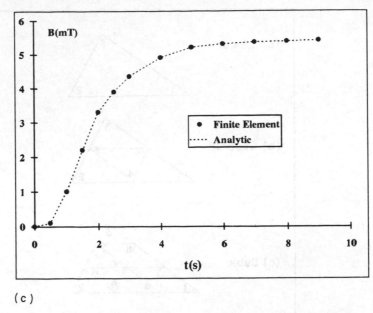

(c)

Fig. 11.4 Transient fields in a ferromagnetic cylinder. (a) Flux lines and mesh, (b) flux density at the centre, (c) flux density versus 'y'.

constitutive relationships at each particular frequency of interest [10]. Thus a Newton–Raphson scheme similar to that presented in Chapter 10 can be applied to the set of eqns (11.15). The general case can be dealt with more rigorously by introducing non-linear iterations at each time step. There are no difficulties in principle, but high computational costs are to be expected.

11.5 SECOND ORDER ELEMENTS

So far only the first order triangular elements defined by eqn (9.1) and Fig. 9.1, section 9.1.1, have been used. These elements are linear in potential and will of course yield a constant field that is discontinuous when traversing from one element to its neighbour. Although these field discontinuities can be smoothed, as described in section 9.4, Figure 9.6, it is often desirable for problems in which the field is changing steeply near conductor surfaces (e.g. skin effect eddy currents) to use a higher order basis function for the potential itself. For example, for planar triangles, instead of using eqn (9.1), the potential may be expressed as

$$u = \alpha_1 + \alpha_2 x + \alpha_3 y + \alpha_4 xy + \alpha_5 x^2 + \alpha_6 y^2. \tag{11.45}$$

There are now six parameters and a quadratic variation in xy. By analogy with the first order triangles, with three nodes for a linear variation, it can be seen that six nodes are now required for completeness—see Fig. 11.5. In order to express eqn (11.45) in terms of potentials at the nodes, i.e. to expresss in terms of basis functions N_i (as in eqn (9.11)), the area coordinates defined by eqn (9.12) can again be used. It is more convenient to

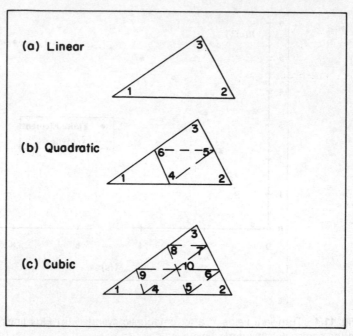

Fig. 11.5 Triangular element family.

use the symbols L_1, L_2 etc. to represent the area coordinates instead of N_i which will be reserved for the basis functions themselves, i.e.

$$u = \sum_{i=1}^{6} N_i u_i \tag{11.46}$$

and

$$L_i = \frac{a_i + b_i x + c_i y}{2S} \tag{11.47}$$

as in eqn (9.12). It should be noted that for the first order case

$$N_i = L_i. \tag{11.48}$$

The explicit forms of the N_i in terms of L_i for the second order case can be obtained directly by setting up the analogous relations to eqn (9.3) but since these will involve equations in six unknowns, the working is tedious and it is better to follow the standard practice developed in finite-element texts, e.g. see Zienkiewicz and Taylor, p. 265 [14], where by using Lagrangian interpolating formulae, it is shown that for quadratic triangles with vertex and mid-side nodes

$$N_i = (2L_i - 1)L_i \tag{11.49}$$

for the vertex nodes, and

$$\left.\begin{array}{l} N_4 = 4L_1 L_2 \\ N_5 = 4L_2 L_3 \\ N_6 = 4L_3 L_1 \end{array}\right\} \tag{11.50}$$

for the mid-side nodes, see Fig. 11.5. These shape functions clearly satisfy the conditions of eqn (9.13), i.e. at node 2, $L_2 = 1$, $L_1 = L_3 = 0$, see Fig. 9.2 and from eqn (11.49) $N_2 = 1$; also at node 5 $L_2 = L_3 = 1/2$ and from eqn (11.50) $N_5 = 1$, and so on. Similar expressions to eqns (11.49) and (11.50) can be derived for cubic and higher basis functions. The Galerkin procedures of Chapter 9 can now be extended to include the second order

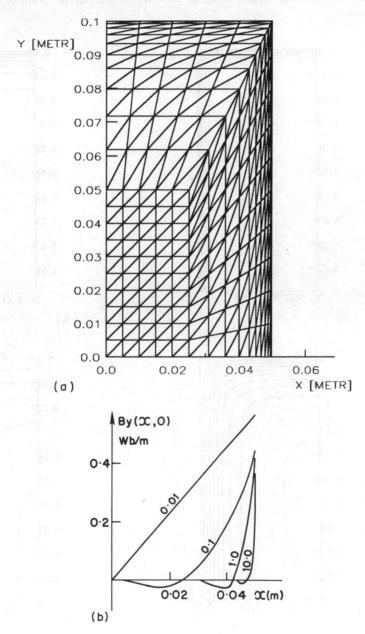

Fig. 11.6 Time harmonic case—rectangular bar problem. (a) Mesh for capturing the skin effect, (b) flux distribution at various frequencies.

basis function of eqns (11.49) and (11.50), the element matrices corresponding to eqns (11.14) and (11.30) for the time-harmonic and transient cases respectively can be obtained by substituting for the shape functions in terms of the area coordinates L_i and then by integrating analytically using the formula of eqn (9.23), since only simple powers are involved. The element matrices are of order 6×6 but this adds nothing new in principle and so the merging and solution procedures are similar to those used for the first order case.

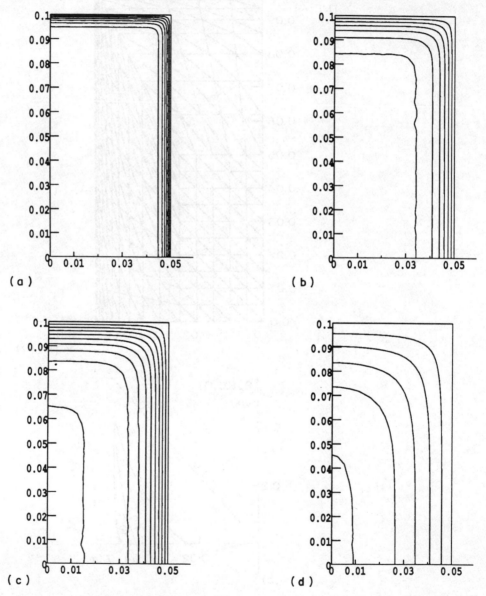

Fig. 11.7 Transient case—rectangular bar problem. (a) $T = 0.1$ s, (b) $T = 0.5$ s, (c) $T = 1.0$ s, (d) $T = 5.0$ s. $\sigma = 10^7$ s/m, $\mu = 1000$, skin depth $= 0.01$ m.

The choice when to use first or second order elements depends upon context. The question, for example, of whether it is more accurate to refine a mesh into more elements or to use the same number with higher order basis functions appears to be problem dependent (Zienkiewicz and Taylor, p. 269 [14]). Practical computation in electromagnetics suggests that for a 'cheap' preliminary examination of the problem, first order elements should be used and then with the same mesh topology a more accurate solution is then obtained using second order elements. The switch from first order to second order should only involve the setting of a single flag in the computer program [3]. The results shown in Figs 11.6 and 11.7 are for the rectangular bar problem with time harmonic and transient excitations respectively, second order elements with mesh refinement near bar edges are used to allow accurate modelling of the pronounced skin effect. The lack of smoothness shown by the innermost contours is because the field here is essentially zero and the 'noise' in the solution, from the discretization error, becomes apparent.

11.6 MODELLING THE EXTERNAL CIRCUIT

So far solutions of the now familiar equation

$$\nabla \times \frac{1}{\mu} \nabla \times \mathbf{A} + \sigma \frac{\partial \mathbf{A}}{\partial t} - J_s \tag{11.51}$$

in two dimensions, where J_s is the source current density, have been considered. This equation on its own is only applicable to a restricted set of cases, namely, those in which the eddy current return path may be considered to be outside the problem domain, implied by symmetry boundary conditions and in which no spatial variation of current density is allowed within the driving conductors, see Fig. 11.8.

Eddy currents in isolated conducting components and those which provide the spatial variation of current density in large source conductors are limited by a potential gradient

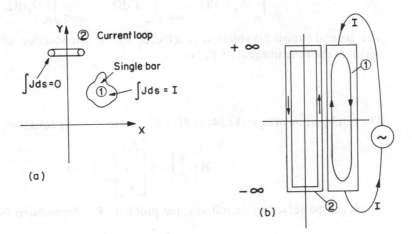

Fig. 11.8 Modelling on external circuit. (a) *XY* plane for 2D systems, (b) long systems, connected at ∞.

along the conductor given by

$$-\int_{\Omega_I} \sigma E \, d\Omega = -\int_{\Omega_I} \left(\frac{\partial A}{\partial t} + \nabla V \right) d\Omega = I, \tag{11.52}$$

where I is the total current flowing, and equals zero for an isolated conductor or equals the amount provided by the power supply in the case of a source conductor, see eqn (1.16). The effect of the potential gradient is a uniform current density over the conducting region, J' given by

$$J' = -\nabla V. \tag{11.53}$$

These effects can be taken into account if eqn (11.51) is replaced by the following two equations which must be solved together [2],

$$\nabla \frac{1}{\mu} \nabla A_z + \sigma \frac{\partial A_z}{\partial t} - J' = 0, \tag{11.54}$$

and

$$\int_{\Omega_J} \left(-\sigma \frac{\partial A_z}{\partial t} - J' \right) d\Omega = \int_{\Omega_J} J_s \, d\Omega. \tag{11.55}$$

In the finite element solution of eqn (11.54) there is now one extra unknown J', for each conducting region. Similarly eqn (11.55) yields one extra equation for each conducting region. Equation (11.54) is solved using the Galerkin weighted residual formulation which leads, in the time harmonic case, to:

$$\left(\frac{1}{\mu} \int_{\Omega_r} \nabla N_i \nabla N_j \, d\Omega + \imath \omega \sigma \int_{\Omega_r} N_i \, d\Omega \right) A_j - \int_{\Omega_r} N_i J' \, d\Omega = 0 \tag{11.56}$$

for the ith equation in the rth conducting region. To maintain a symmetric matrix, eqn (11.55) can be solved by using a unit weight everywhere. For region r this becomes

$$\int_{\Omega_r} N_i A_j \, d\Omega - \frac{1}{\imath \omega \sigma} \int_{\Omega_r} J' \, d\Omega = -\frac{1}{\imath \omega \sigma} \int_{\Omega_r} J_s \, d\Omega. \tag{11.57}$$

The general (transient) solution is achieved by the introduction of another variable F which is the time integral of J', i.e.

$$J' = \frac{\partial F}{\partial t}. \tag{11.58}$$

The combination of eqns (11.54) and (11.55) lead to a matrix equation, see eqn (11.31),

$$\mathbf{R} \begin{bmatrix} A \\ F \end{bmatrix} + \mathbf{S} \begin{bmatrix} \dfrac{\partial A}{\partial t} \\ J' \end{bmatrix} + \mathbf{T} = 0 \tag{11.59}$$

which can be solved as an initial value problem. R is formulated from

$$\mathbf{R} = \frac{1}{\mu} \int_{\Omega} \nabla N_i \nabla N_j \, d\Omega \tag{11.60}$$

with the coefficient of ε being zero the matrix ...

$$ (11.6?) $$

The necessity of applying constraints to the secondary currents must be taken into account in the formulation of the finite-element equations, for where there is a net transport current ...

The effect of these constraints can be seen from the solution for a transformer ... where no current in the secondary winding makes the value of I ... equal ... Figure 11.9(a) shows the output for a solution ... These show the solution without the additional constraints and it can be seen that the flux lines closing at the right-hand side are not balanced currents. Two other features are of the constraints ... Figure 11.9(b) shows the open circuit case, i.e. the zero net current ... windings are useful ... so each part of the winding carrying has a net zero current ... Figure 11.9(c) is the normal situation, balanced non-zero currents in the secondary winding ... and the flux passing through the centre ... Because the secondary winding ... is a continuous loop of steel ... and the ends connected ... current distribution is finite unless this is a two-dimensional model. This example shows that additional conditions can be added to the normal finite-element discretization to allow for important physical constraints.

11.7 MOTIONAL EFFECTS

11.7.1 Modifications to the finite-element scheme

A large number of electromagnetic devices involve moving components having ...

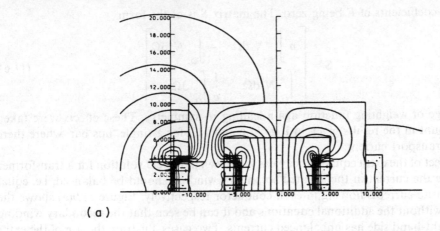

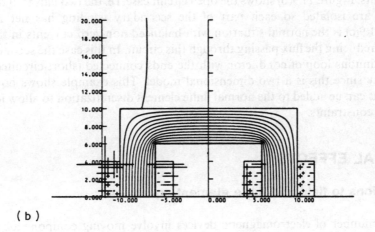

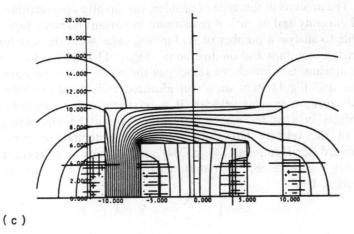

Fig. 11.9 Steady state a.c. solutions for a transformer. (a) No constraints applied to secondary currents, (b) open solution, zero net current in secondary, (c) the normal situation solution.

with the coefficients of F being zero. The matrix S is of the form

$$S = \begin{bmatrix} \sigma \int_{\Omega_r} N_i \, d\Omega & -\int_{\Omega_r} N_i \, d\Omega \\ \int_{\Omega_r} N_i \, d\Omega & \dfrac{1}{\sigma} \int_{\Omega_r} d\Omega \end{bmatrix} \tag{11.61}$$

the change of weighting function again leading to symmetry. These effects were taken into account in the results given in Figs 11.6 and 11.7 for a single 'bus bar' where there is a net transport current.

The effect of the extra equations can easily be seen in the 2D solution for a transformer [2] where the current in the isolated secondary winding should be balanced, i.e. equal and opposite currents flowing in each conductor respectively. Figure 11.9(a) shows the solution without the additional equations and it can be seen that the secondary winding on the right-hand side has unbalanced currents. Two cases illustrate the use of the extra constraints. Figure 11.9(b) shows the open circuit case, i.e. the two halves of the secondary winding are isolated so each part of the secondary winding has net zero current. Figure 11.9(c) is the normal situation with balanced non-zero currents in the secondary winding reducing the flux passing through this circuit. In this case the secondary winding is a continuous loop of conductor with the ends connected (short-circuited) artificially at infinity since this is a two-dimensional model. This example shows how additional equations can be added to the normal finite element discretization to allow for important physical constraints.

11.7 MOTIONAL EFFECTS

11.7.1 Modifications to field and finite element equations

A large number of electromagnetic devices involve moving components, e.g. rotating and linear machines, eddy current brakes and probes for non-destructive testing of materials. The analysis of this type of problem can involve considerable complexity and numerical difficulty and as such it remains an important research topic [11]. However it is possible to analyse a number of 2D limiting cases when the relative velocity of the moving parts is constant and unidirectional. Figure 11.10(a) shows an idealized model of a linear machine is in which the velocity of the 'rotor' is constant and directed in the x direction; and Fig. 11.10(b) shows an idealized cylindrical machine with the rotor rotating at constant angular velocity. It is relatively easy to include these motional velocity effects in a finite-element formulation using the Galerkin procedure. The inclusion of the velocity term into the vector potential equation (eqn (11.51)) is straightforward (see Hammond [4], p. 234)[†]; and it leads in all regions moving with a fixed velocity $\mathbf{v}$ with respect to a stationary inertial frame, to an expression for the current density $\mathbf{J}$:

$$J = \sigma\left(-\frac{\partial \mathbf{A}}{\partial t} + \mathbf{v} \times \nabla \times \mathbf{A} \right) + \mathbf{J}_s, \tag{11.62}$$

[†]The force on moving charge, Q, across a magnetic field is given by $Q\mathbf{v} \times \mathbf{B}$ thus $\mathbf{E} = \mathbf{E}_s + \mathbf{v} \times \mathbf{B}$.

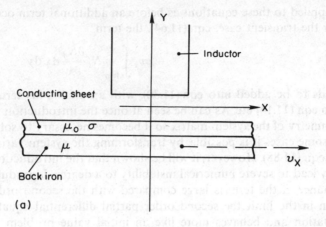

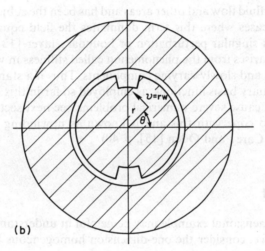

Fig. 11.10 Idealized models involving motion. (a) A single-sided linear induction motor, (b) machine rotating with angular velocity ω.

Accordingly eqn (11.51) for linear materials becomes:

$$\nabla^2 \mathbf{A} + \mu\sigma\mathbf{v} \times \nabla \times \mathbf{A} = \mu\sigma\frac{\partial \mathbf{A}}{\partial t} - \mu\mathbf{J}_s. \tag{11.63}$$

For 2D problems of the type shown in Fig. 11.10(a), eqn (11.63) reduces to:

$$\frac{1}{\mu}\nabla^2 A_z - \sigma v_x \frac{\partial A_z}{\partial x} = \sigma\frac{\partial A_z}{\partial t} - J_{sz}, \tag{11.64}$$

and for the type shown in Fig. 11.10(b) to:

$$\frac{1}{\mu}\nabla^2 A_z + \sigma r\omega\left(\sin\theta\frac{\partial A_z}{\partial x} - \cos\theta\frac{\partial A_z}{\partial y}\right) = \sigma\frac{\partial A_z}{\partial t} - J_{sz}. \tag{11.65}$$

In eqn (11.65) ω is the constant angular velocity at a radius r. If the Galerkin procedure is applied to these equations as before an additional term occurs involving the velocity. For the transient case, eqn (11.64), the term

$$- \sigma v_x \int_{\text{elem}} N_i \frac{\partial N_j}{\partial x} \, dx \, dy$$

needs to be added into eqn (11.30), with an analogous term for the steady-state case into eqn (11.14) etc. As can be seen at once the introduction of these terms destroys the symmetry of the system matrix so it becomes necessary to solve for the complete matrix. In some cases it is possible, by transforming the system variables, to restore symmetry, see eqn (11.88). However, it will be shown that the introduction of the velocity term itself may lead to severe numerical instability to a degree depending upon its magnitude. For instance, if the term is large compared with the second order derivative terms $(\nabla^2 A_z)$ then in the limit the second order partial differential equation becomes a first order equation and behaves more like an initial value problem. This phenomenon is well known in fluid flow and other areas and has been the subject of considerable investigation [12]. In cases where this term dominates the field equations limit to a problem class known as singular perturbation or boundary layer [13]. In these initial value cases a difficulty arises from the phenomenon called stiffness in which solutions have simultaneously fast and slowly varying components. Thus if a standard weighting method is used with ordinary basis functions, as employed so far in this book, the fast transient components will cause severe numerical problems, see next section. A special method has been derived to cope with this and is known as upwinding (Zienkiewicz and Taylor [14], Vol. 2 or Carey and Oden [15], p. 40).

11.7.2 Upwinding

A one-dimensional example may be helpful in understanding both the problem and the solution, i.e. consider the one-dimension homogeneous form of eqn (11.64):

$$\frac{d^2 A}{dx^2} - m \frac{dA}{dx} = 0 \quad \text{in} \quad 0 < x < 1 \tag{11.66}$$

with boundary conditions

$$A(0) = 1, \quad A(1) = 0. \tag{11.67}$$

This example, for illustrative purposes only, would approximate to an infinite slab moving through a stationary y directed field, where m is given by:

$$m = \sigma \mu v_x. \tag{11.68}$$

The analytic solution to this is

$$A(x) = \frac{\exp mx - \exp m}{1 - \exp m}. \tag{11.69}$$

Hence if m is large this solution is essentially unity inside the slab falling through a boundary layer near $x = 1$ to equal the value $A(1) = 0$ on the boundary: Thus if a standard finite element mesh solution is used with say mesh size $h = 1/5$ employing

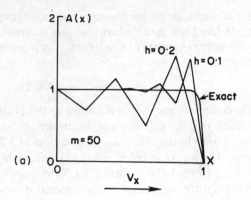

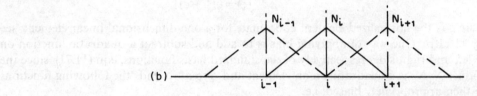

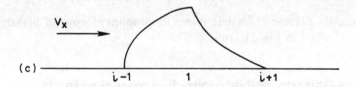

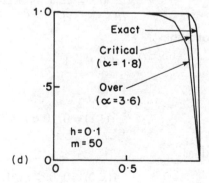

Fig. 11.11 One-dimensional velocity effects. (a) Instability due to velocity term, (b) standard one-dimensional elements, (c) Petrov–Galerkin upwind elements, (d) the effect of upwind elements.

linear basis functions, large oscillations can occur depending on the value of m, see Fig. 11.11(a). It has been established (see also, Carey and Oden [15], p. 40, for more details) that for linear elements of size h the following condition should hold if oscillations are to be avoided:

$$mh = \sigma \mu v_x h < 2, \tag{11.70}$$

where the dimensionless number m is known as the Peclet or magnetic Reynolds number and the product, mh, the cell Reynolds number. Note that the boundary layer in the example corresponds to the skin effect, see section 11.3.2. To avoid the need of specifying large numbers of elements to represent the boundary layer, the standard basis functions can be weighted toward the 'upwind' or 'upstream' direction, i.e. if the direction of motion is to the right (positive x) then the 'upwind' direction is to the left, see Fig. 11.11(c). Suitable shape functions for the one-dimensional case are given by Carey and Oden [15] (p. 285).

Thus let
$$\left. \begin{aligned} N_1(\xi) &= \tfrac{1}{2}(1 - \xi) \\ N_2(\xi) &= \tfrac{1}{2}(1 + \xi) \end{aligned} \right\}, \tag{11.71}$$
and

where ξ is the normalized element coordinate for a one-dimensional linear element, see Fig. 11.11(b). One way of applying bias is to add and subtract a quadratic function on the left and right sides respectively of the standard basis functions, eqn (11.71); since the product $N_1 N_2$ is a quadratic on an element and zero at its ends the following functions are then appropriately biased; i.e.

$$\begin{aligned} N_1(\xi) &= N_1 - \alpha N_1 N_2 = N_1(1 - \alpha N_2) \\ N_2(\xi) &= N_2 + \alpha N_1 N_2 = N_2(1 + \alpha N_1), \end{aligned} \tag{11.72}$$

where α is a factor which determines the amount of 'upwind' bias desired. These functions are sketched in Fig. 11.11(c).

11.7.3 Petrov–Galerkin method—one-dimensional example

As an example consider again the one-dimensional form of eqn (11.63), i.e.

$$\frac{1}{\sigma \mu} \frac{\partial^2 A}{\partial x^2} = \frac{\partial A}{\partial t} + v_x \frac{\partial A}{\partial x} - \frac{J_s}{\sigma} \tag{11.73}$$

with boundary conditions

$$A(0, t) = 1$$

and
$$\left. \begin{aligned} & \\ \frac{\partial A}{\partial x}(1, t) &= 0 \quad 0 < t < T \end{aligned} \right\} \tag{11.74}$$

and initial conditions

$$A(x, 0) = 0; \ 0 \leqslant x \leqslant 1. \tag{11.75}$$

The relative size of v_x and $\sigma \mu$ determine the importance of velocity (advection in fluids) and diffusion terms. Thus the weighted residual expression corresponding to eqn (11.73) is given by

$$\int_0^1 W_i \frac{\partial A}{\partial t} dx + v_x \int_0^1 W_i \frac{\partial A}{\partial x} dx - \frac{1}{\sigma\mu} \int_0^1 W_i \frac{\partial^2 A}{\partial x^2} dx = \frac{1}{\sigma} \int_0^1 W_i J_s dx \qquad (11.76)$$

and integrating by parts to ensure C_0 continuity eqn (11.76) becomes:

$$\int_0^1 W_i \frac{\partial A}{\partial t} dx + v_x \int_0^1 W_i \frac{\partial A}{\partial x} dx + \frac{1}{\sigma\mu} \int_0^1 \frac{\partial W_i}{\partial x} \frac{\partial A}{\partial x} dx = \frac{1}{\sigma} \int_0^1 W_i J_s dx \qquad (11.77)$$

if $W_i(0) = 0$ and eqns (11.74) and (11.75) are satisfied. Introducing the finite element approximation

$$A(x, t) = \sum_j^n A_j(t) u_j(x) \qquad (11.78)$$

and use the standard Galerkin weighting functions defined by eqn (11.71), $u_i = N_i$, eqn (11.77) becomes

$$\sum_{j=1}^n \left[\left(\int_0^1 N_i N_j dx \right) \frac{\partial A_j}{\partial t} + v_x \left(\int_0^1 N_i \frac{\partial N_j}{\partial x} dx \right) A_j + \frac{1}{\sigma\mu} \left(\int_0^1 \frac{\partial N_i}{\partial x} \frac{\partial N_j}{\partial x} dx \right) A_j \right]$$

$$= \frac{1}{\sigma} \int_0^1 N_i J_s dx, \qquad (11.79)$$

or, in matrix form, eqn (11.79) has the canonical form (see eqn (11.31))

$$S \frac{dA}{dt} + v_x R_1 A + \frac{1}{\sigma\mu} R_2 A = F. \qquad (11.80)$$

If instead of the standard Galerkin weighting functions the biased 'upwind' functions of eqns (11.72) are used, the weighting functions in the x coordinate frame become

$$\tilde{N}_i = N_i + \alpha\chi = \begin{cases} \dfrac{x - x_i}{h_i} + \alpha \dfrac{(x - x_{i-1})(x_i - x)}{h_i^2}, & x_{i-1} \leqslant x \leqslant x_i \\ \dfrac{x_{i+1} - x}{h_{i+1}} + \alpha \dfrac{(x - x_i)(x_{i+1} - x)}{h_{i+1}^2}, & x_i \leqslant x \leqslant x_{i+1}. \end{cases} \qquad (11.81)$$

If eqn (11.81) is used, i.e. $u_i = \tilde{N}_i$, eqn (11.77) now becomes

$$\sum_{j=1}^n \left\{ \left[\int_0^1 (N_i N_j + \alpha\chi_i N_j) dx \right] \frac{\partial A_j}{\partial t} + v_x \left[\int_0^1 \left(N_i \frac{\partial N_j}{\partial x} + \alpha\chi_i \frac{\partial N_j}{\partial x} \right) dx \right] A_j \right.$$

$$\left. + \frac{1}{\sigma\mu} \left[\int_0^1 \left(\frac{\partial N_i}{\partial x} \frac{\partial N_j}{\partial x} + \alpha \frac{\partial \chi_i}{\partial x} \frac{\partial N_j}{\partial x} \right) dx \right] A_j \right\} = \frac{1}{\sigma} \int_0^1 N_i J_s dx. \qquad (11.82)$$

Thus eqn (11.80) becomes modified to

$$\left(S \frac{dA}{dt} + v_x R_1 A + \frac{1}{\sigma\mu} R_2 A \right) + \alpha \left(\tilde{S} \frac{dA}{dt} + v_x \tilde{R}_1 A + \frac{1}{\sigma\mu} \tilde{R}_2 A \right) = F. \qquad (11.83)$$

The element contributions can readily be worked out by using appropriate values of N_i for each node according to eqn (11.81). After merging the contributions for each node on a uniform mesh the nodal equations when $J_s = 0$ are:

$$\left(\frac{1}{h^2} + \frac{\alpha m}{6h} + \frac{m}{2h}\right) A_{i-1} - \left(\frac{2}{h^2} + \frac{\alpha m}{3h}\right) A_i + \left(\frac{1}{h^2} + \frac{\alpha m}{6h} + \frac{m}{2h}\right) A_{i+1} = 0. \qquad (11.84)$$

The solution to this equation is [11]

$$A_i = C_1 + C_2 \frac{1 + (1 + (\alpha/3))mh/2}{1 + ((\alpha/3) - 1)mh/2} \qquad (11.85)$$

with C_1 and C_2 constants and is oscillating if

$$\alpha < 3 - \frac{6}{mh} \qquad (11.86)$$

hence the minimal upwinding to suppress oscillations is

$$\alpha = 3 - \frac{6}{mh}. \qquad (11.87)$$

See also eqn (11.70) and Fig. 11.11(d). The analytic solution to eqn (11.66) can also be used to symmetrize the matrices by making in eqn (11.64) the substitution:

$$A = a(x, y) \exp mx. \qquad (11.88)$$

If this is carried out the first derivative term is eliminated and the problem reduces to the standard cases already considered, note that this procedure will only be possible if the velocity is unidirectional in x or y.

11.7.4 Extension to two dimensions

The shape functions used above for one-dimensional line elements can be readily extended to two-dimensional quadrilateral elements, see Zienkiewicz and Taylor Vol. 2, [14], in which two-dimensional weighting functions are derived by simply using the appropriate products of such one-dimensional functions. Triangular elements can pose problems if conforming elements are required, one approach is to merely coalesce two nodes of a quadrilateral element using the limiting expressions of the bi-linear functions [16]. However, this solution may not give any improvement in stability, nevertheless, provided the velocity is parallel to a mesh line as in the example shown in Fig. 11.12(b) simple upwinding in this direction appears to work well. The problem is still subject to considerable research and new ideas are continually under examination, for example the following formulae for applying the up-wind correction using triangular elements has been proposed [17]:

$$\tilde{N}_i = N_i + \tau_i v_i \frac{\mathrm{d}N_i}{\mathrm{d}x} \qquad (11.89)$$

where

$$\tau_i = \frac{1}{2}\left(\frac{S_{ii}}{\max j |R_{1,ij}| + R_{2,ii}}\right). \qquad (11.90)$$

The matrices are those defined in eqn. (11.79) and eqn (11.80).

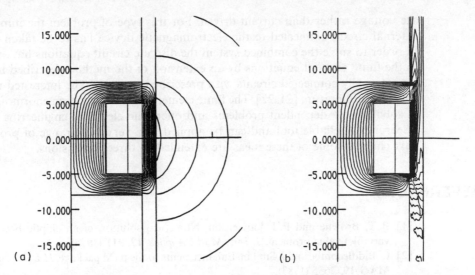

Fig. 11.12 Pipe inspection tool flux plots. (a) Tool stationary, (b) tool moving downwards at 1 m/s.

11.7.5   Examples

Some examples of problems involving motion have been reported in the literature, for example results for a two-dimensional idealization of a linear induction tachometer [18] with a relative velocity of 1.2 m/s which is well below the limit for the need to apply upwinding and good results were obtained. However, in a problem relating to an electromagnetic brake, the investigators [19] encountered oscillations and used upwinding to achieve satisfactory results. In this case the mesh chosen for the moving conductors was regular and parallel to the direction of motion so a simple upwinding technique could be used. The example shown Fig. 11.12 is an idealized model for a inspection tool used for detecting flaws in gas pipes [20]. The result shown in Fig. 11.12(a) is the flux map when the tool is stationary, the field was assumed to be rotationally symmetric about the left-hand axis. The flux 'wake' is clearly shown in Fig. 11.12(b) in which the inspection tool is moving downward at a velocity of 1 m/s, the onset of instability, in this case, becomes critical at 10 m/s and although it is reported that upwinding works well for linear materials it may fail for non-linear problems.

11.8   SUMMARY

This chapter has concentrated on numerical methods for time-dependent problems both for the harmonic and general transient fields. The treatment has been confined entirely to the two-dimensional plane xy symmetry fields using differential operators. The extensions needed for axisymmetry problems are very straightforward in the time domain and only require the methods already developed in Chapter 9 for the space discretization. an important topic, not included in this chapter, concerns electromagnetic devices that are voltage rather than current driven. For this type of problem the impedance of the external circuit connected to the electromagnetic device has to be taken into account. In order to solve the combined system the discrete circuit equations have to be coupled to the finite element equations by an extension of the method described in section 11.6 for externally connected circuits with prescribed currents. The interested reader should consult the literature [21, 22]. The finite element method has been enormously successful in solving time dependent problems and offers the electrical engineering designer and researcher a reliable tool and can be applied to a very wide range of problems. In the next chapter some of these ideas are extended to three dimensions.

REFERENCES

[1] B. T. Browne and P. J. Lawrenson, Numerical solution of an elliptic bounday (complex variable) value problem, *J. Inst. Maths Applics*, **17**, 311 (1976).

[2] C. Biddlecombe and J. Simkin, Enhancements to the pe2d package, *IEEE Trans. on magnetics* **MAG-19**, 2635 (1983).

[3] Vector Fields Ltd, 24 Bankside, Kidlington, Oxford OX5 1JE, *TOSCA, GFUN, CARMEN, ELEKTRA, BIM2D, OPERA, AND PE2D User Manuals* (1990)

[4] P. Hammond, *Applied Electromagnetism.* Oxford: Pergamon Press (1971).

[5] T. Preston and C. Riley, Derivation and implementation of a boundary method for the evaluation of eddy current effects in turbine-generator rotors, *Tech Rep.*, GEC Plc., Stafford, UK, 1981. GEC Report NEP270, Feb.

[6] T. Preston, *Three-dimensional eddy currents in turbine-generators.* PhD thesis, University of London, ICST (1983).

[7] E. M. Deeley and J. Xiang, Improved surface impedance methods for two and three dimensional problems, *IEEE Trans, MAG-24* (1988).

[8] O. C. Zienkiewicz and K. Morgan, *Finite Elements and Approximation Method.* New York: John Wiley (1983).

[9] C. S. Biddlecombe, *Two Numerical Methods for the Solution of 2-D Eddy Current Problems.* PhD thesis, University of Reading (1978).

[10] S. Williamson *et al.*, Solution of two-dimensional non-linear field problems with sinusoidal excitation sources using first order finite elements, *IEEE Trans. on Magnetics*, **19**(6) 2433 (1983).

[11] I. Christie *et al.*, Finite element methods for 2nd order differential equations with significant first derivatives, *IJNME*, **10**, 1389 (1977).

[12] I. Christie *et al.*, Finite element methods for 2nd order differential equations with significant first derivatives, *IJNME*, **11**, 131–44 (1977).

[13] E. P. Doolan *et al.*, *Uniform Numerical Methods for Problems with Initial and Boundary Layers.* Dublin: Boole Press (1960).

[14] O. C. Zienkiewicz and R. Taylor, *The Finite Element Method, 4th Edition, Volumes 1 and 2.* Maidenhead: McGraw-Hill (1991).

[15] G. F. Carey and J. T. Oden, *Finite Elements and Computational Aspects, Vol 3.* New York: Prentice-Hall (1984).

[16] P. S. Huyakorn, Solution of steady state, convective transport equation using an upwind finite element scheme, *Appl. Math. Modelling 1* (1977).

[17] A. Muyukami, An implementation of the streamline-upwind/Petrov-Galerkin method for linear triangular elements, *Comp. Meth. Appl. Mechanics and Eng.*, **49**, 357 (1985).

[18] D. Rodger and J. F. Eastham, A formulation for low frequency eddy current solutions, *IEEE Trans. on Magnetics*, **MAG-19**, 2443 (November 1983).

[19] J. Bigeon *et al.*, Finite element analysis of an electromagnetic brake, *IEEE Trans. on Magnetics*, **MAG-19**(6) (1983).

[20] J. Simkin, Modelling ferromagnetic materials for electromagnetic non destructive testing simulation, in *Third National Seminar on Non-Destructive Evaluation of Ferromagnetic Materials* (Houston, Texax, USA), pp. 109–113, Western Atlas International, Inc. (1988).

[21] S. Williamson *et al.*, Analysis of large induction motors-a combined fields and circuit approach, *IEEE Trans. on Magnetics*, **MAG-21**(6) (1985).

[22] P. Belforte *et al.* A finite element computation procedure for electromagnetic fields under different supply conditions, *IEEE Trans. on Magnetics*, **MAG-21**(6) (1985).

THREE-DIMENSIONAL PROBLEMS

12.1 INTRODUCTION

In extending computational techniques for electromagnetic fields from two-dimensional (2D) to three dimensions (3D) additional complexities arise, not only because of the 3D geometries but also from the physical nature of the field itself which is a vector quantity, see Chapter 1. The construction of suitable meshes is now a formidable task. This is especially true if a differential method of solution is employed since, for this case, a grid of mesh points spanning the entire problem space is now required. However, all considerations of geometric modelling and mesh generation will be left until Chapter 13. The vector nature of the electromagnetic field implies that at each node there will be at least three unknown quantities to compute, which means that there will be a threefold increase in the size of the system matrix over that required for scalar unknowns. This in turn means a considerable escalation in computer costs. To illustrate the increase, suppose that a 2D problem approximated by just 100 nodes gives satisfactory results, is extended to 3D. It now seems reasonable that the 3D problem will need 1000 nodes if all dimensions have the same degree of geometric complexity. If it is assumed that the solution process depends on the square of the number of degrees of freedom, then there will be a 100-fold increase in computing power for a scalar field problem, and for a vector field problem this will further increase to 900 times. In practice the actual increase will be less since solution times using iterative methods are not so pessimistic (see section 13.4.2). Nevertheless, 3D solutions will be very expensive and should only be undertaken after simpler models have been explored, e.g. the 2D cross-section for the end region for an electrical machine, see Fig. 8.2, would be a suitable exploratory limit. In this chapter three-dimensional solutions will be introduced by considering differential formulations for time-invariant problems and it will be shown that in this instance a formulation can be derived in terms of a scalar potential thus avoiding some of the increase in computer costs mentioned above. This is followed by a brief treatment of solving integral equations for three-dimensional static problems using the magnetization vector as the unknown variable. The second part of the chapter deals with eddy current effects. The 3D time harmonic and transient case will be considered using the magnetic vector potential but since this subject is still the subject of the relevant literature should be consulted, see for example references [1–8]. Finally a brief treatment of methods using a field vector (e.g. **H**) will be given together with the concept of 'edged-based' elements.

12.2 SCALAR POTENTIAL FORMULATIONS

The scalar potential formulations have already been introduced in section 8.2.1, where it was shown that the Poisson equation can represent quite generally electrostatic and magnetostatic problems (see also Table 8.1 and eqn (8.1)). These results hold for 3D geometries without modification. The electrostatics case is straightforward, but, because of the non-conservative character of magnetic fields from current sources, the choice of potential type, for magnetic problems, may be important.

12.2.1 The reduced scalar potential approach

In section 1.2.2 it was shown that the magnetic field intensity could be separated into two parts (see eqn (1.22)), i.e:

$$\mathbf{H} = \mathbf{H}_m + \mathbf{H}_s \tag{12.1}$$

$\mathbf{H}_s$ is the source field produced by the current carrying conductors and $\mathbf{H}_m$ the field produced by magnet dipoles, induced or permanent. It was shown (eqn (1.23)) that $\mathbf{H}_m$ could be represented by the gradient of a scalar potential ϕ which satisfies a Poisson equation of the type,

$$\nabla \cdot \mu \nabla \phi = \nabla \cdot \mu \mathbf{H}_s. \tag{12.2}$$

Furthermore, it was also shown that $\mathbf{H}_s$ could be derived from the Biot–Savart law by evaluating:

$$\mathbf{H}_s = \frac{1}{4\pi} \int_\Omega \mathbf{J}_s \times \nabla\left(\frac{1}{R}\right) d\Omega, \tag{12.3}$$

where $R = |\mathbf{r}' - \mathbf{r}|$ is the distance from the source point $\mathbf{r}'$ to the field point $\mathbf{r}$. In many cases, this can be integrated to give an analytic expression for $\mathbf{H}_s$; for complicated current paths, the expression can be integrated by a combination of analytic and numerical quadrature [9, 10, 11, 12]. Frequently in practice the permanent magnet sources can be represented by a modified constitutive equation of the form,

$$\mathbf{B} = \mu(\mathbf{H})(\mathbf{H} - \mathbf{H}_c), \tag{12.4}$$

where μ is a non-linear function of $\mathbf{H}$, which is, in general, a tensor quantity (see Chapter 1, and also Chapter 10). $\mathbf{H}_c$ is the coercive field for the material [13], but in soft magnetic materials is normally assumed to be zero. Whilst direct solutions of eqn (12.2) are possible, in magnetic materials, the two parts of the field $\mathbf{H}_m$ and $\mathbf{H}_s$ tend to be of similar magnitude but in opposite direction, so that cancellation occurs in computing the field intensity H, giving a loss in accuracy [14]. This loss is particularly severe when μ is large. This effect is clearly shown in Fig. 12.9—a more detailed discussion on field cancellation will be given later, see section 12.2.8. It was shown in Chapter 1 that the total field $\mathbf{H}$ can be represented by the gradient of a scalar potential providing there are no current sources—see section 1.2.2, and for these regions this potential, known as the total scalar potential, satisfies:

$$\nabla \cdot \mu \nabla \psi = \nabla \cdot \mu \mathbf{H}_c. \tag{12.5}$$

It is clear then that the total scalar potential can be used to avoid cancellation errors, but unfortunately it cannot represent the whole problem, since in regions where there are currents this potential is multivalued (Stratton [13], p. 141).

12.2.2 The two-potential approach

As already suggested in section 1.2.2 the total scalar potential and reduced scalar potential can be combined to avoid the cancellation associated with reduced potentials and yet allow the inclusion of electric currents. In regions that contain currents, the reduced scalar potential should be used, and elsewhere the total scalar potential. The solutions can then be coupled at the interfaces of the regions. For convenience the main steps are repeated here, thus, consider a two-region problem as shown in Fig. 12.1. In region 1 (Ω_k) there are volumes of magnetic material but no conductors and the field is equal to the gradient of the total scalar potential which satisfies eqn (12.5), on the other hand, in region 2 (Ω_j) there are conductors present, but no permeable materials, and the field can be represented by the reduced scalar potential which satisfies eqn (12.2). At the interface Γ_l between the two regions the normal component of the magnetic induction (**B**) and the tangential components of the field intensity (**H**) must be continuous. Thus, if $\tilde{n}$ is the outward normal from Ω_k, and $\tilde{t}$ a tangent direction to Γ_l, then

$$\mathbf{B}_n = \mu_1 \left[-\frac{\partial \psi}{\partial n} - H_{cn} \right]_1 = \mu_2 \left[H_{sn} - \frac{\partial \phi}{\partial n} \right]_2 \tag{12.6}$$

and

$$\mathbf{H}_t = \left(-\frac{\partial \psi}{\partial t} \right)_1 = \left(-\frac{\partial \phi}{\partial t} + H_{st} \right)_2. \tag{12.7}$$

Equation (12.7) can be integrated over any path on Γ_l to give an integral relationship between the potentials at two points (A and B):

$$\psi_A - \psi_B = \phi_A - \phi_B + \int_A^B \mathbf{H}_s \cdot \tilde{t} \, ds, \tag{12.8}$$

where $\mathbf{H}_s$ is obtained explicitly from eqn (12.3), see Fig. 12.2.

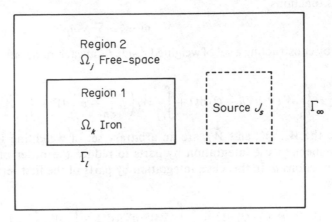

Fig. 12.1 Potential domains and topology.

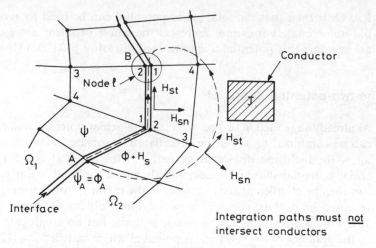

Fig. 12.2 Source scalar potentials and integration paths.

12.2.3 Finite element solutions

In order to introduce the basic steps in the formulation of 3D problems using the finite element method it is convenient to specify a model prblem of the Poisson type as follows:

$$\nabla \cdot \mu \nabla \Phi = Q \; : \; \mathbf{r} \in \Omega, \tag{12.9}$$

where Φ is a scalar potential, either reduced ϕ or total ψ subject to boundary conditions

$$\Phi = \Phi \; : \; \in \Gamma_1 \tag{12.10}$$

$$\mu \frac{\partial \Phi}{\partial n} = \bar{p} \; : \; \in \Gamma_2 \tag{12.11}$$

where $\Gamma = \Gamma_1 + \Gamma_2$ on the surface of a domain Ω. The imposition of eqns (12.10) and (12.11) ensure that solutions to eqn (12.9) are unique—see Appendix 6. As with the two-dimensional problems considered in Chapter 9 the solution is approximated by a set of basis functions

$$\Phi = \tilde{u} = \sum N_i u_i \tag{12.12}$$

and by constructing a set of weighted residuals at each node, see section 9.1.2, the residual is

$$R_i = \int_\Omega W_i(\nabla \cdot \mu \nabla \tilde{u} - Q) \, d\Omega + \int_{\Gamma_2} \bar{W}_i \left(\mu \frac{\partial \tilde{u}}{\partial n} - \bar{p} \right) d\Gamma + \int_{\Gamma_1} \bar{\bar{W}}_i(\tilde{u} - \bar{\Phi}) \, d\Gamma = 0, \tag{12.13}$$

there the W_i, $\bar{W}_i$, and $\bar{\bar{W}}_{ii}$ are an arbitrary set of weighting functions. As before it is convenient to use integration by parts to reduce the order of continuity required for the functions $\tilde{u}$. In this case, integration by parts of the first term in the above equation gives;

$$\int_\Omega W_i(\nabla \cdot \mu \nabla \tilde{u} - Q) \, d\Omega = -\int_\Omega \nabla W_i \cdot \mu \nabla \tilde{u} \, d\Omega + \int_\Gamma W_i \mu \frac{\partial \tilde{u}}{\partial n} \, d\Gamma - \int_\Omega W_i Q \, d\Omega \tag{12.14}$$

and by choosing, $\bar{W}_i = -W_i$ to eliminate the normal gradient term along the boundary Γ_2, eqn (12.13) becomes:

$$R_i = -\int_\Omega \nabla W_i \cdot \mu \nabla \tilde{u} \, d\Omega - \int_\Omega W_i Q \, d\Omega + \int_{\Gamma_2} W_i \bar{p} \, d\Gamma$$

$$+ \int_{\Gamma_1} W_i \mu \frac{\partial \tilde{u}}{\partial n} \, d\Gamma + \int_{\Gamma_1} \bar{W}_i (\tilde{u} - \bar{\Phi}) \, d\Gamma = 0. \tag{12.15}$$

The Galerkin method requires the W_i to be identified with the basis functions, i.e.

$$W_i = N_i. \tag{12.16}$$

As the N_i are functions, local to elements, containing the nodal parameter u_i, eqn (12.15) defines a set of algebraic equations based on the nodes. The boundary condition on Γ_1 is usually enforced and therefore the appropriate integrals are eliminated (see section 9.3). The problem has now been reduced to the solution of a set of linear equations of the form:

$$K_{ij} u_i = C_i, \tag{12.17}$$

and

$$K_{ij} = \int_\Omega \nabla N_i \cdot \mu \nabla N_j \, d\Omega \tag{12.18}$$

the matrix **K** being sparse and symmetric for this particular choice of weighting functions. The right-hand side vector C_i is given by:

$$C_i = \int_\Omega N_i Q \, d\Omega. \tag{12.19}$$

12.2.4 The two-scalar potential formulation

In order to avoid cancellation errors two regions are constructed as in Fig. 12.1. Region $1\,(\Omega_k)$ consists of those parts of the probem which can be represented by the total scalar potential, i.e. volumes of zero current density and region $2\,(\Omega_j)$ consists of the volumes containing currents which can be represented by the reduced potential. Furthermore, without too much loss of generality region 2 can be assumed to have constant permeability (usually free space with $\mu = \mu_0$). This topology conforms to a large range of problems encountered in practice. Each region can be subdivided into finite elements, except that the problem will now not be completely defined, since on the interface boundary Γ_l between the two regions both ψ and ϕ are unknown. This indeterminancy can be resolved by application of the interface conditions, eqn (12.6) and (12.7). For simplicity the contribution from hard magnetic materials (i.e. right-hand side term in eqn (12.5)) will be omitted although it is perfectly straightforward to include these effects [15], see also Chapter 10. Applying the weighted residual method and integration by parts to each region independently,

$$R_1 = -\int_{\Omega_k} \nabla W_i \cdot \mu \nabla \psi \, d\Omega + \int_{\Gamma_l} W_i \mu \frac{\partial \psi}{\partial n} \, d\Gamma \tag{12.20}$$

and

$$R_2 = -\int_{\Omega_j} \nabla W_i \cdot \mu \nabla \phi \, d\Omega + \int_{\Gamma_I} W_i \mu \frac{\partial \phi}{\partial n} \, d\Gamma, \tag{12.21}$$

μ_2 has been removed, since it is constant in the region Ω_j from which it follows that the term containing H_s vanishes, since $\operatorname{div}(\mathbf{H}_s) = 0$. Since the total residual is to be set to zero (i.e. $R_1 + R_2 = 0$), and making use of the interface conditions (eqns (12.6) and (12.8)) the sum of eqns (12.20) and (12.21) become

$$\int_{\Omega_k} \nabla W_i \cdot \mu \nabla \psi \, d\Omega + \int_{\Omega_j} \nabla W_i \cdot \mu \nabla \phi \, d\Omega = -\int_{\Gamma_I} W_i \left[-\mu \frac{\partial \psi}{\partial n} + \frac{\partial \phi}{\partial n} \right] d\Gamma$$

$$= -\int_{\Gamma_I} W_i \mathbf{H}_s \cdot \hat{\mathbf{n}} \, d\Gamma. \tag{12.22}$$

In deriving eqn (12.22) it should be noticed that in eqns (12.20) and (12.21) the normal directions over the interface Γ_b, for the two regions, are of opposite sign, and that it has been assumed that there is continuity of the weight functions between Ω_k and Ω_j. After applying the Galerkin method of eqn (12.22) with a finite-element discretization, the coefficient matrix is identical to eqn (12.18) with appropriate permeability. At the interface Γ_I, either ψ or ϕ can be eliminated by eqn (12.8). Eliminating ϕ results in a right-hand side term for a node on the interface that is given by:

$$\mathbf{C} = \mathbf{Kg} - \mathbf{h} \tag{12.23}$$

where K is the element matrix, eqn (12.18), and

$$\mathbf{g} = \int_0^{t_1} H_{st} \, dt \tag{12.24}$$

$$\mathbf{h} = \int_{t_1}^{t_2} N_l H_{sn} \, dt, \tag{12.25}$$

see Fig. 12.2.

12.2.5   The reduced scalar potential formulation

If cancellation is not expected (see section 12.2.8) then eqn (12.2) can be solved over the entire problem space. Some care is needed at interfaces between regions of differing permeability, i.e. applying the weighted residual method and integration by parts to the two regions independently leads to,

$$R_1 = -\int_{\Omega_k} \nabla W_i \cdot \mu_1 \nabla \phi \, d\Omega + \int_{\Gamma_I} W_i \mu_1 \frac{\partial \phi}{\partial n} \, d\Gamma \tag{12.26}$$

and

$$R_2 = -\int_{\Omega_j} \nabla W_i \cdot \mu_2 \nabla \phi \, d\Omega + \int_{\Gamma_I} W_i \mu_2 \frac{\partial \phi}{\partial n} \, d\Gamma, \tag{12.27}$$

and thence combining eqns (12.26), (12.27) by the continuity conditions (ϕ and $\mu(\partial \phi / \partial n) + H_{sn}$) leads to an additional term in the weighted residual expression. Thus the total residual becomes

$$R = - \int_{\Omega_k} \nabla W_i \cdot \mu_1 \nabla \phi \, d\Omega - \int_{\Omega_j} \nabla W_i \cdot \mu_2 \nabla \phi \, d\Omega + \int_{\Gamma_l} W_i H_{sn} (\mu_1 - \mu_2) \, d\Gamma, \qquad (12.28)$$

It should be noted that the source fields have to be computed at all mesh points in the permeable regions.

12.2.6 3D elements and shape functions

Using this approach a finite-element subdivision is made in each region, and a standard assembly procedure is followed. The contribution to the matrix K from an element is of the form of (cf. eqn (12.18))

$$k_{ij} = \int_{\text{elem}} \mu \left[\frac{\partial N_i}{\partial x} \frac{\partial N_j}{\partial x} + \frac{\partial N_i}{\partial y} \frac{\partial N_j}{\partial y} + \frac{\partial N_i}{\partial z} \frac{\partial N_j}{\partial z} \right] d\Omega. \qquad (12.29)$$

The contribution to the right-hand sides from an element with a node l on the interface Γ_l is

$$C_l = k_{li} g_l, \qquad (12.30)$$

where g_l is given by eqn (12.24). If the element possesses an edge that forms part of Γ_l, its contribution to the right-hand side will also include h, i.e.

$$C_i = k_{li} g_l = h_l, \qquad (12.31)$$

where h_l is given by eqn (12.25), see Fig. 12.2.

The type of element to be used can be selected from several possible choices. The 2D triangular family discussed in Chapter 9 can be generalized to a 3D family of tetrahedra—this is very straightforward and will not be pursued here. An alternative to the tetrahedron is the hexahedron; this element has been widely used by many practitioners of the finite element method with considerable success (Zienkiewicz and Taylor [16], p. 150). A suitable choice of basis function for the 8-node hexahedron element—see Fig. 12.3—is,

$$u = \alpha_1 + \alpha_2 x + \alpha_3 y + \alpha_4 z + \alpha_5 xy + \alpha_6 yz + \alpha_7 zx + \alpha_8 xyz. \qquad (12.32)$$

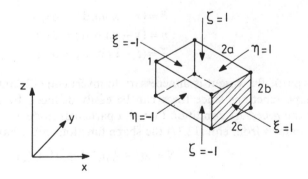

Fig. 12.3 Normalized coordinates for a regular hexahedra (cube)—eight nodes.

Equation (12.32) is the 3D generalized of eqn (9.2) for quadrilateral elements and represents a 'bilinear' variation of the unknown u. The eight terms in eqn (12.32) correspond to the eight corner nodes of a hexahedron, these are sufficient to describe a hexahedra with plane facets. Higher order basis functions are readily available by extending the number of terms to include quadratic terms (20 nodes) and cubic terms (32 nodes) thus allowing curvilinear hexahedral elements—see Fig. 12.4 for examples. Note that the order of variation implied by eqn (12.32) is higher than linear because of the second degree terms, xy etc.; but it is not a *complete* polynomial of degree 2 which has ten terms without the term in xyz which is a cubic term. Equation (12.32) in matrix notation is

$$
\begin{bmatrix} u_1 \\ u_2 \\ \cdot \\ \cdot \end{bmatrix} = \begin{bmatrix} 1 & x_1 & y_1 & z_1 & x_1 y_1 & y_1 z_1 & z_1 x_1 & x_1 y_1 z_1 \\ 1 & x_2 & y_2 & z_2 & x_2 y_2 & y_2 z_2 & z_2 x_2 & x_2 y_2 z_2 \\ \cdot & \cdot & \cdot & \cdot & \cdot & \cdot & \cdot & \cdot \\ \cdot & \cdot & \cdot & \cdot & \cdot & \cdot & \cdot & \cdot \end{bmatrix} \begin{bmatrix} \alpha_1 \\ \alpha_2 \\ \cdot \\ \cdot \end{bmatrix} \tag{12.33}
$$

or concisely as

$$ \mathbf{u}_i = \mathbf{C}\boldsymbol{\alpha} \tag{12.34} $$

where the vectors $\mathbf{u}_i$ and $\boldsymbol{\alpha}$ refer to the element vertices and formally

$$ \boldsymbol{\alpha} = \mathbf{c}^{-1}\mathbf{u}_i \tag{12.35} $$

if the inverse of the matrix $\mathbf{C}$ exists. Furthermore from eqn (12.32) and eqn (12.35), u at any general point on an element is given by

$$ u = [1, x, y, z, xy, yz, zx, xyz]\boldsymbol{\alpha} = \mathbf{P}\boldsymbol{\alpha} = \mathbf{PC}^{-1}\mathbf{u}_i. \tag{12.36} $$

Thus the shape functions N_i, defined in the usual way, i.e.

$$ u = \sum N_i u_i \tag{12.37} $$

implies from eqn (12.36) that

$$ \mathbf{N} = \mathbf{PC}^{-1}. \tag{12.38} $$

It is very convenient to use normalized coordinates in deriving the shape functions. A suitable set of coordinates is shown in Fig. 12.3, in which

$$
\begin{aligned}
\xi &= (x - x_c)/a, \mathrm{d}\xi = \mathrm{d}x/a \\
\eta &= (y - y_c)/b, \mathrm{d}\eta = \mathrm{d}y/b \\
\zeta &= (z - z_c)/c, \ \mathrm{d}\zeta = \mathrm{d}z/c.
\end{aligned} \tag{12.39}
$$

In this particular case it is unnecessary to invert eqn (12.33) to obtain explicit forms for the shape functions since these can be easily deduced by inspection, i.e. in order to satisfy the necessary condition that at a particular node j, $N_j = 1$ and $N_k = 0$, for $k \neq j$ which follows from eqn (12.37) the shape functions must have the form

$$ N = \lambda(1 + \xi_0)(1 + \eta_0)(1 + \zeta_0), \tag{12.40} $$

where

$$ \xi_0 = \xi\xi_i, \eta_0 = \eta\eta_i, \zeta_0 = \zeta\zeta_i. \tag{12.41} $$

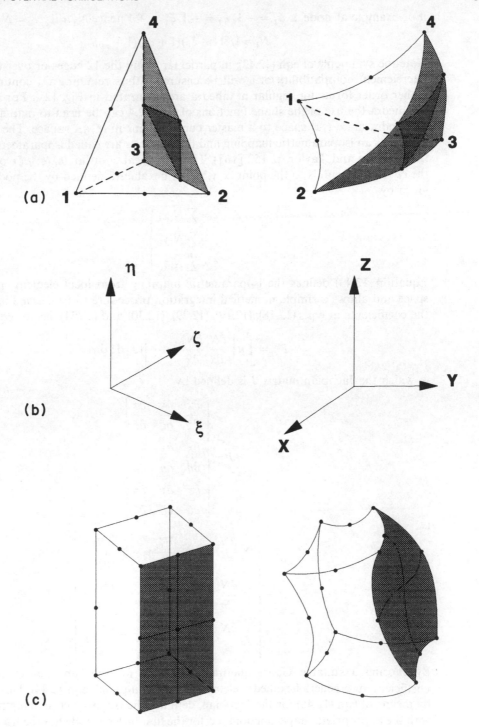

Fig. 12.4 Three-dimensional second order iso-parametric elements. (a) Tetrahedra 10 nodes, (b) local and global coordinates, (c) hexahedra 20 nodes.

For example at node 1: $\xi_1 = -1, \eta_1 = -1, \zeta_1 = +1$ hence $N_1 = 1, N_2 = N_{3...} = 0$, etc.

$$N_i = 1/8(1 + \xi_0)(1 + \eta_0)(1 + \zeta_0). \tag{12.42}$$

Note the symmetry of eqn (12.42); in particular along the 12 edges or over the six faces interelement compatibility of u will be ensured—thus retaining C_0 continuity. Some higher order forms for regular hexahedra are illustrated in Fig. 12.4. For non-regular hexahedra the use of the shape functions of Fig. 12.4 can be used to map a curvilinear hexahedra in (x, y, z) space to a master cube element in (ξ, η, ζ) space. The mapping is known as an isoparametric mapping and the elements are called isoparametric elements (Zienkiewicz and Taylor, p. 150 [16]). Thus the local coordinate, (ξ, η, ζ), of a point in the cube corresponds to the point (x, y, z) in a hexahedra defined by the nodes (x_i, y_i, z_i) given by

$$\left.\begin{array}{l} x = \sum N_i x_i \\[4pt] y = \sum N_i y_i \\[4pt] z = \sum N_i z_i. \end{array}\right\} \tag{12.43}$$

Equation (12.43) defines the isoparametric mapping from local element space to real space and allows a simple numerical integration procedure to be derived in evaluating the coefficients in eqns (12.18), (12.19), (12.29), (12.30) and (12.31), i.e. for eqn (12.18)

$$K_{ij} = \int \mu \left[\frac{\partial N_i}{\partial x} \frac{\partial N_j}{\partial x} + \cdots \right] |J| \, d\xi \, d\eta \, d\zeta, \tag{12.44}$$

in which the Jacobian matrix J is defined by

$$J = \begin{bmatrix} \dfrac{\partial x}{\partial \xi} & \dfrac{\partial y}{\partial \xi} & \dfrac{\partial z}{\partial \xi} \\[10pt] \dfrac{\partial x}{\partial \eta} & \dfrac{\partial y}{\partial \eta} & \dfrac{\partial z}{\partial \eta} \\[10pt] \dfrac{\partial x}{\partial \zeta} & \dfrac{\partial y}{\partial \zeta} & \dfrac{\partial z}{\partial \zeta} \end{bmatrix} \tag{12.45}$$

i.e.

$$J = \begin{bmatrix} \sum \dfrac{\partial N_i}{\partial \xi} x_i & \sum \dfrac{\partial N_i}{\partial \xi} y_i & \sum \dfrac{\partial N_i}{\partial \xi} z_i \\[10pt] \sum \dfrac{\partial N_i}{\partial \eta} x_i & \sum \dfrac{\partial N_i}{\partial \eta} y_i & \sum \dfrac{\partial N_i}{\partial \eta} z_i \\[10pt] \sum \dfrac{\partial N_i}{\partial \zeta} x_i & \sum \dfrac{\partial N_i}{\partial \zeta} y_i & \sum \dfrac{\partial N_i}{\partial \zeta} z_i \end{bmatrix} \tag{12.46}$$

By chosing a suitable Gauss quadrature scheme for the cube master element the coefficient for a genera hexahedra element can be computed (eqn 12.44). This is achieved by means of eqn (12.46) for the Jacobian, defining the isoparametric mapping, together with the appropriate shape functions, i.e. for the first order 8-node hexahedra, eqn (12.42) is used. For the mapping to be possible the Jacobian determinant must satisfy

$$|J| > 0 \tag{12.47}$$

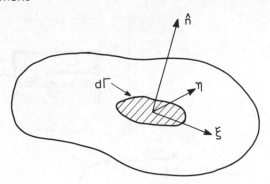

Fig. 12.5 Surface integration.

and eqn (12.47) can be used to check against gross mapping distortions in the computer program. If surface integrations are required, as in eqn (12.31), and if the same isoparametric transformations are used special care must be taken, since the domain of integration will be in only two of the three local coordinates and eqn (12.44) will then be inappropriate. In this case the fundamental magnitudes of first order from differential geometry should be used (Bourne and Kendall [17], p. 145), i.e. if the domain of integration is as shown in Fig. 12.5 for an element surface facet then the element of surface area is given by:

$$d\Gamma = \sqrt{EG - F^2}\, d\xi\, d\eta, \tag{12.48}$$

where

$$E = \left(\frac{\partial x}{\partial \xi}\right)^2 + \left(\frac{\partial y}{\partial \xi}\right)^2 + \left(\frac{\partial z}{\partial \xi}\right)^2$$

$$F = \left(\frac{\partial x}{\partial \xi}\frac{\partial x}{\partial \eta}\right) + \left(\frac{\partial y}{\partial \xi}\frac{\partial y}{\partial \eta}\right) + \left(\frac{\partial z}{\partial \xi}\frac{\partial z}{\partial \eta}\right) \tag{12.49}$$

$$G = \left(\frac{\partial x}{\partial \eta}\right)^2 + \left(\frac{\partial y}{\partial \eta}\right)^2 + \left(\frac{\partial z}{\partial \eta}\right)^2$$

and eqn (12.48) is evaluated for each surface Gauss point by using eqn (12.49) and eqn (12.43).

12.2.7 Examples from magnetostatics

The method described in the previous section forms the basis of the computer code TOSCA [18] which provides solutions for Laplace and Poisson problems and uses the two potential formulation for the magnetostatic case with conductor sources. As well as solving magnetostatic problems with hard or soft magnetic materials, the code can also be used for electrostatic solutions, including the treatment of volume charges. All of the examples presented in this section were solved using the research version of this code [15].

Example 1: Particle beam 'bending' magnet The geometry of a Rutherford Appleton Laboratory type 1 bending magnet is shown in Fig. 12.6(a). These magnets are fabricated

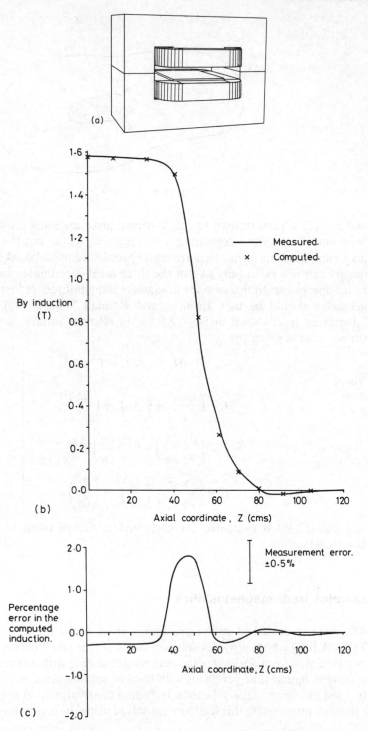

Fig. 12.6 Particle physics Type 1 bending magnet (5200 8-node hexahedra elements). (a) Geometry of the model, (b) flux density as a function of axial position, (c) error in computed solution.

from two different types of steel and the fields have been measured for a range of current excitation [19]. Figure 12.6(b) shows the variation in field along the axis of the magnet, measured and computed, and Fig. 12.6(c) the error in the computed solution as a percentage of the central field. Note that the departure from linearity of its central field is 17% for a current of 450 A. The case illustrated has a maximum error where the field

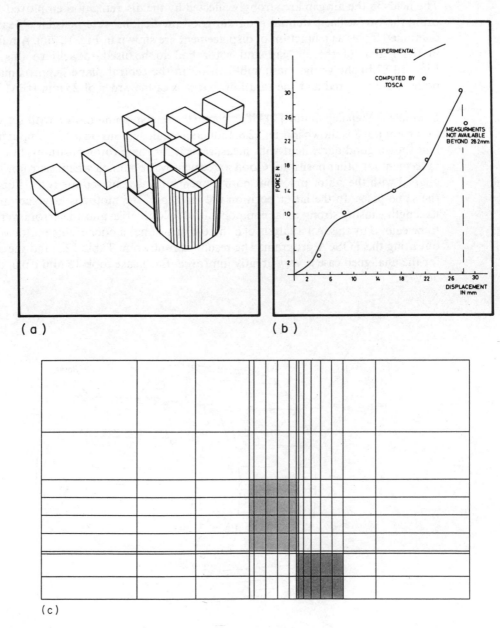

(a)   (b)

(c)

Fig. 12.7 Vernier linear stepping motor. (a) Geometric model, (b) computed forces and measurement, (c) mesh subdivision in central plane.

is changing rapidly and further mesh here should increase the accuracy providing that the material properties and dimensions are sufficiently defined—also note that the measurement error is of the order 0.5%.

Example 2: Vernier linear motor The motor has a small air gap and its fields are strongly influenced by saturation of the iron yoke [20]. The geometry is shown in Fig. 12.7(a). The fields in the air gap are strongly affected by the discretization employed. However, using the two scalar potential program good predictions were possible. Measured and computed forces as a function of displacement are shown in Fig. 12.7(c). A refined mesh in the region of the air gap and rotor had to be used to achieve this accuracy. Figure 12.7(b) shows the mesh sub-division in the central plane. Approximately 8000 nodes were required and the solution time was of the order of 25 min (IBM 360/195).

Example 3: Stepping motor [21] The geometry of a stepping motor with six stator and four rotor poles is shown in Fig. 12.8. Inductance calculations on a 2D computer program had shown good agreement with measurement for some rotor positions, but very poor agreement for other positions. Good agreement was obtained when the rotor teeth were aligned with the stator poles but poor agreement was achieved for rotor slots opposite the same poles. In the latter position the air gap of the motor is large compared with its length causing strong axial components of field to arise and the errors were thought to be caused by this. An analysis of a three-dimensional model of this motor was carried out using the TOSCA program. The results are shown in Table 12.1 and the agreement for the unaligned case is very greatly improved. Each case took 15 min c.p.u. time (IBM 360/75).

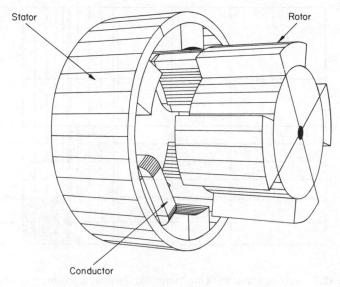

Fig. 12.8 Geometry of a stepping motor.

Table 12.1 Computed and measured inductances for the stepping motor.

Rotor poles aligned with stator poles

Current (Amps)	Measured inductance (mH)	Computed 2D model	inductances (mH) 3D (TOSCA)
4	111	111	116
8	66	66.5	68

Rotor slots aligned with stator poles

Current (Amps)	Measured inductance (mH)	Computed 2D model	inductances (mH) 3D (TOSCA)
4	19.8	13.2	20.6
8	19.8	13.2	20.5

12.2.8 Discussion on field cancellation

In this chapter the use of combined potentials, reduced and total, have been recommended in order to minimize cancellation errors. The effectiveness of single reduced potential methods depend critically on the sub-space of functions used to interpolate the coil fields H_s, and in particular the relation between these functions and the finite element potential solution space. Implementations of the reduced potential method, eqn (12.2), usually use exact evaluation of H_s at all points by analytic integration of the Biot–Savart expression for the field from a defined current distribution, eqn (12.3). To some extent the whole problem of cancellation can be circumvented [22]. To achieve this the space of functions used to interpolate H_s should be the same as that describing the gradients of the reduced potentials. If the two parts of the total field are not supported on the same space of functions then the cancellation between the two parts gives rise to large oscillations in the total field over each element. In order to demonstrate these effects Fig. 12.9(b) and (c) show the fields in a simple model problem (see Fig. 12.9(a)), Fig. 12.9(b) shows the expected field variation compared with the reduced potential results (which are plotted as targets) using the exact H_s from Biot–Savart and in addition the detail of the field variation is shown across two elements by lines joining the targets. This shows the gross fluctuations across elements when the two parts of the field are represented on very different spaces of functions. In Fig. 12.9(c) the H_s field is approximated by centroid values, the wild variations have been removed, but the solution accuracy has been spoiled by the poor interpolation of H_s. Slightly better results are obtained using the bi-linear interpolation of H_s. The results are compared against an 'exact' solution to the problem obtained by using the combined potential method [15].

In all cases it is possible to reduce the cancellation errors to almost insignificant levels by representing H_s on the space of functions that support the gradients of the reduced potentials. In this case the cancellation only causes a problem when the differences become sensitive to computer rounding errors and the calculated fields are the same as

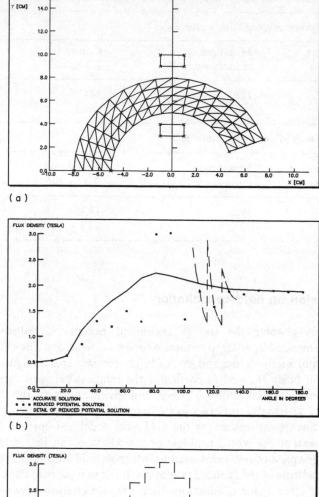

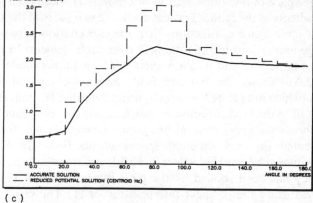

Fig. 12.9 Field cancellation errors.   (a) C-Magnet model problem used to demonstrate cancellation effects—only iron elements and conductors are displayed. The external region is similarly discretized out to 80 cm. (b) Comparison of methods using the reduced with total/reduced potentials. (c) Coil fields approximated by value at element centroid.

those obtained in the combined total and reduced potential method, except for the variations expected for the reason described below, i.e. accuracy of interpolation of $\mathbf{H}_s$. Although using some ingenuity it is possible to solve the cancellation problems arising with reduced potentials it is still not a practical method for 3D magnetostatic fields. The integration for $\mathbf{H}_s$ is expensive for complex coils and when using a reduced potential method these integrations must be performed over all non-linear domains, whereas with the combined reduced and total potential methods these values need only be evaluated in the reduced potential regions. The number of $\mathbf{H}_s$ evaluations can easily be decreased by an order of magnitude using a reasonable combination of reduced and total potential spaces. Furthermore, using numerical basis function to support $\mathbf{H}_s$ implies that the elements must be capable of interpolating $\mathbf{H}_s$ with good accuracy, otherwise as shown in Fig. 12.9(c) the solution suffers. The result of this is that the finite elements must, to some extent, be shaped to fit the coils. This can be almost intractible for a complex coil system, and is completely unnecessary with the combined total and reduced potential method. For most practical applications it can be said that, from a numerical perspective, it is better to compute the small difference quantity directly, i.e. in the case $-\nabla\phi + \mathbf{H}_s = -\nabla\psi$, instead of subtracting two large numbers which at best is inelegant and expensive but also susceptible to error. Nevertheless, on occasion, difficulties may be experienced when the current sources are not balanced (i.e. total net current is not zero). In this case the line integral of field along a closed path inside a connected 'iron' circuit is not zero and the above reasoning advocating the calculation of the difference quantity directly may not apply and it should be possible to obtain good results using the reduced scalar potential.

12.3 INTEGRAL METHODS FOR THREE-DIMENSIONAL PROBLEMS

The generalization of the boundary element method to three-dimensional geometries is relatively straightforward and is amply covered in the literature [23, 24]. However, the alternative formulations given in section 9.8 have the advantage that the normal derivative term present in BEM is not required. Furthermore, if magnetization is used as the unknown variable a very simple formulation specifically appropriate to electromagnetics can be derived and will be considered here.

12.3.1 Magnetization and polarization formulation

Integral equations involving the magnetization can be readily derived, thus by taking the gradient of eqn (9.94) the field at a point due to a region of iron supporting a magnetization $\mathbf{M}$ is given by

$$\mathbf{H}_m = \frac{1}{4\pi}\nabla\int_\Omega \mathbf{M}\cdot\nabla\left(\frac{1}{R}\right)\mathrm{d}\Omega, \tag{12.50}$$

and by eqns (9.92), (9.95) the following integral equation for the magnetization $\mathbf{M}$ results:

$$\mathbf{M}(\mathbf{r}) = \chi(\mathbf{r})\left[\mathbf{H}_s - \frac{1}{4\pi}\nabla\int_\Omega \mathbf{M}(\mathbf{r}')\cdot\nabla\left(\frac{1}{R}\right)\mathrm{d}\Omega\right]. \tag{12.51}$$

This equation may be easily solved by applying a 'moment method' with 'point matching', see section 8.4.1 and eqn (8.64). In its simplest form point matching imposes the constraint that the iron regions are discretized into n elements each having constant magnetization, i.e.

$$\mathbf{M}(\mathbf{r}') = \sum_{i=1}^{n} \mathbf{M}_i \delta_{ij} \tag{12.52}$$

where

$$\delta_{ij} = \begin{cases} 0 & j=i \\ 1 & j \neq i \end{cases}. \tag{12.53}$$

Substitution of eqn (12.52) into eqn (12.51) for each element in turn yields the discretized form of the integral equation:

$$\mathbf{M}_i = \chi_i \left(\mathbf{H}_{si} - \sum_{j=1}^{n} \mathbf{M}_j \mathbf{G}_{ij} \right), \tag{12.54}$$

or

$$\sum_{j=1}^{n} \left(\frac{\delta_{ij}}{\chi_j} + \mathbf{G}_{ij} \right) \mathbf{M}_j = \mathbf{H}_{si}. \tag{12.55}$$

Thus

$$\mathbf{AM} = \mathbf{H}_s \tag{12.56}$$

and, as usual, the continuous equation has been reduced to a set of linear algebraic equations characterized by the matrix $\mathbf{A}$ (see Fig. 12.10).

This technique is an example of a point-matching method in which the integral equation is satisfied at a set of discrete points. In this case it is assumed that the magnetization vector is a constant over each element volume. The matrix coefficient depends upon the choice of element shape and is obtained by integration, thus in eqn (12.55) the matrix coefficient is given by:

$$\mathbf{G}_{ij} = \frac{1}{4\pi} \nabla \int_{elem} \mathbf{e} \cdot \nabla \left(\frac{1}{R} \right) d\Omega \tag{12.57}$$

where $\mathbf{e}$ is a unit vector parallel to $\mathbf{M}$. $\mathbf{G}$ is an example of a tensor of order 2 also termed a dyadic (Bourne and Kendall [17], p. 205). Electric polarization problems can be treated in the same way by replacing $\mathbf{M}$ by $\mathbf{P}$ the polarization vector. For prisms, tetrahedra and polyhedra the integrations can be carried out in closed form [25]. The field from the current sources is obtained by integrating the Biot–Savart Law relation over the volume of the conductors, see eqn (1.24). In fact these quantities are readily available in computer codes for an extensive range of basic conductor elements including straight blocks and curved sectors and many higher order circuits using these elements as building blocks [18]. For most 3D problems the matrix in eqn (12.55) is too large to be held in main memory and so is stored as a direct access disk dataset. Since the matrix has no convenient properties such as symmetry or sparseness, the solution method used is Gaussian elimination with back substitution—see section 9.4. The matrix is

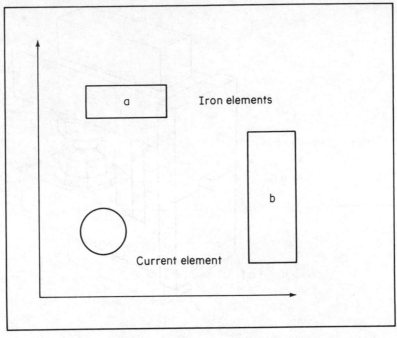

$$\left(\begin{array}{cc} \left(C_{aa}-\dfrac{1}{\chi_a}\right)M_a & +\,C_{ab}M_b \quad =-H_{sa} \\[3ex] C_{ba}M_a & \left(C_{bb}-\dfrac{1}{\chi_b}\right)M_b=-H_{sb} \end{array} \right).$$

Fig. 12.10 Integral equation algorithm with only two elements.

partitioned so that only a fraction of the whole matrix need be in main memory at any time. For non-linear problems, i.e. cases where χ is a function of the magnetizing field, a suitable procedure is to treat the system of equations as quasi-linear and use a simple iterative scheme, i.e.

$$\mathbf{M}^{(k+1)} = \mathbf{A}^{-1}(\mathbf{M}^{(k)} \cdot \mathbf{H}_s). \tag{12.58}$$

However a Newton–Raphson scheme could be used with advantage. Two-dimensional limiting cases for both xy and rz problems are straightforward [25, 26].

Examples [27],

- (a) Polarized target magnet. A computer generated picture of a magnet designed for the detection of polarized particles is shown in Fig. 12.11(a). The truncated conical pole tip is made from a special cobalt-steel, the cylindrical pole and return yokes are fabricated from various other magnet steels. The geometric description of the model used to represent this magnet is shown in Fig. 12.11(a), where for example it can be seen that the cylindrical parts are approximated by an octagonal prism. The rectangular channel through the back yoke was introduced to provide a low field

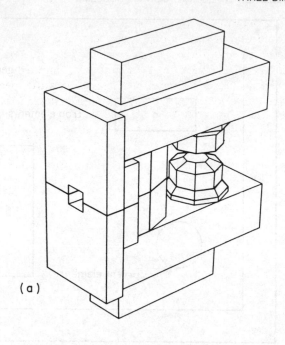

(a)

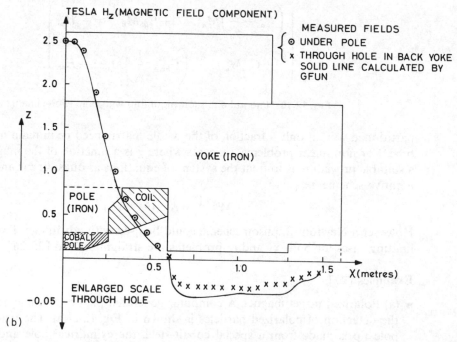

(b)

Fig. 12.11 Polarized target magnet. (a) Computer generated picture of the PEM pole and yoke, (b) comparison with measurements.

path for a beam of charged particles to enter a polarized target placed in the good high field region under the pole tip. The measured field distribution for this magnet is compared with the values calculated in Fig. 12.11(b).

- (b) CERN spectrometer model magnet. A computed generated picture of this magnet, designed as a 1/10 scale prototype for a forward spectrometer magnet, is shown in Fig. 12.12(a). Field measurements along the beam axis are compared with the calculated fields in Fig. 12.12(b). Results show that a high precision (0.10%) on magnetic field prediction, using the integral equation magnetization method, is possible for two-dimensional non-linear problems, providing accurate data of the material properties is known [26]. For three-dimensional magnets, however, the same order

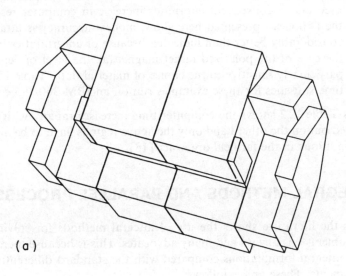

(a)

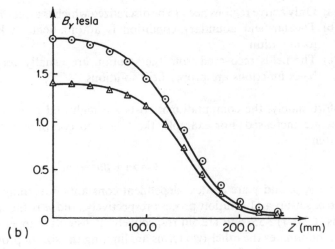

(b)

Fig. 12.12 CERN particle spectrometer. (a) Computer generated picture of the geometry, (b) comparison with measurements.

Table 12.2 Computing statistics.

Number of elements	Probable accuracy (%)	Magnetization Core (Kbytes)	Magnetization CPU Time(s)	Fields at a point Core (Kbytes)	Fields at a point CPU Time(s)
32	10	210	20	210	0.30
100	5	720	360	210	1.00
200	1	720	1800	210	2.00

of precision is not possible without a substantial increase in the number of elements used and a consequent enormous increase in computer resources. Nevertheless, as the two cases presented here show, modest accuracy is attainable and may well be considerably better than indicated because of uncertainties in material properties. In the case of the polarized target magnet (a), the level of field in the hole was found particularly sensitive to the choice of material data. Table 12.2 gives some computer time statistics for these examples run on an IBM 360/195 computer.

As Table 12.2 shows the computer time increases rapidly with the number of elements used but on the other hand only the active regions have to be meshed and the boundary conditions in the far field are exact [18].

12.4 INTEGRAL METHODS AND PARALLEL PROCESSING

As the literature shows the use of integral methods for solving electromagnetic field problems has always had many advocates. This is because there are several advantages of integral formulations compared with the standard differential approach using finite elements. These are as follows:

(a) Only active regions need to be discretized which is an enormous advantage in 3D.
(b) The far-field boundary condition is automatically taken into account by the formulation.
(c) The fields recovered from the solution are usually very smooth since the local basis functions are proper field solutions.

Unfortunately, the computational costs are high and rapidly escalate as the problem sizes are increased. For example, the time t to compute a complete solution can be written,

$$t = \alpha n + \beta n^2 + \gamma n^3, \qquad (12.59)$$

where α, β, and γ are context dependent constants governing the source computation, matrix set-up and solution process respectively, and n is the number of unknowns, see eqns (1.24), (12.55) and (12.56) respectively. It usually happens that long before the n^3 term dominates the other two terms are limiting the size of problem that can be handled for a given computer. However, these terms are essentially parallel operations, and furthermore, a significant degree of parallelism can be exploited in the solution process

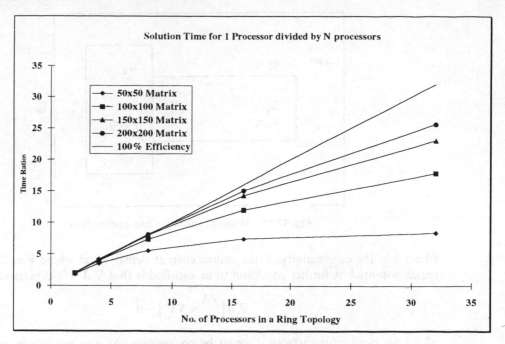

Fig. 12.13 Efficiency of integral equation solution.

itself. Thus the use of parallel hardware should extend the range of applicability of integral methods considerably.

Figure 12.13 shows results from a recent investigation that has implemented the **BIM** of section 9.8 for magnetostatics problems [28]. In this work the parallelism was achieved using an array of 32 transputers (a recently developed low cost 'chip' which is a complete computer on a single chip, with a floating point processor, memory storage, and communications hardware) [29].

12.5 MAGNETIC VECTOR POTENTIAL FORMULATIONS

The discussion now returns to the solution of 3D eddy current and time-dependent field problems using differential operators. In this and the following sections the basic defining equations, uniqueness conditions and the preliminary steps that lead to the discretized forms are covered. But since the expressions of the finite element coefficients for this case are essentially a generalization of the expressions previously derived for the scalar potential the details will be omitted. In section 1.2.1 the magnetic vector potential was introduced and the defining equations for the quasi-static electromagnetic fields were derived, see eqns (1.13), (1.16) and (1.17) and Fig. 12.14.

For convenience these equations are repeated here, i.e.

$$\nabla \times \frac{1}{\mu} \nabla \times \mathbf{A} + \sigma \left(\frac{\partial \mathbf{A}}{\partial t} + \nabla V \right) = \mathbf{J}_s, \tag{12.60}$$

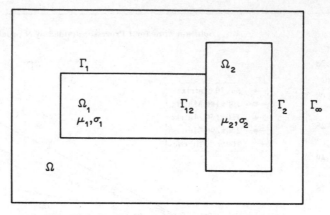

Fig. 12.14 Model problem for two conductors.

where σ is the conductivity, $\mathbf{J}_s$ the source current density, and where V is the electric scalar potential. A further condition to be satisfied is that $\nabla \cdot \mathbf{J} = 0$, expressed as

$$\nabla \cdot \sigma \left(\frac{\partial \mathbf{A}}{\partial t} + \nabla V \right) = 0 \tag{12.61}$$

whilst on conductor surfaces, J_n must be continuous. As was previously emphasized (see section 1.3) these two equations are not independent, since the second arises as a consequence of taking the divergence of the first.

In all the methods to be discussed, the solution to the above equations will be required in conducting regions. In all non-conducting regions (unless specifically mentioned otherwise), it will be assumed that the standard substitution of $\mathbf{H} = -\nabla \phi$ will be made, requiring the solution of $\nabla \cdot \mu \nabla \phi = 0$.

There now follows an overview of the main implementations that have been reported in the literature involving magnetic vector potentials (A). The formulations will be categorized according to the gauge method employed.

12.5.1 Coulomb gauge

- Modified vector potential. The simplest approach is to use a 'modified' vector potential [30] whereby a substitution is made in eqn (12.60)

$$\frac{\partial \mathbf{A}^*}{\partial t} = \frac{\partial \mathbf{A}}{\partial t} + \nabla V$$

leading to the new governing equation

$$\nabla \times \frac{1}{\mu} \nabla \times \mathbf{A}^* + \sigma \frac{\partial \mathbf{A}^*}{\partial t} = \mathbf{J}_s. \tag{12.62}$$

Taking the divergence of eqn (12.62) shows that the method has the implied gauge $\nabla \cdot \sigma \mathbf{A}^* = \nabla \cdot \mathbf{J} = 0$. In the discretized form (i.e. finite elements), this condition will only be satisfied in a weak sense (along with the implication that $\mathbf{J} \cdot \mathbf{n} = 0$ on conductor boundaries). However, this formulation will lead to a unique solution if the product $\omega\sigma > 0$ (for the time harmonic case). Therefore, the formulation is not valid in regions where $\sigma = 0$, or when the frequency $\omega = 0$.

- Coulomb gauge explicitly assigned. An alternative approach is to impose $\nabla \cdot \mathbf{A} = 0$ directly in eqn (12.60), which now requires $\nabla \cdot \mathbf{J} = 0$ as a second equation, leading to the following coupled equations [31],

$$\nabla \times \frac{1}{\mu}\nabla \times \mathbf{A} - \nabla\frac{1}{\mu}\nabla \cdot \mathbf{A} + \sigma\left(\frac{\partial \mathbf{A}}{\partial t} + \nabla V\right) = \mathbf{J}_s \qquad (12.63)$$

$$\nabla \cdot \sigma\left(\frac{\partial \mathbf{A}}{\partial t} + \nabla V\right) = 0. \qquad (12.64)$$

In addition, to ensure a unique solution, it is necessary to explicitly impose $\mathbf{A} \cdot \mathbf{n} = 0$ on the external boundaries to all conductors, see Reference [31].

It can be seen that when σ is constant, V will also be constant (and could be zero for example), leading to the situation where $\mathbf{A}$ is identical to the modified potential $\mathbf{A}^*$ above (for the continuous case). When μ is constant, the first two terms in eqn (12.63) can be combined to form the vector Laplacian operator, see vector identities Appendix A. However, in the discretized form, this is not precisely true, and the resulting matrix is slightly different to that formed were the Laplacian operator used.

The two equations above can be made symmetric if V is replaced by a new variable $\partial\bar{V}/\partial t$. J_n continuity can also be shown to be implied as a natural condition (although in a weak sense only for non-constant σ since $\bar{V}$ is no longer constant). Care must be taken then at zero frequency, since eqn (12.64) is badly structured. In fact $\bar{V}$ should be zero anyway, so can be imposed by forcing the diagonal terms associated with eqn (12.64) to be non-zero. For regions where $\sigma = 0$ this equation does not exist, and eqn (12.63) also reduces to a simpler form. Advantages are that it is straightforward to implement, $\omega = 0$ can be solved (but care over generating the matrix associated with the second equation above), $\sigma = 0$ can be handled (second equation will not exist), and problems involving σ_1/σ_2 can also be solved. However $\mathbf{A} \cdot \mathbf{n} = 0$ must be applied explicitly on conductor outer boundaries (this requires effort to apply to curved boundaries) also a Dirichlet condition on $\bar{V}$ must be applied somewhere if there are no symmetry planes (this being an arbitrary point) and the equations may become badly conditioned at very low frequencies and some loss in accuracy has also been reported at high frequencies [31].

12.5.2 Lorentz gauge

The Lorentz gauge was introduced in section 1.2.1 and can be written as follows,

$$\nabla \cdot \mathbf{A} = -\mu\sigma V. \qquad (12.65)$$

In using this gauge the scalar potential V is not removed from the formulation and thus it leaves a flexibility that can be used to handle the pathological cases of the modified vector potential formulation. The interface condition arising from (12.65) is continuity of $\mathbf{A}$ which immediately removes the multi-valued nature of $\mathbf{A}$ on conductor/conductor interfaces. Also the governing equation for $\mathbf{A}$, which for linear materials and static conditions reduces to the vector Laplacian is,

$$\nabla \times \left(\frac{1}{\mu} \nabla \times \mathbf{A} \right) = -\sigma \frac{\partial \mathbf{A}}{\partial t} + \sigma \nabla \left(\frac{1}{\mu\sigma} \nabla \cdot \mathbf{A} \right). \tag{12.66}$$

Uniqueness

By examining the uniqueness of **A** with the Lorentz gauge applied a range of possible boundary conditions can be deduced to assure uniqueness [32, 33]. Consider a two conductor problem as depicted in Fig. 12.14. The material properties for conductor 1 (Ω_1) are μ_1, σ_1, which are the permeability and magnetic conductivity respectively. Similarly for conductor 2 (Ω_2), which abuts conductor 1, the material properties are μ_2, σ_2. The external conductor surfaces are denoted by Γ_1, Γ_2 respectively and their common interface by Γ_{12}. The governing equations for the three distinct regions are as follows:

Regions 1, 2

$$\nabla \times \frac{1}{\mu_j} \nabla \times \mathbf{A} + \sigma_j \frac{\partial \mathbf{A}}{\partial t} + \sigma_j \nabla V = \mathbf{0} \tag{12.67}$$

$$\nabla \cdot \mathbf{A} = -\mu_j \sigma_j V, \tag{12.68}$$

where $j = 1$ or 2.

Region 3

This is the space that surrounds the two conductors (Ω), in which it is assumed that no current flows, and which possesses a permeability μ,

$$\nabla \cdot \mu \nabla \phi = 0. \tag{12.69}$$

As is customary it is assumed that there are two solutions in each region which lead to the following difference solutions for each region respectively, $\mathbf{A}_1 = \mathbf{A}_1^1 - \mathbf{A}_1^2$, $\mathbf{A}_2 = \mathbf{A}_2^1 - \mathbf{A}_2^2$, $V_1 = V_1^1 - V_1^2$, $V_2 = V_2^1 - V_2^2$, and $\tilde{\phi} = \phi_1 - \phi_2$ (cf. Appendix 6). Following the normal uniqueness procedure equations involving the five difference quantities, $(\mathbf{A}_1, V_1)$, $(\mathbf{A}_2, V_2)$ and $\tilde{\phi}$, for each region respectively, are constructed by integrating eqns (12.67), (12.68) and (12.69) in terms of difference quantities which, after applying vector divergence theorems and the standard interface conditions, lead to:

Regions 1, 2

$$\int_{\Omega_j} \left[\frac{|\nabla \times \mathbf{A}_j|^2}{\mu_j} + \sigma_j \mathbf{A}_j \cdot \frac{\partial \mathbf{A}}{\partial t} - V_j \nabla \cdot \sigma_j \mathbf{A}_j \right] d\Omega$$

$$+ \int_{\Omega_j \cup \Gamma_{12}} \sigma_j V_j \mathbf{A}_j \cdot \mathbf{n}_j \, d\Gamma \tag{12.70}$$

$$+ \int_{\Gamma_j} (\nabla \tilde{\phi} \times \mathbf{n}_j) \cdot \mathbf{A}_j \, d\Gamma + \int_{\Gamma_{12}} \frac{[(\nabla \times \mathbf{A}_j) \times \mathbf{A}_j] \cdot \mathbf{n}_j}{\mu_j} \, d\Gamma = 0.$$

Region 3

$$\int_{\Omega} \mu |\nabla \tilde{\phi}|^2 \, d\Omega - \int_{\Gamma_1} (\nabla \times \mathbf{A}_1 \cdot \mathbf{n}_1) \tilde{\phi} \, d\Gamma - \int_{\Gamma_2} (\nabla \times \mathbf{A}_2 \cdot \mathbf{n}_2) \tilde{\phi} \, d\Gamma = 0, \qquad (12.71)$$

where $\mathbf{n}_i$, $i = 1$ and 2 are the outward normals from regions 1 and 2 respectively. the exterior boundary has been taken to be at infinity and thus no contribution is made to the equations since the potential and all derivatives of the potential are zero.

The last two integrals in (12.71) may be transformed to produce,

$$\int_{\Omega} \mu |\nabla \tilde{\phi}|^2 \, d\Omega + \int_{\Gamma_1} (\nabla \tilde{\phi} \times \mathbf{A}_1) \cdot \mathbf{n}_1 \, d\Gamma - \oint_{P_{\Gamma_1}} \tilde{\phi} \mathbf{A}_1 \cdot \mathbf{dl}$$

$$+ \int_{\Gamma_2} (\nabla \tilde{\phi} \times \mathbf{A}_2) \cdot \mathbf{n}_2 \, d\Gamma - \oint_{P_{\Gamma_2}} \tilde{\phi} \mathbf{A}_2 \cdot \mathbf{dl} = 0 \qquad (12.72)$$

where P_{Γ_i}, $i = 1$ or 2 is the perimeter of each facet that comprises Γ_i, $i = 1$ or 2. The last two integrals of (12.72) vanish because over the conductor both $\tilde{\phi}$ and tangential $\mathbf{A}_i$, $i = 1$ or 2 are continuous.

Adding eqns (12.70) and (12.72) the terms in $\nabla \tilde{\phi}$ cancel and by substituting the Lorentz gauge condition, eqn (12.68), and assuming that σ is invariant with respect to both time and space the following is derived,

$$\int_{\Omega_1} \left(\frac{|\nabla \times \mathbf{A}_1|^2}{\mu_1} \times \frac{1}{2} \frac{\partial(\sigma_1 |\mathbf{A}_1|^2)}{\partial t} + \frac{(\nabla \cdot \mathbf{A}_1)^2}{\mu_1} \right) d\Omega$$

$$+ \int_{\Gamma_1 \cup \Gamma_{12}} \sigma_1 V_1 \mathbf{A}_1 \cdot \mathbf{n}_1 \, d\Gamma$$

$$+ \int_{\Omega_2} \left(\frac{|\nabla \times \mathbf{A}_2|^2}{\mu_2} + \frac{1}{2} \frac{\partial(\sigma_2 |\mathbf{A}_2|^2)}{\partial t} + \frac{(\nabla \cdot \mathbf{A}_2)^2}{\mu_2} \right) d\Omega$$

$$+ \int_{\Gamma_2 \cup \Gamma_{12}} \sigma_2 V_2 \mathbf{A}_2 \cdot \mathbf{n}_2 \, d\Gamma + \int_{\Omega} \mu |\nabla \tilde{\phi}|^2 \, d\Omega = 0. \qquad (12.73)$$

It is clear from (12.73) that at this stage uniqueness is not assured (which will only be the case if the integrands are either squared quantities or vanish when boundary conditions are known). The boundary integrals on surfaces where either V or $\mathbf{A} \cdot \mathbf{n}$ are known become zero but on surfaces where neither are known the integrals remain. It is the surfaces then where neither V nor $\mathbf{A} \cdot \mathbf{n}$ are known on which some extra condition is required in order to define a unique solution. Now for the Lorentz gauge neither V nor $\mathbf{A} \cdot \mathbf{n}$ are known on the bounding surfaces of conductors. Moreover it is clear that on such surfaces there is freedom to choose either $\mathbf{A} \cdot \mathbf{n}$ or $\partial V / \partial n$ since no current flows out of the conductor. This is expressed by the condition $\mathbf{J} \cdot \mathbf{n} = 0$, which in turn yields,

$$\sigma \frac{\partial(\mathbf{A} \cdot \mathbf{n})}{\partial t} + \sigma \frac{\partial V}{\partial n} = 0. \qquad (12.74)$$

Furthermore by choosing $\mathbf{A} \cdot \mathbf{n}$, the equation in V is driven by a different boundary condition thus modifying the distribution of V to relate to the implied $\mathbf{A}$. A suitable choice that reflects both the physics of the situation and the requirement to define a unique system is to relate $\mathbf{A} \cdot \mathbf{n}$ to $\partial V / \partial t$ as follows,

$$\frac{\partial(\mathbf{A} \cdot \mathbf{n})}{\partial t} = k^2 \sigma \frac{\partial V}{\partial t}, \tag{12.75}$$

where k^2 is a positive constant. When (12.75) is integrated with respect to time and the constant of integration chosen to be zero the following condition ensues,

$$\mathbf{A} \cdot \mathbf{n} = k^2 \sigma V. \tag{12.76}$$

It is clear that condition (12.76) produces a squared contribution to eqn (12.73) and this just leaves the integral over the common surface between the two conductors, namely Γ_{12}, to be considered.

If it is assumed that $V_1 = V_2 = V$ on Γ_{12} then the remaining integral from (12.73) has the form,

$$\int_{\Gamma_{12}} V[\sigma_1 \mathbf{A}_1 \cdot \mathbf{n}_1 + \sigma_2 \mathbf{A}_2 \cdot \mathbf{n}_2] \, d\Gamma. \tag{12.77}$$

On Γ_{12}, $\mathbf{J} \cdot \mathbf{n}$ is not zero but continuous, and again it can be seen that there is flexibility in the choice of $\mathbf{A} \cdot \mathbf{n}$ since whatever condition is chosen defines the weighted jump in $\partial V / \partial n$. This becomes clear when the $\mathbf{J} \cdot \mathbf{n}$ condition is expressed in terms of $\mathbf{A}$ and V, where it is assumed that $\mathbf{A}$ is continuous across Γ_{12}, i.e.

$$(\sigma_1 - \sigma_2) \frac{\partial(\mathbf{A}_1 \cdot \mathbf{n}_1)}{\partial t} = -\left[\sigma_1 \left(\frac{\partial V}{\partial n_i} \right)_1 + \sigma_2 \left(\frac{\partial V}{\partial n_i} \right)_2 \right]. \tag{12.78}$$

Condition (12.78) lends itself to a number of choices, for example if $\sigma_1 > \sigma_2$, then

$$\frac{\partial(\mathbf{A}_1 \cdot \mathbf{n}_1)}{\partial t} = k^2 (\sigma_1 - \sigma_2) \frac{\partial V}{\partial t}$$

which on integrating becomes

$$\mathbf{A} \cdot \mathbf{n} = k^2 (\sigma_1 - \sigma_2) V, \tag{12.79}$$

where the arbitrary constant is set to zero as in eqn (12.76). Utilizing the above condition the final uniqueness equation becomes, where it has been assumed that $\sigma_1 > \sigma_2$,

$$\int_{\Omega_1} \left(\frac{|\nabla \times \mathbf{A}_1|^2}{\mu_1} + \frac{1}{2} \frac{\partial(\sigma_1 |\mathbf{A}_1|^2)}{\partial t} + \frac{(\nabla \cdot \mathbf{A}_1)^2}{\mu_1} \right) d\Omega$$

$$+ \int_{\Omega_2} \left(\frac{|\nabla \times \mathbf{A}_2|^2}{\mu_2} + \frac{1}{2} \frac{\partial(\sigma_2 |\mathbf{A}_2|^2)}{\partial t} + \frac{(\nabla \cdot \mathbf{A}_2)^2}{\mu_2} \right) d\Omega$$

$$+ \int_{\Gamma_2} \frac{(\mathbf{A}_2 \cdot \mathbf{n}_2)^2}{(k_2)^2} \, d\Gamma + \int_{\Gamma_1} \frac{(\mathbf{A}_1 \cdot \mathbf{n}_1)^2}{(k_1)^2} \, d\Gamma$$

$$+ \int_{\Gamma_{12}} \frac{1}{k^2} (\mathbf{A}_1 \cdot \mathbf{n}_1)^2 \, d\Gamma + \int_{\Omega} \mu |\nabla \tilde{\phi}|^2 \, d\Omega = 0. \tag{12.80}$$

Equation (12.80) shows that the above conditions produce an unique solution for $\mathbf{A}$ even in the time limiting case of magnetostatic conditions. This is one area where the formulation based on the modified magnetic vector potential fails.

In time varying situations uniqueness for $\mathbf{A}$ is proved by choosing initial conditions for $\mathbf{A}, V$ and ϕ and then integrating (12.80) with respect to time. This is because the integrands not involving time are all positive producing a positive quantity which is then together with the squared terms in $\mathbf{A}$ and $\mathbf{A} \cdot \mathbf{n}$ equal to zero. Unique $\mathbf{A}$ leads to unique V and ϕ by reference to (12.68) and (12.72) with the usual requirement that the potential ϕ be specified at one point.

The Lorentz gauge formulation leads to an unique solution for abutting conductors without requiring any multi-valued nature in $\mathbf{A}$, V or ϕ, unlike the modified magnetic vector potential, where $\mathbf{A}$ becomes multi-valued on the mutual conductor surfaces.

Lorentz gauge formulations

As just intimated the Lorentz gauge formulation produces unique solutions in situations where the modified magnetic vector potential formulation breaks down either explicitly or practically. Although therefore this formulation has some advantages it is also evident that three unknowns per mesh point within conductors has been exchanged for four. The question arises as to whether the impact of such a change can be minimized?

The basic components of the Lorentz gauge formulation are the following,

$$\nabla \times \mathbf{H} = \mathbf{J} \tag{12.81}$$

$$\mathbf{J} = -\sigma \frac{\partial \mathbf{A}}{\partial t} - \sigma \nabla V \tag{12.82}$$

$$\nabla \cdot \mathbf{A} = -\mu \sigma V \tag{12.83}$$

$$\mathbf{A} \cdot \mathbf{n} = k^2 \sigma V; \text{ on exterior boundaries} \tag{12.84}$$

$$(\mathbf{A} \cdot \mathbf{n}) = k^2 (\sigma_1 - \sigma_2) V; \text{ on conductor/conductor interfaces} \tag{12.85}$$

but these components can be mixed together in a number of ways. Three formulations that each produce a symmetric matrix are detailed below:

Case 1

Combine eqns (12.81) and (12.82) to produce the standard equation in $\mathbf{A}$ and V, namely,

$$\nabla \times \left(\frac{1}{\mu} \nabla \times \mathbf{A} \right) = -\sigma \frac{\partial \mathbf{A}}{\partial t} - \sigma \nabla V \tag{12.86}$$

This equation is then integrated over Ω the solution domain in the usual weighted residual form and the term in V is then transformed utilizing the identity $\nabla \cdot (\sigma \mathbf{N} V) \equiv \sigma \mathbf{N} \cdot \nabla V + V \nabla \cdot (\sigma \mathbf{N})$, where $\mathbf{N}$ is a vector shape function, to produce,

$$\int_\Omega \left[\nabla \times \mathbf{N} \cdot \left(\frac{1}{\mu} \nabla \times \mathbf{A} \right) \right] d\Omega + \int_\Omega \sigma \mathbf{N} \cdot \frac{\partial \mathbf{A}}{\partial t} d\Omega$$

$$- \int_\Omega (\nabla \cdot \sigma \mathbf{N}) V \, d\Omega$$

$$- \int_\Gamma \left[\mathbf{N} \times \left(\frac{1}{\mu} \nabla \times \mathbf{A} \right) \right] \cdot \mathbf{n} \, d\Gamma + \int_\Gamma (\sigma V \mathbf{N} \cdot \mathbf{n}) \, d\Gamma = 0. \tag{12.87}$$

This equation is solved with the weighted residual form of (12.83),

$$\int_\Omega \sigma U \nabla \cdot \mathbf{A}\, d\Omega + \int_\Omega \mu \sigma^2 U V\, d\Omega = 0, \tag{12.88}$$

where U is a scalar shape function, and the conditions (12.84) and (12.85) have been applied where relevant.

This formulation always produces a non-zero V even in magnetostatic problems. This is because $\nabla \cdot \mathbf{A}$ is never identically zero in discretized form, and therefore eqn (12.88) must produce non-zero values for V. This can also be related to the fact that $\mathbf{A} \cdot \mathbf{n}$ is not zero as the formulation demands for magnetostatic problems. This discrepancy can be thought of as a modelling error, and indeed, as the mesh is refined the V values decay albeit very slowly. Alternatively, eqn (12.84) shows that V and $\mathbf{A} \cdot \mathbf{n}$ are related and their relative magnitudes can be varied by the choice of k^2. If k^2 is chosen to be large ($> 10^6$) then the values for V are comparable with the rounding error of the machine for single precision arithmetic. However, the larger k^2 chosen the more ill-conditioned the matrix becomes, this is simply because the dominant term in the matrix (for zero frequency cases) becomes increasingly the $(\nabla \times \nabla \times)$ operator which is singular in character.

Case 2

This approach is very similar to the above except on this occasion the weighted residual form of (12.86) stands and the weighted residual form of (12.88) is transformed with the identity, $\nabla \cdot (\sigma U \mathbf{A}) \equiv \nabla(\sigma U) \cdot \mathbf{A} + \sigma U \nabla \cdot \mathbf{A}$ to produce,

$$\int_\Omega \nabla(\sigma U) \cdot \mathbf{A}\, d\Omega + \int_\Omega \mu \sigma^2 V U\, d\Omega + \int_\Gamma \sigma U \mathbf{A} \cdot \mathbf{n}\, d\Gamma = 0. \tag{12.89}$$

This formulation responds in a very similar way to the one outlined in the last section with non-zero V in magnetostatic cases; but a V that is adjustable in magnitude according to the choice of k^2.

The previous two formulations both solve for $\mathbf{A}$ and V simultaneously and thus in a way maximize the impact of moving from three to four unknowns per mesh point within the conductor.

Case 3

This approach reflects the classical motivation for the Lorentz gauge which was to retain the scalar V but to produce two equations, one for $\mathbf{A}$ and one for V. To achieve this the weighted residual form of eqn (12.66) is constructed and the interface term in $1/(\mu\sigma)\nabla \cdot \mathbf{A}(= V)$ is simply replaced by condition (12.84) to produce a single equation in $\mathbf{A}$ (this is for a single conductor region only—otherwise condition (12.85) would be used as well and the equation would become more complex),

$$\int_\Omega \left[\nabla \times \mathbf{N} \cdot \left(\frac{1}{\mu} \nabla \times \mathbf{A} \right) \right] d\Omega \times \int_\Omega (\nabla \cdot \sigma \mathbf{N})\left(\frac{1}{\mu\sigma} \nabla \cdot \mathbf{A} \right) d\Omega$$

$$+ \int_\Omega \sigma \mathbf{N} \cdot \frac{\partial \mathbf{A}}{\partial t}\, d\Omega$$

$$- \int_\Gamma \left[\mathbf{N} \times \left(\frac{1}{\mu} \nabla \times \mathbf{A} \right) \right] \cdot \mathbf{n}\, d\Gamma + \int_\Gamma \frac{\mathbf{A} \cdot \mathbf{n}}{k^2} \mathbf{N} \cdot \mathbf{n}\, d\Gamma = 0. \tag{12.90}$$

After solving (12.90) for $\mathbf{A}$, $\nabla \cdot \mathbf{J} = 0$ can be solved for V by utilizing (12.82). The advantage in solving $\nabla \cdot \mathbf{J} = 0$ for V is that in magnetostatic cases V is identically zero or a chosen constant since it is related to the time derivative of $\mathbf{A}$.

The above formulation minimizes the impact of the move from solving for the modified magnetic vector potential alone to solving for both $\mathbf{A}$ and V within the conductors. This is achieved by constructing a sequential solution procedure that solves for $\mathbf{A}$ first and then uses this solution to solve for the resulting V. This has the advantage that if only the magnetic field is required the second equation for V does not need to be calculated and therefore the resulting computation time is comparable with that for the solution of the modified magnetic vector potential. However the operators are different for the two cases, in fact since the Lorentz operator is closer to the Laplacian it should exhibit better behaviour than the $(\nabla \times \nabla \times)$ operator that is used in the solution of the modified magnetic vector potential. Corroboration of this statement is found in the fact that the Lorentz formulation actually produces a non-singular matrix (for reasonable choices of k^2) and therefore a solution in the magnetostatic limit.

12.5.3 Examples

It can be seen from the theoretical development that case 3 described above is superior to cases 1 and 2, and therefore apart from the results for the zero frequency problem all other results pertain to case 3 alone.

Zero frequency limit

To show that all three Lorentz formulations do indeed limit to stable solutions a very simple problem of a 2 metre conducting cube in a 0.5 Tesla uniform field is convenient. Table 12.3 shows results for two meshes, for a range of k^2 values. The correct solution is obtained for large k^2, independent of mesh even for points near the corner. For small values of k^2 the effect of discretization is apparent and as well small non-zero values of V are produced by cases 1 and 2 but not case 3.

Bath cube problem

This and the following problem were first introduced by the University of Bath, UK and are standard bench-marks for eddy current computer code validation and have recently been adopted by the international workshop group TEAM (Testing Electromagnetic Analysis Methods) [34]. The results [35] for the bath cube [36] using the Lorentz gauge formulation are very similar to those produced by the modified vector potential formulation (CARMEN code) as should be expected since for this problem the two formulations are essentially equivalent.

Bath plate problem

This bench-mark introduces a new complexity because the configuration is multiply connected, see Fig. 12.15, and consists of a conducting plate with holes, driven by a current carrying coil (TEAM problem 3—Bath plate [37]). A comparison of all the

Table 12.3 Conducting block in 0.5 Tesla at zero frequency

Fields on block diagonal for two meshes

k^2	Case 1			Case 2			Case 3		
	10^{-6}	1	10^6	10^{-6}	1	10^6	10^{-6}	1	10^6
$\mathbf{r}_1$	-0.511	-0.513	-0.5	-0.562	-0.561	-0.5	-0.520	-0.518	-0.5
	-0.497	-0.497	-0.5	-0.497	-0.497	-0.5	-0.496	-0.495	-0.5
$\mathbf{r}_2$	-0.458	-0.460	-0.5	-0.416	-0.418	-0.5	-0.433	-0.438	-0.5
	-0.497	-0.497	-0.5	-0.496	-0.497	-0.5	-0.495	-0.495	-0.5
$\mathbf{r}_2$	-0.637	-0.620	-0.5	-0.580	-0.578	-0.5	-0.591	-0.597	-0.5
	-0.489	-0.496	-0.5	-0.494	-0.494	-0.5	-0.484	-0.492	-0.5
$\mathbf{r}_4$	-0.210	-0.315	-0.5	-0.447	-0.447	-0.5	-0.197	-0.335	-0.5
	-0.417	-0.480	-0.5	-0.511	-0.511	-0.5	-0.392	-0.482	-0.5

$\mathbf{r}_i = (0.125, 0.125, 0.125) + (0.25, 0.25, 0.25) * (i - 1); \ i = 1, 2, 3, 4$

Details of finite element mesh used

Mesh	Nodes	Elements	Conductor elements
1	125	64	8
2	301	192	64

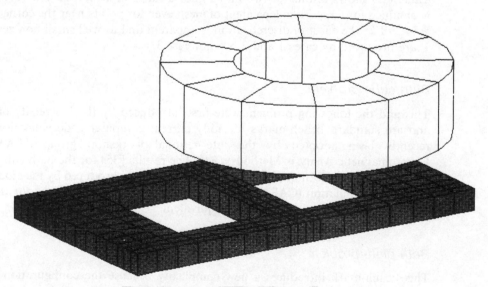

Fig. 12.15 The bath plate problem.

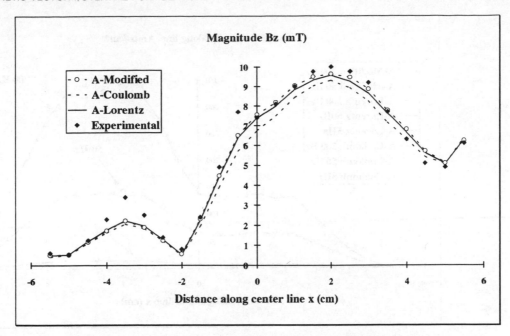

Fig. 12.16 Bath plate: fields at centre line for three methods at 200 Hz.

above methods have been carried out [38] using this problem. The methods compared are the modified vector potential formulation, the Coulomb gauge, and the Lorentz gauge formulations. The results for the Lorentz gauge method were obtained using the ELEKTRA code [18]. The modified vector potential formulation required the use of low conductivity air in the hole, in order to make the problem simply connected. The Coulomb and Lorentz formulations used zero conductivity vector potential in the hole.

The modulus of the field along a predefined line is shown in Fig. 12.16 for the asymmetric case at a drive frequency of 200 Hz. As a further test, the frequency was reduced to 50 Hz, 5 Hz, 0.5 Hz and the DC limit of 0 Hz, and the current along a line interior to the conducting block plotted in Fig. 12.17. The modified vector potential algorithms only just converged at 5 Hz, and did not converge for any frequencies lower than this. The Coulomb and Lorentz methods all succeeded in solving all cases, including the DC limit. For this particular example, the Coulomb method tended to converge quicker than the Lorentz (75 iterations versus 100 iterations), and there were fewer equations for the Coulomb problem due to the boundary conditions (4339 Coulomb equations compared with 4441 Lorentz equations). Figures 12.18 and 12.19 show comparisons of the amplitude and phase for the Lorentz and the modified vector potential formulations and the experimental results.

Segmented torus problem

This problem introduces a multiply-connected configuration involving two abutting materials where the conductivities differ by an order of magnitudes, see Fig. 12.20(a).

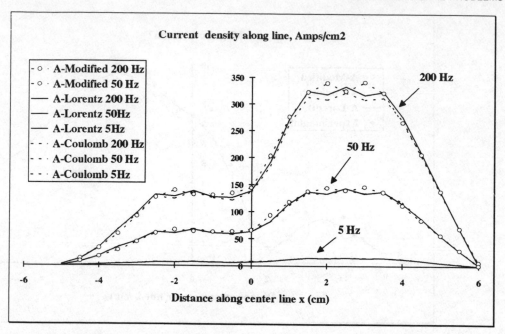

Fig. 12.17 Bath plate: current along line interior to conducting block at 200 Hz, 50 Hz and 5 Hz.

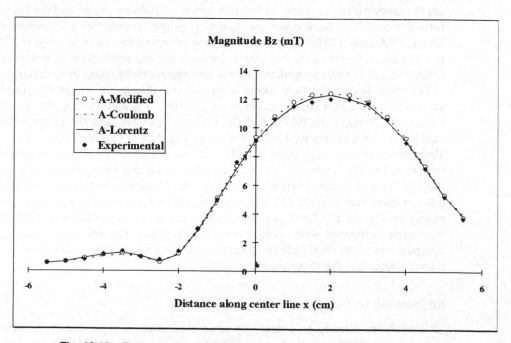

Fig. 12.18 Field amplitude comparison for the bath plate problem (50 Hz).

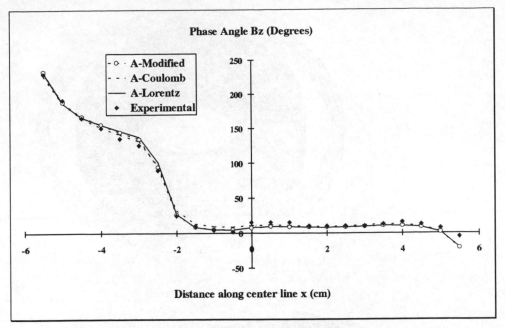

Fig. 12.19 Phase angle comparison for the bath plate problem (50 Hz).

The result shown in Fig. 12.20(b) compares well with those, using methods involving fields rather than potentials (see next section), reported at the TEAM workshop [39].

This last problem exemplifies advantages of the Lorentz formulation which can be summarized as follows. It is simple to implement, has the minimum number of degrees of freedom for field computations despite a very small secondary (Laplacian) problem to be solved to recover the currents, also there is no need to explicitly set $\mathbf{A}.\mathbf{n} = 0$ on conductor boundaries, and $\omega = 0$, $\sigma = 0$ can be solved (i.e. the second equation is no longer present). Furthermore, problems involving σ_1/σ_2 can be solved without the need of introducing multiple values of the unknowns at material interfaces. However, a Dirichlet condition on V must be applied somewhere if there are no symmetry planes (this being an arbitrary point).

12.5.4 Ungauged solutions

The eqns (12.60) and (12.61) have also been solved directly, where it has been shown that the use of appropriate types of elements can imply a unique solution [40]. The equations that arise are not symmetric, but can be made so by replacing V by $\partial \bar{V}/\partial t$ (as in the Coulomb method above).

In looking at the uniqueness of a solution, it was shown in [40] that $\bar{V}$ must be specified at two or more points. If there are planes of symmetry over which $\bar{V}$ is specified, then one other point must also have $\bar{V}$ set. However, as in the Coulomb method above, the continuity of J_n is implied by the formulation itself. This approach has the advantage that $\omega = 0$ is allowed, but care must be taken over constructing the matrices; $\sigma = 0$ may

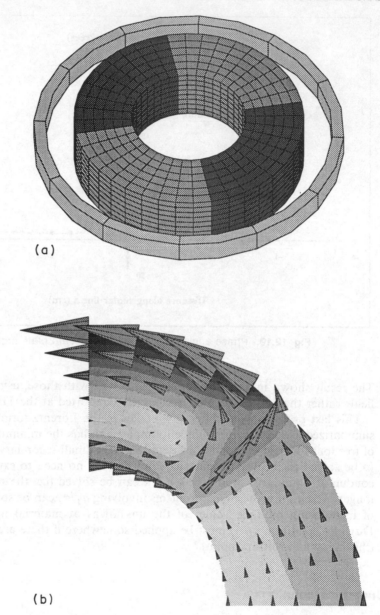

Fig. 12.20 Current vectors/contours for the segmented torus problem. (a) Geometric model showing the drive coil surrounding eddy-conductor, (b) one quadrant showing current flow.

also be allowed. However, the disadvantages are significant, i.e. $\bar{V}$ must be specified at *two* points if there are no symmetry planes, and at one other point if there are symmetry planes; the choice of elements to be used in the discretization is important, since brick elements may work, but tetrahedral elements will not.

12.6 COMMENTS ON VECTOR POTENTIAL FORMULATIONS

A series of formulations has been reviewed, all based on the magnetic vector potential, and which form a 'family' of methods. They range from ungauged formulations, to methods with an implied gauge, and methods in which a gauge is added explicitly (either Coulomb or Lorentz).

The ability to use any shape of element in the discretization (including a mixture of hexahedra, tetrahedra and pentahedra) implies that some form of gauging should be imposed. The ungauged method fails on this count, although the modified vector potential (which does not depend on the element type to gauge the problem) has proven very successful.

To solve general 3D problems, the ability to model DC is essential (both the Coulomb and Lorentz methods work well for DC problems). Other useful features such as zero conductivity (for multiply-connected problems) are also well handled by these two methods.

A disadvantage of solving for **A** and V is that the current is a function of ∇V, thus implying a loss of accuracy due to the required differentiation. This can be overcome if a higher order element is used to model the V equations in the system. This can be achieved very easily for example using the Lorentz implementation.

There may be occasions when it is necessary to model the free space regions by the vector potential instead of the scalar potential. This can arise when the coupling terms between the vector and scalar regions lead to abrupt changes in the magnitude of the matrix coefficients. In these circumstances convergence problems may be encountered. It is sometimes difficult to avoid this especially when very narrow regions are involved [33], or more critically when the interface between vector and scalar is at the iron–air boundary with large and abrupt changes in permeability. One solution to this problem is to extend the vector potential region to occupy the entire space and to accept the higher computational costs involved, this an extension of the standard 2D formulations discussed in Chapter 11 where both the Coulomb and Lorentz gauges are satisfied automatically. Alternatively, the vector region may be extended into the free space region to force the coupling to be on surfaces where there is no change in permeability. This procedure is, in any case, necessary for multiply-connected problems where the 'hole' in a torus, for example, has to be a vector region, see the bath plate and segmented torus examples above (Fig. 12.15 and 12.20). The Lorentz gauge formulation applies to both of these situations naturally [33]. Furthermore, the Lorentz gauge in its general form, which includes the displacement current term, can be used to solve problems over the entire frequency spectrum [41].

12.7 H FORMULATION USING EDGE ELEMENTS

The field equation in terms of the field intensity **H** is given by eqns (1.33), i.e.

$$\nabla \times \nabla \times \mathbf{H} + \sigma \frac{\partial}{\partial t}(\mu \mathbf{H}) = 0 \tag{12.91}$$

however in the global region where $\sigma = 0$ and $\nabla \times \mathbf{H} = 0$, eqn (12.91) limits to eqn (1.28) as before. A direct application of classical finite elements, using nodal basis functions,

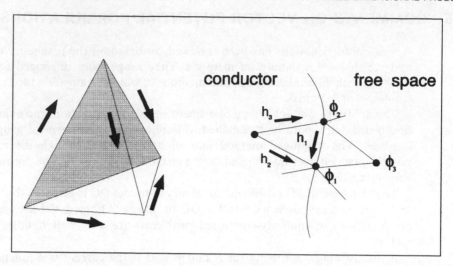

Fig. 12.21 Edge elements.

in the conducting regions will lead to difficulties. This is because the physical interface conditions, eqns (1.11) and (1.12) do not emerge naturally when a Galerkin weighted residual method is applied. The solution to this problem is to use the edge variable elements which enforce the correct interface conditions. Edge variables for tetrahedral elements are defined by

$$h(\mathbf{r}) = \mathbf{a} + \mathbf{b} \times \mathbf{r}, \tag{12.92}$$

where $\mathbf{r}$ is the position vector and $\mathbf{a}$ and $\mathbf{b}$, respectively, are vectors dependent on the geometry of the element. The basis function expansion is then given by

$$\mathbf{H} = \sum \mathbf{h}_e(\mathbf{r})\mathbf{H}_e, \tag{12.93}$$

where $\mathbf{h}_e$ is the vector basis function for edge e, and $\mathbf{H}_e$ is the value of the field along an element edge, see Fig. 12.21. The functions, eqn (12.93) have the following useful properties. $\nabla \cdot \mathbf{h} = 0$ and $\nabla \times \mathbf{h} = 2b$. Most importantly, the edge variable ensures that the tangential component of $\mathbf{H}$ is continuous whilst allowing for the possibility of a discontinuity in the normal component. In non-conducting regions where the field is modelled by a scalar potential standard nodal elements can be used, see Fig. 12.21. At the interface between conducting and non-conducting regions the nodal basis functions couple elegantly to the edge variables, i.e. $h_1 = (\phi_2 - \phi_1)/l$ [42]. In an important implementation of this method [43] the scalar region has been solved using a boundary integral equation allowing the infinite space to be modelled.

12.8 SUMMARY

In this chapter some of the most widely used methods for solving 3D statics and low frequency field problems have been presented. For the statics case the scalar potential formulations have proved to be very effective and economic but care must be taken in the selection of potential type (reduced or total?) since cancellation effects between the computed fields and source fields may introduce unwanted errors.

For eddy current problems vector unknowns have to be calculated and the most widely used algorithms are based on the magnetic vector potential. In this chapter the problem of gauging has been examined in some detail and it has been suggested that the Lorentz gauge limit for low frequency fields yields a robust formulation able to compute problems having both simply and multiply-connected geometries, inhomogeneous conductivities and permeabilities, and can be reduced to the static limit if required. In all of the above formulations pathological problems arise from time to time. For example the discretized system may have an ill-conditioned matrix which causes convergence difficulties. This can happen when two types of potential are involved (e.g. $A - \phi$ formulation) as previously discussed, see section 12.6. In such cases it may be necessary to extend the vector potential regions to include free space (air) in order to avoid abrupt changes in matrix coefficients. A computer code that provides this option has recently been developed, see for example ELEKTRA [18]. This remark serves to remind the reader that the subject is still very much a research area and the references cited should be consulted. In this context it should be emphasized that up to quite recently most numerical schemes using the finite element method have been based on the nodally based shape functions, see section 12.2.6, and as was pointed out in section 12.7 edge variables offer certain advantages and are now becoming popular.

The development of computer solutions of three-dimensional field problems is still at a relatively early stage and in this chapter it has only been possible to present a brief outline of some basic techniques. However, the methods described are all of general validity as regards geometry and, in the case of the finite element method are at least extendable to include transient nonlinear problems. It is now evident that many designers of electrical devices are dependent on three-dimensional solutions and regard computational methods as indispensible for the design of magnets which involve the accurate predictions of magnet field shapes (focusing of particle beams and resonance imaging for medical diagnostics), inductance and force calculation in rotating machines and loss calculation in transformers. The methods of solution using finite elements, developed in this chapter for time invariant fields, offers a near optimal and highly accurate approach to field computation. However, further improvements in efficiency are to be expected as faster methods for the numerical solution of algebraic equations become available, see Chapter 13. So far as eddy-current fields are concerned, the situation is less well established and remains a topic very much for research. Nevertheless considerable progress has been made, particularly in special applications. Indeed in the electrical machine industry, for power generation, the art of three-dimensional computation has been extended in special cases to include coupled electrical, mechanical and thermal phenomena [44].

For further information in 3D field computation the reader should consult the proceedings of the major conferences on the computation of magnetic fields, e.g. COMPUMAG, INTERMAG, and MAGNET TECHNOLOGY, see references. A major barrier for users of these methods arises in the data generation stage—the establishing of correct geometric models and finite element meshes is a formidable task in 3D and electromagnetics problems are among the more difficult. This subject, as well as the software environment a designer needs, in order to use to best effect the computational algorithms discussed in this book effectively will, be discussed in the next chapter.

REFERENCES

[1] J. Simkin (ed.), *Compumag Oxford Proceedings* (Chilton, Oxfordshire) (1976).

[2] J. C. Sabonnadiere (ed.), *Compumag Grenoble Proceedings* (ERA 524 CNRS, Grenoble, France) (1979).

[3] L. Turner (ed.), *Compumag Chicago Proceedings, IEEE Trans. on Magnetics*, **MAG-18**(2) (1982).

[4] G. Molinari (ed.), *Compumag Genoa Proceedings, IEEE Trans. on Magnetics*, **MAG-19**(6) (1983).

[5] W. Lord (ed.), *Compumag Fort Collins Proceedings, IEEE Trans. on Magnetics*, **MAG-21**(6) (1985).

[6] K. Richter (ed.), *Compumag Graz Proceedings, IEEE Trans. on Magnetics*, **MAG-24**(1) (1988).

[7] K. Miya (ed.), *Compumag Tokyo Proceedings, IEEE Trans. on Magnetics*, **MAG-26**(2) (1990).

[8] R. Martone (ed.), *Compumag Sorrento Proceedings, IEEE Trans. on Magnetics*, **MAG-28**(2) (1992).

[9] C. J. Collie, Magnetic fields of a class of 3-dimensional circuits, *Tech. Rep. RL 73–049*, Rutherford Laboratory (1973).

[10] L. Urankar, Vector potential and magnetic field od a current-carrying finite arc segment in analytic form, *IEEE Trans. on Magnetics*, **MAG-16** and **MAG-18** (1980 and 1984).

[11] N. J. Diserens, A search for faster magnetic field routines for curved conductors, *IEEE Trans. on Magnetics*, **MAG-19**(6) (1983).

[12] M. R. G. Drago and G. Secondo, Inductances and high accuracy field computation in ironless structures, in *Int. Symp. on Electromag. Fields in El. Eng., 20–22 September, Lodz, Poland*, pp. 9–12 (1989).

[13] J. A. Stratton, *Electromagnetic Theory*. New York: McGraw-Hill (1941).

[14] J. Simkin and C. W. Trowbridge, On the use of the total scalar potential in the numerical solution of field problems in electromagnets, *IJNME*, **14**, 423 (1979).

[15] J. Simkin and C. W. Trowbridge, Three dimensional non-linear electromagnetic field computations using scalar potentials, *Proceedings of the IEE*, **127**(6), 368–374 (1980).

[16] O. C. Zienkiewicz and R. Taylor, *The Finite Element Method, 4th Edition, Volumes 1 and 2*. Maidenhead: McGraw-Hill (1991).

[17] D. E. Bourne and P. C. Kendall, *Vector Analysis and Cartesian Tensors*. London: Van Nostrans Reinhold (1977).

[18] Vector Fields Ltd, 24 Bankside, kidlington, Oxford OX5 1JE, *TOSCA, GFUN, CARMEN, ELEKTRA, BIM2D, OPERA, and PE2D User Manuals* (1990).

[19] P. J. S. Ritchie and B. G. Loach, *Nimrod Beam Line Equipment Data Handbook* (1968).

[20] J. W. Finch, Department of Electrical Engineering, Newcastle University, Private communication (1979).

[21] P. J. Lawrenson *et al.*, Analysis of switched reluctance stepping motor, Private Communication (1979).

[22] J. Simkin, Magnetostatic fields in 3-d, *VECTOR Electromagnetics Newsletter*, **4**(2), (1988).

[23] C. A. Brebbia, *The Boundary Element Method for Engineers*. Printec Press, 2nd edition (1980).

[24] I. M. Mayergoyz et al., Boundary Galerkin's method for 3-d finite element electromagnetic field computation, *IEEE Trans. on Magnetics*, **MAG-19**(6) (1983).

[25] M. J. Newman, L. R. Turner, and C. W. Trowbridge, G-FUN: an interactive program as an aid to magnet design, in *Proc. Int. Conf. on Magnet Tech.* (**MT** 4), (Y. Winterbottom, ed.), pp. 617–626, Brookhaven National Laboratory (1972).

[26] A. Armstrong *et al.*, New developments in the magnet design computer program g-fun, in *Fifth Conference on Magnet Technology* (N. Sacchetti *et al.*, eds), (Cas. Postale 70–00044 Frascati, Rome, Italy), p. 168, Laboratori Nazionali del CNEN (1975).

[27] C. W. Trowbridge, *Applications of Integral Equation Methods for the Numerical Solution of Magnetostatic and Eddy Current Problems.* Chichester: Wiley (1979).

[28] C. F. Bryant and C. W. Trowbridge, Specification of the boundary demonstrator using transputers, Tech. Rep., Vector Fields Ltd, 24 Bankside, Kidlington, Oxford OX5 1JE, 1988, Esprit (1051), ACCORD/WP5/VFL/DEL/001/291.88/CWT Project Report.

[29] Transputer technical information. INMOS Ltd, 1985, PO Box 424, Bristol, BS99 7DD, UK.

[30] C. R. I. Emson and J. Simkin, An optimal method for 3D eddy currents, *IEEE Transactions on Magnetics*, **19**, 2450 (November 1983).

[31] O. Biro, Coulomb gauged vector potential formulation, *Proceedings of the Eddy Current Workshop, Capri* (October 1988).

[32] C. F. Bryant, C. R. I. Emson and C. W. Trowbridge, A comparison of Lorentz gauge formulations in eddy current computations, in *Compumag Conference on the Computation of Electromagnetic Fields, Tokyo, September 1989 Proceedings, IEEE Trans. on Magnetics*, **MAG-26**(2) (1990).

[33] C. F. Bryant, C. R. I. Emson and C. W. Trowbridge, A general purpose 3-D formulation for eddy currents using the Lorentz gauge, in *Intermag Conference, Brighton, April 1990 Proceedings, IEEE Trans. on Magnetics*, **MAG-28**(5) 2373–2375 (September 1990).

[34] L. R. Turner *et al.*, Papers on bench-mark problems for the validation of eddy current codes, *COMPEL*, vol. 7 (March/June 1982).

[35] C. R. I. Emson, J. Simkin and C. W. Trowbridge, Further developments in three dimensional eddy current analysis, *IEEE Trans. on Magnetics*, **21**, 2231 (November 1985).

[36] J. A. M. Davidson and M. J. Balchin, 3-D field calculations by network methods, *IEEE Trans. on Magnetics*, **MAG-19** (November 1983).

[37] A. Bossavit, Results for benchmark problem 5, the bath cube experiment: an aluminium block in an alternating field, *COMPEL*, vol. 7, No. 1 and 2. (Mar/June 1988).

[38] C. R. I. Emson, C. F. Bryant and C. W. Trowbridge, Finite element solutions of general 3-D eddy current problems using a family of magnetic vector potential formulations, in *3DMAG, International Symposium on 3-D Electromagnetic Fields, Okayama, Japan, September 1989 Proceedings*, pp. 41–44, *COMPEL*, Vol. 9, Supp. A (1990).

[39] Z. Ren and A. Razek, Calculation of 3-d eddy currents using electric field formulation, in *Team Workshop and Meeting on the Applications of Eddy Currents Compuations* (EDF, Research and Development Division, 92141 Clamart, France) (1989).

[40] A. Kameari, Three dimensional eddy current calculation using finite element method with A-V in conductor and Ω in vacuum, *IEEE Trans. on Magnetics*, **MAG-24** (January 1988).

[41] C. W. Trowbridge, C. F. Bryant and C. R. I. Emson, Some recent developments in electromagnetic field computation, in *MAFELAP 1990, 7th Conference on The Mathematics of Finite Elements and Applications, April 1990*, London: Academic Press (1990).

[42] A. Bossavit and J. C. Verite, A fixed FEM-BIEM method to solve 3D eddy current problems, *IEEE Trans. on Magnetics*, **MAG-18**, 431 (March 1982).

[43] A. Bossavit and J. C. Verite, The TRIFOU code: solving the 3D eddy currents problem by using **h** as state variable, *IEEE Trans. on Magnetics*, **MAG-19**, 2465 (November 1983).

[44] T. W. Preston and A. B. J. Reece, The contribution of the finite element method to the design of electrical machines: an industrial viewpoint, *IEEE Trans. on Magnetics*, **MAG-19**, 2375 (November 1983).

13

ELECTROMAGNETIC SOFTWARE
ENVIRONMENT

13.1   INTRODUCTION

Electromagnetic analysis is an important factor in the solution of a very broad range of industrial and research problems. The 'user community' includes designers of equipment working in the following application areas:

- Particle physics
- Electron optics
- Fusion technology
- Non-destructive testing and evaluation
- Electrical machines
- Medical systems—MRI magnets
- Computer industry
- Recording industry
- Audio industry
- Communications industry
- Transport systems
- Robotics
- Components industry
- Industrial processes
- Electromagnetic compatibility
- Power transmission,

etc.

In this chapter some of the practical aspects of computer-aided engineering are explored with a particular emphasis upon the software environment. A designer needs to build a computer model of his problem, execute a field solution and analyse his results in a meaningful way. These three stages will be considered in some detail, mainly by focusing attention on established ideas, but some recent development, which are likely to remain important in the future, are also included. Two related topics, error analysis and optimization, will also be considered because of their impact on future developments.

13.2   ELEMENTS OF A CAD SYSTEM

Before considering the components of a practical software environment for electromagnetics CAD it may be useful to try and give an overall picture of the requirements and to give some indication of the compromises involved.

An idealization of the design process in which a designer iterates towards a satisfactory solution by the heuristic approach is shown in Fig. 13.1. There are three conceptual stages:

- A data input pre-processor for defining geometric and material data.
- A solution processor for solving the equations numerically.
- A post-processor for examining and extracting the results, i.e. fields, gradients, integrals and forces etc.

All three processors should access a common database. The pre- and post-processors should be interactive, utilizing graphics techniques, and should reside in either a multi-user machine environment or a cluster of networked single-user workstations. For large problems, the solution processors may reside on a mainframe with file transfer between locally networked or perhaps widely distributed machines.

In Fig. 13.2 the components are specified in a little more detail; the user can be imagined as controlling the pre- and post-processing phases at a graphics terminal using heuristic techniques to optimize his design. Because the problem is to be solved numerically the model has to be discretized—i.e. meshes of elements have to be generated to conform to the data model—and the accuracy of the result will depend on the level of discretization, see section 9.4. In the ideal system shown in Fig. 13.2(b) this discretization process takes place inside the solution processor ('solver'), which can

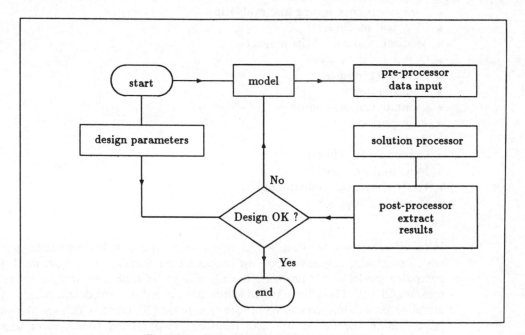

Fig. 13.1   Flow diagram for heuristic design.

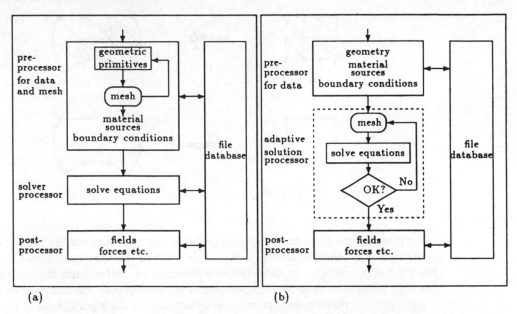

Fig. 13.2 Components of CAD system (a) normal system, (b) ideal system.

generate a mesh adaptively to achieve a specified accuracy. It is a current limitation that this ideal has not been satisfactorily met, at least for three-dimensional systems, despite considerable research that is under way.

Systems available today usually follow the scheme outlined in Fig. 13.2(a) in which the pre-processor includes a mesh generation stage. The crux of the problem is to try to achieve a 'clean interface' between the geometric model for the problem and the internally discretized model, and this requires an algorithmic method for mesh generation.

13.3 PRE-PROCESSING AND MESH GENERATION

In the past, two main classes of method have been used for mesh generation: semi-automatic and manual, though, fully automatic has long been a research topic. Many examples of these two approaches are described in the literature with several implementations available commercially [1].

13.3.1 Semi-automatic

The semi-automatic method allows the user to build the entire mesh from a number of regions which are defined by simple geometric primitives, for example a two-dimensional mesh may be constructed by modelling the geometry by a set of curvilinear quadrilaterals as shown in Fig. 13.3. Note particularly that the elements are automatically generated within each primitive, the number and density can be controlled by a subdivision at

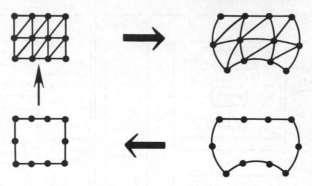

Fig. 13.3   Semi-automatic isoparametric meshing.

region boundaries. The ideas of iso-parametric mapping can be readily used to construct an algorithm for this purpose [4]. With this method the user has to ensure manually that continuity of element subdivision is maintained across inter-region boundaries, although some systems will check for violations and display the mismatches [2].

With good terminal ergonomics, i.e. graphics editing via a puck, mouse, joystick, etc. the basic primitives can be rapidly edited and very complicated meshes efficiently constructed. See for example, Fig. 13.4, where the basic primitives for an electrical machine are shown as well as the completed mesh, it should be noticed how all mirror and anti-symmetric reflections and rotations have been specified as part of the data for each basic primitive.

13.3.2   Manual

The manual method of generating a finite element mesh implies that individual elements are entered sequentially and at the basic level this is indeed an option but by using powerful editing software and the high graphical interaction rates of a single user computer [5] the user can build elements from points; he can then delete, move, copy and replicate, elements; he can even build macros to define a set of elements which can be labelled for future or immediate use. The MAGNET program [3] is an elegant example of this approach which offers, at the basic level, editing and replication facilities so that, after entering a few elements, large meshes can be constructed fairly rapidly. Elements can be subdivided singly (TRISECT operation) or in pairs (BISECT operation), but this can, in some circumstances, lead to element shapes with large aspect ratios. However the quality of the final mesh can subsequently be improved by using the Delaunay algorithm, see section 3.4. This method appeals to many, because it can be satisfying to build a model which subsequently works well, and some users can become very efficient at creating economical models in which they have combined physical insight (and artistry) to achieve excellent results.

For two-dimensional cases this technique is well matched to existing graphics media (graphics displays are fundamentally two-dimensional) but, unfortunately, the extension to 3D is inelegant. This difficulty, however, can be overcome to some extent by combining ideas from both the semi-automatic and manual methods, or by subdividing the geometry

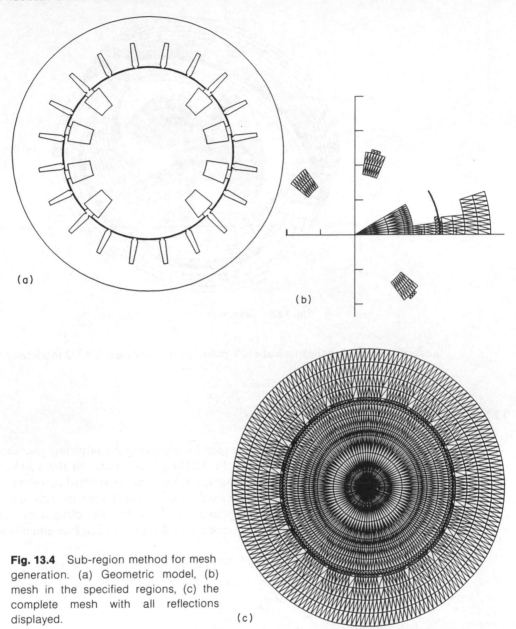

Fig. 13.4 Sub-region method for mesh generation. (a) Geometric model, (b) mesh in the specified regions, (c) the complete mesh with all reflections displayed.

into parallel slices which are essentially 2D. The latter procedure is very suitable for problems when the degree of complexity is greater in a particular plane, e.g. an electrical machine where say the xy plane contains the profile of rotor and stator with several slots and poles and small air gaps. The electromagnetics pre- and post-processor system OPERA [2] has been used for this type of model, see also Fig. 13.5 for an example used in the design of the OSCAR cyclotron [6]. As can be seen in the latter case the method has been generalized to allow rotations and curvilinear transformations with

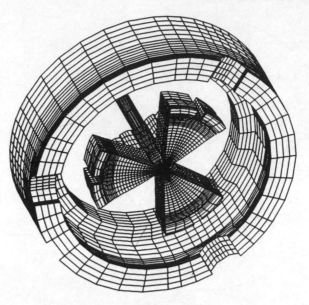

Fig. 13.5   Geometry of the 'Oscar' cyclotron.

the facility to modify the mesh at each 'plane' thus allowing a full 3D modelling flexibility [2].

13.3.3   Mapping methods

A number of schemes have been proposed for achieving a completely automatic mesh; one such scheme was developed for the TRIM [7] program. In the TRIM (Triangle Mesh) code the user in effect has to imagine a rubber sheet stretched across the geometry of the problem; the sheet has a mesh of fixed topology ruled upon its surface so a regular mesh is distorted to produce an irregular mesh. Providing the distortions are not too large this process is equivalent to solving a coupled pair of Laplace equations, i.e.

$$\frac{\partial^2 x}{\partial u^2} + \frac{\partial^2 x}{\partial v^2} = 0 \tag{13.1}$$

$$\frac{\partial^2 y}{\partial u^2} + \frac{\partial^2 y}{\partial v^2} = 0 \tag{13.2}$$

where the point $P(x, y)$ in the xy plane is mapped on to the point $Q(u, v)$ in the uv plane. The method is illustrated for a simple problem in Fig. 13.6.

The model shown in Fig. 13.6(a), representing the actual problem space, is transformed to the simpler model shown in Fig. 13.6(b) by continuous deformation. The regular mesh in u–v space is then used to discretize eqns (13.1) and (13.2) in turn, first with the appropriate boundary conditions for the x-coordinates of the real model and secondly with those for the y-coordinates. The x and y solutions are the u–v 'equipotentials' which

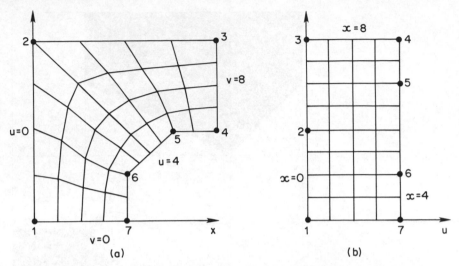

Fig. 13.6 Laplacian mesh generation.

can then be plotted in x–y space as in Fig. 13.6(a) to produce the mesh. The method works quite well for convex regions but near a concave boundary the mesh lines crowd together (e.g., segment 5–6 in Fig. 13.6(a)) and, for sharp corners, can fall outside the boundary.

Another variant on this is to actually stretch points and lines in space by means of direct interaction with a model displayed on a terminal [5]; this is possible using recent work stations which have sufficient power to allow 'rubber-band' graphics in real time, see for example Fig. 13.7 in which a node of a regular mesh is moved together with its connecting lines to confirm to the geometric model.

13.3.4 Automatic mesh generation

Considerable recent work has been directed toward purely automatic grid generation in which the user specifies the geometric model for his problem together with additional information to define the degree of mesh refinement that he estimates will be necessary in order to satisfy a given accuracy of a finite element solution. The important question of accuracy and element size will be left until after a method of forming points to produce good intrinsic element shape has been considered, see section 4.

An illustration of the mesh generation problem is shown diagrammatically in Fig. 13.8, the question being 'what is the optimum way of connecting n points in a plane to form triangles of acceptable shape?'; clearly, the solution offered in Fig. 13.8(a) has a number of triangles with large obtuse angles and consequently the mesh is far from regular. On the other hand the mesh in Fig. 13.8(b) appears to be well proportioned and is a good solution, in fact it may be considered optimum in that the triangles are as near to equilateral as is possible in a special sense which will be referred to later. The number of triangles and edges in the mesh are the same in both cases. This is quite general since it will always depend upon the number of bounding and internal nodes only. Although,

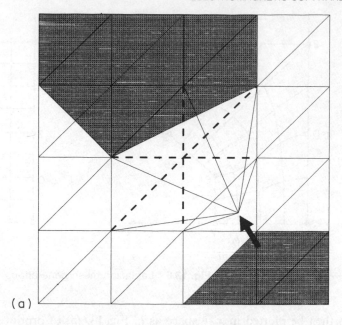

(a)

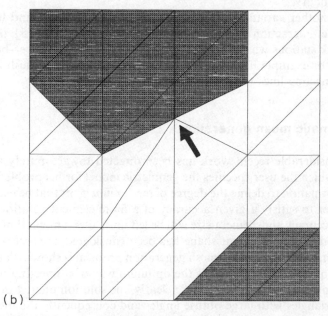

(b)

Fig. 13.7 Rubber-band interaction (a) Modifying a mesh by direct screen interaction, (b) the final position of the shifted node is on the region boundary.

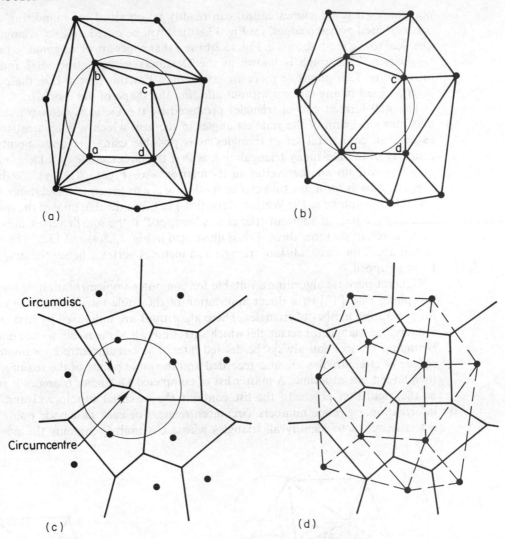

Fig. 13.8 Graphs constructed on a set of points. (a) Points forming a 'bad' mesh, (b) points forming a 'good' mesh, (c) Dirichlet tessellation, (d) Delaunay triangulation.

for the simple case illustrated the optimum mesh could be constructed by inspection, for most practical problems an algorithmic procedure is desirable. Several more or less equivalent algorithms have been developed and the literature is quite extensive [8]. For example, the covering of space by polygonal tiles is a classic problem solved by Dirichlet who constructed a tessellation which is based on the idea of nearest neighbours [9]. Thus for a finite set of given points the ith Dirichlet region is the set of all points in the plane that is closer to the ith point than to any other given point. In Fig. 13.8(c) the Dirichlet tessellation is shown and can easily be constructed geometrically by drawing the bisectors of line segments connecting the points, with the bisections connecting to form a set of non-overlapping polygons. A dual of the Dirichlet tessellation (or Voronoi polygon as it is sometimes called) can readily be obtained by connecting grid points across shared polygon edges, see Fig. 13.8(d). It will be noted that the triangles are now identical to the mesh shown in Fig. 13.8(b) and that edges are orthogonal to the Dirichlet tessellation. This graph is known as the Delaunay triangulation [10] and is unique provided no four points or more are co-circular. For this special case these points can be connected in any fashion without affecting the shape of the triangles.

The well formed grid of triangles produced by the Delaunay construction actually maximizes the sum of the smallest angles in the grid which is equivalent to achieving as near an equilateral set of triangles as is possible using the given points. Another property of the Delaunay triangulation is that the circumcircle of a Delaunay triangle may not contain another vertex in its interior—see Fig. 13.8(a) and 13.8(b) and this suggests a basis for a test that can be used in order to generate a Delaunay mesh. This procedure is known as the Watson algorithm [11], and the diagonal of the quadrilateral formed by a pair of adjacent triangles is 'swapped' if the fourth vertex falls inside the circumcircle of the other three. This is illustrated in Fig. 13.8(a) and 13.8(b) by the quadrilateral abcd; the circumdisk of triangle abd includes vertex c hence the diagonal needs to be swapped.

General purpose algorithms, suitable for computer implementation, using either the 'swapping rule' [12] or a direct application of the circle test [11] can be constructed for generating meshes of triangles. These algorithms are initialized by first establishing a bounding triangle (or rectangle) which surrounds all of the nodes to be added—those bounding vertices can always be deleted later. The circumcentre coordinates and the square of circumradius are also recorded and the node points of the required mesh are then added one at a time. A master list of completed Delaunay triangles is maintained as the algorithm proceeds; the list contains the essential topological and geometric information, e.g. node numbers, circumcentre etc. For each new node point added the list is searched to identify all triangles whose circumdisk contains the new point. It

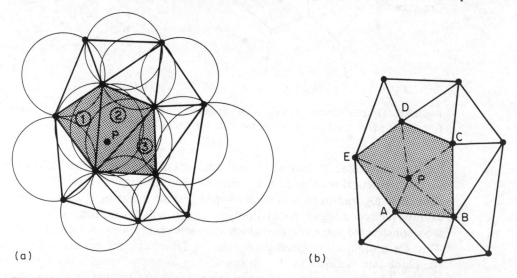

(a)   (b)

Fig. 13.9 Insertion of a new node in a Delaunay triangulation. (a) New node at point P inside three circumdisks, (b) insertion polygon ABCDE formed from three deleted triangles and new mesh completed.

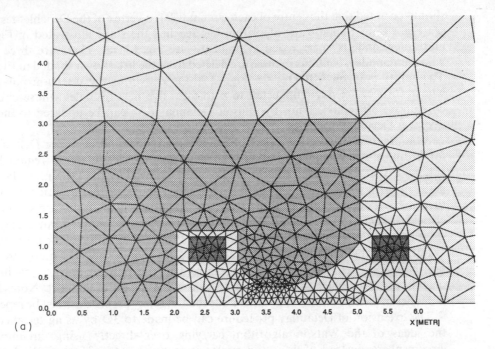

(a)

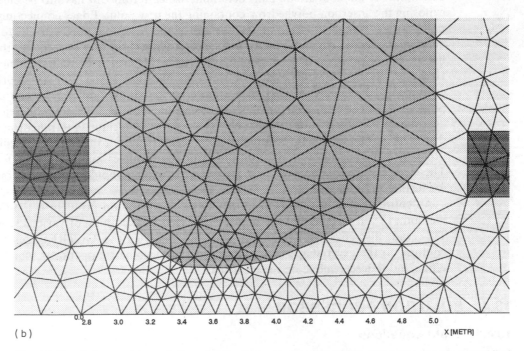

(b)

Fig. 13.10   Delaunay triangulation of a dipole electromagnet. (a) Discretization density controlled by specifying the subdivision along region boundaries with additional points added to ensure no triangle larger than a specified size, (b) mesh detail under the pole.

should be noted that the centre of each circumdisk is a vertex of the Dirichlet tessellation, see Fig. 13.8(c). Suppose the algorithm has reached the stage illustrated in Fig. 13.8(b) and a new node is to be added at point p—see Fig. 13.9(a). There are three triangles whose circumdisk contains p, these are deleted from the list, shown shaded in Fig. 13.9(a). The empty polygon ABCDE, see Fig. 13.9(b) is known as the insertion polygon [11] whose vertices are now connected to p to form five new triangles whose circumdisks contain no other nodes thus the master Delaunay disk can be updated to include the five new Delaunay triangles.

As a practical illustration Fig. 13.10 shows the model and resulting Delaunay mesh for a dipole electromagnet. Initially the boundary of each material region has been subdivided into a number of nodes by the user, followed by the Delaunay algorithm to generate a triangular mesh connecting all of those nodes, further additional nodes have then been added automatically by trisection to ensure all triangles are smaller than a critical dimension depending upon their distance from a prescribed boundary. At each stage the additional nodes would be checked by further application of the Delaunay test.

Further improvements may be also used to regularize the mesh by moving non-boundary defining nodes to the centroid of the polygon formed by its nearest neighbours. This polygonal region is known as the region of support. Note, however, if this lies in the circumdisk of another node the Delaunay procedure must be repeated.

The extension of Delaunay procedure can be made to 3D by using tetrahedra [11], the ideas of the Watson algorithm carrying over directly using circumballs and circumspheres instead of discs and circles respectively. At each stage of addition of a new node the union of circumballs determine which tetrahedra have to be deleted thus exposing the 'insertion polyhedron' containing the new point. Edges connecting the new point to all triangular facets of the surface of the insertion polyhedron are created, defining tetrahedra which fill the empty space. Adding these new elements to the master list as before produces a new Delaunay mesh in 3D which contains the new point. However, several problems exist with this generalization of the Watson algorithm, for example accumulated computer truncation error can cause ambiguity when an added point is very close to a circumsphere causing an incorrect decision on rejection or acceptance. A more serious problem occurs when a valid Delaunay tetrahedron becomes distorted; this arises when the fourth vertex becomes close to the plane containing the other three vertices despite the well-proportioned triangles making up the four faces. These 'slivers' can be removed and the interested reader should consult reference [11] for further information.

An entirely different approach to the 3D mesh generation problem is the OCTREE method [13], in this method the model is replaced by the union of various sized cubes. This crude discretization is then improved by modifying the 'OCTREE' description to conform to the exact shape of the object utilizing geometric operators which deform, refine and indeed remove octree elements.

13.4 SOLUTION PROCESSOR

13.4.1 Field equations

The heart of any electromagnetics software system is the solution processor, see Fig. 13.2. This processor must have all the necessary algorithms to discretize the governing field

Table 13.1 Summary of field of formulations

	Eddy current field formulations	
Method	Defining equation	Gauge
$\mathbf{H} - \phi$	$\nabla \times \nabla \times \mathbf{H} = -\mu\sigma(\partial\mathbf{H})/\partial t$	
$\mathbf{A} - V - \phi$	$\nabla \times (1/\mu)\nabla \times \mathbf{A} = -\sigma(\partial/(\partial t)\mathbf{A} + \nabla V)$	$\nabla \cdot \mathbf{A} = 0$
	$\nabla \cdot \sigma(\partial/(\partial t)\mathbf{A} + \nabla V) = 0$	$\nabla \cdot \mathbf{A} = -\mu\sigma V$
$\mathbf{A}^* - \phi$	$\nabla \times (1/\mu)\nabla \times \mathbf{A}^* = -\sigma(\partial\mathbf{A}^*)/\partial t$	$\nabla \cdot \sigma\mathbf{A}^* = 0$
$\mathbf{T} - \phi$	$\nabla \times (1/\sigma)\nabla \times \mathbf{T} = -\mu(\partial/\partial t)(\mathbf{T} - \nabla\phi)$	$\mathbf{T} \cdot \mathbf{u} = 0$
	$\nabla \cdot \mu(\mathbf{T} - \nabla\phi) = 0$	

Free space in all cases ϕ *satisfies* $\nabla^2\phi = 0$.
Conductors
    H Gauge implied by formulation
        Edge elements appropriate in conductor regions
        Nodal elements in free-space regions
  A − V Gauge needs to be applied, Lorentz or Coulomb?
    A* Gauge implied by the formulation
    T − φ Gauge applied (can be used to reduce the number of components)

equations (see Table 13.1), to generate the matrix coefficients and to solve the linear and non-linear algebraic equations.

There are many omissions in Table 13.1 since the criterion for inclusion was to include techniques that appear at the moment to be capable of some degree of generality. For example, in two dimensions robust procedures are widely available in computer codes [1]. Similarly 3D non-linear magnetostatics problems can be handled by robust codes with efficient pre- and post-processing [14]. The situation with 3D eddy currents is not as stable with only rather restricted functionality. However, substantial progress has been made not only with the techniques of finite elements etc. but also by applying special methods to specific problems, these often involve integral methods based on circuit analogues [15, 16, 17] and it is interesting that a sharper understanding of the connection between these approaches and the methods described here is emerging [18, 19]. One of the most significant events in recent years has been the series of international workshops on eddy current code validation which grew out of the fusion technology programme in the United States and sponsored by the Compumag conferences (TEAM) [20]. A number of bench-mark problems have been defined and solved by a mix of special and general methods. The results of the first twelve problems have now been published [20], with solutions for another group of more advanced problems underway. It is interesting that many researchers are now using these problems to help them in developing new methods as well as to validate existing computer codes.

Which method is best? It is not possible to give a definitive answer at this early stage of validation but the following points may be worth considering [1].

(a) The results from the eddy-current workshop so far have shown that good results were obtained for a wide range of methods. Both for the general methods discussed in this book and for special purpose methods with limited functionality.

(b) Special methods often give good results more economically, but general methods that allow a reduction in the number of components also can be efficient.
(c) The development of special methods needs expensive resources, and they often have no life beyond their original purpose. In order to minimize this investment developers are considering integrating special methods into general purpose pre- and post-processing environments.
(d) Apart from their intrinsic research interest, general methods are needed by designers who have not the resources or expertise to develop special techniques.

13.4.2 Methods for solving the algebraic system of equations

The discretization schemes (finite element, finite differences etc.) invariably reduce the governing equations to the canonical form,

$$K(u)u = Q, \tag{13.3}$$

where K is a matrix of coefficients depending upon the defining equation (integral or differential), discretization and numerical method (e.g. finite elements etc.). In general Q is a function of position and defines the sources (conductors, charges, etc.) and u is the dependent solution variable (potential, field, etc.). For non-linear problems the matrix K also depends upon u. There is a vast literature on this topic [21, 30] stimulated largely by the enormous demand made by engineers and scientists to solve large and complex problems. The computer's ability to solve large real problems is often limited by the efficiency of the linear-equation solution techniques as well as the intrinsic power of the computer used; this is particularly true for three-dimensional problems where a large escalation in the system matrix size is to be expected. By way of illustration, suppose that the geometry of a 2D problem can be sufficiently well modelled by 20×20 node points, e.g. a long dipole magnet where the end effects have been ignored. If now the analysis is to be extended into the third dimension with a similar degree of complexity the discretization will now require $20 \times 20 \times 20$ node points, i.e. an increase from 400 to 8000. Furthermore, if u is a vector quantity with three components, as in the eddy current case, the number of unknowns in the problem will increase to $8000 \times 3 = 24\,000$. Thus the need for an efficient solver is paramount.

Methods for solving eqn (13.3) can be classified into two types, direct methods and iterative methods; for example when solving integral equations the matrix K will be fully populated and Gaussian elimination is often used, this is a direct method and, to quote from the text by Strang (p. 265 [22]), 'is a perfect algorithm, except perhaps if the particular problem has special properties—as almost every problem has!' In fact with finite element solutions of differential equations, see Table 13.1, the matrix is sparse with a context dependent pattern, and although direct elimination methods can be successfully used, a self-correcting iterative method like Gauss–Seidel [23] may be more efficient.

In the remainder of this section a broad review of the more practical results of linear algebra will be given together with some further variations on iterative methods. For more details and mathematical rigour the interested reader is urged to consult the literature, for example the review by Jacobs [21] and the texts by Strang [22], Wait [24] and Varga [23].

13.4.3 Gaussian elimination revisited and remarks on conditioning

It will be realized that Gaussian elimination is a method of reducing an $n \times n$ system of linear equations to upper triangular form, see for example, section 9.4. The system of simultaneous linear equations, eqn (13.3) can be written in full as:

$$\left.\begin{array}{l} k_{11}u_1 + k_{12}u_2 + \cdots + k_{1n}u_n = q_1 \\ k_{21}u_1 + k_{22}u_2 + \cdots + k_{2n}u_n = q_2 \\ \vdots \qquad \vdots \qquad\qquad \vdots \qquad \vdots \\ k_{n1}u_1 + k_{n2}u_2 + \cdots + k_{nn}u_n = q_n \end{array}\right\} \tag{13.4}$$

and for Gaussian elimination when the coefficient k_{11} is non-zero a multiple k_{i1}/k_{11} of the first equation is subtracted from the ith equation for $i = 2,\ldots,n$, with the same operations on the right-hand sides. The effect is to eliminate u_i, from all equations below the first, so that a coefficient matrix is obtained with zeros in the first column below k_{11} and all elements from the second row downwards, with the right-hand sides, are modified. The element k_{11} is called the pivot and the ratios $m_{i1} = k_{i1}/k_{11}$ for $i = 2, 3, \ldots, n$ are called the row multipliers.

This procedure can be repeated with the diagonal element in the second equation of the modified matrix as pivot and if this is non-zero u_2 can then be eliminated from eqns $3,\ldots,n$. This process is further repeated until u_{n-1} is eliminated from the nth equation at which point an upper triangular system is obtained. The full solution is then recovered by a final back substitution. Thus Gaussian elimination is a direct method which terminates after a fixed number of steps so it is possible to calculate the number of arithmetic operations or 'operation count' for the whole process. It is shown in the literature (Strang [22], p. 5) that this number is given by $1/3n^3$ for the elimination and $n(n-1)/2$ for the back substitution. Thus most of the work performed is in the elimination process, overwhelmingly so, e.g. for a system of 100×100 this amounts to over 98% of the arithmetic.

Not all matrices **K** of course will lead to a unique solution of eqn(13.4), only non-singular matrices in which

$$\mathbf{KK}^{-1} = \mathbf{I}. \tag{13.5}$$

In fact, the Gaussian elimination method will break down if the matrix is singular and it is a strength of the method that singular matrices can be detected to prevent catastrophic effects. Clearly if a particular pivot is zero then the multiplier cannot be formed and the elimination cannot proceed; however if, in this case, there is a lower row with a non-zero element in the same column position, i.e. if k_{ii} is zero but there is an element k_{li} non-zero for $l > i$, then equations i and l may be interchanged and k_{li} may be used as a pivot, on the other hand if k_{ii} is zero and all elements in the ith column below are zero then no possible pivot can be found. In this case interchanging of columns would only postpone the problem to a later stage so the method has broken down because the matrix **K** is singular. Even though an upper triangular system **U** could be obtained the (i, i)th position of the matrix **U** would be zero and the back substitution would then fail on division by zero. In fact a pivot with the exact value zero is unlikely to arise in practice with real numbers because of rounding error but the occurrence of a pivot which is close to zero may cause difficulties in finite precision arithmetic.

It is important to prevent large growth in the modified matrix elements and an effective way of doing this is to choose as pivot the largest element in the pivotal sub-column. Thus at the ith elimination stage the lower rows are scanned to find the element of largest absolute value; this is then brought up to the pivotal position by interchange of equations. The multipliers calculated from this pivot necessarily satisfy $|m_{li}| \leqslant 1$ for $l = i + 1, \ldots, n$. This is the strategy of partial pivoting. It is also a good idea to perform some form of scaling on the original matrix if the coefficients vary considerably in size but care is needed not to corrupt the partial pivoting strategy. A reasonable compromise is to use row equilibration where each row of the original matrix is scaled so that

$$\max_{j} |k_{ij}| = 1. \tag{13.6}$$

The Gaussian elimination method is equivalent to obtaining triangular factors of the matrix $\mathbf{K}$; so if $\mathbf{K}$ is non-singular it can be written $\mathbf{LU} = \mathbf{K}$, where $\mathbf{U}$ is the upper triangular matrix formed by the elimination process. The lower triangular matrix $\mathbf{L}$ has each element set to unity on the diagonal and below the diagonal they are the multipliers m_{ij}. It should be noted that the main diagonal entries of $\mathbf{U}$ are the pivots.

In the case of a symmetric matrix $\mathbf{K}$, e.g. the system matrix from a finite element discretization, other important considerations may be helpful. If, for example, a symmetric matrix has the property

$$\mathbf{u}^T \mathbf{K} \mathbf{u} > 0 \tag{13.7}$$

for all $u \neq 0$ then $\mathbf{K}$ is said to be *positive definite* then it can be shown that $\mathbf{K}$ is non-singular and can be factorized into the form

$$\mathbf{K} = \mathbf{LL}^T, \tag{13.8}$$

where $\mathbf{L}$ is a lower triangular matrix with positive diagonal elements [25]. In addition it can also be shown that pivoting is not required for positive definite matrices. A method based on this $\mathbf{LL}^T$ factorization is known as Cholesky decomposition and is widely used [25]; it is efficient, stable and easy to program.

After computing an approximate solution of eqn (13.4) $\mathbf{u}^{(i)}$, some or all the components of $\mathbf{u}^{(i)}$ contain errors usually in the last few significant digits. The residual error can be defined as

$$\mathbf{r}^{(i)} = \mathbf{q} - \mathbf{K} \mathbf{u}^{(i)}, \tag{13.9}$$

and in general will be non-zero. This error can be improved by a technique known as *iterative refinement* where an elimination scheme

$$\mathbf{u}^{(i)} = \mathbf{q} - \mathbf{K} \mathbf{u}^{(i)}, \tag{13.10}$$

is defined with $\delta \mathbf{u}$ the solution of

$$\mathbf{K} \delta \mathbf{u} = \mathbf{r}^{(i)}. \tag{13.11}$$

It should be noted that the elimination process is only done once to obtain the upper triangular matrix $\mathbf{U}$ and the successive rhs's (e.g. $\mathbf{r}^{(n)}$) can be computed by the cheap process of backward substitution. The algorithm is as follows:

$$\left. \begin{array}{r} \mathbf{r}^{(k)} = \mathbf{q} - \mathbf{K} \mathbf{u}^{(k)} \\ \delta \mathbf{u}^{(k)} = \mathbf{K}^{-1} \mathbf{r}^{(k)} \\ \mathbf{u}^{(k+1)} = \mathbf{u}^{(k)} + \delta \mathbf{u}^{(k)} \\ k = 1, 2 \ldots \end{array} \right\} \tag{13.12}$$

A difficulty in solving eqn (13.4) occurs when small perturbations in matrix coefficients or rhs sources can produce a large change in the solution. In these circumstances the equations are said to be *ill-conditioned* and usually means that the computed solution is likely to be a poor approximation to the true solution. Iterative refinement with a higher precision calculation of the residuals can virtually eliminate this computational error but the results may be misleading because of the inherent errors in the basic data. What is needed is a quantitative measure of how data errors can affect the solution and such a measure is provided by the condition number of a linear system. This is achieved using vector and matrix norms as follows.

The norm of a vector is defined as

$$|\mathbf{u}| = (\mathbf{u}^T, \mathbf{u}), \tag{13.13}$$

and for a matrix as

$$|\mathbf{A}| = \max_{u \neq 0} \frac{|\mathbf{A}\mathbf{u}|}{|\mathbf{u}|}. \tag{13.14}$$

A simpler norm based on the maximum value is also used, i.e.

$$|\mathbf{u}|_\infty = \max_{1 \leqslant i \leqslant n} |\mathbf{u}| \tag{13.15}$$

and

$$|\mathbf{A}|_\infty = \max_{1 \leqslant i \leqslant n} \sum_{j=1} |k_{ij}| \tag{13.16}$$

these non-negative real numbers provide a measure of the sizes of vectors and matrices. A simple expression for the approximate bound for the relative perturbation of the solution of eqn (13.4), $\delta\mathbf{u}$, induced by perturbations in the data $\delta\mathbf{K}$ and $\delta\mathbf{q}$ can be derived in terms of vector and matrix norms, i.e.

$$\frac{|\delta\mathbf{u}|}{|\mathbf{u}|} \leqslant |\mathbf{K}| \cdot |\mathbf{K}^{-1}| \left\{ \frac{|\delta\mathbf{q}|}{|\mathbf{q}|} + \frac{|\delta\mathbf{K}|}{|\mathbf{K}|} \right\}. \tag{13.17}$$

The factor on the right of the inequality,

$$\text{cond}(\mathbf{K}) = |\mathbf{K}| \cdot |\mathbf{K}^{-1}| \tag{13.18}$$

is called the condition number of $\mathbf{K}$. If cond($\mathbf{K}$) is small ($u \sim 1$), data and rounding errors will not affect the solutions using a stable method like Gaussian elimination unduly. If, on the other hand, cond($\mathbf{K}$) is large, data errors may be magnified and produce a large error in the solution irrespective of the computational method employed. Similarly it can be shown that if $\mathbf{u}_1$ is an incorrect solution, then

$$\frac{|\mathbf{u} - \mathbf{u}_1|}{|\mathbf{u}|} \leqslant \text{cond}(\mathbf{K}) \cdot \frac{|\mathbf{r}|}{|\mathbf{q}|}. \tag{13.19}$$

Although the bound on the error $\mathbf{u} - \mathbf{u}_1$ depends directly on the condition number it will also depend on the elementary rounding error of the computer, the relative precision ε. It is shown by Forsythe and Moler [26] that the relation between these quantities is very roughly

$$\text{cond}(\mathbf{K}) \sim \frac{1}{n\varepsilon} \cdot \frac{|\mathbf{u} - \mathbf{u}_1|}{|\mathbf{u}|}. \tag{13.20}$$

For the full matrices that arise in the solution of integral equations direct methods such as Gaussian elimination or 'Cholesky' are very effective and can be used with confidence provided due attention is paid to scaling, pivoting and error analysis, however, because of the $\sim n^3$ operation count the computing time will escalate and ultimately, of course, impose an upper limit to the size of problem that can be solved by the computer in question, note however, the very effective improvements that can be expected when parallel computers are used, see section 12.4. For example a boundary element integral equation with a 1000 nodes took approximately 30 minutes running on an IBM 360 195 serial computer [27] and a problem ten times as large would therefore take 500 hours! Also the number of storage locations increases prohibitively though in this case 'backing store' can be used to extend the range. It can be seen then that integral equation methods are limited to problems of small and medium scale ($n < 1000$), at least with current serial hardware; on the other hand differential methods like conventional finite elements, lead to sparse matrices where most of the elements are zero. In the latter case, provided the bandwidth (see section 9.4) is kept small, large systems can be analysed by direct methods and if compact storage methods are used, see Fig. 9.5, the operation count goes as $m^2 n$, m being the bandwidth. On the other hand for large sparse matrices iterative methods have many advantages.

13.4.4 Iterative methods

The important SOR iterative scheme based on the Gauss–Seidel method was presented in section 10.2.2. This scheme requires the specification of a relaxation factor, thus for Laplace equation in a square region, with a square mesh of m subdivisions as in a finite difference scheme, it can be shown that the optimum relaxation factor is given by

$$w_0 = \frac{2}{1 + \sin \pi/m}. \tag{13.21}$$

Unfortunately for practical problems no simple formula can be derived and so the optimum factor is found by heuristic methods. For fuller discussion the reader should consult the text by Varga [23]. Thus, although the SOR method will deal with the largest of problems the choice of relaxation factor to accelerate the convergence militates against its use to some extent.

Conjugate gradient method—CG

An attractive alternative technique for solving large sparse systems which does not involve problem dependent parameters is the conjugate gradient method (Reid, 1971) [24]. In what follows it is convenient to use the conventional notation

$$(\mathbf{x}, \mathbf{y}) = \mathbf{x}^T \mathbf{y}, \tag{13.22}$$

where $\mathbf{x}$ and $\mathbf{y}$ are column vectors respectively, also instead of using superscripts to denote iteration numbers *subscripts* will be used. The problem of solving the system of equations defined by eqn (13.4) is equivalent to minimizing the function $F(u)$ given by

$$F(u) = \sum_j \left(\sum k_{ij} u_j - q_i \right)^2 \tag{13.23}$$

since,

$$\frac{\partial F}{\partial u_j} = -2 \sum_i \left(\sum_j k_{ij} - q_i \right) k_{il} = 0 \tag{13.24}$$

for a minimum implies that

$$\left(\sum_j k_{ij} - q_i \right) = 0 \tag{13.25}$$

for all i. In terms of matrices eqn (13.23) may be written as

$$F = \mathbf{r}^T \mathbf{r} = (\mathbf{Ku} - \mathbf{q})^T(\mathbf{Ku} - \mathbf{q}) \tag{13.26}$$

and if instead of solving eqn (13.4) directly the solution vector u may be regarded as the point in multi-dimensional space where F attains its minimum. This point may be located by starting at some initial approximation u_D and taking a succession of steps in suitable directions which progressively reduces the value of $\mathbf{r}^T\mathbf{r}$. If $\mathbf{K}$ is positive definite then $(\mathbf{r}, \mathbf{Kr}) > 0$ and it is convenient to use a related function to minimize, namely

$$F = (\mathbf{r}, \mathbf{Kr}). \tag{13.27}$$

Since $\mathbf{K}^{-1}$ is also positive definite and hence $F > 0$ and the minimum value, $F(\mathbf{u}) = 0$, is attained as before when $\mathbf{r} = 0$. In the diagram in Fig. 13.11 the current approximation $\mathbf{u}_n$ with $F(\mathbf{u}_n) > 0$ is to be improved by finding the point on a line defining the current search direction $\mathbf{P}_n$.

The appropriate point will be the point that minimizes $F(\mathbf{u})$ along the line P_n. Now any point along the search line can be determined by using the vector equation of the line, i.e.

$$\mathbf{u} = \mathbf{u}_n + \alpha \mathbf{P}_n, \tag{13.28}$$

and the particular point on the line at which $F(\mathbf{u})$ is minimum is denoted by

$$\mathbf{u}^* = \mathbf{u}_n + \alpha^* \mathbf{P}_x. \tag{13.29}$$

The residual vector $\mathbf{r}$ at the point $\mathbf{u}$ in Fig. 13.11 can be expressed by using eqns (13.28) and (13.29),

$$\mathbf{r} = \mathbf{r}^* + \mathbf{K}(\alpha^* - \alpha)\mathbf{P} \tag{13.30}$$

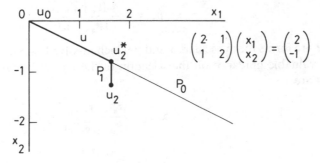

Fig. 13.11 Graphical illustration of the conjugate gradient method.

and hence

$$\mathbf{K}^{-1}\mathbf{r} = \mathbf{K}^{-1}\mathbf{r}^* + (\alpha^* - \alpha)\mathbf{P}. \tag{13.31}$$

Now in order to derive an expression for F, see eqn (13.27), it is necessary to post-multiply the transpose of eqn (13.30) by eqn (13.31), i.e.

$$F(\mathbf{u}) = F(\mathbf{u}^*) + 2(\alpha^* - \alpha)\mathbf{P}_n^T\mathbf{r}^* + (\alpha^* - \alpha)^2\mathbf{P}_n^T\mathbf{K}\mathbf{P}_n \tag{13.32}$$

for a symmetric matrix. The last term in eqn (13.32) is always positive since $\mathbf{K}$ is positive definite and if the second term is zero it follows that: $F(\mathbf{u}^*) < F(\mathbf{u})$. Thus provided

$$(\alpha^* - \alpha)(\mathbf{P}_n, \mathbf{r}^*) = 0 \tag{13.33}$$

for α the point $\mathbf{u}^*$ is the required minimum point along the line. This condition will be satisfied for $(\mathbf{P}_n, \mathbf{r}^*) = 0$, now

$$\mathbf{r}^* = \mathbf{q} - \mathbf{K}\mathbf{u}^* = \mathbf{q} - \mathbf{K}(\mathbf{u}_n + \alpha^*\mathbf{P}_n) = \mathbf{r}_n - \alpha^*\mathbf{K}\mathbf{P}_n \tag{13.34}$$

so from eqns (13.33) and (13.34) the unidirectional minimum is attained when

$$\alpha^* = \alpha_n = \frac{(\mathbf{P}_n, \mathbf{r}_n)}{(\mathbf{P}_n, \mathbf{K}\mathbf{P}_n)} \tag{13.35}$$

and a new approximation to the solution is given by

$$\mathbf{u}_{n+1} = \mathbf{u}_n + \alpha_n\mathbf{P}_n \tag{13.36}$$

with

$$\mathbf{r}_{n+1} = \mathbf{r}_n - \alpha_n\mathbf{K}\mathbf{P}_n. \tag{13.37}$$

It now remains to chose the search directions $\mathbf{P}_0$, $\mathbf{P}_1$, $\mathbf{P}_2$,... etc. It is usual to take $\mathbf{P}_0 = \mathbf{r}_0$, and for $n > 0$ to define $\mathbf{P}_n$ as a combination of the current residual vector $\mathbf{r}_n$ and previous direction $\mathbf{P}_{n-1}$, i.e.

$$\mathbf{P}_n = \mathbf{r}_n + \beta_{n-1}\mathbf{P}_{n-1}. \tag{13.38}$$

In the *conjugate gradient* method the scalar parameter β is chosen so that $\mathbf{P}_n$ and $\mathbf{P}_{n-1}$ are conjugate with respect to the matrix $\mathbf{K}$, i.e.

$$(\mathbf{P}_n, \mathbf{K}\mathbf{P}_{n-1}) = 0 \tag{13.39}$$

hence from eqns (13.38) and (13.40)

$$\beta_{n-1} = -\frac{(\mathbf{r}_n, \mathbf{K}\mathbf{P}_{n-1})}{(\mathbf{P}_{n-1}, \mathbf{K}\mathbf{P}_n)}. \tag{13.40}$$

There are forms for the scalars α and β which involve less computation, Reid (1971) [24] gives valuable guidance on the selection of these and discusses errors. The alternative forms are

$$\alpha_{n-1} = \frac{(\mathbf{r}_{n-1}, \mathbf{r}_{n-1})}{(\mathbf{P}_{n-1}, \mathbf{K}\mathbf{P}_n)} \tag{13.41}$$

$$\beta_{n-1} = \frac{(\mathbf{r}_n, \mathbf{r}_n)}{(\mathbf{r}_{n-1}, \mathbf{r}_{n-1})}. \tag{13.42}$$

The reason for the particular choice of search directions, eqn (13.40), is that theoretically, this ensures convergence on a finite number of steps (equal to the number of equations). However in practice the CG method is not direct because of rounding errors and the method is regarded as iterative for which a terminating criterion is needed. Because of a cumulated rounding error the residual calculated from eqn (13.37) will be slightly different from the true residual calculated by eqn (13.30). A measure of this difference is given by

$$d = |\mathbf{q} - \mathbf{K}\mathbf{u}_n - \mathbf{r}_n|^2, \tag{13.43}$$

where d is a squared norm which will increase with n (number of iterations) due to rounding errors; the computed residual however, $|\mathbf{r}_n|^2$, decreases gradually as the solution is approached. When $d \sim |\mathbf{r}_n|^2$ the rounding error level in $\mathbf{r}_n$ has been reached and the iteration should be terminated.

Preconditioning and ICCG method

In recent years the CG method has been very successful in solving electromagnetics problems, particularly for both two and three-dimensional static fields for which the $\mathbf{K}$ matrix is both positive definite and symmetric. This success has largely come about by an improvement to the basic CG method which is known as preconditioning [21]. For example, the system $\mathbf{K}\mathbf{u} = \mathbf{q}$ can be transformed to $\mathbf{H}v = \mathbf{p}$ by using an approximate Cholesky factorization, $\mathbf{K} \sim \mathbf{L}\mathbf{L}^T$, with,

$$\mathbf{v} = \mathbf{L}^T\mathbf{u} \tag{13.44}$$

$$\mathbf{p} = \mathbf{L}^{-1}\mathbf{q} \tag{13.45}$$

and,

$$\mathbf{L}^{-1}\mathbf{K}(\mathbf{L}^T)^{-1}\mathbf{v} = \mathbf{p}. \tag{13.46}$$

If the lower triangular matrix $\mathbf{L}$ has the same sparsity pattern as the lower triangular part of $\mathbf{K}$ it is found that the computational effort to solve the transformed system, eqn (13.46), is considerably less than for the original system [21]. This incomplete Cholesky preconditioning with conjugate gradients is known as ICCG [28].

13.4.5 A comparison of methods

The relative merits of direct and iterative methods for a standard problem are shown in Table 13.2.

The comparisons shown in Table 13.2 should only be used as a rought guide; the superiority of ICCG for systems that are symmetric and positive definite on the one hand and where there has been no special ordering of elements to minimize bandwidth on the other is clear. However, other methods can also be competitive if attention is given to node ordering strategies, e.g. in nested disection [29] or alternatively to superposition of meshes as in multi-grid [30]. The choice of methods for solving the fully populated systems encountered in integral equations would appear to be severely limited since, in general, the matrix has very little structure to exploit and therefore a direct Gaussian elimination method, with operation count $\sim n^3$, would seem optimal;

Table 13.2 Comparison of methods for sparse systems (times in arbitary units).

n[1]	m[2]	GE[3]	GS[4]	SOR[5]	$ICCG$[6]
36	8	1	2	1	1
121	13	3	30	11	5
256	18	12	175	41	13
441	23	33	619	123	28
676	28	80	—	268	49
961	33	152	—	536	70
2025	—	—	—	—	192

Notes
1. The timing tests are for solutions of the Poisson equation over a unit square [23]. A uniform mesh of first order triangles with n points was used to discretize the problem.
2. The system matrix bandwidth is related to the edge subdivision by $m = s + 2$.
3. A symmetric band solver using Gaussian elimination. The operation count is proportional to $m^2 n$ and the results show a good fit to this dependence.
4. The basic Gauss–Seidel method needs far too many iterations to be competitive.
5. SOR decreases the number of iterations but finding an optimum value or the relaxation factor is difficult. This method is widely used for finite difference discretizations with tri-diagonal systems, where the matrix coefficients are trivial and can be cheaply calculated *in-situ*, this approach is also suited to non-linear problems since the latest solution vector can be used as a starting value for the next non-linear iteration. The operation count goes as $m \times n \times I$, where I is the number of iterations, the results in the table were obtained after converging to 1×10^{-5}.
6. The pre-conditioned conjugate gradient methods are very popular and the results here suggest why—the operation count goes approximately $n \log n$ and is largely independent of bandwidth!

however, as the literature shows, simple iterative methods can be very effective for cases when the matrix is full but is also diagonally dominant [31, 32]. Finally, it must be understood, that the results given here apply only to serial computers and it is to be expected that for some algorithms which have intrinsic parallelism, e.g. integral equations, the use of concurrent computers will change the conclusions outlined here significantly [33].

13.5 ERROR ESTIMATION

Up to now the important subject of error estimation in finite element analysis has been largely ignored and it is probably true to say it has very largely been neglected by industrial practitioners. In fact large design departments use specialists in finite element or other forms of analysis because experience is essential when reliable results are required within tight time schedules. The errors in approximate numerical solutions are due to three primary causes, truncation or discretization errors, round-off error in precision and approximations in the mathematical model of the real problem. Discretization error, only, is considered here. It is due to the incomplete satisfaction of the governing equations and their boundary conditions, and is introduced by the trial function approximation.

Translation of the physical problem into a computational model has been made much easier by the development of sophisticated interactive pre-processors for preparing geometric and mesh data. With these tools design engineers can quickly become proficient at the creation of finite element grids for example, and their existing application experience provides the engineering science background needed to appreciate the physical modelling. However, the quality of the results obtained from existing finite element software is critically dependent on the user defined models and assessment of the likely error is partly by experience and partly heuristic.

The traditional method of assessing accuracy is to re-solve the problem with increased discretization or alternatively by using higher order finite elements. A rough rule of thumb is that when the dimensions of all the finite elements are halved in size the errors with low order elements will be reduced by approximately a factor of four. Unfortunately the computer costs of carrying out this type of checking quickly become prohibitive, especially with three-dimensional problems. Invariably analysts must resort to local refinement of particular portions of the model. This is time consuming and relatively unreliable since judgement has to be used to decrease the number of possibilities.

Error estimation methods for finite element procedures have been an active area of research. Two basic approaches have been reported [34, 35], the first based on mathematical analysis of the difference between the discrete approximation and the continuum equation it supports, the second using a combination of solutions that by their nature have the property of providing upper and lower bounds for system quantities such as energy. Some of these methods are briefly described in the next section as well as a 'cheap' alternative base on interpolation theory [36].

13.5.1 Error estimators for finite element calculations

Most static problems in electromagnetics are defined by the Poisson equation. For example the following two forms have been frequently used:

$$\nabla \cdot \frac{1}{\mu} \nabla A_z = -J_z, \tag{13.47}$$

for plane two-dimensional problems, and

$$\nabla \cdot \mu \nabla \phi = -\nabla \cdot (\mu H_s) \tag{13.48}$$

$$\nabla \cdot \mu \nabla \psi = 0 \tag{13.49}$$

for three-dimensional problems.

The current density J_s and conductor field H_S are known sources and μ the material permeability. Equation (13.47), in terms of the vector potential $\mathbf{A}$, is usually known as the standard formulation and eqns (13.48) and (13.49), in terms of the reduced scalar potential ϕ and total scalar potential ψ, is known as the complementary formulation. The ideas of energy functionals are closely related to the variational method and provide an alternative approach to field computation. It has been shown for instance that numerical solutions of eqn (13.47) give an upper bound and eqns (13.48) and (13.49) a lower bound [35]; a result that has been very effectively used for error estimation.

For linear problems eqns (13.47), (13.48) and (13.49) can be written in generic form,

$$Lu = g, \tag{13.50}$$

where L is a linear differential operator. In the finite element method u is locally approximated by,

$$\hat{u} = \sum N_i a_i, \tag{13.51}$$

where the N_i are the element shape functions and the a_i are unknown parameters to be determined. The shape functions are usually low order polynomials in x and y and the parameter a_i the approximate solution at the element nodes.

The local error in a computed solution at some level of mesh refinement can be expressed formally by;

$$\mathbf{e} = \mathbf{u} - \hat{\mathbf{u}}. \tag{13.52}$$

Unfortunately, eqn (13.52) requires the exact solution $\mathbf{u}$ and in any case gives no information on how well the original equation, eqn (13.50), is satisfied. For this reason it is usual to introduce a global measure defined as the energy in the error, the so called energy norm [34],

$$|\mathbf{E}|^2 = \int_\Omega \mathbf{e}^T L \mathbf{e} \, d\Omega \tag{13.53}$$

which can easily be related to the domain residual,

$$\mathbf{R} = L\hat{u} - \mathbf{g} \tag{13.54}$$

which indicates how well the original equation is satisfied by the appropriate solution [37]. Thus from eqns (13.52), (13.53) and (13.54), becomes

$$|\mathbf{E}|^2 = \int_\Omega \mathbf{e}^T \mathbf{R} \, d\Omega. \tag{13.55}$$

The domain residual has two parts; first, the contribution arising from the distributed residual interior to each element and secondly, the step contribution arising at element interfaces since the approximation eqn (13.51) does not allow derivative continuity. In fact this contribution can be used as an error estimate in itself [38] by calculating the step change in normal flux in crossing an element edge. Unfortunately the local error $\mathbf{e}$ can only be estimated by additional computation, for example in one technique a local approximation to the error at a node is made by solving an auxiliary problem satisfying eqn (13.50) over the local region of support, i.e.

$$|\mathbf{E}|^2 = \sum_i \int_{\Omega_i} |\nabla(\mathbf{u} - \mathbf{u}_s|^2 \, d\Omega. \tag{13.56}$$

where u_s is the auxiliary solution and the domain Ω_i only includes those elements sharing node i, see Fig. 13.12. This error norm has been worked out for a number of problems and element type and several formulae are available in the literature in which dependence on the characteristic parameter h defining the element size are given—error estimators in this class are known as h type.

An alternative approach is to estimate $\mathbf{e}$ by noting that as the element size h decreases

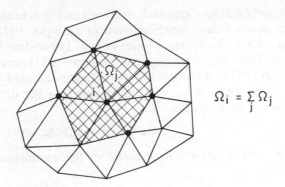

$$\Omega_i = \sum_j \Omega_j$$

Fig. 13.12 Finite elements in the region of support.

to zero, the error in the polynomial approximation of an arbitrary function tends in the limit to be one order higher than the approximation function itself [37]. Thus if a hierarchical addition of a single shape function is added to eqn (13.51) the error in the vicinity of the additional term is given by:

$$e \sim N_{n+1} a_{n+1} \tag{13.57}$$

an approximation to a_{n+1} can be determined using a weighted residual method and the energy norm is calculated as,

$$|\mathbf{E}|^2 = \mathbf{K}_{n+1,n+1}^{-1} \left[\int_\Omega N_{n+1} \mathbf{R} \, d\Omega \right]^2, \tag{13.58}$$

where $\mathbf{K}$ is the additional sub-matrix. The literature contains many examples of the hierarchical shape function method, and when used as a method of refinement it is known as the p method, as opposed to the h method above, and can lead to a faster rate of convergence [37]. See Fig. 13.13 for an illustration of hierarchical shape functions.

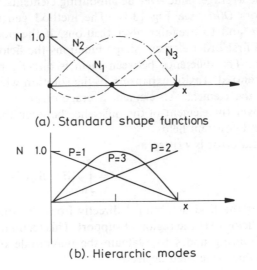

(a). Standard shape functions

(b). Hierarchic modes

Fig. 13.13 One-dimensional quadratic shape functions.

An entirely different approach to error analysis is to use the concept of standard and complementary forms already referred to, i.e. eqn (13.47) which gives an upper bound and eqn (13.48) which gives a lower bound. These ideas have been successfully applied to the solution of electromagnetic problems [35] and to error analysis and mesh adaptation [39] in which by solving both the standard and complementary problems for the same mesh an error estimate based on the difference between the two field solutions can be postulated, i.e.

$$|e|^2 = (\mathbf{B}_C - \mathbf{B}_S) \cdot (\mathbf{B}_S - \mathbf{B}_C), \tag{13.59}$$

where $\mathbf{B}_C$ and $\mathbf{B}_S$ are the local fields for the complementary and standard problems respectively.

13.5.2 Local error estimation using interpolation theory

All of the methods introduced so far require additional computation to a greater or lesser extent. However, an alternative approach is possible which, whilst retaining the basic idea embodied in eqn (13.55), avoids the solution or the estimation of an auxiliary problem [36].

The solution to eqn (13.47) using first order finite elements with linear shape functions will have nodal potential errors given by,

$$e_i = O(h^2) \tag{13.60}$$

and these nodal values will be exact if the real solution is a polynomial with terms less than cubic. As is well known, taking the derivatives of potentials from the shape functions is a poor method of evaluating the field since this leads to constant field values. This gives errors in the fields $O(h)$. Interpolation theory suggests that the most reasonable approach would be to use, what in one dimension turns out to be central differences, to calculate the potential derivatives at the nodes, or by the equivalent procedure of taking the average value over neighbouring elements. This gives nodal derivatives that have errors $O(h^2)$, see Fig. 13.14. The method generalizes to more than one space dimension and to irregular discretizations. If the nodal field values are interpolated using the first order element shape functions the field and potential solutions are not consistent. The difference between them is directly related to the cubic terms in an element centred. Taylor expansion of the solution which uses the next level neighbour nodes of the element. An element average error can be evaluated by computing the integral over the element of the difference between the nodally interpolated fields and the element constant fields.

The local error is written as

$$e = \int_{\Omega_i} (B - B_s) \, d\Omega, \tag{13.61}$$

where B is the field determined directly from the shape function and B_s the nodally averaged fields over the region of support. This technique uses the information provided by neighbouring nodes to evaluate the magnitude of the higher order terms in the solutions that have been neglected.

The following two problems demonstrate this error estimation method [36]. The first

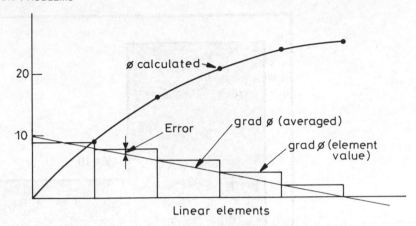

Fig. 13.14 One-dimension interpolation theory.

is a simple potential problem, the L-shaped section shown in Fig. 13.15(a). This problem is relevant in this context because it exhibits a singular solution at the corners. Figure 13.15(b) shows the actual error and the error computed from the estimator along the line joining the internal and external corners of the L-shaped section. It should be observed that the estimator bounds the actual error.

The second example is a rotationally symmetric solenoid. The fields from the coil were computed by adaptive integration to an accuracy of 1 in 10^6 for comparison with the finite element solution. The coil's inside diameter was 100 mm and it had a 30 mm square cross-section. Figure 13.16 shows the error in the computed flux density at (10, 10) as a function of the discretization size, and compares this with the predicted error estimator for the solution (see the error bars).

This subject has received a significant amount of attention recently. The interested reader is recommended to study the work carried out by the group at the University of Geneva where several of the error estimators proposed in the literature are surveyed and compared as well some new procedures of their own [40].

13.6 OPEN BOUNDARY PROBLEMS

A major consideration in any electromagnetic field analysis is the placement of the far-field boundary. In many cases the natural boundary of a magnet is essentially at infinity although in practice the presence of remote objects and their possible effect on the field shape will have to be taken into account. Special care will be needed with all methods based on solving the differential form of the defining equations. A number of approaches to this problem have been described in the literature [41]:

- Terminating the field at a sufficiently large distance. This of course begs the question and at the very least will require several solutions in order to achieve confidence. A useful technique here is to solve two problems at each trial with Dirichlet and Neumann boundary conditions respectively in order to bound the solution.
- To use special finite element basis functions which have the correct asymptotic

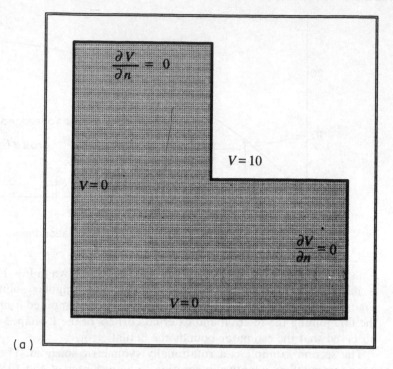

(a)

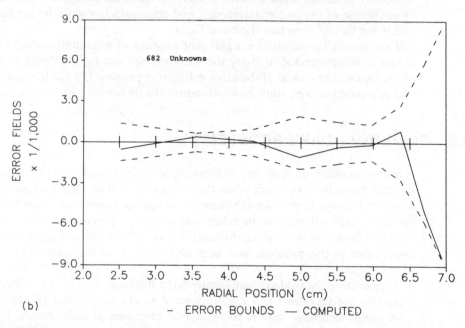

(b)

Fig. 13.15 L-shaped example for checking interpolation error estimator. (a) Mesh, (b) computed and actual field errors along the line (0,0) to (5,5).

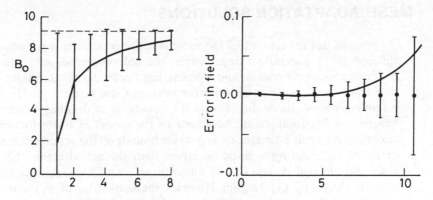

Fig. 13.16 Convergence and errors of the field in a short solenoid.

behaviour for large distances [42]. The major advantage of using special finite elements is that the matrix size and bandwidth are not significantly increased, but on the other hand some knowledge of how the field decays is needed in order to fix scaling parameters. This approach has been successfully implemented in a number of cases [41].

- A number of methods based on the idea of a super global element to model the exterior region. The method of recursive ballooning [43] in which the global element is generated iteratively by successively adding concentric rings of 'scaled elements' to embrance the finite element model. At each step the adjacent nodes between rings are removed so there are only nodes on the original and far boundaries. Boundary conditions at the far boundary can now be applied. This method has been applied very successfully to 2D geometries where it has been shown that the distance to the outer boundary increases gemetrically (e.g. after several iterations the distance is measured in light-years!). The method is not so effective in 3D.

  A major difficulty in these global element methods is that the matrix structure is strongly. affected and that the bandwidth will increase to a point where the matrix becomes essentially full and an advantage of the differential approach disappears.

- Mapping techniques have also been extensively used to transform the exterior infinite space to a finite space. The classical Kelvin transformation (see Kellogg [44], p. 232) is an example of this in which the transformation $rr' = a^2$, where r and r' are the inverse points with respect to a sphere of radius a. An inversion in a sphere is one-to-one except that the centre of the sphere of inversion has no corresponding point. The neighbourhood of the origin maps into a set of points at a large distance—into an infinite domain. This transformation has been used in a number of finite element systems to model the infinite domain in which the exterior space to a sphere surrounding the actual model is solved as an interior problem by means of the Kelvin transformation. The nodes of the two spaces, now bounded by spheres, are connected by forcing their solution values to be identical [45], [46], [47].

- The need for special methods is obviated totally if integral methods are used, see sections 9.8 and 12.4. This is, of course, one of the major advantages of this approach and can be recommended for small to medium sized problems or larger if computers with parallel architectures are available.

13.7   MESH ADAPTATION SOLUTIONS

As pointed out in section 13.2 the most effective procedure is to include the costly and difficult mesh generation stage within the solution processor. The development of automatic mesh generators, and efficient, fast linear algebra techniques makes this ideal achievable now that *a posteriori* error estimator can be derived. The development of a reliable error estimator then allows the possibility of deciding where additional nodes (degrees of freedom) should be added to the model in order to improve the overall accuracy. An error estimator must provide bounds on the actual error, and this estimated error, to be meaningful, must be larger than the actual error; that the interpolated estimator defined by eqn (13.61) for example, can satisfy this requirement for actual cases is shown by Fig. 13.15(c). However the knowledge of an error estimator in itself is not the only criterion for refinement of the mesh [37], it is also necessary if optional meshes are to be constructed to know whether the addition of a particular degree of freedom will actually reduce the error, such an index is known as a refinement or correction indicator.

To illustrate the process of a solver using adaptive mesh techniques some results for a 'tape head' magnet results, using a development code [2], are shown in Fig. 13.17. The three models show 'snap-shot' results of the mesh at various stages of adaptation. In this case the triangulations were constructed using the Delaunay method described in section 13.3.4 and the equations at each stage were solved using the conjugate gradient method, pre-conditioned by the incomplete Cholesky factorization (the ICCG method). The significant advantages of this latter method for adaption are: (a) that storage requirements only increase linearly with node count, (b) it is a reliable and stable solution scheme that does not require a banded matrix structure thus not needing expensive node renumbering and (c) it is an iterative technique and so it can use an approximate solution as a starting vector.

If the error estimator defined by eqn (13.61) is used to decide whether to add an additional node at a particular point, the adaption will not, in itself, identify possible orthogonality to the error in the extra degree of freedom, hence the mesh is not optimal [37] but it does achieve the requested precision in the field results.

13.8   POST-PROCESSING AND OPTIMIZATION

The post-processing (see section 13.2) element of the CAD system should allow the conversion of the numerical results of a problem solution (potentials) to useful and meaningful engineering quantities, e.g. at the lowest level fields, gradients, forces etc. up to design parameters like stored energy, inductance joule heat etc. and many others. The post-processor may, for example, involve smoothing techniques to refine the quality of the output data so that the gradient values required for forces can be reliably computed. The post-processor, as well as the pre-processor, is at the engineering interface and is a key element in the loop of the heuristic design procedure, see Fig. 13.1. The post-processor is, to a large extent, subject dependent unlike the pre-processor and solver which have a lot of commonality with other areas, e.g. thermal, fluid and stress analysis, all of these need to model the geometry of a particular object and are often described by Poisson-like equations.

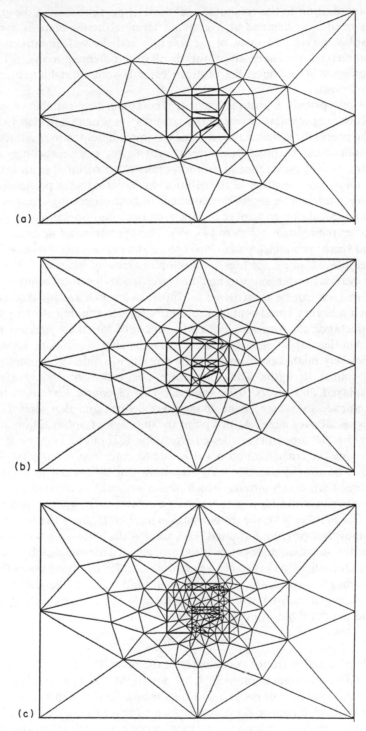

Fig. 13.17 Adaptive meshing of a 'tape head' magnet. (a) Starting mesh, (b) after two iterations, (c) after four iterations.

In addition to the computational tasks of post-processing the engineer needs to view his results on demand at a graphics terminal hence there is a need for facilities for display curves and maps of the solution and derived quantities. Indeed a number of commercial packages allow the display of potential maps, fields, forces along line segments, at this level the post-processor is somewhat limited and is often termed a post-viewer.

Conceptually a post-processor is a computer tool capable of manipulating the field solutions (potentials) in much the same way as a hand calculator manipulates numbers. The processor is interactive, it responds immediately to user requests. Ideally it includes colour raster graphics hardware to provide the engineer with graphical output in the form of field plots, zone plots and graphs. Also required is an interactive input device to allow interrogation of the solution by 'pointing' at a particular area of the display. A very interesting recent development of post-processing ideas has been to extend the 'hand calculator' concept to include vector calculus operations on general field quantities rather than single numbers [48, 49]. The operations of addition, multiplication, inner and vector product, gradient, divergence and curl are made available with 'programming facilities' allowing the user to develop 'macros' of his own. Thus point quantities can be evaluated over contours and areas which can then if necessary be integrated. A library of macros can be constructed for future use which are invoked by simple commands; such a library could contain procedures for determining stored energy, forces, torques, inductance, etc. An example of a post-processor with these facilities is shown in Fig. 13.18.

Another aspect of post-processing is to be able to compare the solutions for a number of closely related problems, each having slightly different geometry, material properties or source strengths. Thus the vector post-processor should allow differences to be displayed and errors to be assessed. This of course leads directly back to the input pre-processor where design changes are defined and the 'what if?' heuristic procedure begins all over again. At this point the question of optimization arises—mere trial and error is no substitute for design creativity. It is of course, the engineer's proper role to provide the creativity and not to waste too much time on 'what if?' experiments or, at least, he should consider if an algorithm can be devised in order to provide answers rather than 'knob turning' which misses or could never reach a solution in time—the number of states for a binary choice escalates as 2^n where n is a number of variables.

Put another way, the process of design often requires the determination of boundary values that produce a desired field, i.e. it is the solution to the inverse problem that is needed. As an example consider the problem of determining a boundary shape to produce a uniform field in the gap of an electromagnet. In the region under the pole, see Fig. 13.19, the magnetic field will, in particular, depend upon the geometry, material and current sources generally but will depend upon the four quantities $(\omega_1, \omega_2, d_1, d_2)$ the shim widths and depths. Thus at point P in Fig. 13.19

$$B_p = B(\omega_1, \omega_2, d_1, d_2, x, y), \tag{13.62}$$

where x and y are the coordinates of the point P.

The optimization problem is to determine the values of ω, d etc. that produce a constant value of B_p inside a domain. One way of doing this is to solve the field equations by finite elements or boundary elements for a range of values of ω etc. using a CAD system which includes pre- and post-processing facilities to speed up all the steps. To do this suppose only the two limiting values of each shim parameter are explored, i.e.

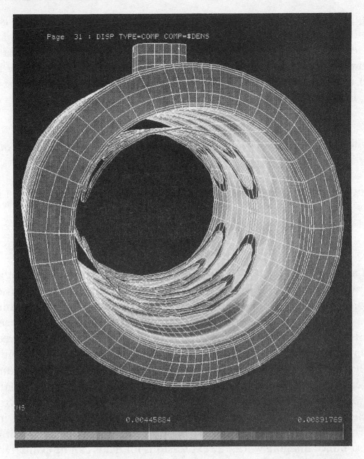

Fig. 13.18 An example of post-processing. Magnetic resonance imaging (MRI) 'body scanner' gradient coils, showing the current density induced on the inner surface of the cryostat.

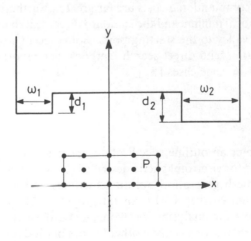

Fig. 13.19 An optimization of field correcting 'shims'.

eight states, thus there will be $2^8 = 256$ cases to run! This clearly is unfair because trends will be observed on the way and experience will eliminate many of the trials, nevertheless a lot of computation and time is to be expected. In cases like this automatic optimization should be considered; there are several methods available in the literature but in order to illustrate the ideas the well tried least squares method will be described.

Equation (13.62) may be written

$$B_p = B(\mathbf{g}, \mathbf{r}), \tag{13.63}$$

where $\mathbf{g}$ is a vector of geometric points to be determined, i.e. in the dipole magnet example,

$$\mathbf{g} = (\omega_1, \omega_2, d_1, d_2) \tag{13.64}$$

and $\mathbf{r}$ defines the field point. Suppose the useful field region in Fig. 13.19 can be represented by a rectangular grid containing n points then the sum

$$F(\mathbf{g}) = \sum_i |B(\mathbf{g}, \mathbf{r}_i) - B(\mathbf{g}, \mathbf{o}_i)|^2 \tag{13.65}$$

is a measure of the departure of the field from its value at the origin. Thus for a perfectly uniform field the objective function $F(\mathbf{g})$ in eqn (13.65) will be identically zero so the computational problem is to determine the vector $\mathbf{g}$ that minimizes $F(\mathbf{g})$ and this will be the best that can be done with the number of degrees of freedom in $\mathbf{g}$ available (in this case four). A better solution may be possible with more degrees of freedom, i.e. with perhaps additional shims. By using standard procedures for minimizing functions of several variables [50], providing, for any given set $\mathbf{g}$, B can be computed either by finite elements or boundary integrals (boundary integrals are particularly useful in this case), the design problem can be solved. Computer routines for minimizing functions are more efficient if derivatives $\partial F / \partial g$ are available; for many situations this is not a practical proposition; however, in the case of optimizing the shape and position of coils in magnetic resonance imaging work, the expressions for fields can be evaluated including derivatives thus speeding up the calculations [2]. In practice not all values of the vector $\mathbf{g}$ will be admissible this means that in general the function F has to be minimized subject to a number of constraints, fortunately there are a number of routines that allow constraints though the computational costs are far greater. Furthermore the objective function may have many local minima and the standard direct search methods [51] will only determine the nearest valley to the starting point. Some recent work, using stochastic procedures in combination with direct search methods, has reported that near-global optima can be obtained in some cases [52].

13.9 SUMMARY

In this chapter an outline specification has been given for the main components of a CAD system for electromagnetic design analysis. Particular attention has been given to automatic mesh generation, solution methods and error estimation. It is possible to provide design engineers with an integrated CAD system which should enrich their array of tools for analysing electromagnetic devices. Whether this will improve the quality of design depends upon other factors but it does offer rapid methods of checking out untried ideas and a further reduction in the cost of prototype testing. As hardware becomes more powerful the range and size of problem that can be handled will increase. Of particular interest are the new developments in parallel processors and transputers which will enable solutions of large problems [33] achievable today on large mainframes only, to be carried out on small work-stations within a few years. On the other hand, basic algorithm development in techniques for linear and non-linear algebra, optimization theory and discretization techniques will in due course extend the frontiers of computing to include complex coupled problems, involving the interaction of electromagnetic, thermal and stress phenomena. Automatic mesh generating and self-adaptive schemes will progress to the point where 3D calculations can be carried out to prescribed tolerances.

Other computational techniques, not covered here are beginning to have an impact; these include the incorporation of design rules into knowledge base systems and the use of declarative languages to establish solutions to design problems not amenable to classical analysis [53]. These considerations are becoming increasingly important as engineers begin to use the techniques of artificial intelligence to interrogate and make inferences by using the vast store of knowledge and past-experience data amassed on electrical devices of all types.

A final remark to offset this rather exciting prospect for the future is to say that the results of computer simulations will only be meaningful and robust if the physical modelling is well understood and appropriate to the problem in hand; and in the end this will depend upon the judgement and inventiveness of the engineer. True success will come if these qualities are not orthogonal.

REFERENCES

[1] G. Molinari, *Classification of Software Available and Criteria of Approach in Industrial application of Electromagnetic Computer Codes*, Kluwer Academic Publishers, Dordrecht (1990).

[2] Vector Fields Ltd, 24 Bankside, Kidlington, Oxford OX5 1JE, *TOSCA, GFUN, CARMEN, ELEKTRA, BIM2D, OPERA, and PE2D* (1990).

[3] Infolytica Corporation, Suite 430, 1500 Stanley Street, Montreal, Canada, H3A 1R3, *The MAGNET Program* (1988).

[4] O. C. Zienkiewicz and R. Taylor, *The Finite Element Method, 4th Edition, Volumes 1 and 2*. Maidenhead: McGraw-Hill (1991).

[5] J. Simkin and C. W. Trowbridge, Electromagnetics CAD using a single user machine, *IEEE Trans. on Magnetics*, **MAG-19**, 2655 (1983).

[6] M. Kruip, Oxford Instruments plc, Eynsham, Oxon, UK, Private Communication (1989).

[7] A. A. Winslow, Numerical solution of the quasi-linear poisson equation in a non-uniform triangular mesh, *J. Comput. Phys.*, **1**, 149 (1971).

[8] C. L. Lawson, Software for CI surface interpolation, *Mathematical Software III* (Ed. R. Rice), p. 161 (1977).

[9] G. L. Dirichlet, Uber die reduction der positiven quadratischen forman mit drei unbestimmten ganzen zahlen, *J. Reine Angew. Math.*, **40**(3), 209 (1850).

[10] B. Delaunay, Sur la sphere vide, *Izvestiya Akademii Nauk, USSR, Math. and Nat. Sci. Div, No 6*, p. 793 (1934).

[11] J. C. Cavendish *et al.*, An approach to automatic 3d fe mesh generation, *IJNME, 21*, p. 329 (1985).

[12] Z. Cendes *et al.*, Magnetic field computation using Delaunay triangulation and complementary finite element methods, *IEEE Trans. on Magnetics*, **MAG-19** (1983).

[13] M. S. Shephard, Automatic and adaptive mesh generation, *IEEE Trans. on Magnetics*, **MAG-21**(6), 2484 (1985).

[14] C. W. Trowbridge, Electromagnetic computing: the way ahead, *IEEE Trans. on Magnetics*, **MAG-24**, 13 (January 1988).

[15] J. A. M. Davidson and M. J. Balchin, 3-D field calculations by network methods, *IEEE Trans. on Magnetics*, **MAG-19** (November 1983).

[16] D. W. Weissenburger and U. R. Christenson, A network mesh method to calculate eddy currents on conducting surfaces, *IEEE Trans. on Magnetics*, **MAG-18** (March 1982).

[17] L. R. Turner and R. J. Lari, Applications and further developments of the eddy current program EDDYNET, *IEEE Trans. on Magnetics*, **MAG-18** (1982).

[18] A. Bossavit and J. C. Verite, The TRIFOU code: solving the 3D eddy currents problem by using **h** as state variable, *IEEE Trans. on Magnetics*, **MAG-19**, 2465 (November 1983).

[19] R. Albanese and G. Rubinacci, Integral formulation for 3D eddy current computation using edge elements, *IEE Proceedings*, **135**, 457 (September 1988).

[20] L. R. Turner *et al.*, Papers on bench-mark problems for the validation of eddy current codes, *COMPEL*, **7** (March June 1988) and **9** (September 1990).

[21] D. A. H. Jacobs, A review of recent developments in the solution of large systems of equations, *CERL RD/L/N/126/78* (1978).

[22] G. Strang, *Linear Algebra and its Applications*. New York: Academic Press (1976).

[23] R. S. Varga, *Matrix Iterative Analysis*. New Jersey: Prentice-Hall (1962).

[24] J. K. Reid, *On the Method of Conjugate Gradients for the Solution of Large Sparse Systems of Equations*, New York, Academic Press, (1971).

[25] D. A. H. Jacobs, Preconditioned conjugate gradient methods for solving systems of algebraic equations, *CERL RD/L/N/193/80* (1980).

[26] G. F. Forsythe and C. B. Moler, *Computer Solution of Linear Algebraic Systems*. Prentice-Hall (1967).

[27] A. Armstrong *et al.*, The solution of 3D magnetostatic problems using scalar potentials, in *Compumag Grenoble* (J. C. Sabbonadiere, ed.), (Grenoble, France), Laboratoire d'Electrotechnique, ENSEGP (1978).

[28] J. A. Meijerink and V. der Vorst, An iterative solution method for systems of which the coefficient matrix is a symmetric in matrix, *Maths Comp.*, **31**, 148 (1977).

[29] A. George, Nested disection of a regular finite element mesh, *SIAM J. Num. Anal.*, **10**, 345 (1980).

[30] I. S. Duff, *Sparse Matrices and their Uses*. New York: Academic Press (1981).

[31] W. R. Hodgkins and J. F. Waddington, The solution of 3-D induction heating problems using an integral equation method, *IEEE Trans. on Magnetics*, **MAG-18**(2), 431 (1982).

[32] J. Bettes, Iterative methods for fully populated systems of equations, Private Communication (1985).

[33] C. F. Bryant and C. W. Trowbridge, Specification of the boundary demonstrator using transputers, Tech. Rep., Vector Fields Ltd, 24 Bankside, Kidlington, Oxford OX5 1JE, (1988). Esprit (1051), ACCORD/WP5/VFL/DEL/001/291.88/CWT Project Report.

[34] I. Babuska and W. C. Rheinboldt, Error estimates for adaptive finite element computations, *SIAM, J. Numer. Anal.*, **15**(4) (1978).

[35] P. Hammond and J. Penman, Calculations of eddy current by dual energy methods, *Proc. IEE*, **125**(7) (1978).

[36] C. S. Biddlecombe, J. Simkin, and C. W. Trowbridge, Error analysis is finite element models of electromagnetic fields, *IEEE Trans., Intermag. Arizona* (1986).

[37] O. C. Zienkiewicz and A. W. Craig, Adaptive mesh refinement and a posteriori error estimation for the p version of the finite element method, *SIAM Workshop on Adaptive Computational Methods, Univ. of Maryland, USA* (1983).

[38] C. W. Trowbridge, Status of electromagnetic field computation, *Proc. 9th Magnet Technology Conference, Zurich, SIN*, p. 707 (1985).

[39] J. Penman and M. D. Grieve, An approach to mesh generation, *IEEE Trans. on Magnetics*, **MAG-21**(6) (1985).

[40] P. Girdinio *et al.*, Local error estimates for adaptive mesh refinement, *IEEE Trans. on Magnetics*, **MAG-24**, 299 (January 1988).

[41] C. R. I. Emson, Methods for the solution of open-boundary electromagnetic problems, *IEEE Proceedings, Pt A*, **135**, 152 (March 1988).

[42] P. Bettes, Infinite elements, *Int. J. Number. Methods Eng.*, 53–64 (1977).

[43] P. P. Silvester *et al.*, Exterior finite elements for 2-dimensional field problems with open boundaries, *IEEE Proceedings*, **118**, 1743–1747 (1971).

[44] O. D. Kellogg, *Foundations of Potential Theory*. New York: Dover Publications (1954).

[45] R. Albanese and G. Rubinacci, Solution of three dimensional eddy current problems by integral and differential methods, *IEEE Trans. on Magnetics*, **MAG-24**, 98 (January 1988).

[46] Q. Xiuying and N. Guangzeng, Electromagnetic field analysis in boundless space by finite element method, *IEEE Trans. on Magnetics*, **MAG-24**(1) (1988).

[47] E. M. Freeman and D. A. Lowther, An open boundary technique for axisymmetric and three dimensional magnetic and electric field problems, *IEEE Trans. on Magnetics*, **MAG-25**, 4135–4137 (1989).

[48] D. Lowther *et al.*, A stack configured vector calculator for electromagnetic field evaluation, *IEEE Trans. on Magnetics*, **MAG-18**(2) (1982).

[49] C. S. Biddlecombe and C. P. Riley, Post processing of 3-D electromagnetic field calculations, *IEEE Trans. on Magnetics*, **MAG-24** (1988).

[50] A. G. A. M. Armostrong *et al.*, Automated optimisation of magnet design using a boundary integral method, *IEEE Trans. on Magnetics*, **MAG-18**(2) (1982).

[51] R. Fletcher, *Practical Methods of Optimisation*, 2nd edition, John Wiley, Chichester (1991).

[52] J. Simkin, C. W. Trowbridge, Optimizing electromagnetic devices combining direct search methods with simulated annealing. *IEEE Transactions on Magnetics*, **MAG-28** (March 1992).

[53] C. Emson *et al.*, Towards an environment for the Integration of analysis and design, *Computer-Integrated Manufacturing Systems*, **3**(4) (November 1990).

APPENDIX 1

A1.1 POTENTIAL AND FLUX FUNCTIONS

To facilitate analysis, a *potential function* ψ is defined such that the *change* in this function between any two points is proportional to the *change in potential* between them. Its value at any point, with respect to some origin (of potential), is a direct measure of the value of the potential there and, in addition, a line joining points having the same value of potential function is an equipotential line.

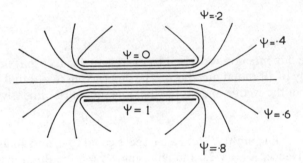

Consider the field of two charged conducting plates. Let $\psi = 0$ represent the value of the potential on one plate and $\psi = 1$ that on the other, so that there is unit difference of potential between the plates. Equipotential lines can be drawn in the space round the plates, representing $\psi = $ constant, for values of ψ between 0 and 1. For example, $\psi = 0.1$ represents a line joining points differing in potential from that of the lower potential plate by one tenth of the potential difference between the plates.

In a similar manner a *flux function* ϕ is defined such that $\phi = $ constant defines a flux line; and two lines $\phi = \phi_0$ and $\phi = \phi_0 + n$ have n units of flux passing between them.

From the above definitions it is seen that differences in flux and potential function represent quantities of flux and potential differences (the zeros of both functions being chosen arbitrarily). Lines drawn for constant values of potential and flux functions, so chosen that ψ and ϕ change in equal steps, form a field map in which the regions enclosed between the intersecting lines are curvilinear squares. In a uniform field these curvilinear squares become exactly square.

Since the potential and flux functions are orthogonal one function can be derived from the other by use of the equation

$$\left(\frac{dy}{dx}\right)_{\phi = \text{constant}} = -1 \Big/ \left(\frac{dy}{dx}\right)_{\psi = \text{constant}}.$$

However, the equations relating ψ and ϕ directly are most simply developed, not from the above equation or its equivalent in other coordinate forms, but by the use of complex variable theory.

As an example of the forms of flux and potential functions consideration is again given to the field of a line charge. It is shown that, in general, the difference in potential between the ends of a contour C is $\int_C E \, dl$, and so

$$\psi = K_1 \int_C E \, dl + K_2, \tag{A1.1}$$

where K_1 and K_2 are chosen arbitrarily. For a line charge $E = q/2\pi\varepsilon_0\varepsilon r$ and so

$$\psi = K_1 \int_C \frac{q \, dr}{2\pi\varepsilon_0\varepsilon r} + K_2 \tag{A1.2}$$

$$= \frac{K_1 q}{2\pi\varepsilon_0\varepsilon} \log r + K_2.$$

Thus choosing $K_1 = 1$ and $K_2 = 0$ simplicity gives

$$\psi = \frac{q}{2\pi\varepsilon_0\varepsilon} \log r.$$

The flux function is derived by expressing the fact that flux is emitted from the charge equally in all radial directions. The flux leaving in a radial wedge is proportional to the angle of the wedge and so the flux function varies linearly with θ, and has the form

$$\phi = K_3 \theta + K_4. \tag{A1.3}$$

The total flux emitted from a charge is equal to q, and so as θ changes by 2π, ϕ changes by q. Hence, $K_3 = q/2\pi$ and choosing $K_4 = 0$ the flux function may be written

$$\phi = \frac{q\theta}{2\pi}. \tag{A1.4}$$

It will be noted that the flux and potential functions for the field of two charged concentric circular conductors are the same as those for the field of a line large, both fields having circular equipotential lines and straight radial flux lines.

A1.2 THE MAGNETIC FIELD OF LINE CURRENTS

The field strength due to a line current acts tangentially to circles centred on the current, and the work done in moving a magnetic pole once round the current in any closed path is constant. The work done is the product of force and the distance moved, which is $2\pi r$ for a circular path for radius r centred on the current; hence the field strength, the force per unit pole, is given in magnitude by

$$H = \frac{i}{2\pi r}.$$

(A1.5)

The flux density B is given by

$$B = \frac{\mu\mu_0 i}{2\pi r},$$

(A1.6)

and, since it also acts tangentially, the magnetic flux lines for a line current are circles having the conductor as centre.

Flux function

From the last equation it is seen that the flux passing between two points at radii r_1 and r_2 is

$$\int_{r_1}^{r_2} \frac{\mu\mu_0 i}{2\pi r} \, dr,$$

which equals

$$\frac{\mu\mu_0 i}{2\pi} [\log r]_{r_1}^{r_2}.$$

The flux function is proportional to this, and is usually expressed in a form independent of $\mu\mu_0 i$ as

$$\phi = \frac{1}{2\pi} \log r.$$

(A1.7)

Magnetic potential of a line current

The gradient of the potential is in the direction the vector **B**, and so equipotential lines are radial. Therefore, because of the symmetry of the field, as a point moves round the current its potential changes in direct proportion to the change in its angular position θ with respect to the current. The change in potential for one complete revolution is i, and so the change in potential for movement through an angle θ is given by $\theta i/2\pi$. Because the potential changes continuously with rotation about the current, it is multi-valued at a point.

The field of a current-carrying conductor of circular section

To demonstrate the use of vector potential in a simple case, the field of an infinitely long straight conductor of relative permeability μ and of circular section, carrying current of uniform density J, is examined. This field has circular symmetry, and consequently the field vectors do not vary with movement at constant radius r about the centre of the conductor. Because of this it may be described without reference to $\mathbf{A}$, and this is first done.

When a unit pole is moved round a circle concentric with the conductor, the work done is equal to the current enclosed. Therefore, when the circle is inside the conductor, the field strength there, H_1, is expressed by

$$2\pi r H_1 = \pi r^2 J,$$

and so

$$H_1 = \frac{Jr}{2}. \tag{A1.8}$$

Again, when the circle is outside the conductor, the field strength there, H_2, is given by

$$2\pi r H_2 = \pi a^2 J,$$

a being the radius of the conductor, and so

$$H_2 = \frac{Ja^2}{2r}. \tag{A1.9}$$

Thus it is seen that the field strength is proportional to the radius inside the conductor and inversely proportional to the radius outside, where it is identical with that of a line current situated at the centre of the conductor.

To determine the form of the vector potential function describing this field, consider a rectangular path in an axial plane, having two sides at radius r and $(r + \delta r)$ parallel to the direction of the current, and two sides unit distance apart. By definition the line integral of $\mathbf{A}$ taken once round this rectangle is equal to the flux linking the rectangle. Thus as the vector potential is constant at any given radius, the change in the vector potential function from radius r to $(r + \delta r)$ is equal to $B\delta r$. Inside the conductor, where $A = A_1$, the flux density is, from eqn (A1.8) $\frac{1}{2}\mu\mu_0 Jr$, and so

$$dA_1 = \tfrac{1}{2}\mu\mu_0 Jr\,dr,$$

which, when integrated, gives

$$A_1 = \tfrac{1}{4}\mu\mu_0 Jr^2 + C_1.$$

Since only changes in vector potential are specified by curl $A = B$ the origin of A_1 is arbitrary and, for convenience, the constant C_1 can be made zero to give

$$A_1 = \tfrac{1}{4}\mu\mu_0 Jr^2. \tag{A1.10}$$

Outside the conductor, where $A = A_2$, the flux density is, from eqn (A1.9) $\mu_0 Ja^2/2r$, and so

$$dA_2 = \tfrac{1}{2}\mu_0 \frac{Ja^2}{r}\,dr,$$

which, when integrated, makes

$$A_2 = \tfrac{1}{2}\mu_0 Ja^2 \log r + C_2.$$

Now the vector potential is continuous across the boundary of the conductor and so, when $r = a$, $A_1 = A_2$. Hence

$$\tfrac{1}{4}\mu\mu_0 Ja^2 = \tfrac{1}{2}\mu_0 Ja^2 \log a + C_2,$$

giving

$$C_2 = \tfrac{1}{4}\mu_0 J a^2 (\mu - 2\log a).$$

Thus substituting for C_2 yields

$$A_2 = \tfrac{1}{4}\mu_0 J a^2 \left[\mu + 2\log\left(\frac{r}{a}\right) \right]. \tag{A1.11}$$

In this simple example the boundary is a flux line, and an expression for the field strength is simply obtained. In general, however, it is necessary first to obtain a solution in terms of vector potential, satisfying Poisson's equation and the boundary conditions (see section 4.1).

A1.3 CONJUGATE FUNCTIONS

The application of complex variable theory to the analysis of fields makes possible the simple solution of many problems which would otherwise be difficult or impossible. The real and imaginary parts of any continuous, regular[†] function of a complex variable are called *conjugate functions* and it is shown that both are solutions of Laplace's equation. Further, the flux and potential function for any field are conjugate functions, and they may be combined together in a single function of a complex variable and so handled with facility.

Laplace's equation

Consider a complex variable z defined by

$$z = x + jy. \tag{A1.12}$$

Let a complex variable t, defined by

$$t = u + jv, \tag{A1.13}$$

be any continuous regular function of z; that is, let

$$t = F(z). \tag{A1.14}$$

Differentiating this equation partially with respect to x gives

$$\frac{\partial t}{\partial x} = \frac{\partial F(z)}{\partial z}\frac{\partial z}{\partial x};$$

but

$$\frac{\partial z}{\partial x} = 1,$$

and so

$$\frac{\partial t}{\partial x} = F'(z). \tag{A1.15}$$

[†]A continuous single-valued function is regular if the partial derivatives $\partial u/\partial x$, $\partial v/\partial x$, $\partial u/\partial y$, $\partial v/\partial y$ exist, are continuous and satisfy the Cauchy–Riemann equations.

Further, differentiating this equation again with respect to x gives

$$\frac{\partial^2 t}{\partial x^2} = \frac{\partial}{\partial x} F'(z)$$

$$= \frac{\partial F'(z)}{\partial z} \frac{\partial z}{\partial x},$$

and so

$$\frac{\partial^2 t}{\partial x^2} = F''(z). \tag{A1.16}$$

Also, differentiating eqn (A1.14) partially with respect to y gives

$$\frac{\partial t}{\partial y} = F'(z) \frac{\partial z}{\partial y};$$

but

$$\frac{\partial z}{\partial y} = j,$$

and so

$$\frac{\partial t}{\partial y} = j F'(z). \tag{A1.17}$$

Differentiating again with respect to y gives

$$\frac{\partial^2 t}{\partial y^2} = \frac{\partial}{\partial y} j F'(z)$$

$$= j F''(z) \frac{\partial z}{\partial y},$$

and so

$$\frac{\partial^2 t}{\partial y^2} = - F''(z). \tag{A1.18}$$

Finally, combining eqns (A1.16) and (A1.18)[†] gives

$$\frac{\partial^2 t}{\partial x^2} + \frac{\partial^2 t}{\partial y^2} = 0. \tag{A1.19}$$

Thus any regular function of a complex variable obeys eqn (A1.19), which will be recognized as Laplace's equation.

Equation (A1.19) may be written as two equations, in terms of the real and imaginary parts of t, by noting that differentiation of eqn (A1.13) gives

$$\frac{\partial^2 t}{\partial x^2} = \frac{\partial^2 u}{\partial x^2} + j \frac{\partial^2 v}{\partial x^2}$$

and

$$\frac{\partial^2 t}{\partial y^2} = \frac{\partial^2 u}{\partial y^2} + j \frac{\partial^2 v}{\partial y^2}.$$

[†]This assumes that $F'(z)$ and $F''(z)$ are unique; that is, that $F(z)$ is regular.

The resulting equations are

$$\frac{\partial^2 u}{\partial x^2} + \frac{\partial^2 u}{\partial y^2} = 0 \qquad (A1.20)$$

and

$$\frac{\partial^2 v}{\partial x^2} + \frac{\partial^2 v}{\partial y^2} = 0, \qquad (A1.21)$$

and they shown that not only a function, but also the real and imaginary parts of any regular function of a complex variable obey Laplace's equation.

Cauchy–Riemann equations

Consider again eqn (A1.13). When differentiated with respect to x it gives

$$\frac{\partial t}{\partial x} = \frac{\partial u}{\partial x} + j\frac{\partial v}{\partial x},$$

and combining this with eqn (A1.15) gives

$$F'(z) = \frac{\partial u}{\partial x} + j\frac{\partial v}{\partial x}. \qquad (A1.22)$$

Similarly it is seen from eqns (A1.13) and (A1.17) that

$$jF'(z) = \frac{\partial u}{\partial y} + j\frac{\partial v}{\partial y}. \qquad (A1.23)$$

Then eliminating $F'(z)$ between eqns (A1.22) and (A1.23) yields

$$j\frac{\partial u}{\partial x} - \frac{\partial v}{\partial x} = \frac{\partial u}{\partial y} + j\frac{\partial v}{\partial y}, \qquad (A1.24)$$

and, since the real and imaginary parts on both sides of the equation must be equal, then

$$\frac{\partial u}{\partial x} = \frac{\partial v}{\partial y} \qquad (A1.25)$$

and

$$\frac{\partial u}{\partial y} = -\frac{\partial v}{\partial x}. \qquad (A1.26)$$

There are the Cauchy–Riemann equations and they are satisfied by the real and imaginary parts of any regular function of a complex variable.

Flux and potential functions as conjugate functions

A constant value of t defines two curves in the z-plane, $u(x, y) = $ constant and $v(x, y) = $ constant; and since t may take an infinite number of values these equations define two families of curves. Let $u = u_0$ and $v = v_0$ define one curve from each family and consider the intersection of the curves at point $z = z_0$. Taking first the curve for $u = u_0$, the slope is dy/dx and, since u remains constant as x and y change,

$$\frac{\partial u}{\partial x}\,\delta x + \frac{\partial u}{\partial y}\,\delta y = 0,$$

and, in the limit,

$$\left(\frac{dy}{dx}\right)_{z_0} = \left(\frac{-\partial u/\partial x}{\partial u/\partial y}\right)_{z_0}. \tag{A1.27}$$

Similarly, the slope of the curve $v = v_0$ is given by

$$\left(\frac{dy}{dx}\right)_{z_0} = \left(\frac{-\partial v/\partial x}{\partial v/\partial y}\right)_{z_0}. \tag{A1.28}$$

From eqns (A1.27) and (A1.28) the product of the slopes is

$$\left(\frac{\partial u}{\partial x}\bigg/\frac{\partial u}{\partial y}\right)_{z_0}\left(\frac{\partial v}{\partial x}\bigg/\frac{\partial v}{\partial y}\right)_{z_0},$$

which, from the Cauchy–Riemann equations (A1.25) and (A1.26), is equal to -1. Therefore the two curves intersect at right angles and this proves that families of curves corresponding to constant values of conjugate functions are orthogonal.

Because conjugate functions are solutions of Laplace's equation and are orthogonal functions they may be used to represent flux and potential functions. Consider the flux and potential functions describing a field in the (x, y) plane, and let

$$\varphi = f_1(x, y) = f_1(z) \quad \text{and} \quad \psi = f_2(x, y) = f_2(z).$$

Then, since these can be represented as conjugate functions, they may be combined together in a single function of a complex variable, $w(z)$, where

$$w(z) = f_1(z) + jf_2(z).$$

This function, w, is called the *complex potential function* and it is of fundamental importance in the use of complex variable theory for the solution of field problems; in terms of φ and ψ it is

$$w = \varphi + j\psi. \tag{A1.29}$$

It should be emphasized that all the above quantities are merely numbers having no dimensions. However, in a particular problem it is often convenient to choose a scale constant so that the value of either φ or ψ gives directly a quantity of flux or a value of potential difference. Note that φ and ψ can be interchanged in eqn (A1.29) when convenient.

It is pointed out earlier that φ and ψ can be derived from each other, and it is now evident that, being conjugate functions, they are related by the Cauchy–Riemann equations. Hence, from these equations, the relationships are

$$\frac{\partial \varphi}{\partial x} = \frac{\partial \psi}{\partial y} \tag{A1.30}$$

and

$$\frac{\partial \psi}{\partial x} = -\frac{\partial \varphi}{\partial y}. \tag{A1.31}$$

These are based on the expression of the field in cartesian coordinates, but they may be expressed equally in the forms appropriate to other coordinate systems. For example, in circular cylinder coordinates the equations are

$$\frac{\partial \varphi}{\partial r} = \frac{1}{r}\frac{\partial \psi}{\partial \theta} \tag{A1.32}$$

and

$$\frac{\partial \psi}{\partial r} = -\frac{1}{r}\frac{\partial \varphi}{\partial \theta}. \tag{A1.33}$$

Simple examples of the use of conjugate functions

To demonstrate the use of conjugate functions and the complex potential function, two fields treated earlier in terms of purely real functions are considered.

Charged concentric cylinders. It is shown earlier, that, for the field between charged concentric cylinders, the potential function is given by

$$\psi = \frac{q}{2\pi\varepsilon_0\varepsilon} \log r \tag{A1.34}$$

and the flux function by

$$\varphi = \frac{q}{2\pi} \theta. \tag{A1.35}$$

In order that φ and ψ can be combined to give the complex potential function w, it is necessary that they are expressed with appropriate scale factors by multiplying one of them by a constant factor; in this case it is convenient to multiply the potential function by $\varepsilon_0\varepsilon$ to give

$$\psi = \frac{q}{2\pi} \log r. \tag{A1.36}$$

The complex potential function then becomes [interchanging ψ and φ as compared with eqn (A1.29)]

$$w = \psi + j\varphi = \frac{q}{2\pi} (\log r + j\theta),$$

which may be expressed in terms of the complex variable t, where $t = r \exp(j\theta)$, as

$$w = \frac{q}{2\pi} \log t. \tag{A1.37}$$

This represents the field of two charged concentric cylinders (or of a line charge), centred about the origin of the t-plane, and substitution for the coordinates of a point in the t-plane gives the values of flux and potential function there, to a scale determined by the form of eqn (A1.37). This equation gives quantities of flux directly in the m.k.s. system of units, but to make the solution dimensionless it is convenient to use the form

$$w = \frac{1}{2\pi} \log t. \tag{A1.38}$$

The field map can be obtained by writing this equation in the form

$$t = \exp(2\pi\psi + 2\pi j\varphi),$$

and by substituting values of ψ and φ.

The field of a line current

To show the way in which the flux and potential functions may be derived from each other, the field of a line current in complex-variable form is considered. From eqn (A1.7) the flux function is

$$\varphi = \frac{1}{2\pi} \log r, \tag{A1.39}$$

and the Cauchy–Riemann equation (A1.32) may be rewritten

$$\psi = \int r \frac{\partial \varphi}{\partial r} \delta\theta.$$

Hence, differentiating eqn (A1.39) and substituting gives, in general,

$$\psi = \frac{1}{2\pi} \theta + f(r).$$

From symmetry, it is apparent that ψ is a function of θ only and so $f(r)$ is equal to an arbitrary constant, which may be ignored, and so the potential function becomes

$$\psi = \frac{1}{2\pi} \theta. \tag{A1.40}$$

φ and ψ can, of course, be combined to form

$$w = \frac{1}{2\pi} \log t, \tag{A1.41}$$

which is identical with eqn (A1.38). Therefore the complex potential functions for the fields of a line charge and a line current can be expressed by identical functions, for the fields of a line charge and a line current vary in identical ways, but with the flux and potential functions interchanged.

APPENDIX 2
THE FIELD INSIDE A HIGHLY PERMEABLE RECTANGULAR CONDUCTOR

Let the boundary of the rectangular conductor have sides $2a$ and $2b$, and let its centre be at the origin of the (x, y) plane. It is shown in Fig A2.1 together with a plot of flux lines inside the conductor for the case $b = 2a$. The solution for the field requires the determination of a vector potential function A which satisfies Poisson's eqn (4.17), and which has a constant (but unknown) value on the boundary of the rectangle (assuming $\mu = \infty$), that is, satisfying the boundary conditions,

$$B_x = \frac{\partial A}{\partial y} = 0, \quad \text{when } x = -a \text{ and } a$$

and

$$B_y = -\frac{\partial A}{\partial x} = 0, \quad \text{when } y = -b \text{ and } b.$$

(A2.1)

To form a suitable function A, which is a solution of Poisson's equation, the method is the same as that used in the solution of an inhomogeneous ordinary differential equation; it is formed as the sum of the solution to the corresponding homogeneous equation plus a suitable particular integral of the inhomogeneous one. In the present case of Poisson's equation the homogeneous equation is that of Laplace,

$$\frac{\partial^2 A}{\partial x^2} + \frac{\partial^2 A}{\partial y^2} = 0,$$

---

†There are many important fields the analyses of which are identical with those of the magnetic fields of infinitely permeable, current-carrying conductors. These include the following: most notably, the Saint Venant problem of torsion in a linear isotropic prism, the temperature distribution in a prism in which heat is generated uniformly and the boundary temperature is everywhere constant, the motion of a non-viscous liquid of uniform vorticity circulating in a fixed prism, and the lateral displacement of a uniformly loaded homogeneous membrane, supported on a horizontal contour by a uniform edge tension.

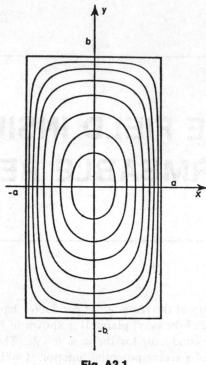

Fig. A2.1

which, from eqn (3.72), has the general solution

$$A = \sum_{m=1}^{\infty} (c_m \sin mx + d_m \cos mx)(g_m \sinh my + h_m \cosh my) + k_1 + k_2 x + k_3 y.$$

As, however, the desired solution for A inside the rectangle must be an even function of x and y, it is necessary to take only

$$A = \sum_{m=1}^{\infty} d_m \cos mx \cosh my. \tag{A2.2}$$

The particular integral of eqn (4.17) must also be an even function of x and y, and the simplest is

$$A = -\tfrac{1}{2}\mu\mu_0 J x^2. \tag{A2.3}$$

The complete solution of eqn (4.17) is then given by the sum of eqns (A2.2) and (A2.3) (adjusting the value of d_m) as

$$A = -\tfrac{1}{2}\mu\mu_0 J\left[x^2 + \sum_{m=1}^{\infty} d_m \cos mx \cosh my \right], \tag{A2.4}$$

in which the constants m and d_m are to be determined from the boundary conditions, eqn (A2.1). By differentiating eqn (A2.4) with respect to y, the first boundary condition gives

$$\sum_{m=1}^{\infty} d_m \sinh my \cos ma = 0,$$

which requires that

$$m = \frac{n\pi}{2a},$$

where n is an odd integer. Substituting for m in eqn (A2.4) and using the second boundary condition gives

$$2x - \sum_{n=1,3,\ldots}^{\infty} \frac{n\pi}{2a} d_m \cosh \frac{n\pi b}{2a} \sin \frac{n\pi x}{2a} = 0. \qquad (A2.5)$$

To obtain from this equation the values of the constants d_m it is necessary to expand $2x$ in a form directly equatable with the infinite series, i.e. as a Fourier series in terms of $\sin(n\pi x/2a)$ in the range $0 < x < a$. The nth coefficient C_n of this series is

$$C_n = \frac{2}{a} \int_0^a 2x \sin \frac{n\pi x}{2a} \, dx$$

$$= \frac{(-1)^k 16a}{(2k+1)^2 \pi^2}$$

and thus, from eqn (A2.5)

$$C_n = \frac{n\pi}{2a} d_m \cosh \frac{n\pi b}{2a}$$

or

$$d_m = \frac{(-1)^k 32a^2}{(2k+1)^3 \pi^3 \cosh(2k+1)\pi b/2a},$$

where k is any integer (including zero) and $m = 2k + 1$. Finally, therefore, substituting for d_m and m in eqn (A2.4), the vector potential function describing the field inside the conductor is

$$A = -\frac{1}{2} \mu \mu_0 J \left[x^2 + \sum_{k=0}^{\infty} \frac{(-1)^k 32a^2}{(2k+1)^3 \pi^3 \cosh(2k+1)\pi b/2a} \right.$$

$$\left. \times \cosh \frac{(2k+1)\pi y}{2a} \cos \frac{(2k+1)\pi x}{2a} \right]. \qquad (A2.6)$$

The term $(2k+1)^3$ in the denominator makes the series rapidly convergent (three or four terms give an accuracy much better than 1 per cent), and therefore convenient for computation. The expressions for B_x and B_y are simply found by the appropriate differentiations of eqn (A2.6). In the denominator they contain the term $(2k+1)^2$, and so they too are rapidly convergent. B_x has its maximum value at the middle of the sides $y = \pm b$ and B_y its maximum at the middle of the sides $x = \pm a$.

APPENDIX 3
TABLE OF TRANSFORMATIONS

This is a list of transformation equations and corresponding boundary shapes, which are of particular interest in the solution of electric and magnetic field problems. The transformations relate the planes of z and t, where

$$z = x + jy$$

and

$$t = u + jv.$$

In the case of transformations to the upper half-plane (which throughout is taken to be that of t) only the z-plane is shown, but the corresponding points of the t-plane are marked *inside* the mapped region.

Transformations to the unit circle can be obtained by combining those to the infinite straight line with the bilinear transformation.

A3.1 TRANSFORMATIONS TO THE UPPER HALF-PLANE

1.1 Circles

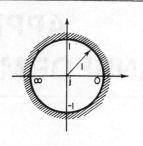

$$z = \frac{j - t}{j + t}.$$

(Exterior of circle on lower half-plane)

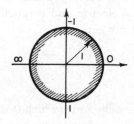

$$z = \frac{j + t}{j - t}.$$

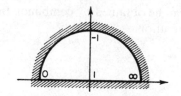

$$z = \frac{\sqrt{t} - 1}{\sqrt{t} + 1}.$$

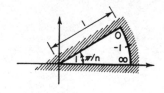

$$z = \left(\frac{\sqrt{t} - 1}{\sqrt{t} + 1} \right)^{1/n}$$

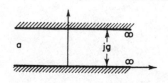

1.2 Straight-line segments

One defining vertex

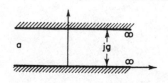

$$z = \frac{g}{\pi} \log (t - a).$$

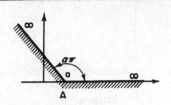

$$z + A = (t - a)^{\alpha}, \quad \alpha \neq 0.$$

Two defining vertices

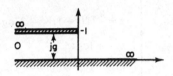

$$z = \frac{g}{\pi}(1 + t + \log t).$$

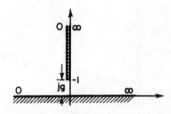

$$z = \frac{g}{2}(t^{1/2} - t^{-1/2}).$$

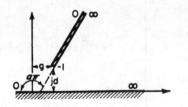

$$z = S\left(\frac{t^{1-\alpha}}{1-\alpha} - \frac{t^{-\alpha}}{\alpha}\right), \quad \begin{array}{l} \alpha \neq \pi, \\ \neq 0. \end{array}$$

S and position of origin evaluated from equivalence of $z = g + jd$ and $t = -1$.

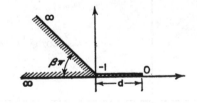

$$z = d\,[2 - \beta)(t + 1)^{1-\beta}$$
$$- (1 - \beta)(t + 1)^{2-\beta}$$

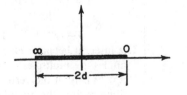

$$z = d\,\frac{1 - t^2}{1 + t^2}.$$

($t = j$ corresponds with $z = \infty$).

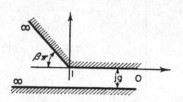

$$z = -\frac{g}{\pi} \int \frac{(1-t)^\beta}{t}\, dt.$$

Representable in elementary functions if $\beta = p/q$, where $0 < p < q$, and p and q are integers.

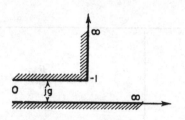

$$z = \frac{g}{\pi} \{ 2(t+1)^{1/2} - 2 \log[(t+1)^{1/2} + 1] + \log t.$$

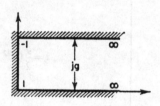

$$z = \frac{g}{\pi} \cosh^{-1} t.$$

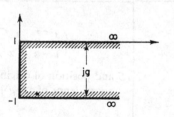

$$z = \frac{g}{\pi} [t\sqrt{t^2 - 1} - \log(t + \sqrt{t^2 - 1})].$$

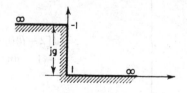

$$z = \frac{g}{\pi} (\cosh^{-1} t - \sqrt{t^2 - 1}).$$

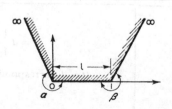

$$z = S \int t^{[(\alpha/\pi) - 1]} (t - 1)^{[(\beta/\pi) - 1]}\, dt;$$
$$S = \frac{l}{(-1)^{[(\beta/\pi) - 1]}} \, \Gamma \, \frac{(\alpha + \beta)/\pi}{\Gamma(\alpha/\pi)\Gamma(\beta/\pi)}.$$

The integral becomes elliptic when the angles are multiples of either $\pi/3$, $\pi/4$, or $\pi/6$. When all the vertices occur in the finite region, the boundary is triangular.

Three defining vertices

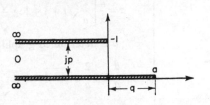

$$z = S[\tfrac{1}{2}t^2 + (1-a)t - a\log t] + k.$$

S and k are determined graphically.

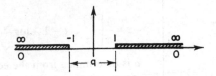

$$z = \tfrac{1}{4}q(t + t^{-1}).$$

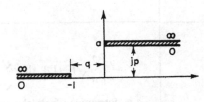

$$z = S\left[t + (1-a)\log t + \frac{a}{t} \right] + k.$$

S and k are determined graphically.
See p. 171.

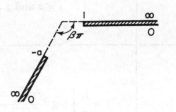

$$z = \frac{t^{2-\beta}}{2-\beta} + \frac{(a-1)}{(1-\beta)}t^{1-\beta} + \frac{a}{\beta}t^{-\beta}.$$

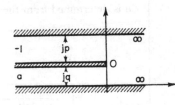

$$z = \frac{p}{\pi}\left[\log(t+1) + \frac{q}{p}\log\left(t - \frac{q}{p} \right) \right] - \frac{q}{\pi}\log\frac{q}{p}.$$

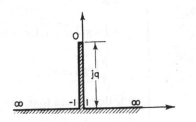

$$z = q\sqrt{t^2 - 1}.$$

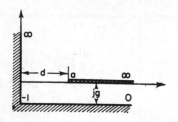

$$z = \frac{g}{\pi}\left[\frac{2}{a}\sqrt{t+1} + \log\frac{\sqrt{t+1}+1}{\sqrt{t+1}-1}\right];$$

a is determined graphically from

$$\frac{d}{g}\pi = \frac{2}{a}\sqrt{a+1} + \log\frac{\sqrt{a+1}+1}{\sqrt{a+1}-1}.$$

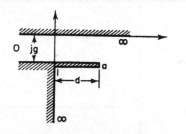

$$z = \frac{g}{\pi}\left[\frac{2j}{a}\sqrt{t-1} - \log\frac{\sqrt{t-1}-j}{\sqrt{t-1}+j}\right].$$

a is determined from the equivalence of

$$t = a \text{ and } z = d - jg.$$

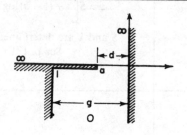

$$z = \frac{g}{\pi}\left[\frac{2}{a}\sqrt{t-1} + j\log\frac{\sqrt{t-1}-j}{\sqrt{t+1}+j}\right].$$

a is determined from the equivalence of

$$t = a \text{ and } z = -d.$$

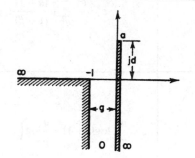

$$z = \frac{jg}{\pi}\left[\log\frac{\sqrt{t+1}-1}{\sqrt{t+1}+1} - \frac{2}{a}\sqrt{t+1}\right].$$

a is determined from the equivalence of

$$t = a \text{ and } z = jd.$$

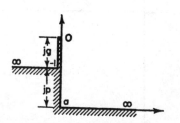

$$z = \frac{p}{\pi}\frac{2}{(a-1)}\left\{\sqrt{(t+1)(t-a)} + \frac{(a-1)}{2}\right.$$

$$\left. \times \left[\log(2\sqrt{(t+1)(t-a)} + 2t + 1 - a)\right]\right\}$$

$$- \frac{p}{\pi}\log(1+a).$$

a is determined from the equivalence of

$$t = 0 \text{ and } z = j(p+g).$$

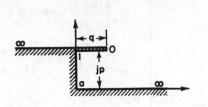

$$z = \frac{p}{\pi}\frac{2}{(a-1)}\left\{\sqrt{(t-1)(t-a)} + \frac{(a-1)}{2}\right.$$

$$\left. \times \left[\log(2\sqrt{(t-1)(t-a)} + 2t - 1 - a)\right]\right\}$$

$$- \frac{p}{\pi}\log(a-1).$$

a is determined from the equivalence of

$$t = 0 \text{ and } z = q + jp.$$

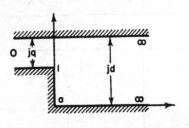

$$z = \frac{d}{\pi}\left[\cosh^{-1}\frac{2t-(a+1)}{a-1}\right.$$

$$\left. - \cosh^{-1}\frac{(a+1)t-2a}{(a-1)t}\right]$$

$$a = \left(\frac{d}{g}\right)^2.$$

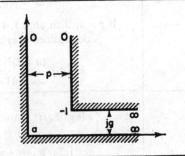

$$z = \frac{2g}{\pi}\left[\frac{p}{g}\tan^{-1}\frac{pu}{g} + \frac{1}{2}\log\left(\frac{1+u}{1-u}\right)\right];$$

$$u^2 = \frac{1-(g/p)^2}{t+1}.$$

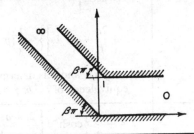

$$z = S\int\frac{1}{t}\left(\frac{1-t}{a+t}\right)^\beta dt.$$

Representable in terms of elementary
functions if $\beta = p/q$, where $0 < p < q$
and p and q are integers. See Köber,
p. 156.

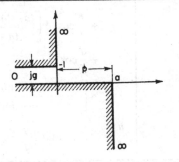

$$z = -\frac{jg}{\pi\sqrt{a}}\left[\frac{(a+1)R}{R^2-1} + (1-a)\tanh^{-1}R\right.$$

$$\left. + j\sqrt{a}\log\frac{(R\sqrt{a}-j)}{(R\sqrt{a}+j)}\right]; \quad R = \sqrt{\frac{t+1}{t-a}},$$

$$a = 1 + 2p/g \pm \sqrt{(1+2p/g)^2 - 1}.$$

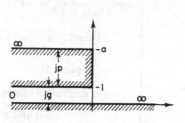

$$z = \frac{g}{\pi}\left[\frac{(a+1)}{\sqrt{a}}\tanh^{-1}R + \frac{(a-1)}{\sqrt{a}}\right.$$

$$\left.\times \frac{R}{(1-R^2)} + \log\frac{(R\sqrt{a}-1)}{(R\sqrt{a}+1)}\right];$$

$$R = \sqrt{\frac{t+1}{t+a}},$$

$$a = -1 + 2k^2 \pm 2k\sqrt{k^2-1}, \quad k = 1 + p/g.$$

Four defining vertices

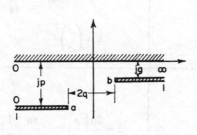

$$z = \frac{g}{\pi}\left[ab\log t + (1-ab)\log(t-1)\right.$$

$$\left. + (1+a+b+ab)\frac{1}{(t-1)}\right] + k;$$

$$ab = p/g,$$

If $p = g$, then $ab = 1$, $k = -jg$ and a
is determined graphically from

$$\log a + \frac{(a+1)^2}{a(a-1)} = -\frac{q}{g}\pi.$$

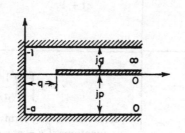

$$z = \frac{1}{\pi}\left\{g\log\left[2\sqrt{(t+a)(t+1)} + 2t + a + 1\right]\right.$$

$$+ p\log\left[\frac{2\sqrt{a}\sqrt{(t+a)(t+1)}}{t}\right.$$

$$\left.\left. + \frac{2a}{t} + a + 1\right]\right\} - \frac{1}{\pi}(p+g)\log(1-a);$$

$gb = \sqrt{a}\,p$, and a is determined from the
equivalence of $t = b$, $z = q$.

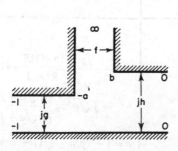

$$z = -\frac{jf}{\pi}\left\{\cosh^{-1}\left[\frac{2t+a-b}{a+b}\right]\right.$$

$$+ \frac{h}{f}\cos^{-1}\left[\frac{(a-b)t-2ab}{(a+b)t}\right]$$

$$+ \frac{g}{f}\cos^{-1}$$

$$\left.\times\left[\frac{2(a-1)(b+1)(b-a+2)(t+1)}{(a+b)(t+1)}\right]\right\};$$

$$\frac{h}{f} = \sqrt{ab}, \quad \text{and} \quad \frac{g}{f} = \sqrt{(b+1)(a-1)}.$$

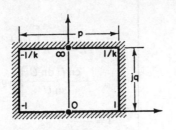

$$t = \operatorname{sn}\left(\frac{2K}{p}z, k\right);$$

$$\frac{K'}{K} = \frac{2q}{p}.$$

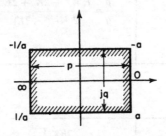

$$z = \frac{p}{E - k'^2 K}$$

$$\times \left\{ E' - k^2 K - j\left[Z(\beta) + \frac{\beta}{K}(E - k'^2 K) \right] \right\};$$

$$k \operatorname{sn} \beta = -\sin\left[2\tan^{-1}(-t)\right]$$

$$\frac{p}{g} = \frac{E - k'^2 K}{E' - k^2 K'}.$$

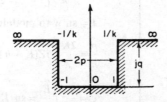

$$z = \frac{p}{E} E(t, k);$$

$$\frac{q}{p} = \frac{K' - E'}{E}.$$

Five defining vertices

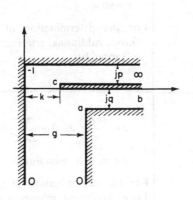

$$z = \frac{1}{\pi}\left\{ p\cosh^{-1}\left[\frac{2t + 1 - a}{1 + a}\right]\right.$$

$$+ q\cos^{-1}\left[\frac{(1 - a)t - 2a}{(1 + a)t}\right] + q\cosh^{-1}$$

$$\left. \times \left[\frac{2(1 + b)(b - a) + (1 + 2b - a)(t - b)}{(1 + a)(t - b)}\right]\right\};$$

$$\frac{c\sqrt{a}}{b} = \frac{g}{p},$$

$$\frac{c - b}{b}\sqrt{\frac{b - a}{1 + b}} = \frac{q}{p},$$

and the third relationship between the constants is determined by substituting for $t = c$, $z = k$.

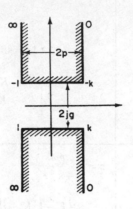

$$z = \frac{jp}{2E - k'^2 K}\left[U\left(k'^2 - \frac{2E}{K} \right) - 2Z(U) \right.$$

$$\left. - \frac{\operatorname{cn} U \operatorname{dn} U}{\operatorname{sn} U} \right] - g;$$

$$t = \frac{1}{\operatorname{sn} U},$$

$$\frac{p}{g} = \frac{K'k'^2 - 2K' + 2E'}{2(Kk'^2 - 2E)}.$$

Six defining vertices

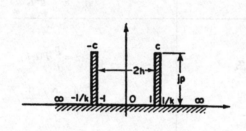

$$z = \frac{2hK'}{\pi}\left[Z(w) + \frac{\pi w}{2KK'} \right];$$

$$t = \operatorname{sn} w \text{ to modulus } k,$$

$$\frac{p}{h} = \frac{2K'}{\pi}\left[jZ(K + jv) + \frac{\pi v}{2KK'} \right],$$

$$c = \frac{1}{K}\sqrt{\frac{E'}{K'}} = \operatorname{sn}(K + jv).$$

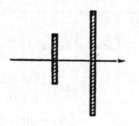

$$z = S\left[Z(w) + \frac{\pi w}{2KK'} + \frac{\operatorname{cn} w \operatorname{dn} w}{c + \operatorname{sn} w} \right] + \text{const.};$$

$$t = \operatorname{sn} w.$$

For the determination of the constants see
Love, Additional references of
Chapter 10.

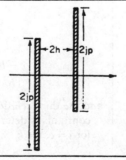

$$z = \frac{2hK'}{\pi}\left[Z(w) + \frac{\pi w}{2KK'} + jBk \operatorname{sn} w \right];$$

$$t = \operatorname{sn} w \text{ to modulus } k.$$

For the determination of the constants see
Love, Additional references of
Chapter 10.

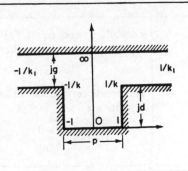

$$z = -\frac{2g}{\pi}\left[\frac{\operatorname{sn}\alpha\,\operatorname{dn}\alpha}{\operatorname{cn}\alpha}U + \Pi(U,\alpha,k)\right];$$

$t = \operatorname{sn}\alpha$ to modulus k,

$k_1 = k\operatorname{sn}\alpha$,

$$\frac{p}{g} = \frac{4K}{\pi}\left[Z(\alpha) - \frac{\operatorname{sn}\alpha\,\operatorname{dn}\alpha}{\operatorname{cn}\alpha}\right],$$

$$\frac{d}{g} = \frac{2K'}{\pi}\left[Z(\alpha) - \frac{\operatorname{sn}\alpha\,\operatorname{dn}\alpha}{\operatorname{cn}\alpha}\right] + \frac{\alpha}{K}.$$

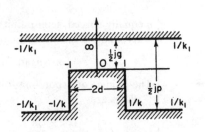

$$z = \frac{p}{\pi}\frac{k^2\operatorname{sn}\alpha\,\operatorname{cn}\alpha}{\operatorname{dn}\alpha}\left[U - \frac{\operatorname{dn}\alpha}{k^2\operatorname{sn}\alpha\,\operatorname{cn}\alpha}\Pi(U,\alpha)\right];$$

$t = \operatorname{sn} U$ to modulus k,

$k_1 = k\operatorname{sn}\alpha$,

$$\frac{g}{p} = \frac{2K'}{\pi}\left[\frac{k^2\operatorname{sn}\alpha\,\operatorname{cn}\alpha}{\operatorname{dn}\alpha} - Z(\alpha)\right] + 1 - \frac{\alpha}{K}$$

$$\frac{d}{p} = \frac{K}{\pi}\left[\frac{k^2\operatorname{sn}\alpha\cdot\operatorname{cn}\alpha}{\operatorname{dn}\alpha} - Z(\alpha)\right],$$

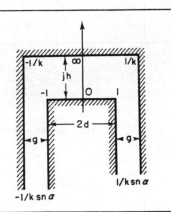

$$z = U - \frac{\operatorname{dn}\alpha}{k^2\operatorname{sn}\alpha\,\operatorname{cn}\alpha}\Pi(U,\alpha,k);$$

$t = \operatorname{sn} U$,

$\alpha = K - \frac{1}{2}jK'$,

$$d = K\left(\frac{1+k}{2k} - \frac{\pi}{4kK}\right),$$

$$g = \frac{\pi}{2k},$$

$$h = \frac{K'(1+k)}{2k}.$$

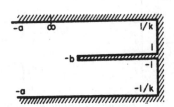

$$z = kC(1 - bk\sin\delta)\frac{\operatorname{sn}\delta}{\operatorname{cn}\delta\,\operatorname{dn}\delta}\Pi(\gamma,\delta,k)$$

$$+ \frac{kbC\lambda}{a} - \frac{jC(a-b)k^2\operatorname{sn}^2\delta}{\operatorname{cn}\delta\,\operatorname{dn}\delta}$$

$$\times \sin^{-1}\sqrt{\left(\frac{a^2-1}{1-k^2}\frac{1-k^2t^2}{a^2-t^2}\right)}.$$

For the evaluation of the constants see
E.P.Adams, Electric distributions on
cylinders, *Proc. Am. Phil. Soc.* **125**, 11
(1936).

n defining vertices

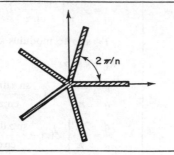

n equally spaced, equal, finite intersecting plates terminating on the unit circle.

$$z = [\cos(n\tan^{-1}t)]^{2/n}.$$

The points corresponding to the ends of the plates are

$$t = \tan\frac{(2k-1)\pi}{2n}, \quad k = 1, 2, 3, \ldots n.$$

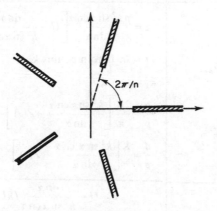

n equally spaced, semi-infinite plates, terminating on the unit circle.

$$z = [\cos(n\tan^{-1}t)]^{-2/n}.$$

The points corresponding to the ends of the plates are

$$t = \tan\frac{(2k-1)\pi}{2n}, \quad k = 1, 2, 3, \ldots n.$$

1.3 Straight-line segments and curves

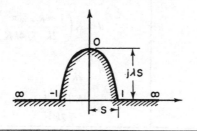

Semi-ellipse and straight line.

$$z = S(t + \lambda\sqrt{t^2 - 1}).$$

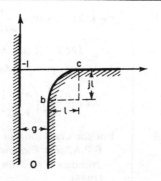

$$z = \frac{-2g}{\pi(\sqrt{b} + \lambda\sqrt{c})}\left\{\sqrt{t-b}\right.$$

$$-\sqrt{b}\tan^{-1}\sqrt{\left(\frac{t-b}{b}\right)} + \lambda\left[\sqrt{t-c}\right.$$

$$\left.\left.-\sqrt{c}\tan^{-1}\sqrt{\left(\frac{t-c}{c}\right)}\right]\right\}.$$

b, c, and λ are determined from the equivalence of $t = -1$ and $z = 0$ and from the desired shape of the rounded corner.

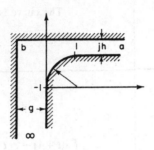

$$2 = -j\sqrt{\frac{a+1}{b-a}}\log\frac{1+\sqrt{\left(\frac{b-a}{a+1}\right)\left(\frac{2}{b-1}\right)}}{1-\sqrt{\left(\frac{b-a}{a+1}\right)\left(\frac{2}{b-1}\right)}}$$

$$+2j\tan^{-1}\sqrt{\frac{2}{b-1}-\left(\frac{b+1}{b-1}\right)}$$

$$\times\log\frac{1+\sqrt{\frac{2}{b+1}}}{1-\sqrt{\frac{2}{b+1}}}+2\sqrt{\left(\frac{b+1}{b-1}\right)\left(\frac{a-1}{b-a}\right)}$$

$$\times\tan^{-1}\sqrt{\left(\frac{b-a}{a-1}\right)\left(\frac{2}{b+1}\right)};$$

$$h=\pi\left\{\sqrt{\frac{a+1}{b-a}}+\sqrt{\left(\frac{b+1}{b-1}\right)\left(\frac{a-1}{b-a}\right)}\right\},$$

$$g=\pi\left\{1+\sqrt{\frac{b+1}{b-1}}\right\}.$$

Choice of constants gives a rounded corner
with almost uniform curvature.

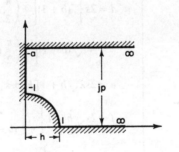

$$z=\frac{2p}{\pi(1+\lambda)}\left[\tan^{-1}\sqrt{\left(\frac{t-1}{t+a}\right)}\right.$$

$$\left.+\lambda\tanh^{-1}\sqrt{\left(\frac{t+1}{t+a}\right)}\right].$$

λ and a are determined from the equivalence
of $t=-1$ and $z=jb$, and $t=1$, $z=b$. The
curve approximates closely to a circle.

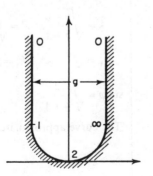

$$z=jg\frac{(K'-E')}{E}+\frac{1}{2}g(1-j);$$

$$k=\sqrt{t}.$$

The curve is a semicircle.

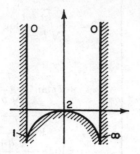

$$z = jg\frac{K'}{K} - \frac{1}{2}g(1+j);$$

$$k = \sqrt{t}.$$

The curve is a semicircle.

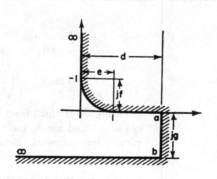

$$z = S\int\frac{\lambda(t+1) + \sqrt{t-1}}{\sqrt{[(t-a)(t-b)]}}\,dt;$$

$$g = 2s\left[\lambda\sqrt{b+1}\,E\left(\frac{\pi}{2},k\right) + \sqrt{(b-1)}\right.$$

$$\left.\times E\left(\frac{\pi}{2},k'\right)\right],$$

$$d = 2s\left\{\lambda(b+1)\left[F\left(\frac{\pi}{2},k_1\right) - E\left(\frac{\pi}{2},k_1\right)\right]\right.$$

$$\left.+ \sqrt{(b-1)}\left[F\left(\frac{\pi}{2},k_1'\right) - E\left(\frac{\pi}{2},k_1'\right)\right]\right\},$$

$$e = 2s\lambda\sqrt{b+1}\left[F(\varphi,k_1) - E(\varphi,k_1,)\right],$$

$$f = 2s\sqrt{b-1}\left\{\sqrt{\left[\frac{2(b+1)}{(b-1)(a+1)}\right]}\right.$$

$$\left.- E(\varphi,k')\right\},$$

$$k = \sqrt{\frac{b-a}{b+1}}, \quad k' = \sqrt{\frac{b-a}{b-1}},$$

$$k_1 = \sqrt{1-k^2}, \quad k_1' = \sqrt{1-k'^2},$$

$$\sin\varphi = \sqrt{\frac{2}{a+1}}$$

The curve approximates to a circular arc.

A3.2 OTHER TRANSFORMATIONS

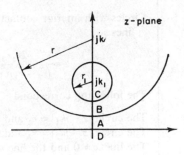

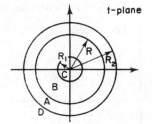

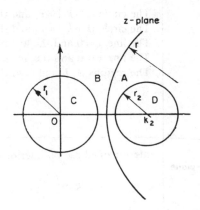

Non-concentric circles (interior) and straight line on concentric circles.

$$2 = \frac{jp_1(R_2 + t)}{R_2 - t};$$

$$p_1 = \sqrt{k_1^2 - r_1^2}, \quad k_1 > r_1 > 0.$$

The following correspond:

The line $y = 0$ and the circle $|t| = R_2$.
The circle $|z - jk_1| = r_1$ and the circle

$$|t| = R_1 = \frac{R_2 r_1}{|k_1| + \sqrt{k_1^2 - r_1^2}}.$$

The circle $|z - jk| = r$ and the circle $|t| = R$ where

$$k = p_1 \frac{R_2^2 + R^2}{R_2^2 - R^2}$$

and

$$r = \frac{2p_1 R_2 R}{|R_2^2 - R^2|}.$$

The regions A, B, C and D in the two planes.

Non-concentric circles (exterior) on concentric circles.

t-plane as above. $z = \dfrac{a(r_1 t - b R_1)}{r_1 t - a R_1};$

$ab = r_1^2, (k_2 - a)(k_2 - b) = r_2^2$ and

$$\frac{R_2}{R_1} = \frac{r_2}{r_1} \frac{b}{(k_2 - b)}, \quad k_2 > 0.$$

The following correspond:

The circle $|z| = r_1$ and the circle $|t|R_1$.
The circle $|z - k_2| = r_2$ and the circle $|t| = R_2$.
The circle $|z - k| = r$ and the circle $|t| = R$,

$$\text{where } k = \frac{r_1^2(R_1^2 - R^2)}{bR_1^2 - aR^2},$$

$$\text{and } r = \frac{r_1 R_1 R |b - a|}{|aR^2 - bR^2|}.$$

The regions A, B, C, and D in the two planes.

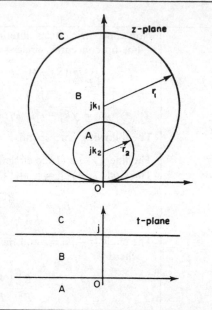

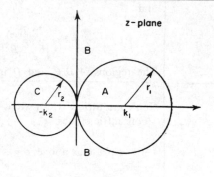

Circles with interior contact on straight lines.

$$z = \frac{2r_1 r_2}{(r_2 - r_1)t - r_2 j}.$$

The following correspond:

The circle $|z - jk_1| = r_1$ and the line $v = 0$.
The circle $|z - jk_2| = r_2$ and the line $v = j$.
The line $x = 0$ and the line $u = 0$.
The line $y = 0$ and the line $v = jr_2/(r_2 - r_1)$

The regions A, B, and C in the two planes.

Circles with exterior contact on straight lines.

$$z = \frac{2r_1 r_2}{j(r_1 + r_2)t + r_2}.$$

The following correspond:

The circle $|z - k_1| = r_1$ and the line $v = 0$.
The circle $|z + k_2| = r_2$ and the line $v = j$.
The line $x = 0$ and the line $v = r_2/(r_1 + r_2)$.
The line $y = 0$ and the line $u = 0$.
The regions A, B and C in the two planes.

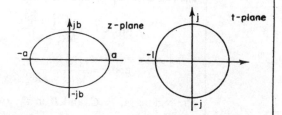

Interior of ellipse on interior of unit circle:

$$t = \sqrt{k}\,\text{sn}\left(\frac{2K}{\pi}\sin^{-1}\frac{z}{\sqrt{a^3 - b^2}}\right)$$

to modulus k;

$$\left(\frac{a - b}{a + b}\right)^2 = \exp\left(-\frac{\pi K'}{K}\right).$$

APPENDIX 4
USEFUL VECTOR IDENTITIES
AND INTEGRAL THEOREMS

In the formulae below U and V are scalar and $\mathbf{F}$ and $\mathbf{G}$ are vector functions of position.

A4.1 VECTOR IDENTITIES

The following relationships involving differentiation of vectors have been frequently used in the text and are here stated without proof, see for example [1].

$$\nabla(U + V) = \nabla U + \nabla V \tag{A4.1}$$

$$\nabla \cdot (\mathbf{F} + \mathbf{G}) = \nabla \cdot \mathbf{F} + \nabla \cdot \mathbf{G} \tag{A4.2}$$

$$\nabla \times (\mathbf{F} + \mathbf{G}) = \nabla \times \mathbf{F} + \nabla \times \mathbf{G} \tag{A4.3}$$

$$\nabla(UV) = U\nabla V + V\nabla V \tag{A4.4}$$

$$\nabla \cdot (V\mathbf{F}) = V\nabla \cdot \mathbf{F} + \mathbf{F} \cdot \nabla V \tag{A4.5}$$

$$\nabla \times (V\mathbf{F}) = V\nabla \times \mathbf{F} + (\nabla V) \times \mathbf{F} \tag{A4.6}$$

$$\nabla \cdot (\mathbf{F} \times \mathbf{G}) = \mathbf{G} \cdot \nabla \times \mathbf{F} - \mathbf{F} \cdot \nabla \times \mathbf{G} \tag{A4.7}$$

$$\nabla \times (\mathbf{F} \times \mathbf{G}) = (\mathbf{G} \cdot \nabla)\mathbf{F} - (\mathbf{F} \cdot \nabla)\mathbf{G} + \mathbf{F}\nabla \cdot \mathbf{G} - \mathbf{G}\nabla \cdot \mathbf{F} \tag{A4.8}$$

$$\nabla(\mathbf{F} \cdot \mathbf{G}) = (\mathbf{G} \cdot \nabla)\mathbf{F} + (\mathbf{F} \cdot \nabla)\mathbf{G} + \mathbf{G} \times \nabla \times \mathbf{F} + \mathbf{F} \times \nabla \times \mathbf{G} \tag{A4.9}$$

$$\nabla \times \nabla V = 0 \tag{A4.10}$$

$$\nabla \cdot \nabla \times \mathbf{F} = 0 \tag{A4.11}$$

A4.2 INTEGRAL THEOREMS

The above identities can be used to establish a number of useful integral relationships involving vector fields. They are primarily based on the divergence and circulation theorems which are quoted first, see for example [2].

$$\int_\Gamma \nabla\cdot\mathbf{F}d\Omega = \int_\Gamma \mathbf{F}\cdot d\Gamma \text{ (Gauss's or divergence theorem)} \tag{A4.12}$$

$$\int_\Gamma \nabla\times\mathbf{F}\cdot d\Gamma = \oint \mathbf{F}\cdot d\mathbf{l} \text{ (Stokes's or circulation theorm)} \tag{A4.13}$$

$$\int_\Omega \nabla V d\Omega = \int_\Gamma V d\Gamma \tag{A4.14}$$

$$\int_\Omega \nabla\times\mathbf{F}d\Omega = -\int_\Gamma \mathbf{F}\times d\Gamma \tag{A4.15}$$

$$\int_\Omega (\nabla U\cdot\nabla V + U\nabla^2 V)\,d\Omega = \int_\Gamma U\nabla V\cdot d\Gamma \text{ (Green's first theorem)} \tag{A4.16}$$

$$\int_\Omega (U\nabla^2 V - V\nabla^2 U)\,d\Omega = \int_\Gamma (U\nabla V - V\nabla U)\cdot d\Gamma \text{ (Green's second theorem)} \tag{A4.17}$$

The divergence theorem eqn (A4.12) can be used to derive the generalization of the integration by parts rule used frequently in this book, i.e. for the product ($U\mathbf{F}$) the divergence theorem can be written as:

$$\int_\Omega \nabla\cdot(U\mathbf{F})\,d\Omega = \int_\Gamma U\mathbf{F}\cdot\mathbf{n}\,d\Gamma, \tag{A4.18}$$

and using the divergence of a scalar-vector product rule, eqn. (A4.5), and rearranging leads to

$$\int_\Omega U\nabla\cdot\mathbf{F}d\Omega = \int_\Gamma U\mathbf{F}\cdot\mathbf{n}\,d\Gamma - \int_\Omega (\nabla U\cdot\mathbf{F})\,d\Omega. \tag{A4.19}$$

Finally, a vector form of Green's theorem can be derived using eqn (A.7) as follows:

$$\int_\Omega \mathbf{F}\cdot\nabla\times\mathbf{G}\,d\Omega = \int_\Omega \mathbf{G}\cdot\nabla\times\mathbf{F}\,d\Omega - \int_\Omega \nabla\cdot(\mathbf{F}\times\mathbf{G})\,d\Omega \tag{A4.20}$$

and by applying the divergence theorem to the second term of the r.h.s. in eqn (A4.20) to get

$$\int_\Omega \mathbf{F}\cdot\nabla\times\mathbf{G}\,d\Omega = \int_\Omega \mathbf{G}\cdot\nabla\times\mathbf{F}\,d\Omega - \int_\Gamma (\mathbf{F}\times\mathbf{G})\cdot\mathbf{n}\,d\Gamma. \tag{A4.21}$$

Equation (A4.21) is used extensively in deriving the discretized equations for three-dimensional eddy current analysis, where for example, $\mathbf{F}$ is equated to the vector weight function $\mathbf{N}$ and $\mathbf{G} \equiv (1/\mu)\nabla\times\mathbf{A}$.

REFERENCES

[1] D. E. Bourne and P. C. Kendall, *Vector Analysis and Cartesian Tensors*, London: Van Nostrand Reinhold (1977).
[2] J. A. Stratton, *Electromagnetic Theory*. New York: McGraw-Hill (1941).

APPENDIX 5
QUADRATURE RULES FOR NUMERICAL INTEGRATION

The numerical quadrature rules needed for evaluating element matrices have been exhaustively studied and are liberally reported in the literature, see for example the text by Carey and Oden [1], (Vol. III, p. 340) the original work by Hammer, Marlowe and Stroud [2], the book by Stroud and Secrest [3], and for more recent work on triangles, see the paper by Cowper [4].

Table A5.1 gives some limited information on Gaussian quadrature points and weights for triangles.

The points and weights for one-dimensional quadrature are given in Table A5.2, these can be generalized for two and three-dimensional integrations over the unit square by tensor multiplication.

Table A5.1 Gauss quadrature for triangle.

Quadrature points and weights			
Order	Points	Area coordinates	Weights
Linear $O(h^2)$	1	$\xi_1 = \frac{1}{3}(1,1,1)$	1
Quadratic $O(h^3)$	3	$\xi_1 = \frac{1}{2}(1,1,0)$	
		$\xi_2 = \frac{1}{2}(0,1,1)$	$\frac{1}{3}$
		$\xi_3 = \frac{1}{2}(1,0,1)$	
Cubic $O(h^3)$	4	$\xi_1 = \frac{1}{3}(1,1,1)$	$-\frac{27}{48}$
		$\xi_2 = \frac{1}{5}(3,1,1)$	
		$\xi_3 = \frac{1}{5}(1,3,1)$	$+\frac{25}{48}$
		$\xi_4 = \frac{1}{5}(1,1,3)$	

Table A5.2 One-dimensional quadrature.

Legendre–Gauss quadrature points and weights		
Order	ξ_j	ω_j
1	0	2
2	$\pm\sqrt{\frac{1}{3}}$	1
3	0	$\frac{8}{9}$
	$\pm\sqrt{\frac{3}{5}}$	$\frac{5}{9}$
4	$\pm$ 0.33998 10435 84856	0.65214 51458 62546
	$\pm$ 0.86113 63115 94053	0.34785 48451 37454
5	0	0.56888 88888 88889
	$\pm$ 0.53846 93101 05683	0.47862 86704 99366
	$\pm$ 0.90617 98459 38664	0.23692 68850 56189
6	$\pm$ 0.23861 91860 83197	0.46791 39345 72691
	$\pm$ 0.66120 93864 66265	0.36076 15730 48139
	$\pm$ 0.93246 95142 03152	0.17132 44923 79170

REFERENCES

[1] G. F. Carey and J. T. Oden, *Finite Elements and Computational Aspects, Vol 3.* New York: Prentice-Hall (1984).

[2] P. C. Hammer, O. P. Marlowe, and A. H. Stroud, Numerical integration over simplexes and cones, *Math. Tables Aids Comp*, **10**, 130–137 (1956).

[3] A. H. Stroud and D. Secrest, *Gaussian Quadrature Formulas.* Englewood Cliffs, N. J. Prentice-Hall (1966).

[4] E. R. Cowper, Gaussian quadrature formulae for triangles, *Int. J. Num. Meth. Eng.*, **7**, 405–408 (1973).

APPENDIX 6
UNIQUENESS OF SCALAR POTENTIAL

In addition to the defining equations, see section 1.3 it is necessary to prescribe boundary conditions otherwise non-unique solutions are possible and the problem will not be properly defined. This can be seen by a simple application of the integral theorems (see Appendix 4) to the generic form of the Poisson equation. Suppose the equation is valid in a domain Ω bounded by a surface Γ, i.e.

$$\nabla \cdot \kappa \nabla u = q. \tag{A6.1}$$

Thus if the solution to eqn (A6.1) is not unique then there will be at least two solutions u_1 and u_2 and their difference is given by $U = u_1 - u_2$ and $\nabla \cdot \kappa \nabla U = 0$. Therefore, by Green's first theorem (see Appendix 4),

$$\int_\Omega U \nabla^2 U \mathrm{d}\Omega + \int_\Omega \nabla U \cdot \nabla U \mathrm{d}\Omega = \int_\Gamma U \frac{\partial U}{\partial n} \mathrm{d}\Gamma. \tag{A6.2}$$

The first term vanishes by definition leaving

$$\int_\Omega (\nabla U)^2 \mathrm{d}\Omega = \int_\Gamma U \frac{\partial U}{\partial n} \mathrm{d}\Gamma. \tag{A6.3}$$

and accordingly if the r.h.s. of eqn (A6.3) is zero then, since the integrand of the l.h.s. is a squared quantity. ∇U must be zero and the field is unique. The r.h.s. will vanish if either U or $\partial U/\partial n$ are zero on the boundary. This is achieved by specifying $u = u_1 = u_2$, known as the Dirichlet boundary condition, or by specifying $\partial U/\partial n = \partial u_1/\partial n = \partial u_2/\partial n$, known as the Neumann boundary condition. In the latter case u still needs to be specified at one point at least in order to ensure uniqueness of potential since it is the field only that is made unique by causing the r.h.s. of eqn (A6.3) to vanish. Similar arguments apply to vector fields as was shown in section 12.5.2 and for a complete treatment the reader should consult the book by Kellogg [1].

REFERENCE

[1] O. D. Kellogg, *Foundations of Potential Theory*. New York: Dover Publicatons (1954).

BIBLIOGRAPHY

Abraham, M., and Becker, R., *The Classical Theory of Electricity and Magnetism*, Blackie, London and Glasgow (1937).

Allen, D. N. de G., *Relaxation Methods*, McGraw-Hill, London, New York, and Toronto (1954).

Batemann, H., *Partial Differential Equations of Mathematical Physics*, Dover, New York (1944).

Beckenbach, E. F., *Modern Mathematics for the Engineer*, McGraw-Hill, New York (1956).

Bewley, L. V., *Two-dimensional Fields in Electrical Engineering*, Dover, New York (1963).

Bourne, D. E. and Kendall, P. C., *Vector Analysis and Cartesian Tensors*, Van Nostrand Reinhold, London (1977).

Bowman, F., *Introduction to Elliptic Functions with Applications*, English Universities Press, Suffolk (1953).

Brebbia, C. A., *The Boundary Element Method for Engineers* (2nd Edition), Printec Press (1980).

Buchholz, H., *Elektrische und magnetische Potentialfelder*, Springer, Berlin (1957).

Buckingham, R. A., *Numerical Methods*, Pitman, London (1957)

Byerly, W. E., *An Elementary Treatise on Fourier Series and Spherical, Cylindrical and Ellipsoidal Harmonics*, Ginn, Boston (1893).

Byrd, P., and Friedman, M. D., *Handbook of Elliptic Integrals for Engineers and Physicists*, Springer, Berlin (1954).

Carey, G. F. and Oden, J. T., *Finite Elements and Computational Aspects, Vol 3*, Prentice Hall, New York (1984)

Carslaw, H. S. and Jaegar, J. G. *Conduction of Heat in Solids* (2nd Edition), Clarendon Press, Oxford (1959).

Carter, G. W., *The Electromagnetic Field in its Engineering Aspect*, Longmans, London, New York, and Toronto (1954).

Churchill, R. V., *Fourier Series and Boundary Value Problems*, McGraw-Hill, New York (1941).

Copson, E. T., *An Introduction to the Theory of Functions of a Complex Variable*, Clarendon Press, Oxford (1935).

Crandall, S. H. *Engineering Analysis*, McGraw-Hill, New York (1956).

Doolan, E. P., *et al. Uniform Numerical Methods for Problems with Initial and Boundary Layers*, Boole Press, Dublin (1960).

Duff, I. S., *Sparse Matrices and their Uses*, Academic Press, New York (1981).

Edwards, J., *A Treatise on the Integral Calculus*, Macmillan, London (1921).

Ferraro, V. C. A., *Electromagnetic Theory*, Athlone Press, London (1956).

Finlayson, B. A., *The Method of Weighted Residuals and Variational Principles*, Academic Press, New York (1972).

Fletcher, R., *Practical Methods of Optimization* (2nd Edition), Wiley Interscience, Chichester (1991).

Forsythe, G. E., and Wasow, W. R., *Finite Difference Methods for Partial Differential Equations*, Wiley, New York and London (1960).

Frank, P., and Mises, R. V., *Die Differential- und Integralgleichungen der Mechanik und Physik*, 2 vols., Vieweg, Braunschweig (1935).

Gibbs, W. J., *Conformal Transformations in Electrical Engineering*, Chapman and Hall, London (1958).

Grinter, L. E., *Numerical Methods of Analysis in Engineering*, Macmillan, New York (1949).

Grover, F. W., *Inductance Calculations*, van Nostrand, New York (1946).

Hague, B., *The Principles of Electromagnetism Applied to Electrical Machines*, Dover, London (1962).

Hammond, P., *Applied Electromagnetism*, Pergamon Press, Oxford (1971).

Hammond, P., *Energy Methods in Electromagnetism*, Clarendon Press, Oxford (1981).

Harrington, R. F., *Field Computation by Moment Methods*, MacMillan, New York (1968).

Jeans, J., *The Mathematical Theory of Electricity and Magnetism*, University Press, Cambridge (1951).

Jennings, W., *First Course in Numerical Methods*, Macmillan, London (1964).

Kantorovich, L. V. and Krylov, V. I., *Approximate Methods of Higher Analysis, Groningen* (1958).

Karman, T. V., and Burgers, J. M., *General Aerodynamic Theory*, vol. II, Springer, Berlin (1935).

Kellogg, O. D., *Foundations of Potential Theory*, Dover Publications, New York (1952).

Kelvin, Lord, *Reprint of Papers on Electrostatics and Magnetism*, Macmillan, London (1872).

King, L. V., *Numerical Evaluation of Elliptic Functions and Elliptic Integrals*, Cambridge (1924).

Köber, H., *Dictionary of Conformal Representation*, Dover, New York, (1952).

Koppenfels, W. von, and Stallmann, F., *Praxis der konformen Abbildung*, Springer, Berlin (1959).

Lanczos, C., *Applied Analysis*, Pitman, London (1957).

Legendre, A. M., *Mémoires sur les Transcendantes élliptiques* (1793).

Legendre, A. M., *Exercices de calcul intégral* (1811).

Lowther, D. and Silvester, P. P., *Computer Aided Design in Magnetics*, Springer-Verlag, Berlin (1986).

Maxwell, J. C. A dynamical theory of the electromagnetic field. *Roy. Soc. Trans.*, **155** 459–512 (1864).

Maxwell, J. C., *A Treatise on Electricity and Magnetism*, 2 vols., Clarendon Press, Oxford (1892).

McLachlan, N. W., *Bessel Functions for Beginners* (2nd edition), Clarendon Press, Oxford (1955).

Meyer, H. A., *Symposium on Monte Carlo Methods*, Chapman and Hall, London (1956).

Milne, W. E., *Numerical Solution of Differential Equations*, Wiley, New York; Chapman and Hall, London (1953).

Miline-Thomson, L. M., *Theoretical Hydrodynamics*, Macmillan, London (1938).

Moon, F. C., *Magneto-Solid Mechanics*. Wiley Interscience, New York (1984).

Moon, P., and Spencer, D. E., *Field Theory for Engineers*, van Nostrand, New York (1961).

Moore, J. et al. *Moment Methods in Electromagnetics*, Research Studies Press, Chichester (1983).

Morse, P. M., and Feshbach, H., *Methods of Theoretical Physics*, 2 vols., McGraw-Hill, New York (1953).

Moullin, E. B., *Principles of Electromagnetism*, University Press, Oxford (1955).

Ollenderf, F., *Technische Elektrodynamik*, Band 1, *Berechnung magnetischer Felder*, Springer, Vienna (1952).

Papoulis, A., *The Fourier Integral and Its Applications*, McGraw-Hill, New York (1962).

Ralston, A., and Wilfe, H. S., *Mathematical Methods for Digital Computers*, Wiley, New York, London (1960).

Reid, J. K., *On the Method of Conjugate Gradients for the Solution of Large Sparse Systems of Equations*, Academic Press, New York (1971).

Richtmeyer, R. D., *Difference Methods for Initial Value Problems*, Wiley-Interscience, New York (1957).

Seely, S., *Introduction to Electromagnetic Fields*, McGraw-Hill, New York (1958).

Shaw, F. S., *An Introduction to Relaxation Methods*, Dover, New York (1953).

Silvester, P., *Modern Electromagnetic Fields*, Prentice Hall, New York (1968).

Silvester, P. P. and Ferrari, R. L., *Finite Elements for Electrical Engineers*, Cambridge University Press, Cambridge (1990).

Smith, G. D., *Numerical Solution of Partial Differential Equations*, Oxford (1965).

Smythe, W. R., *Static and Dynamic Electricity* (3rd Edition), McGraw-Hill, New York (1968).

Sokolnikoff, I. S. and E. S., *Higher Mathematics for Engineers and Physicists*, McGraw-Hill, New York and London (1941).

Southwell, R. V., *Relaxation Methods in Theoretical Physics*, University Press, Oxford (1946).

Stoll, R. L. *The Analysis of Eddy-Currents*, Clarendon Press, Oxford (1974).

Strang, G., *Linear Algebra and its Applications*, Academic Press, New York (1976).

Stratton, J. A., *Electromagnetic Theory*, McGraw-Hill, New York (1941).

Stroud, A. H. and Secrest, D., *Gaussian Quadrature Formulas*, Prentice Hall, Englewood Cliffs, NJ (1966).

Tannery, J., and Molk, J., *Elements de la théorie des fonctions élliptiques*, Gauther-Villars, Paris (1893).

Tegolpoulos, J. A. and Kriezis, E. E. *Eddy-Currents in Linear Conducting Media*, Elsevier, Amsterdam (1985).

Todd, J., *Survey of Numerical Analysis*, McGraw-Hill, New York (1968).

Varga, R. S., *Matrix Iterative Analysis*, Prentice-Hall, London (1962).

Vitkovitch, D. (editor), *Field Analysis: Experimental and Computational Methods*, Von Nostrand, London (1966).

Wachspress, E. L., *Iterative Solution of Elliptic Systems*, Prentice-Hall (1966).

Wait, R., *The Numerical Solutions of Algebraic Equations*, Wiley, Chichester (1979).

Walsh, J., *Numerical Analysis: An Introduction*, Academic Press, London, New York (1966).

Weber, E., *Electromagnetic Fields*, vol. 1, *Mapping of Fields*, Wiley, New York, London (1960).

Whittaker, E. T., and Watson, G. N., *A Course of Modern Analysis*, University Press, Cambridge (1920).

Zienkiewicz, O. C. and Morgan, K., *Finite Elements and Approximation Method*, Wiley, New York (1983).

Zienkiewicz, O. C. and Taylor, R. I., *The Finite Element Method*, (4th Edition), *Volumes 1 and 2*, McGraw-Hill, Maidenhead (1991).

Zworykin, V. K., Morton, G. A., Ramberg, E. G., Hillier, J., and Vance, A. W., *Electron Optics and the Electron Microscope*, Wiley, New York (1945).

Tables of elliptic functions

Byrd, P. F., and Friedman, M. D., *Handbook of Elliptic Integrals for Engineers and Physicists*, Springer, Berlin, Göttingen, and Heidelberg (1954).

Dwight, H. B., *Tables of Integrals and other Mathematical Data*, Macmillan, New York (1934).

Jahnke, E., and Emde, F., *Table of Functions with Formulas and Curves*, Dover, New York (1945).

Miline-Thomson, L. M., *Jacobian Elliptic Function Tables*, Dover, New York (1950).

Pearson, K., *Tables of Complex and Incomplete Elliptic Integrals*, London (1934).

Peirce, B. O., *A Short Table of Integrals*, Ginn, Boston (1929).

Spenceley, G. W., and Spenceley, R. M., *Smithsonian Elliptic Function Tables*, Smithson, Washington (1947).

Additional bibliography

Abraham, M., and Becker, R., *The Classical Theory of Electricity*, Blackie, London (1932).

Bieberbach, L., *Einführung in die konforme Abbildung* (Sammlung Göschen 768), 5 Aufl., Berlin (1956).

Bieberbach, L., *Lehrbuch der Funtionentheorie*, 2 vols.; reprint by Chelsea Publishing Co., New York, 1945; originally published by Teubner, Leipzig.

Byerly, W. E., *Fourier's Series and Spherical, Cylindrical and Ellipsoidal Harmonics*, Ginn, Boston (1902).

Caratheodory, C., *Conformal Representation*, Cambridge (1932).

Chaffee, E. L., *Theory of Thermionic Vacuum Tubes*, McGraw-Hill, New York (1933).

Courant, R., and Hilbert, D., *Methoden der mathematischen Physik*, vol. I, 1931, vol. II, 1937, Springer, Berlin.

Crandall, S. H., *Engineering Analysis*, McGraw-Hill, New York (1956).

Cullwick, E. G., *The Fundamentals of Electromagnetism*, Macmillan, New York (1939).

Eck, B., *Einführung in die technische Strömungslehre*, vol. I, *Theory*, 1935; vol. II, *Laboratory Methods*, 1936, Springer, Berlin.

Evans, G. C., *The Logarithmic Potential, Discontinuous Dirichlet, and Newmann Problems*, American Mathematical Society, vol. VI, New York (1927).

Fourier, J. B. J., *Théorie analytique de la chaleur* (translated by A. Freeman), Dover, New York (1955).

Goursat, E., *Cours d' analyse mathématique*, A. Hermann, Paris (1910, 1911).

Hammond, P., *Applied Electromagnetism*, Pergamon, Oxford (1971).

Handbuch der Physik, vol. 12, *Theorien der Elektrizidät, Elektrostatik*, 1927; vol. 15, *Magnetismus, elektromagnetisches Feld*, 1927, Springer, Berlin.

Heine, E., *Anwendungen der Kugelfunktionen*, Berlin (1881).

Hobson, E. W., *Spherical and Ellipsoidal Harmonics*, Cambridge (1931).

Hurwitz, A., *Vorlesungen über allgemeine Funktionentheorie*, Springer, Berlin (1929).

Karapetoff, V., *The Electric Circuit*, McGraw-Hill, New York (1910).

Karapetoff, V., *The Magnetic Circuit*, McGraw-Hill, New York (1910).

Karman, Th. V., and Burgers, J. M., *General Aerodynamic Theory, Perfect Fluids*, vol. II of *Aerodynamic Theory* (edited by W. F. Durand), Springer, Berlin (1935).

Kellogg, O. D., *Foundations of Potential Theory*, Springer, Berlin (1929).

Korn, A., *Lehrbuch der Potentialtheorie*, Dummler, Berlin (1899).

Lamb, H., *Hydrodynamics*, 6th edn., Cambridge University Press, England (1932).

Lance, G. N., *Numerical Methods for High Speed Computers*, Iliffe, London (1960).

Livens, G. H., *The Theory of Electricity*, Cambridge University Press, London (1926).

Macrobert, T. M., *Spherical Harmonics*, E. P. Dutton, New York (1927).

Macrobert, T. M., *Functions of a Complex Variable*, Macmillan, London (1954).

Mason, M., and Weaver, W., *The Electromagnetic Field*, Chicago University Press (1929).

Miller, K. S., *Partial Differential Equations in Engineering Problems*, Prentice-Hall, New York (1953).

Milne-Thomson, L. M., *Theoretical Aerodynamics*, Macmillan, London (1952).

Murnaghan, F. D., *Introduction to Applied Mathematics*, Wiley, New York (1948).

Muskhelishvili, N. G., *Some Basic Problems of the Mathematical Theory of Elasticity* (translated by J. R. M. Radok), P. Neordhoff Ltd., Groningen-Holland (1953).

Myers, L. M., *Electron Optics*, Van Nostrand, New York (1939).

Nehari, Z., *Conformal Mapping*, McGraw-Hill, New York, Toronto, and London (1952).

Neville, H. H., *Jacobian Elliptic Functions*, Clarendon Press, Oxford (1951).

Ollendorff, F., *Potentialfelder der Elektrotechnik*, Springer, Berlin (1932).

Peirce, B. O., *Newtonian Potential Function*, Ginn, Boston (1902).

Pierpoint, J., *Functions of a Complex Variable*, Ginn, Boston (1914).

Poloujadoff, M., *The Theory of Linear Induction Machines*, University Press, Oxford (1980).

Ramsay, A. S., *A Treatise of Hydromechanics*, Bell (1935).

Richter, R., *Elektrische Maschinen*, vol. 1, *Fundamentals and D. C. Machines*, 1924; vol. 2, *Synchronous Machines and Converters*, 1930; vol. 3, *Transformers*, 1932; vol. 4, *Asynchronous Machines*, 1936, Springer, Berlin.

Robinson, P. D., *Fourier and Laplace Transforms*, Rouledge and Keegan Paul, New York (1971).

Rothe, R., Ollendorff, F., and Pohlhausen, K., *Theory of Functions as Applied to Engineering Problems*, Technology Press, Cambridge, Mass. (1933).

Spangenberg, K. R., *Vacumm Tubes*, McGraw-Hill, New York (1948).

Sternberg, W., *Potentialtheorie*, W. de Gruyter, Leipzing (1925).

Strutt, M. J. O., *Moderne Mehrgitter-Elektronenröhren*, vol. 2, Springer, Berlin (1928).

Thom, A., and Apelt, C. J., *Field Computations in Engineering and Physics*, Van Nostrand, New York (1961).

Titchmarsh, E. C., *Theory of Functions*, Oxford (1932).

Walker, M., *Conjugate Functions for Engineers*, Oxford (1933).

Webster, A. G., *Partial Differential Equations*, Teubner, Leipzig (1927).

INDEX

Alternative boundary element methods 274
Ampère's law 4
Analogous fields 2
Anisotropy 304
Application areas 219, 385
Area coordinates 248
Axisymmetric problems 277
Axisymmetry
  improving the accuracy near the axis 281

Basic field equations 4, 397
Basis functions 233, 246
Bilinear transformation 127
  applied to concentric circles 130
  applied to parallel straight lines, the
    doublet 129
  mapping properties 128
  the cross-ratio 132
Biot—Savart law 6, 76
Boundary conditions 4, 10, 43, 50, 51,
    59—65, 253
  surface impedance 319
Boundary element method (BEM) 9, 261
  convergence 268
  multiple regions 269
  open boundary problems 269
  system of equations 265
  weighted residuals 261
Boundary elements
  coefficients 265
  singular point 263
Boundary integral method 274, 359
Boundary matching methods 66
Boundary value problems
  of the first kind 215
  of the second and mixed kinds 217

Cancellation errors 357
Capacitance
  calculation of 13
  of a parallel plate capacitor 157

of concentric cylinders 137
of two cylindrical conductors 135
of wire between conducting boundaries 27
of wire in rectangular cylinder 29
Cauchy—Riemann equations 127, 427—429
Charge density, on capacitor plates 157
Charge near conducting plane 24, 25
Charge or current near circular boundary
    31—33
Charged boundaries 194, 195
Chebyshev series 211
Circular boundary
  in applied uniform field 37
  transformation to polygon 180—185
  with specified distributions of potential or
    potential gradient 55, 215
Circular cylinder in a uniform field 33
Circulation theorem 454
Collocation 233
Collocation methods 234
Complex potential 120, 430
Complex potential function 429
  of Schwarz 215
Comparison of methods 397
Compumag conferences 343, 381, 397
Computer model 222
Conductor
  circular in slot 100
  effective resistance 102
  inductance of, in slot 92
  rectangular in slot 85
Conductor fields 344
Conductors
  capacitance of two cylindrical 135
  in an infinite, parallel air gap 81
  inside rectangular boundaries 87
  non-magnetic, in air 70
  of infinitely permeable material 433
  of three-dimensional shape 77
  two rectangular, in slot 85
  voltage gradient between 135

Conformal transformation 117
  and field sources 189
  bilinear 127
  choice of corresponding points 158
  choice of origin 124
  conservation of flux and potential 126
  field maps 125
  from a circular to a polygonal boundary
    180
  Joukowski 138
  logarithmic 122
  numerical methods 208
  of curved boundaries 127–148
  of ellipse 134, 145
  of finite boundaries 194
    line current near impermeable boundary
      194
    line current near permeable boundary
      194
    uniform applied field 194
  of infinite boundaries 190
  of the interior of a polygon 149
  of the exterior of a polygon 175
  parametric equations 143, 146
  polygonal boundaries 149–187
  scale relationship between planes 161
  Schwarz–Christoffel 150
  using series 146
Conjugate functions 117, 427
Conservation of flux and potential in
    transformation 126
Constitutive parameter 225
Constitutive relationships 4, 6
Continuity, C_0 249
Convergence rates 309
Coulomb gauge 6, 366
Cross-ratio 132
Current distributions
  and the method of images 80
  solutions of Poisson's equation 69–95
Current doublet 35, 52
Current flow problems 9
Current sheets 66, 73
Curved boundaries 196
Curvilinear polygons 198
Cylinder
  effect of, represented by a doublet 52
  in applied uniform field 33, 50
  permeable, as magnetic screen 50
  premeable, influenced by current

Determinant, for bilinear transformation 128
Diffusion equation 8, 18, 95–112
  embedded circular conductor 101
  general time dependence 107
  harmonic excitation 96–102
  plane-faced block 97

rectangular section conductor 99
solid poles, losses 103
solutions in cartesian coordinates 97, 99,
  103, 107, 110
solutions in cylindrical coordinates 100,
  106, 110
three space dimensions 109
travelling fields 102–107
Dirichlet boundary conditions 225
Dirichlet problem 72, 124, 217, 262, 263
  for interior of unit circle 72
Discretisation 227
Discretisation methods 387
Discretisation schemes 261
Displacement current 4, 313, 379
Distributed currents 69–115
Distributional sources 15
Divergence theorem 454
Double Fourier series 86
Doublet 33
  field at infinity 35
  images of 36
  representation of cylinder 52

Eddy current code validation 397
Eddy current losses 98, 99, 103
Eddy current problems
  3D 365
  comparison of methods 373
  pathological problems 381
Eddy current problems, analytic 95–112
  numerical 313–341
Eddy currents, numerical 313
  skin effect 319
  steady state AC 314
  surface impedance 319
  transient case 320
  ungauged solutions 377
Effective resistance 99, 102
Electric vector potential 7
Electrostatic lens 173
Electrostatics 8
Elements of a C.A.E.-M. system 386
Elliptic function 200
Elliptic integrals 202
  third kind 207
End windings 66, 76–80
Engineering objectives 221
Equivalent pole or charge distributions 12
Error estimation 406
  cheap estimator 410
  hierarchical shape functions 409
  interpolation theory 410
  upper and lower bounds 410
Errors 222
Example problems, numerical
  bath plate 373

bath cube 373
C magnet 277
E-core magnet 317
electromagnetic brake 339
ferromagnetic cylinder 322
high voltage bush 280
L-shape problem 412
laminated structure 306
linear induction tachometer 339
magnetised cylinder 303
objective lens 285
particle bending magnet 353
pipe inspection tool 339
polarised target 3D magnet 362
rectangular bar 234, 238, 252, 268, 299, 327, 328, 433
segmented torus 378
spectrometer magnet 363
stepping motor 355
switched reluctance motor 301
transformer 331
vernier linear motor 355

Faraday's law 4
Field of line current near a permeable corner 192
Field of line current near impermeable plate 191
Fields, analytic
  diffusion and eddy current 95–112
  in circular-section conductor 99
  in conducting cylinders 106, 110
  in conductors embedded in permeable material 100
  in plane-faced block 97
  in rectangular-section conductor 99
  in solid poles 103
  Laplacian
    between windings of a transformer 173
    between unequal, opposite slots 211
    in a contactor 170
    inside the unit circle 215
    maps 125
    near the salient pole-tip 196
    of a charged, conducting plate 178–180
    of a current-carrying conductor of circular section 425
    of a current in a slot 163
    of a current near a permeable corner 164
    of a line charge 123
    of a line current 122, 432
    of a line current and a permeable plate of finite cross-section 194
    of a line current in cylindrical air gap 53
    of a line current influenced by permeable cylinder 31, 48, 141
    of a rectangular salient pole 163
    of a simple electrostatic lens 173
    of concentric cylinders 130
    of current between two infinite, parallel permeable surfaces 159
    of flow round a circular hole 140
    of hyperbolic equipotential boundary 121
    of opposite parallel plates 168
    of two currents inside an infinitely permeable tube 133
    of two cylindrical conductors 135
    of two finite charged plates 204
    of two semi-infinite equipotential planes 124
    outside a charged, conducting boundary of elliptical shape 145
    outside a charged rectangular conductor 194
    three-dimensional 76, 109–112
  Poissonian 69
    between permeable block 81
    due to rectangular conductor in air 70
    inductances 92
    in permeable box 86
    in permeable slot 85
    three dimensional 109–112
    transformer winding forces 90
    transient 107
    with travelling fields 102
    within permeable conductors 70, 433
Finite difference method 223, 227
Finite element
  matrices 251, 255, 259, 317
Finite element method 224, 231, 245, 346
  error estimation 406
Finite elements
  3D 351
  bilinear 349
  coefficients 248
  continuity 249
  edge elements 379
  gross distortions 353
  hexahedra 350
  isoparametric 352
  rectangle 246
  second order 325
  surface integrations 353
  tetrahedra 353
  triangle 246, 326
Finite elements versus finite differences 281
Flow round circular hole 140
Flux density
  on armature surface opposite a slot 167
  sinusoidal distribution in inductor alternator 63
Flux distribution
  at the corner of a transformation core 164

Flux distribution (*cont.*)
    inside machine stators 56
Flux fringing at edge of capacitor plates 155
Force
    acting on a boundary 16, 17, 40
    between parallel rectangular bus-bars 73
    between rotor and stator conductors in a
        cylindrical machine 53
    distribution 17
    in a contactor 170
    on three-dimensional end windings 77
Fourier series
    cartesian coordinates 57, 81, 86, 110
    cylindrical coordinates 44, 110
    double 86
    single 57, 81
Fourier transform and inverse 107–109
Free-space modelling 366, 368, 379
Fringe factor 158
Fröhlich relation 300

Galerkin's method 234, 236, 241, 246, 347
Gauging strategies 366
Gauss's law 4
Gauss's theorem 454
Gauss–Seidl method 294
Gaussian quadrature 455
Geometric model 226
Green's theorems 9
Greens' theorems 454

H, fields 344
H field formulations 8, 379
Harmonic fields 96, 97, 102
Helmholtz equation 96, 99, 104, 108, 110
Heuristic design 386
History of electromagnetic computation 222

Ideal systems 387
Images 21
    complex variable representation 25
    parallel plane boundaries 25
    plane boundaries 22
    single plane boundaries 22
Inductor alternator 62
Inductance 14
    of a transformer winding 90
    of conductor influenced by permeable
        plate 186
    of parallel bus-bars 29
    of rectangular conductor in slot 92
    of three-dimensional circuits 79
Integral equations 9, 226
Integral method
    3D problems 359
Integral methods 224, 260, 359
    computing time statistics

magnetisation method 359
non-linearity 361
parallel processing 364
polarisation method 359
Integral theorems 454
Integration by parts 454
Interconnected regions 66, 94, 107
Interface conditions 4, 10
Intersecting plane boundaries
    four 29
    three 28
    two 27
Interconnecting thin plates 178
Inversion geometrical and complex 131
Isoparametric mapping 352

Jacobian matrix 295, 352
Joukowski transformation 138

Lagrangian interpolating formulae 326
Laminated structures 306
Laplace equation 5, 8, 17, 18, 43–68
    integral solutions 65
    solution in cartesian coordinates 57–65
    solution in circular cylinder coordinates
        45–57
    solution using conjugate functions and
        conformal transformation 116–27
Laplacian 18, 69, 95
Leakage flux of slot 165
Liebmann method 293
Line change near a conducting cylinder 33
Line current
    complex potential function 122, 424
    image of 24
    in slot 163
    influenced by iron cylinder 141
    magnetic field of 424
    near a permeable cylinder 33
    potential function in cartesian coordinates
        58
    potential function in cylindrical
        coordinates 46
    vector potential of 70, 425
Line poles 47, 58
Line sources 15
    infinite array with alternating signs 26, 29
    inside an infinitely permeable tube 133,
        134
    near boundaries 31, 141, 188–194
        finite 194
        infinite 190
        mixed impermeable/permeable 191
Linear algebra 255, 291, 295, 398
    comparisons 405
    conjugate gradients 402
    Gauss–Seidl 294, 402

Gaussian elimination 255, 399
ICCG 405
iterative methods 402
preconditioning 405
SOR 402
Logarithmic function 122–124, 424
Lorentz force 4
Lorentz gauge 6, 367
examples 373
formulations 367
uniqueness 368
Low frequency limit approximation 4, 313
LU decomposition 293

Magnetic contactor, forces in 170–173
Magnetic scalar potential 6
Magnetic vector potential 5
Magnetostatics 8
3D 344
Mathematical model 224
Matrix assembly 251, 316
Matrix symmetry 251, 317
Maxwell's equations 4
Mesh adaption 414
Mesh generation 387
automatic 391
Delaunay method 394
Dirichlet tesselation 393
manual 388
mapping methods 390
octree method 396
rubber-banding 392
semi-automatic 387
sub-region method 389
Voronoi polygon 393
Watson algorithm 394
Mixed boundary condition 225
Modelling the external circuit 329
Modified Gauss–Seidl 293
Modified vector potential 366
Moment method 240, 359
Motional effects 332
Multiple regions 66, 94, 107
Multiple transformations 125
Multiply connected problems 371, 373, 375, 379

Neumann boundary conditions 225
Newton–Raphson method 295
Non-concentric circular boundaries 132–138
Non-equipotential boundaries 55, 183, 215
Non-linearity 289, 323
Non-magnetic conductors in air 70–80
Normalised coordinates 350
Numerical integration of a complex variable 186
Numerical quadrature rules 455

Ohm's law 4
Open boundary problems 411
ballooning 413
heuristic method 411
infinite elements 413
integral methods 413
Kelvin transformation 413
Optimisation 414

Parabolic boundary 147
Parallel plane boundaries 25, 81, 155
Parallel plate capacitor 155
Parallel processing 365
Parametric equations 143
Partial differential equations 17
Permanent magnets 302
Permeance factors for the field of a current in a slot 167
Petrov–Galerkin method 336
Plate
charged conducting 178
finite, permeable, and line current 183
intersecting thin 178
opposite parallel equipotential 168
Point matching 233
Poisson equation 7, 8, 17, 18, 67–115, 224, 245
axisymmetric form 277
scalar form 278
Rogowski's method 81
Roth's method 86
solution in cartesian coordinates 71–76, 80–95, 112
solution in polar coordinates 70, 110, 112, 425
solutions in double Fourier series 86
solutions in single Fourier series 81
three dimensions 346
Poisson integral 55, 180, 183, 215
Pole-face loss 103
Pole profile, in the inductor alternator for a sinusoidal flux distribution 63
Polygonal boundaries 149–187
circle to polygon 180–185
exterior of polygon 175–180
interior of polygon 149–175
scale relationship 161
with angles not multiples of $\pi/2$ 199
with negative vertex angles 168–170
Polygons
curvilinear 128, 130, 198
with five vertices 173
with four vertices 163, 168, 170
with negative vertex angles 168–170
with rounded corners 196
with three vertices 155, 159, 170

Polygons (*cont.*)
  with two parallel sides 153
  with two vertices 152
Post-processing 414
Post-processor
  function analyser 416
Potential and flux-functions 423
Pre- and post-processing 386
Pre-processing 387

Quasi-static fields 3

Rectangular bus-bar 70
Rectangular conductor
  in a slot 85
  in a rectangular window 87
  in parallel air gap 81
  of highly permeable material 433
Regular function 117, 427
Rogowski electrode 158
Rogowski's method for Poisson equation
  scope of 86
Roth's method 86
  scope 94
Rotationally symmetric problems 277
Rounded corners 196

Saturation effects 289
Scalar integral equation 275
Scalar potential
  cancellation errors 357
  modelling free space (air) 366, 368, 379
  reduced 5, 225, 344, 348
  three dimensions 344
  total potential 7, 225, 345
  two-potential method 345, 347
Schwartz—Christoffel 150
  choice of corresponding points 158
  classification of integrals 185
  evaluation of constants 162
  integrals expressible in terms of non-
    simple functions 185
  integrals expressible in terms of simple
    functions 186
  scale relationship between planes 126, 161
Screening effect of a permeable cylinder 50
Second versus first order elements 329
Separation of the variables 44, 96
  cartesian coordinates 57
  cylindrical coordinates 44
  possible coordinate systems 44
  scope of the method for Laplace's
    equation 65
Series transformation 143, 146
  closed boundaries 146
  open boundaries 147

Shape functions 233, 245, 349
Simple iterative methods 291
Singularly perturbed problems 334
Skin depth 97—99, 106
Solution processor 398
Solution techniques 398
Stages in computing a solution 259
Static fields 8
Stokes' theorem 454
Sub domain collocation 236
Surface pole chemistry 13
System matrix 255
  solution methods 398

Table of transformation 437—451
Team workshops 373, 397
Three-dimensional field solutions, analytic
  76—80, 109—112
Three-dimensional problems, numerical 343
Time dependent formulations 320
Time discretisation 320
Time harmonic case 314
Time marching schemes 320
Transformer windings 87
  forces and inductances 90, 93
Transient field solutions, analytic 107—109
Transputers 365
Travelling fields 102—107
Trial functions 233
Truncation error 229
Two-dimensional case 8

Uniqueness of scalar potentials 457
Uniqueness of vector potentials 368
Upwinding 334
User community 385

Variational method 232
Vector Green's theorem 454
Vector identities 454
Vector potential 69, 79, 81—93, 365
  Coulomb gauge 366
  discussion 379
  Lorentz gauge 367
  modified 366
  multiply connected problems 371
  of line current 70, 126
  three dimensions 365
Virtual work 16
Volume magnetisation 9

Weak form 250
Weighted function 233
Weighted residual method 222, 233
  boundary elements 261
Which weighting function? 239